Cognitive Science and Technology

Series Editor

David M. W. Powers, Adelaide, SA, Australia

More information about this series at http://www.springer.com/series/11554

Chang S. Nam
Editor

Neuroergonomics

Principles and Practice

Springer

Editor
Chang S. Nam
Department of Industrial
and Systems Engineering
North Carolina State University
Raleigh, NC, USA

ISSN 2195-3988 ISSN 2195-3996 (electronic)
Cognitive Science and Technology
ISBN 978-3-030-34783-3 ISBN 978-3-030-34784-0 (eBook)
https://doi.org/10.1007/978-3-030-34784-0

This Springer imprint is published by the registered company Springer Nature Switzerland AG
The registered company address is: Gewerbestrasse 11, 6330 Cham, Switzerland

Foreword

Passion runs deep in the hearts of researchers who strive to understand the connections between people and the systems around them. This passion spans a wide range of scientific fields and has not only shaped our understanding of the interactions between people and the world, but it has also changed the ways we live our lives. Chang S. Nam and his colleagues share this passion, and their research in Neuroergonomics clearly demonstrates this passion. Another thread that runs across time and connects many of these passionate researchers is their desire to understand and solve problems in order to help individuals and society. ***Neuroergonomics: Principles and Practice*** is a publication that seeks to do this for the scientific community and the general reader who is interested in better understanding how Neuroergonomics, and its application, can shape everyday life and emerging applications. As one explores this publication, it is useful to look at some related history and examine some common elements.

In 1959, Charles Darwin showcased his passion in the groundbreaking publication, ***On the Origin of Species***. He outlined natural selection and advanced both the scientific community's and nonspecialist readers' understanding of the interaction between the environment and living organisms. Darwin's work was foundational in advancing our understanding of how changes in the characteristics of a species occur over several generations via the process of natural selection. Understanding evolution has been, and is, helping people solve problems that impact their lives. Evolution is the root of human behavior and is inseparable from the system, the environment, that shapes species.

In 1938, B. F. Skinner showcased his passion in the groundbreaking publication, ***The Behavior of Organisms***, where he advanced a new science based on selection by consequences as the mechanism through which behavior changes during the lifetime of the individual. He outlined the science of behavior and advanced our understanding that behavior is determined by its consequences, be they reinforcements or punishments, which make it more or less likely that the behavior will occur again. Skinner's work has successfully been applied widely in real-world applications because he explored the connection between the environment and individual behavior. Understanding the science of behavior has been, and is, helping people solve problems that impact their lives because most of society's problems involve human behavior. Behavior is the root of human interaction and is inseparable from the system, the environment, that shapes the individual.

In 1960, J. C. R. Licklider showcased his passion in the quintessential publication, ***Man-Computer Symbiosis***, where he put forth the novel notion that a cooperative interaction between people and computers was possible via a new symbiotic partnership paradigm. He outlined how the close coupling of people and electronic partners would enable a partnership that could far exceed the capabilities that a human or computational system could achieve individually. His work set people on the quest to tightly couple human brains and computing machines, so that the resulting partnership could process information that no human brain is capable of alone and process data in a way that no computer is capable of alone. Understanding the symbiotic relationship between people and computers has been, and is, enabling the creation of better decision-making capabilities and is a key component in enabling individuals to more effectively solve complex and fast-paced challenges. Interaction is the root of human–machine symbiosis and is inseparable from the system, the environment, that shapes individuals' brain and behavior.

Today's human–machine symbiotic systems can now be developed at a much deeper level than simply observable behavior between humans and machines. Modern human–machine symbiotic systems can now exploit our understanding of the brain and the associated behavioral correlate. Neuroergonomics is the study of brain and behavior at work and is the combination of the fields of neuroscience and ergonomics. In 2019, 160 years after the publication of Darwin's original publication, Chang S. Nam and his colleagues are showcasing their passion in ***Neuroergonomics: Principles and Practice***, where they seek to solidify the application of neuroscience to ergonomics. Like Charles Darwin, Chang S. Nam and his colleagues are advancing our understanding by explaining Neuroergonomics in a manner that speaks to both the scientific community and nonspecialist readers. Like B. F. Skinner, Chang S. Nam and his colleagues articulate a path forward that will result in the production of real-world applications. Like J. C. R. Licklider, Chang S. Nam and his colleagues put forth ideas and a future direction for the field of Neuroergonomics that will support the design of safer and more efficient systems. Chang S. Nam and his colleagues are advancing our understanding of the relationship between brain function and performance in real-world tasks. Like all of these prior contributors, Chang S. Nam and his colleagues pursue these goals to better help people solve problems that impact their lives.

Neuroergonomics: Principles and Practice has a rich history and is a milestone in the field of Neuroergonomics. In the late 1990s, Raja Parasuraman coined the term Neuroergonomics and spent his career defining and advancing the field. This was driven by his passion to study how humans interact with machines and his passion for cognitive neuroscience. During this same time period, I coined the term Augmented Cognition, and I spent my career defining and advancing this related field. This was driven by my passion to advance the frontier between human–computer interaction, psychology, ergonomics, and neuroscience, with the aim of creating revolutionary human–computer interactions. Therefore, it seems only natural to me that Neuroergonomics and Augmented Cognition have remained tightly coupled. The tight connection was further deepened in 2010 when Raja Parasuraman, the *Father of Neuroergonomics*, accepted the Human Factors and Ergonomics Society's Augmented Cognition Technical Group's Spirit of Innovation Award. Today, Chang S. Nam and his colleagues are providing us this essential publication at a time when the seeds planted over the last few decades have ripened. This publication builds upon this history and is helping to carry the field of Neuroergonomics forward.

For me, it is a wonderful coincidence that I originally met Chang S. Nam while attending the same Human Factors and Ergonomics Society's Augmented Cognition Technical Group that had previously celebrated the *Father of Neuroergonomics*. When I met Chang S. Nam at this conference, our passion for basic and applied research in human factors and ergonomics engineering to advance the science of Human–Computer Interaction quickly drew us together. A key element of that connection was our common, broad perspective on the application of systems and information engineering to human-centered technologies, including brain–computer interfaces and Augmented Cognition-Based Systems. Additionally, our passion for real-world application, particularly in the application of challenges facing operational military personnel, provided the thread to connect our thoughts. Operational military challenges are often particularly useful in tying together innovation, science, and immediate problems that need solving. Today, I particularly resonate with Chang S. Nam and his colleagues who worked on this publication because they address fundamental human factor issues and they emphasize the role of the human nervous system. This work is a key element of Augmented Cognition-Based Systems that sense a multitude of brain states, combined with other behavior and modeling techniques and adapt to users in real time, providing a true symbiosis between the human and computational systems.

It is important to remember that Neuroergonomics is helping to enable practitioners to leverage basic knowledge from brain–computer interfaces and neuroscience to achieve more effective human–systems integration. Neuroergonomics provides practitioners with more effective tools and systems to build Augmented Cognition Systems. The work in this publication details scientific developments that were necessary to create optimal systems, explains the sensors necessary for success, and includes many lessons learned regarding integration into existing and new applications. This knowledge provides the reader with useful information that is beneficial in understanding how to select adaptation strategies and manage adaptation processes. The optimal design of adaptive strategies can significantly enhance human performance and enable the development of effective Augmented Cognition Systems.

Passion is contained throughout this publication and its contents will enable readers to understand the connections between people and the systems around them at a level of detail that cannot readily be found elsewhere. This passion and its associated contents will change the ways we live our lives. Keeping in mind the past, and with hope for the future, enjoy exploring this publication today.

Captain Dylan D. Schmorrow, Ph.D.
United States Navy (Ret.)
e-mail: schmorrow@yahoo.com

Soar Technologies, Inc.
Ann Arbor, MI, USA

Preface

Neuroergonomics is a combination of the Greek words neuro, meaning "relating to nerves or the nervous system," and ergonomics, meaning "the study of work"—the study of brain and behavior at work. Neuroergonomics is an emerging area whose meaning is collectively defined as the study of the human brain function and behavior in relation to behavioral performance in natural environments and everyday settings. The domains impacted by neuroergonomics are varied; research has been conducted in the military, health, workplace, education settings, and so on. Thus, the impact of neuroergonomics is large; however, there are currently few books to provide students, practitioners, and researchers, including those outside of academia, with a single, go-to source containing state-of-the-science information about neuroergonomics. This book provides up-to-date coverage for researchers, students, and practitioners, including those with no formal training in neuroergonomics, to be able to grapple with a synopsis of key findings and theoretical and technical advances from neuroergonomics-related fields.

This book is organized into an introductory chapter with an emphasis on the evolution of neuroergonomics, five main parts, and a conclusion chapter. First part, consisting of four chapters, opens with an introduction to the fundamental components of neuroanatomy and brain function, as well as brain processes neuroergonomists need to understand (chapter "An Introduction to Neuroergonomics: From Brains at Work to Human-Swarm Teaming"). Chapter "Brain Basics in Neuroergonomics" provides a brief overview of some of the aspects of EEG-based experiments that will serve as an entry point and guide to researchers, including those who are new to the field. In addition, two other guides are intended to provide the necessary information for neuroergonomists to be able to understand the effects and usages of two other common neuroimaging methods—functional near-infrared spectroscopy or fNIRS (chapter "The EEG Cookbook: A Practical Guide to Neuroergonomics Research") and transcranial direct current stimulation or tDCS (chapter "Functional Near-Infrared Spectroscopy (fNIRS) in Neuroergonomics") on the brain. Second part introduces three computational approaches applicable to neuroergonomics, such as adaptive control of thought-rational (ACT-R, chapter "Transcranial Direct Current Stimulation (tDCS): A Beginner's Guide for Neuroergonomists"), deep learning techniques (chapter "Adaptive Control of Thought-Rational (ACT-R): Applying a Cognitive Architecture to Neuroergonomics"), and dynamic causal modeling (DCM, chapter "Deep Learning Techniques in Neuroergonomics"), where the history of advancements, its concepts, and applications in neuroergonomics research are described. Third part presents six chapters that discuss various neuroergonomics assessments of cognitive and physical performance in the areas of physical activity and sedentary behavior (chapter "Dynamic Causal Modeling (DCM) for EEG Approach to Neuroergonomics"), psychophysiology (chapter "Physical Activity and Sedentary Behavior Influences on Executive Function in Daily Living"), emotion (chapter "Neuroergonomics and Its Relation to Psychophysiology"), motor skill (chapter ""Hello Computer, How Am I Feeling?", Case Studies of Neural Technology to Measure Emotions"), training (chapter "The Neural Basis of Cognitive Efficiency in Motor Skill Performance from Early Learning to Automatic Stages"), and music proficiency and performance (chapter "Approaches for Inserting Neurodynamics into the Training of Healthcare Teams"). Fourth part illustrates how two emerging fields of brain–computer interface (BCI) and neuroergonomics can be brought together to add tremendous insight to an important issue—enhancing the quality of life for people with severe disabilities, through the design, development, and implementation of a hybrid EEG-functional transcranial Doppler ultrasound (fTCD) BCI (chapter "The Neuroergonomics of Music Proficiency and Performance"), BCI for spinal cord injury rehabilitation (chapter "Hybrid EEG–fTCD Brain–Computer Interfaces"), and BCI-controlled functional electrical stimulation (FES) for hand-grasp rehabilitation (chapter "Brain–Computer Interfaces for Spinal Cord Injury Rehabilitation"). Fifth part presents everyday and emerging applications of neuroergonomics in the areas of car driving (chapter "A Sensorimotor Rhythm-Based Brain–Computer Interface Controlled Functional Electrical Stimulation for Handgrasp Rehabilitation"), driving and navigation (chapter "Neuroergonomics Behind the Wheel: Neural Correlates of Driving"), augmented reality (AR) and virtual reality (VR) (chapter "Fundamentals and Emerging Trends of Neuroergonomic Applications to Driving and Navigation"), information visualization (chapter "Neuroergonomic Solutions in AR and VR Applications"), and trust (chapter "Neuroergonomic Applications in Information Visualization"). This book ends with a series of reflections on the future of, which suggests where to go from here.

This book was motivated by the desire many neuroergonomists have had to further understand the neurocognitive mechanisms and correlates related to perception, memory, attention, and the planning and execution of actions in a variety of contexts. Thus, we hope that our readers will find the information presented in this book timely and useful in guiding their research.

On behalf of the editorial team, I would sincerely like to thank the contributing authors for their professionalism as well as their commitment to the success of this book.

Raleigh, USA Chang S. Nam

Contents

Editor and Contributors

About the Editor

Chang S. Nam is a Professor of Edward P. Fitts Industrial and Systems Engineering at North Carolina State University, USA. He is also an Associate Professor of the UNC/NCSU Joint Department of Biomedical Engineering as well as the Department of Psychology. He received a Ph.D. from the Grado Department of Industrial and Systems Engineering at Virginia Tech in 2003.

Dr. Nam is the author or co-author of more than 100 research publications including journal articles, edited books, book chapters, and conference proceedings. Dr. Nam's research interests center around neuroergonomics, brain–computer interface (BCI) and rehabilitation, neuroadaptive automation in large-scale unmanned aerial vehicles, hyperscanning, cognitive neuroscience, and trust in human–robot interaction. His research has been supported by federal agencies including the National Science Foundation (NSF), the Air Force Research Laboratory (AFRL), and the National Security Agency. Dr. Nam has received the US Air Force Summer Faculty Fellowship Program (AFSFFP) Award, NSF CAREER Award, Outstanding Researcher Award, and Best Teacher Award. He is the Editor of two BCI books—*Brain–Computer Interfaces Handbook: Technological and Theoretical Advances* and *Mobile Brain–Body Imaging and the Neuroscience of Art, Innovation and Creativity*—and is currently working on an edition of *Trust in Human-Robot Interaction: Research and Applications.* Currently, Dr. Nam serves as the Editor-in-Chief of the journal Brain–Computer Interfaces.

Contributors

Hussein A. Abbass School of Engineering and IT, University of New South Wales, Canberra, Australia

Murat Akcakaya University of Pittsburgh, Pittsburgh, USA

Brett Baker The University of Texas at Austin, Austin, TX, USA

Darla Castelli The University of Texas at Austin, Austin, TX, USA

Inchul Choi Edward P. Fitts Department of Industrial & Systems Engineering, North Carolina State University, Raleigh, NC, USA

Jinyoung Choi School of Electrical Engineering and Computer Science, Gwangju Institute of Science and Technology, Gwangju, South Korea

Sanghyun Choo Edward P. Fitts Department of Industrial & Systems Engineering, North Carolina State University, Raleigh, NC, USA

Ian Daly Brain-Computer Interfacing and Neural Engineering Lab, Department of Computer Science and Electronic Engineering, University of Essex, Colchester, UK

Elizabeth Eskander Department of Psychology, North Carolina State University, Raleigh, NC, USA

Bryn Farnsworth von Cederwald iMotions A/S, Copenhagen, Denmark

Jing Feng Department of Psychology, North Carolina State University, Raleigh, NC, USA

Trysha Galloway The Learning Chameleon, Inc., Culver City, CA, USA

Jacob Green Edward P. Fitts Department of Industrial & Systems Engineering, North Carolina State University, Raleigh, NC, USA

Vishesh Hiremath University of Virginia, Charlottesville, VA, USA

Jiali Huang Edward P. Fitts Department of Industrial & Systems Engineering, North Carolina State University, Raleigh, NC, USA

Maarten A. Immink University of South Australia, Adelaide, Australia

Sehyeon Jang School of Electrical Engineering and Computer Science, Gwangju Institute of Science and Technology, Gwangju, South Korea

Sung C. Jun School of Electrical Engineering and Computer Science, Gwangju Institute of Science and Technology, Gwangju, South Korea

Aya Khalaf University of Pittsburgh, Pittsburgh, USA

Inki Kim University of Illinois at Urbana-Champaign, Urbana, IL, USA

Ji-Eun Kim University of Washington, Seattle, USA

Na Young Kim Edward P. Fitts Department of Industrial & Systems Engineering, North Carolina State University, Raleigh, NC, USA

Nayoung Kim Edward P. Fitts Department of Industrial & Systems Engineering, North Carolina State University, Raleigh, NC, USA

Satyam Kumar Department of Electrical and Computer Engineering, University of Texas at Austin, Austin, TX, USA

Tae-Ho Lee Virginia Tech, Blacksburg, USA

Joshua McNiven Royal Welsh College of Music and Drama, Cardiff, UK

Ranjana K. Mehta Department of Industrial and Systems Engineering, Texas A&M University, College Station, TX, USA

Alyssa Merante Department of Biomedical Engineering, North Carolina State University, Raleigh, NC, USA

Alessio Meroni Holst Centre, IMEC, Eindhoven, The Netherlands

Vojkan Mihajlović Holst Centre, IMEC, Eindhoven, The Netherlands

Chang S. Nam Edward P. Fitts Department of Industrial & Systems Engineering, North Carolina State University, Raleigh, NC, USA

Joseph K. Nuamah Department of Industrial and Systems Engineering, Texas A&M University, College Station, TX, USA

Erfan Pakdamanian University of Virginia, Charlottesville, VA, USA

Paruthi Pradhapan Holst Centre, IMEC, Eindhoven, The Netherlands

Nathan Sanders Edward P. Fitts Department of Industrial & Systems Engineering, North Carolina State University, Raleigh, NC, USA

Ervin Sejdic University of Pittsburgh, Pittsburgh, USA

Navid Shahriari Holst Centre, IMEC, Eindhoven, The Netherlands

Thomas J. Smith University of Minnesota, Minneapolis, USA

Ronald Stevens UCLA School of Medicine, Brain Research Institute, Los Angeles, CA, USA;
The Learning Chameleon, Inc., Culver City, CA, USA

Liya Thomas Edward P. Fitts Department of Industrial & Systems Engineering, North Carolina State University, Raleigh, NC, USA

Willem B. Verwey University of Twente, Enschede, The Netherlands

Macie Ware Edward P. Fitts Department of Industrial & Systems Engineering, North Carolina State University, Raleigh, NC, USA

Ann Willemsen-Dunlap JUMP Simulation and Education Center, Peoria, IL, USA

Duncan Williams Digital Creativity Labs (Computer Science Department), University of York, York, UK

Jolanda Witteveen Holst Centre, IMEC, Eindhoven, The Netherlands

David L. Wright Texas A&M University, College Station, USA

Yu Zhang Department of Psychiatry and Behavior Sciences, Stanford University, Stanford, CA, USA

Introduction

An Introduction to Neuroergonomics: From Brains at Work to Human-Swarm Teaming

Hussein A. Abbass

Abstract The field of Neuroergonomics sits at the interface between humans' brains and humans' working environment with an aim to improve the work humans do. In this chapter, I will review the scientific journey that led to the birth of Neuroergonomics. The journey starts from the Biocybernetics and Brain–Computer Interfaces projects in 1960s, followed by work on adaptive aiding in 1970s, and all the way to early 2000s with work on Augmented Cognition and Neuroergonomics. An extension to Neuroergonomics gave birth to Cognitive-Cyber Symbiosis, whereby Neuroergonomics is augmented with artificial intelligence agents that act as relationship managers between the human brain and the information-centric work environment. Some challenges facing these fields today are then discussed using a human-swarm teaming lens.

1 Introduction

Neuroergonomics is the study of the human brain in relation to its contributions to the human's efficiency in a working environment (Parasuraman & Wilson, 2008). The field has grown over decades with foundations that could be traced back to 1960s. The evolution of the field has witnessed a series of concepts and terminologies that started with what appeared to be morphologically different, but equally similar at their foundational cores, dimensions of the field. This chapter will explore these dimensions, starting with Licklider's work.

In his seminal paper "Man-Machine Symbiosis" (Licklider, 1960), Licklider wrote: "A multidisciplinary study group, examining future research and development problems of the Air Force, estimated that it would be 1980 before developments in artificial intelligence make it possible for machines alone to do much thinking or problem-solving of military significance. That would leave, say, five years to develop man-computer symbiosis and 15 years to use it. The 15 may be 10 or 500, but those years should be intellectually the most creative and exciting in the history

H. A. Abbass (✉)
School of Engineering and IT, University of New South Wales, Canberra, Australia
e-mail: h.abbass@adfa.edu.au

C. S. Nam (ed.), *Neuroergonomics*, Cognitive Science and Technology,
https://doi.org/10.1007/978-3-030-34784-0_1

of mankind." Licklider had a vision that humans and machines will work together in harmony. The enabling technology for this vision is artificial intelligence (AI). Today, we have started to live the years that Licklider labeled as "intellectually the most creative". Before we delve into some of these creative ideas and how they are reshaping human's efficiency at work, we will first offer a glimpse of selected history that led to neuroergonomics as we know it today.

2 Before and After Neuroergonomics

The inception of neuroergonomics could be traced back to two ambitious projects: Biocybernetics (Wiener & Schadé, 1964) and Brain–Computer Interfaces (Vidal, 1973) or BCI. Both projects had the intriguing idea of using physiological responses as objective metrics to assess a human's mental load and processing. As stated in the concluding report of the Biocybernetics project: "the objective in the DARPA Biocybernetics Program was to investigate the use of pupillary responses in assessing attentional demands on operators of complex man-machine systems. The results of this project were strongly positive: the task-evoked pupillary response has emerged as an excellent physiological indicator of mental workload." (Beatty, 1979, 1).

The BCI project, led by Vidal, focused on the identification of signal responses in the brain to external stimuli and the opportunity this offers to design a communication channel between the human brain and the work environment. Vidal in his seminal paper asked: "Can these observable electrical brain signals be put to work as carriers of information in man-computer communication or for the purpose of controlling such external apparatus as prosthetic devices or spaceships? Even on the sole basis of the present states of the art of computer science and neurophysiology, one may suggest that such a feat is potentially around the corner." (Vidal, 1973, 157).

Biocybernetics and BCI offered evidence to strengthen the hypothesis that objective indicators could be extracted from human signals, whether they are signals due to brain/cognitive functions or behavioral responses, and could be interpreted into actionable knowledge. Encoding the information residing within these signals offered the missing key to open a door of opportunities to connect humans and machines. It then did not take much time for the work to evolve to a different level of ambitious. A level whereby automation adapts in response to human mental states.

Adaptive Aiding (Rouse, 1988) is a program of research that started in 1974; almost straight after the Biocybernetics and BCI projects. Rouse explains that in a complex system, there are many components interacting. At one point of time, automating one component is essential and useful, while at another point of time, this automation may not be that useful. As the task demands change, the level of aiding and the strategy needed to define the way human and machine interact should also change. He defined three requirements for adaptive aiding:

- As task demands increase, the level of aiding should increase.
- As task demands increase, the interaction between human and machine should be streamlined.
- Variations of aiding and mode of interaction should be initiated by the aiding/automation tool.

These simple rules laid out the foundations for an intertwined journey in three streams of research that need to come together to deliver on the ambitious aim. One stream focuses on collection and interpretation of signals from humans. The second pays attention to the design of metrics to analyze the complexity of a task. The third is centered on algorithmic design of the adaptive logic that uses the actionable information from the first two streams to decide when, what, and how to switch functions or subtasks between humans and machines.

While brain imaging offered insights into the functional topology of the brain, real-time data analysis required more temporal resolution and flexibility in the sensors used to collect the data that other brain imaging techniques such as functional magnetic resonance imaging (fMRI) could not offer. Electroencephalography (EEG) offered more opportunities in this direction with the air traffic control (ATC) domain offering the perfect problem due to the mentally demanding tasks an air traffic controller does. Other techniques in the literature include: functional near-infrared spectroscopy (fNIRS) and transcranial Doppler (TCD), where both sense brain activities, while transcranial direct current stimulation (tDCS) delivers a low electric current of one to two milliamperes for neuromodulation.

Brookings, Wilson, and Swain (1996) conducted one of the early studies in this area, albeit with a small sample size of eight subjects. They experimented with three ATC scenarios that varied traffic volume, traffic complexity, and time criticality. They used 19 EEG channels according to the 10–20 electrode system and analyzed five bands: delta (1.1–3.9 Hz), theta (4.3–7.8 Hz), alpha (8.3–11.9 Hz), beta 1 (12.3–15.8 Hz), and beta 2 (16.2–24.9 Hz). Results for power spectral band values were calculated as the percentage of the total power between 1.1 and 24.9 Hz. They concluded the study with a number of findings including: the percent theta power at central, parietal, one frontal, and one temporal site significantly increased as task difficulty increased, the beta 1 band was sensitive to the traffic conditions at F3, Fz, F4, Cz, T4, and the interaction between traffic and difficulty manipulations was reflected in delta activity sites F3, Fz, F4, and T3. The study demonstrated the differential sensitivity of a variety of workload measures in complex tasks.

By early 2000, sufficient literature demonstrated the plausibility of the premise that EEG measured from the human scalp reveals mental processing information that could be leveraged to adapt automation to the human. However, the reliance on off-line post-analysis meant that the findings are mere academic ones. It was only meaningful to take a step to transform these findings into real-time systems to have any practical significance.

The above motivation gave birth to Augmented Cognition (AugCog) program (Schmorrow & Kruse, 2002). The aim of AugCog was to develop mental-state measurements and tracking technologies. As described by Schmorrow and Kruse (2002), the plan was "In FY 2002, the AugCog program is developing robust, noninvasive, real-time, cognitive state detection technology for measuring the cognitive processing state of the user. In FY 2003, AugCog will be developing and testing integrated multi-sensor interface technologies that will permit human state manipulation." (Schmorrow & Kruse, 2002, 7). Along with a similar timeframe, the literature saw the birth of Neuroergonomics. Probably the first use of the term was in Parasuraman's paper (Parasuraman, 1998), but the real impact and use of the term were more influential in his 2003 paper (Parasuraman, 2003). In that paper, Parasuraman defined neuroergonomics as "the study of brain and behaviour at work." (Parasuraman, 2003, 5).

Interestingly, despite the revolutionary concept of AugCog, the significant majority of research that has been done in the last two decades remained on the course of either off-line analysis or real-time BCI for medical applications (Nam & Nijholt, 2018). Only a few studies looked at real-time EEG analysis using realistically complex operational scenarios (Abbass, Tang, Amin, Ellejmi, & Kirby, 2014). To put it simply, AugCog and Neuroergonomics remain today a largely unexplored territory with significant missed—opportunities that practitioners could easily leverage. The working environment, in practice, has seen few attempts to use real-time EEG-based indicators to influence and shape the environment, let alone the potential for a two-way communication between the human brain and the task the human is operating on. This has been due to some challenging factors; some technological, while others are social.

On the technological level, factors ranged from the lack of robust sensors that could be used in an operational environment without imposing significant discomfort on the human from long usages, to the difficulties in generalizing EEG indicators across subjects. While trends in the EEG signals were consistent, at least among the majority of subjects, each subject requires different parameterizations of their models, let alone that the EEG signal for the same subject could be significantly impacted with caffeine, alcohol, and even time of the day. Social factors included the ethical implications of continuous assessments of the mental states of an operator in a work environment and the lack of compelling business cases to use what some would consider a "fancy" technology that may only be well-needed in safety-critical domains.

The above challenges called for new ways to think about how to transform the experimental results obtained in carefully conditioned environments to operational settings. The complexity of real-time calibration of the models to the human operator, contextual analysis, understanding and assessment of the task, and the level of sophistication required to create a truly symbiotic relationship between the human and the machine, called for a level of sophistication in the design of artificial intelligence (AI) beyond simple forms of adaptive logic.

The Cognitive-Cyber Symbiosis (CoCyS) concept (Abbass, Petraki, Merrick, Harvey, & Barlow, 2016) was then introduced to design this AI agent. CoCyS "revolutionised the adaptation process from a machine adapting to a human to smart adaptive agents (called 'ecookies') that act as autonomous relationship managers between humans and machines." (Abbass, 2019, 163). CoCyS leverages the revolution in AI to reduce the technological challenges faced in Neuroergonomics by automating the process to a sufficient level of autonomy that enables a plug and play environment. Such automation will allow massive productions of the technologies that will ease the financial burdens to create the business case for it. In simple terms, CoCyS is a bridge for AugCog and Neuroergonomics to make the visions created by visionaries and innovators, such as Schmorrow and Parasuraman, over the last 60 years an economically viable reality.

3 Where to from Here?

Neuroergonomics has evolved over the years as being discussed above. With the levels of sophistication seen today in sensor technologies, the complexity of the tasks expected from humans, and the advances made in artificial intelligence, the field needs to start tackling unprecedented challenges. In this section, the author will use his current research to lay out some of the most pertinent challenges in Neuroergonomics. To start with, Fig. 1 depicts the bigger picture of the author's research program in cognitive-Cyber Symbiosis for Human-Swarm Teaming (CoCyS-HST).

HST is one of the tasks that is expected to be most demanding on the human. An air traffic controller would normally be handling 7–9 aircraft at any point of time. In HST, the human could be teaming with other humans and one or more swarms of autonomous vehicles, each consisting of tens of vehicles, if not more.

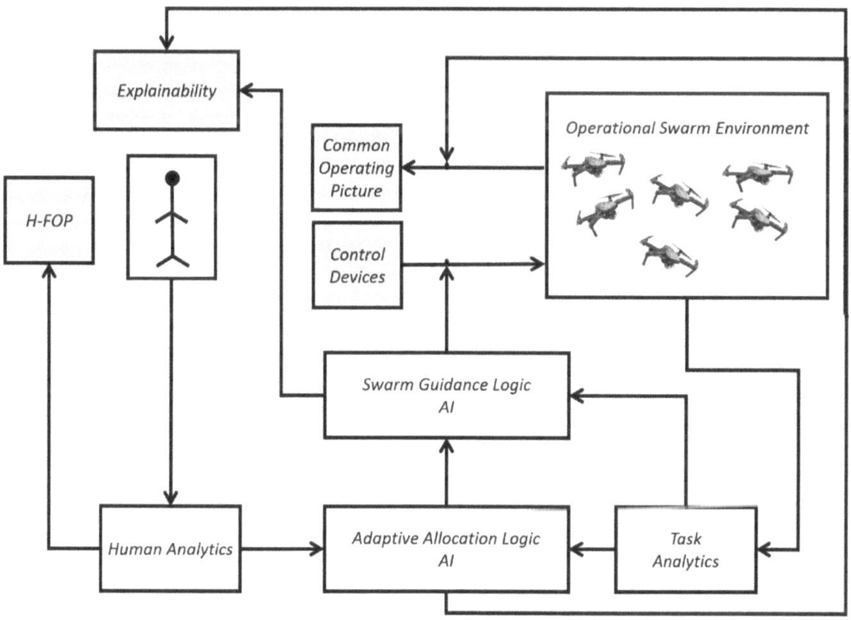

Fig. 1 Cognitive cyber symbiosis for human-swarm teaming

This level of complexity is unlikely to be manageable by a human. The situation calls for an AI to assist the human in managing the complexity and automate swarm guidance. This swarm guidance AI (SGAI) is responsible for autonomous guidance of the swarm in support of the human operator. The level of autonomy in SGAI is controlled by another AI, the adaptive allocation logic AI (AAL-AI), that relies on real-time analytics of the task and the human mental states to determine the optimal allocation of functions between the humans and the SGAI. Such allocations of functions are then communicated directly to SGAI to adapt its level of autonomy, and to the human through the interface depicting the common operating picture of the swarm to adjust human's situation awareness and actions.

To integrate the dynamic allocation of functions between the human and SGAI with the human, both AAL-AI and SGAI need to offer a level of explanation to the human. These explanations are necessary for many reasons. First, they are more likely to improve human's trust in the system. Second, they assist the human to maintain their situation awareness; especially when SGAI changes its level of autonomy which may cause the human to disorient. For example, if SGAI decides to reduce the level of autonomy due to a high-risk decision that needs to be made, the human is suddenly required to integrate more information to make a decision. Take for example a self-driving car that decides to hand control back to the human in a high-risk situation, the human needs to be quick in forming and/or updating the human's situation awareness picture.

The human factors operational picture (H-FOP) is a concept introduced in Ma-Wyatt, Fidock, and Abbass (2018), whereby a human analytic AI offers real-time assessments of human's mental states from a multitude of heterogenous data modalities including EEG, galvanic skin conductance, heart rate, and behavioral data such as facial expressions, mouse movements, and keystrokes.

The neuroergonomics challenges in the above system reside mainly in the H-FOP, human analytic, task analytic, and AAL-AI components. There are classic challenges with significant literature to address them. For example, the first challenge is related to the question of which indicators are useful in this task to evaluate human's attention level, situation awareness, mental load, level of engagement, and fatigue. With reliance on data from different modalities, each modality could contribute some indicators. For example, both EEG and heart rate variability could offer indicators of mental load. The second challenge is related to the design of appropriate fusion functions to integrate information from these diverse modalities. A third challenge revolves around similar lines but for the task itself; that is, how to characterize and assess the complexity of a situation as the task evolves?

We need to switch focus to challenges that have not been discussed sufficiently in the literature. The first of these is real-time calibration of the models based on EEG indicators. This is possibly one of the main obstacles in taking EEG from the lab environment to the real-world. In simple terms, if the system is plug and play, any human should be able to join the system and the system should work with this new human autonomously and with ease.

This form of calibration is very difficult but not impossible. Possible ways to achieve it include the following. One could calibrate against moments of subject relaxation. That is, if the operator needs to take rest every two or three hours, in the rest break, the operator could get trained to spend 2 min with their eyes closed and attempting to relax, for the system to calibrate. A second way is to calibrate against events in the task. The task analytic agent monitors the task in real time and is able to identify certain events such as the sudden appearance of an obstacle. These events could be used for calibration as they are expected to be followed by event-related potentials. A third possibility is to collect data from significantly large number of operators in a variety of task contexts, time of day, with different impacts of different beverages, and sufficient diversity to represent the population of operators. This later case seems possible, and may generalize well on a particular targeted subpopulation, but seems to run a hidden risk if the system maps a subject to the wrong behavior in a critical moment due to a bias in the sample.

Calibration within a subject own experience seems to be the most reliable approach to follow as it is subject centric and does not assume a single parameterization fits-all model. Moreover, a human who is using the same system on a daily basis may not require continuous calibrations as per the suggestions above. The system would have collected enough data to generalize and work well for that particular human.

A second challenge that does not get discussed sufficiently in the literature is the AAL-AI. Most of the literature assumes a very simple rule-based AI taking the form of a recommender system. However, the AAL-AI in complex tasks need to be far more sophisticated (Abbass, 2019). It first needs to be contextually aware of the human, the particular situation the human is faced with, and the overall mission. Second, the AAL-AI needs to anticipate the impact of each decision on human's situation awareness to decide an appropriate protocol for switching functions between the human(s) and SGAI.

Acknowledgements The author would like to acknowledge funding from the Australian Research Council Discovery Grant number DP160102037.

References

Abbass, H. A. (2019). Social integration of artificial intelligence: Functions, automation allocation logic and human-autonomy trust. *Cognitive Computation, 11*(2), 159–171.

Abbass, H. A., Petraki, E., Merrick, K., Harvey, J., & Barlow, M. (2016). Trusted autonomy and cognitive cyber symbiosis: Open challenges. *Cognitive Computation, 8*(3), 385–408.

Abbass, H. A., Tang, J., Amin, R., Ellejmi, M., & Kirby, S. (2014). Augmented cognition using real-time EEG-based adaptive strategies for air traffic control. In *Proceedings of the human factors and ergonomics society annual meeting* (Vol. 58, pp. 230–234). Los Angeles, CA: SAGE Publications.

Beatty, J. (1979). *Concluding report: ARPA biocybernetics project* (Technical Report ADA078523). The Ruth H. Hooker Technical Library, Naval Research Laboratory.

Brookings, J. B., Wilson, G. F., & Swain, C. R. (1996). Psychophysiological responses to changes in workload during simulated air traffic control. *Biological Psychology, 42*(3), 361–377.

Licklider, J. C. (1960) Man-computer symbiosis. *IRE Transactions on Human Factors in Electronics, 1*, 4–11.

Ma-Wyatt, A., Fidock, J., & Abbass, H. A. (2018). Quantifying and predicting human performance for effective human-autonomy teaming. In *Proceedings of the International Conference on Science and Innovation for Land Power, Defence Science and Technology Group*. Australia: Department of Defence.

Nam, C. S., Nijholt, A., & Lotte, F. (2018). *Brain–computer interfaces handbook: Technological and theoretical advances*. Boca Raton: CRC Press.

Parasuraman, R. (1998). *Neuroergonomics: The study of brain and behavior at work*. Washington, DC: Cognitive Science Laboratory. Available online at www.psychology.cua.edu/csl/neuroerg.html.

Parasuraman, R. (2003). Neuroergonomics: Research and practice. *Theoretical Issues in Ergonomics Science, 4*(1–2), 5–20.

Parasuraman, R., & Wilson, G. F. (2008). Putting the brain to work: Neuroergonomics past, present, and future. *Human Factors, 50*(3), 468–474.

Rouse, W. B. (1988). Adaptive aiding for human/computer control. *Human Factors: The Journal of the Human Factors and Ergonomics Society, 30*(4), 431–443.

Schmorrow, D., & Kruse, A. A. (2002). Darpa's augmented cognition program-tomorrow's human computer interaction from vision to reality: Building cognitively aware computational systems. In *Proceedings of the 2002 IEEE 7th Conference on Human Factors and Power Plants, 2002* (pp. 7–7). IEEE.

Vidal, J. J. (1973). Toward direct brain-computer communication. *Annual Review of Bio-physics and Bioengineering, 2*(1), 157–180.

Wiener, N., & Schadé, J. P. (1964). *Progress in biocybernetics*. Amsterdam: Elsevier Publishing Company.

Brain Basics in Neuroergonomics

Bryn Farnsworth von Cederwald

Abstract This chapter provides an introduction to (1) the fundamental components of neuroanatomy and brain function, (2) how brain processes give rise to behaviors that are relevant to study from a neuroergonomic perspective, and (3) how these brain processes can be detected and investigated with neuroimaging methods typically employed in neuroergonomics.

1 Introduction

The brain sits at the center of the human nervous system and is involved in regulating or executing almost every action that the human body produces. It is therefore difficult to overstate the importance of this organ with regards to our physiological functions. Studying the function of human action or thought will ultimately and inevitably lead to the brain. Neuroergonomics is well positioned to facilitate an understanding of this link between mental processes (such as actions or thoughts) and brain function. In order to make this link, however, an understanding must begin at both ends of the process, both with how the action is completed, and the processes that underlie the action. Knowledge of the workings of the brain helps us understand the processes before and during the action. This chapter aims to provide such knowledge, so that the output and the links between them can be understood and discerned with greater precision.

The brain has been studied extensively, particularly in recent history as neuroimaging and molecular methods have both improved and become easier to implement. This has allowed for many breakthroughs to be made and we now have a clearer idea of what the brain is, what it consists of, how it works, develops, and how it learns. An introduction to the current state of understanding of each of these areas will be introduced below. It should, however, be made clear—while great progress has been made (and continues to be made), we are still far from a true, holistic, and complete understanding of brain function.

B. Farnsworth von Cederwald (✉)
iMotions A/S, Copenhagen, Denmark
e-mail: bryn.farnsworth@imotions.com

C. S. Nam (ed.), *Neuroergonomics*, Cognitive Science and Technology,
https://doi.org/10.1007/978-3-030-34784-0_2

The brain in anatomical terms consists of the cerebrum, the cerebellum, and the brain stem. The cerebrum is usually what is pictured when considering the brain—the two-hemisphere structure with multiple folds. The cerebellum is located at the back of the head and outwardly shows multiple fine grooves and furrows. The brain stem consists of several structures that reside within the cerebrum, and is the link between the cerebrum and the spinal cord. Below, we will go through the functioning of each of these structures, and highlight areas that have particular relevance to neuroergonomics.

2 An Introduction to Neuroanatomy

2.1 The Cerebrum

In broadest terms, the cerebrum consists of six brain areas that span two hemispheres. Four of these areas can be seen from the external view of the brain—the frontal, parietal, occipital, and temporal lobes. Two other lobes, the limbic and insular lobes, are found within the cerebrum. Each of these lobes can be said to be the site of exclusive functions, although there is also considerable overlap.

On the outermost point of the cerebrum sits the cerebral cortex (often, and herewith referred to as the cortex), a sheet of neurons roughly 1.5–3 mm thick. The cortex is finely wrinkled, creating the gyri (ridges) and sulci (furrows) that are seen on the surface of the brain. This wrinkling dramatically increases the surface area of the cortex, allowing for many more neurons to reside within this space. Various other areas constitute the remainder of the cerebrum, including the basal ganglia, hippocampus, and the olfactory bulb, among others (Martin, Radzyner, & Leonard, 2012) (Fig. 1).

2.2 The Cortex

The cortex is where the majority of neuronal processing takes place before that signal is transmitted to subcortical areas (and beyond). This is evidenced by the approximately 77 billion neurons (Azevedo et al., 2009) that reside in the cortex. The cortex itself is composed of multiple columns of neurons (Defelipe, Markram, & Rockland, 2012) that contain multiple layers. There are six layers within the column, although the exact composition of the neurons within the layers differs depending on where it is situated. This will be returned to within Sect. 2.1.

Initial processing of visual information is carried out within the cortex, in the occipital lobe (at the back of the head). As light enters the retina, it activates photoreceptors (a sensor of light within the eye) that trigger a cascade of signals that are sent to the visual cortex. Around 50% of the signal is sent in a contralateral fashion (i.e.,

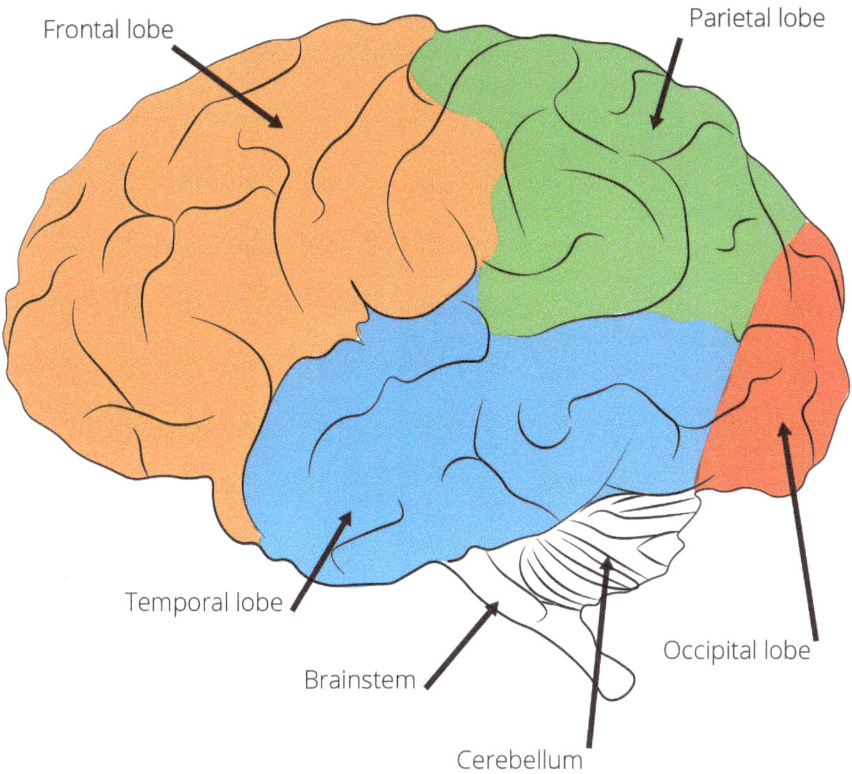

Fig. 1 A gross anatomy map of the brain. The frontal lobe is shown in orange, the temporal lobe in blue, the parietal lobe in green, and the occipital lobe in red. The brain stem and cerebellum are shown without color

light that enters the right eye will ultimately be processed within the left occipital cortex, and vice versa). There are various parts of the occipital cortex that respond to different features of the visual scene, such as edges, movement, color, etc. (Livingstone & Hubel, 1988). This means that certain features of a visual scene will trigger activity within certain regions of a brain, while other features will not.

Similar to the visual cortex, somatosensory information (relating to our body and senses) is also processed in a contralateral fashion (although more so, around 90% of the signals travel to the opposite side of the brain; Elbert et al., 1994). The subdivisions of the somatosensory cortex are further responsible for signals relating to a distinct body part. This is known as a somatotopic arrangement, in which distinct regions are activated for sensory information related to distinct body parts (Feldman & Brecht, 2005). Furthermore, this somatotopic organization appears upside down relative to our body position, with the feet and legs at the very top of our brain, while the face is at the bottom. Interestingly, the areas concerned with somatosensory activity of the face are inverted within this organization, and thus begin at the brow and follow downward to the mouth, after the area concerned with somatosensory activity of the neck (this is shown in Fig. 2). The more sensitive parts of our body (such as our lips) take up a disproportionate amount of space within this cortical somatotopic map, which is in accordance with the increased number of sensory receptors at these physical locations (Lumpkin & Caterina, 2007). The somatosensory cortex is responsible for receiving haptic (touch), proprioceptive (location), nociceptive (pain), and temperature information. Activation of specific brain regions within this area could then relay information about the pain experienced in the arm, or the location of our fingers, etc. These signals generated by the somatosensory cortex can then determine how we respond to our tactile environment. This may trigger a downstream change in activity within the motor cortex, the area largely responsible for organizing our movements.

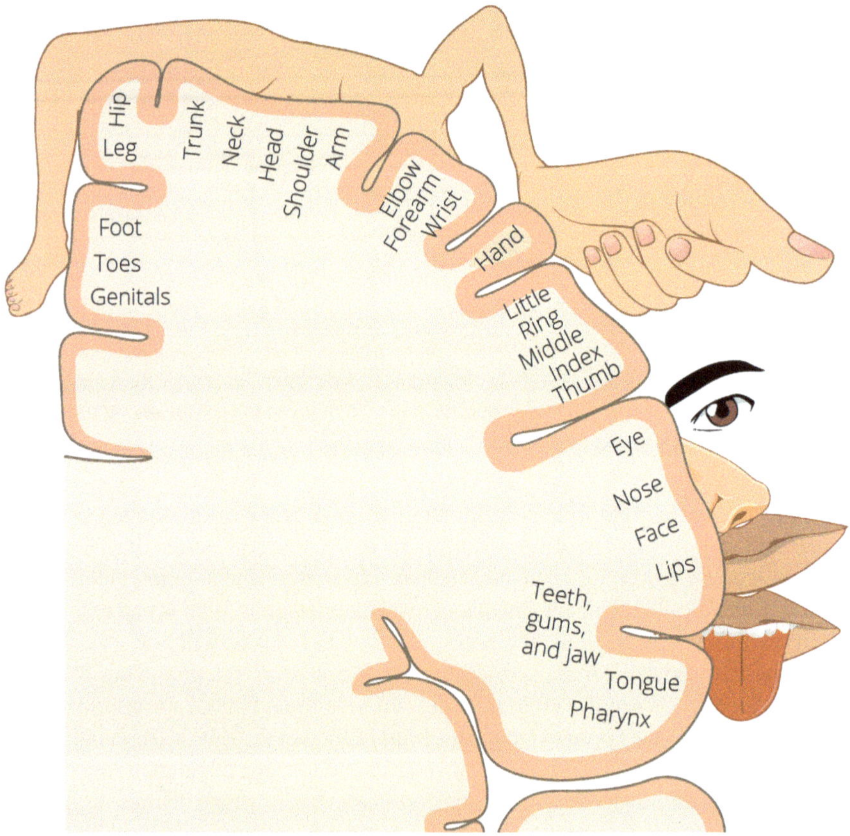

Fig. 2 Somatotopic map of the somatosensory cortex. The primary motor cortex is arranged in a similar manner. Adapted from OpenStax College—CC BY 3.0, https://cnx.org/contents/FPtK1zmh@6.27:KcreJ7oj@5/Central-Processing

The motor cortex consists of three areas—the primary motor cortex, premotor cortex, and the supplementary motor cortex. The primary motor cortex is similar to the somatosensory cortex in that it receives information largely in a contralateral fashion, and that it has a somatotopic map. The activation of certain specific parts of the primary motor cortex can cause bodily movements. This can be completed with the use of transcranial magnetic stimulation (or other stimulatory techniques) to target specific parts of the body, causing, for example, a twitching of a finger. Although further research is needed to delineate the exact roles of the two other areas within the motor cortex, research has suggested that the premotor cortex is involved in the planning of movements prior to their execution, while the supplementary motor cortex is involved in the learning and execution of sequences of movements (Gallese, Fadiga, Fogassi, & Rizzolatti, 1996; Picard & Strick, 2003).

At the front of the cortex sits the prefrontal cortex. While not a sensory processing area per se, this area is highly interconnected with other cortical, subcortical, and brain stem areas (Öngur & Price, 2000). It has also drawn considerable interest due to its apparent role in a range of cognitive control functions, such as attention, decision-making, planning complex behaviors, and regulating social actions (Miller & Cohen, 2001). More commonly referred to as executive function, this comprises a set of cognitive processes essentially related to supervising and controlling behavior (Roberts, Robbins, & Weiskrantz, 2003). A meta-analytic review of literature has shown that the involvement of this area with executive functions isn't necessarily clear-cut, with research ultimately suggesting that the prefrontal cortex is sensitive to executive functions, but isn't necessarily specific to their functioning (Alvarez & Emory, 2006). This means that while the prefrontal cortex is likely involved in executive functions, it's unlikely it operates alone. Multiple regions are likely needed to complete such complex actions as decision-making and behavioral control. Further research will continue to develop a more precise and nuanced understanding of how (and how much) the prefrontal cortex is involved in executive function, although it is clear that the area is central to this process.

2.3 Subcortical Regions

The subcortical regions consist of the basal ganglia, the hippocampus, amygdala, claustrum, and the basal forebrain. Each of these areas resides within the brain, connected to the cortex by white matter tracts (explained in further detail in Sect. 2.2). Each serves a distinct purpose in sensory processing and action selection, often first receiving information from various parts of the cortex (Fig. 3).

The basal ganglia are a set of interconnected structures that are involved in a range of crucial processes, but have largely been considered in the context of voluntary movement control (DeLong, 1990). The regions have also more recently been investigated for their involvement in both cognitive and affective processes (Middleton & Strick, 2000; Paulmann, Pell, & Kotz, 2008). The structures consist of the dorsal striatum (caudate and putamen), ventral striatum (nucleus accumbens

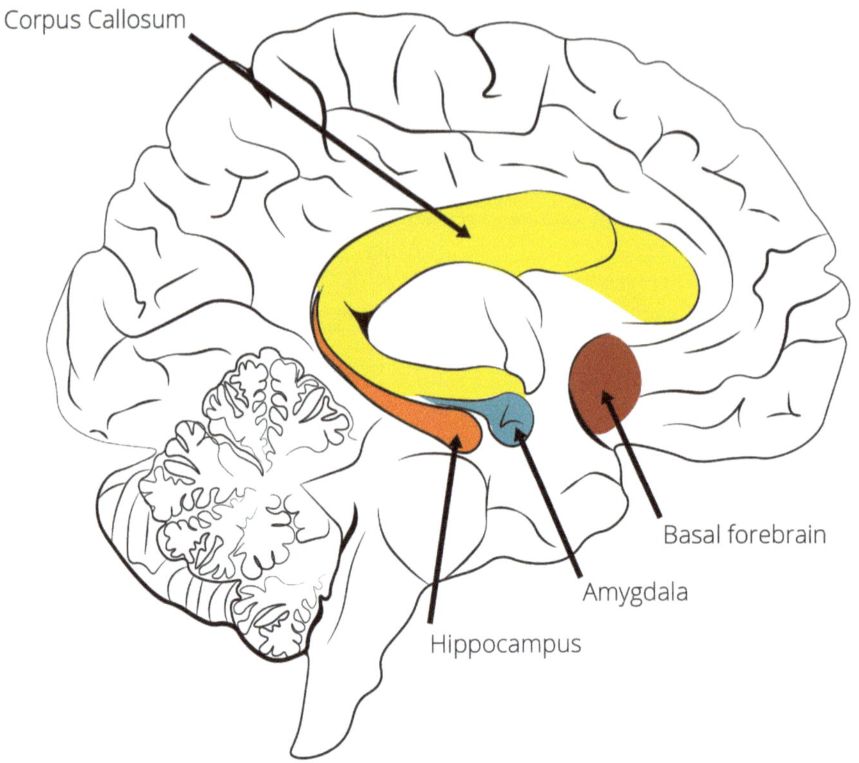

Fig. 3 Selected subcortical regions. The corpus callosum is shown in yellow, the amygdala in blue, the hippocampus in orange, and the basal forebrain in red. All of the highlighted regions, with the exception of the corpus callosum, are found deeper within the brain than the midpoint, and the figure shows their approximate location within the sagittal plane

and olfactory tubercle), globus pallidus (internal and external segments), the subthalamic nucleus, and substantia nigra (pars reticulata and pars compacta) (Nelson & Kreitzer, 2014). Several of these structures are of note due to their direct and identifiable role in movement, and subsequently movement disorders. Parkinson's disease, primarily a disorder in movement initiation, is the result of a drastic reduction in the number of dopaminergic cells in the substantia nigra pars compacta (Burns et al., 1983). Huntington's disease, a disorder that ultimately also affects gait and movement (Hausdorff, Cudkowicz, Firtion, Wei, & Goldberger, 1998), is associated at early stages with damage to the dorsal striatum (other basal ganglia and brain structures ultimately also show damage; Dogan et al., 2013). The nucleus accumbens (within the ventral striatum of the basal ganglia) is often referred to as the "reward center" of the brain, as an area that has been shown to be central in the release of dopamine in the response to rewarding events (Ikemoto & Panksepp, 1999). This association has also implicated the importance of this region (and also the basal ganglia as a whole) in the formation of addiction (Pontieri, Tanda, Orzi, & Chiara, 1996), as well as motivational processes in general. The strengthening of connections through reward processes, such as the release of dopamine within the nucleus accumbens, is essential in promoting sets of behaviors that benefit survival.

The basal ganglia taken together have also been implicated in behavioral disorders such as Tourette's syndrome and obsessive–compulsive disorder both of which exhibit symptoms related to control of actions (Pauls, Towbin, Leckman, Zahner, & Cohen, 1986). While each region of the basal ganglia can be seen to perform different functions, one theory is that taken together the structures form a control circuit of action selection in the brain. This is suggested to function through the culmination of other signals (e.g., from the somatosensory cortex) leading a decision or movement to be acted upon, or not acted upon, dependent on the configuration and processes of the basal ganglia itself (Alexander & Crutcher, 1990; Alexander, DeLong, & Strick, 1986).

The hippocampus, a small structure with a horn-like appearance (originally named for its resemblance to the shape of a seahorse), is located deep in the lower portion of the brain, proximal to the brain stem. The hippocampus initially drew a great deal of interest due to the infamous case of a patient identified as H.M. (later identified to be a man by the name of Henry Molaison). The patient was surgically treated for intractable epilepsy, in which a great portion of the hippocampus was removed. Upon waking, the patient was unable to form any new, explicit memories, confining his memory to only the presurgery events of his life. His working memory (an approximately 30 s short-term memory store) and his procedural memory (an implicit memory of procedures, such as riding a bike) were still intact, implicating the hippocampus a critical component for the formation of explicit memories (Squire, 2009). This is reinforced with the finding that the hippocampus is one of the first brain structures to show damage in Alzheimer's disease (Hyman, Hoesen, Damasio, & Barnes, 1984), of which worsened memory is a central symptom. Further research has also shown the importance of the hippocampus in spatial memory skills (Girardeau, Benchenane, Wiener, Buzsáki, & Zugaro, 2009). While the exact function of the hippocampus in relation to memory formation is still debated (Eichenbaum, 2004), although is clear that it plays a critical role. One theory suggests that the hippocampus works as a central hub in explicit memory formation and recall, in which signals are sent to other regions and the information is stored more locally (Battaglia, Benchenane, Sirota, Pennartz, & Wiener, 2011). This is suggested to be the reason why procedural memories could still be learnt in the case of H.M., as the memory formation bypassed the hippocampus (Squire, 1992).

Much of what we know about the amygdala has emerged in a similar way to that of the hippocampus, through the case study of an individual with a brain lesion. The patient in question, S.M., has Urbach–Wiethe disease, a rare genetic condition that causes a hardening of the medial temporal lobes. In S.M. this led to a destruction of the amygdalae and as a resulting symptom, the loss of the ability to experience fear (largely—a study administering the inhalation of carbon dioxide was able to elicit such a response, but this remains an exception; Feinstein et al., 2013). Other research has highlighted the importance of the amygdala in the memory processing

(Packard, Cahill, & Mcgaugh, 1994), decision-making (Bechara, Damasio, Damasio, & Lee, 1999), and in the formation of emotional responses (particularly, although not exclusively, with fear and anxiety responses, as suggested by the case of S.M.; Adolphs, Tranel, Damasio, & Damasio, 1994). One of the core functions of the amygdala is theorized to be in learning and memory formation in reference to emotional events (Dolcos, Labar, & Cabeza, 2004). This in part accounts for the propensity of the amygdala to more strongly be activated by negative emotional experiences (Ochsner et al., 2004), as these events are likely more instructive about long-term survival. Having a strong memory of a threatening event will reduce the chance that such a threat will be encountered in the future.

The claustrum is a small structure that resides relatively close to the globus pallidus of the basal ganglia. It is a structure with reciprocal connections to most, if not all, areas of the cortex and thalamus (Torgerson, Irimia, Goh, & Horn, 2014). Lesion studies in which only the claustrum is affected are extremely rare—the location of this brain region is such that exclusive damage to this region is unlikely. It has been proposed that such widespread interconnections with other regions suggest that the claustrum could have a crucial role in the generation of the conscious experience (Crick & Koch, 2005). More recent research has suggested that this region could instead play a role in regaining, but not maintaining, consciousness (Chau, Salazar, Krueger, Cristofori, & Grafman, 2015). There is continued speculation about the functionality of this brain region that has yet to be resolved.

The basal forebrain, located in front of and below the striatum of the basal ganglia, is a region of the brain that consists of several structures that are densely connected to the hippocampus, amygdala, and the frontal cortex (Mufson, Ginsberg, Ikonomovic, & Dekosky, 2003). Various studies have indicated that this area plays a role in mediating activity of areas downstream, suggesting a link in moderating attention through links to the frontal cortex, as well as a link in moderating memory processes through links to the hippocampus (Muir, Page, Sirinathsinghji, Robbins, & Everitt, 1993). Overall, it has been suggested that this area plays a role in the "modulation of cortical information processing" (Baxter & Chiba, 1999).

Connecting both hemispheres of the brain is the corpus callosum. While strictly speaking a nerve tract, rather than a brain area, this portion of the brain allows the different hemispheres to communicate with each other. This is particularly crucial for the common contralateral connections that take place for sensory processing (e.g., the visual system receives light in the left eye that is ultimately processed in the right hemisphere, this communication is transferred through the corpus callosum). While clearly a crucial component of the brain, individuals who have had their corpus callosum removed (often in the treatment of otherwise intractable epilepsy) do not necessarily suffer particularly deleterious impacts on their everyday functioning. Known as "split-brain" patients, interesting elements of stimulus perception have been shown in experiments testing their abilities. When a visual stimulus (such as a card showing a number) is presented exclusively to the right visual field, the split-brain patient can respond verbally and with their right hand. However, when this is repeated with an exclusive presentation to the left visual field, the patient does not appear aware of the stimulus (Pinto et al., 2017). This is due to the involvement of the right hemisphere in visual perception of the left visual field and in left-side motoric movements (Sakata & Taira, 1994), while the left hemisphere has an increased involvement in language processing (Vigneau et al., 2006). Early research on split-brain patients therefore suggested the possibility of two conscious hemispheres emerging from the splitting of the brain. Further research has elucidated that there is more nuance to these findings than initially reported, but that perceptual differences across hemispheres do exist (Pinto, Haan, & Lamme, 2017).

2.4 *The Brain Stem*

The brain stem consists of the midbrain, the pons, and the medulla oblongata. All of these structures reside deep within the cerebrum. The midbrain (sometimes referred to as the diencephalon) is at the cusp of the brain stem and cerebrum, often leading to inconsistencies in where this structure is said to be located. The midbrain principally contains three thalamic structures: the thalamus, hypothalamus, and epithalamus.

The principal connections from the brain to the rest of the body pass through the brain stem to the spinal cord, making this set of structures essential for supporting human life at its most fundamental level. As such, the brain stem is central to core physiological functions, such as heart rate, breathing, sleeping, eating, drinking, sexual behavior, and circadian rhythms (Abbott et al., 2013; Ikeda et al., 2016). Motor activity can also be controlled by the brain stem. While fine motoric movements often result as a convergence of activity between the motor cortex, somatosensory cortex, and the cerebellum, the brain stem is able to generate whole-body motoric movements. These whole-body movements may be walking, running, or other positioning, such as a defensive stance (Massion, 1992).

2.5 *The Cerebellum*

The cerebellum, located at the most anterior part of the head, is a finely convoluted structure that consists of almost half of the total neurons of the brain (Azevedo et al., 2009). Originally thought to only have an involvement in the planning and execution of motoric movements, more recent research has expanded this understanding, showing an involvement in various other processes (Buckner, 2013). Connected to the rest of the brain via the brain stem, with connections to the cortex via the thalamus, the cerebellum is often overlooked when considering the brain, but plays a crucial role in various processes.

The traditional notion of the cerebellar involvement in motoric processes is not without validity—the cerebellum is central and critically involved in the coordination and fine-tuning of movement. A certain degree of consensus has formed around the notion that the cerebellum operates internal models of motor movements that help to adjust and refine ongoing motoric processes (Manto et al., 2011). The cerebellum is ideally positioned for this task, due to its relative proximity to the sites of movements, ensuring a faster rate of motoric feedback can be expected compared to signals returning from the motor cortex or basal ganglia (Ramnani, 2006).

While the exact role that the cerebellum plays in cognitive and emotional processes remains largely at a hypothetical level (Koziol et al., 2014), a range of brain imaging research has shown that a variety of links do exist (Buckner, 2013). A central theory is that the function of the cerebellum in cognitive processes operates in a similar manner to its involvement in motor control—in storing models of efficient representations of cognitive behavior (Koziol et al., 2014).

3 A Cellular Understanding of the Brain

The previous section established some of the core anatomical components of the brain. This section will go a level further and discuss the cellular components that make up our central nervous system. There are multiple different cellular types, and beyond that various different subtypes, all of which have unique functions. The main cell type that is discussed within the context of the brain is the neuron (also known as a nerve cell), the cell that sends and receives signals to and from other neurons. There are also glial cells that perform a largely supportive role to the function of neurons (although are also capable of other functions). In addition to these two central nervous system cells, the blood supply to the brain has a central function in regulating energy use, and is often also a measure of brain activity as well. Each of these components will be discussed in further detail below.

3.1 Neurons

Neurons are the principal communicators of the brain. The brain regions discussed above are chiefly composed of regions of densely bundled together neurons. These cells can be split into three components—the cell body, the axon, and the dendrite. While it is beyond the scope of this chapter to dig too deeply into the molecular composition of this cell, it is beneficial to understand how the functions of this cell give rise to changes in brain activity and therefore changes in behavior. In essence, for a typical bipolar neuron, the dendrite is the component of the cell that receives the signal from another neuron or nerve cell, which causes a chain of reactions that involves interactions across the cell body, while the axon sends the signal further (Fig. 4).

The cell body is where the nucleus of the cell resides, which holds the DNA. The transcription (put in more simple terms, the decoding) of the DNA will ultimately determine how this cell develops and is maintained. This has implications for how the cell receives and sends signals. DNA that incorrectly codes for a certain necessary

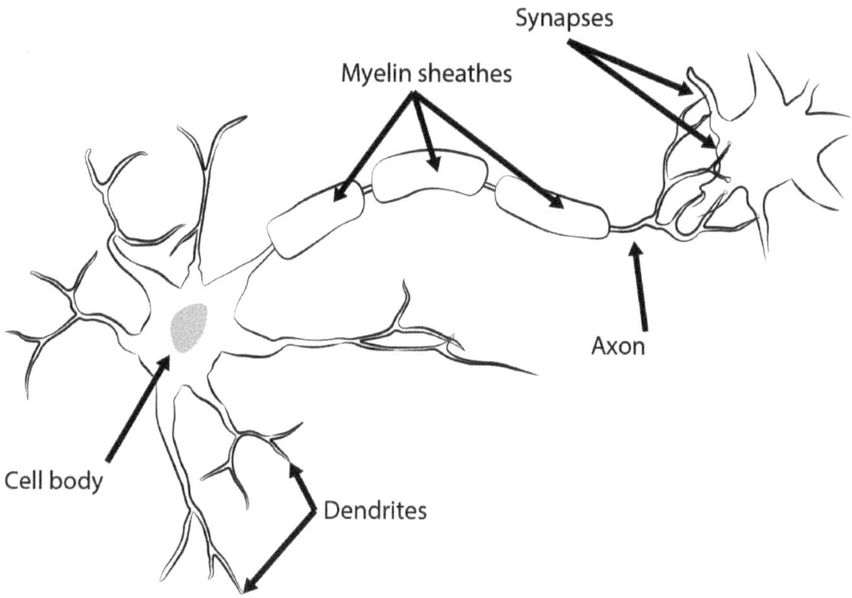

Fig. 4 A basic diagram of a neuron

protein could lead to a malfunction—this may not be of great consequence at the single-cell level, but a critical mass of malfunctioning neurons can lead to neurological or psychiatric disease or disorders. The nucleus is also the site of creation of many neurotransmitters, the molecules that determine how a signal is sent from the neuron to another.

A single neuron's cell body often has thousands of dendrites attached, all of which can receive input from multiple axons. Dendrites are highly branched—such arborization means that neurons can readily receive signals from multiple neurons. The dendrites themselves also include dendritic spines—multiple small protrusions that emerge from each branch. Dendritic spines contain pockets of electrochemical charges that can be readily released in the presence of a signal, extending that signal further within the cell it belongs to. In the presence of repeated signals to the dendrite, the spines can extend—theorized to be a facilitation of signal propagation (Wong & Ghosh, 2002). This process is directly linked to the formation and strengthening of memories (at least in the rat; Leggio et al., 2005). Similarly, if no signal is delivered to a dendritic spine over a long enough time period, the protrusion will diminish and eventually disappear.

The axons, which in most neuronal cells exist as singular protrusions, are transmitters of the neuronal signal. The electrochemical signal is propagated along the length of this extension. Within the brain most axons are wrapped at interspersed intervals, by a fatty (lipid-based) membrane. This insulation (or "myelination") is termed the myelin sheath, which helps to project the signal within the axon at faster speeds than without. The signal is able to essentially jump between the unmyelinated gaps of the axon (referred to as the "nodes of Ranvier"), vastly increasing the rate at which the electrochemical signal can travel. The signal at one node of Ranvier is able to provoke another action potential at the next node. This jumping of the signal between one node of Ranvier to another is known as saltatory conduction. The signal ultimately will arrive at the synapse—the end of the axon. The synapse is where neurotransmitters are released into what is called the synaptic cleft—the gap between the synapse and the subsequent point of action, such as a dendrite. From there the process of signal propagation can continue. The next part of this chapter will explain what that signal actually consists of.

3.2 Action Potentials

An action potential is a signal that is sent by a neuron. This is an electrochemical signal that travels along the membrane of the cell across the axon. In the absence of a signal, however, the cell remains at rest. The so-called resting state of a neuron is negatively charged, with the relative voltage inside the cell compared to the outside at −70 mV. An imbalance of sodium (Na) and potassium (K) ions maintains this disparity. While both sodium and potassium are positively charged ions (cations), a series of pumps in the cell membrane selectively creates an imbalance, with an increased number of potassium cells inside the cell, and an increased number of sodium cells outside of the cell at rest. These pumps use energy to return to and retain this disparity if the imbalance is perturbed.

When a signal arrives at a dendrite, the electrochemical imbalance can be adjusted. An increase of roughly 15 mV (changing the resting potential from −70 to −55 mV), resulting from a sudden ionic change, is often enough to trigger the continuation of the signal across the cell. Amounts less than this may change the voltage differences, but in the absence of another signal that can push the potential over the tipping point, the cell will return to the baseline of −70 mV. If, however, the change passes this threshold, a rapid series of changes occur in the composition of the cell, allowing the signal to propagate across the membrane.

If the inner part of the cell increases to −55 mV relative to outside, sodium channels will open. These are barriers that selectively allow sodium to pass through (referred to as voltage-gated ion channels). Due to the imbalance, the sodium ions will then rapidly enter the cell as the electrochemical attraction pulls them inward. The ions strive for electrochemical equilibrium, and this sudden change will overshoot, changing the imbalance ultimately to a relative potential of +40 mV inside the cell. The change in current affects neighboring areas, creating a domino-like effect of electrochemical changes, causing surrounding cell membranes to adjust in the same manner. This is how the action potential propagates (or "fires") through electrochemical perturbation. This is a rapidly occurring, all-or-nothing response. If the initial change isn't sufficient, the electrochemical potential will simply return to a resting state of −70 mV, or it will trigger the aforementioned changes. This imbalance though, once created, doesn't, however, remain at +40 mV. It's at this point that the sodium channels close, stopping the entrance of sodium into the cell. The potassium channels open, allowing potassium to cross over into the extracellular space, pulled by the attraction of the new electrochemical imbalance (that is in the opposite direction of the cell at rest). The new flooding of potassium into the extracellular space inverts the imbalance once more, and the inside of the cell becomes negatively charged relative to the outside. The potassium channels close once more. The cell, now more negative than at resting state, enters a brief refractory period in which the negative imbalance is so strong that new signals cannot trigger a new imbalance. The cell (or rather, the specific area of the cell) becomes momentarily impervious to outside interference. To return to the final resting state (−70 mV), a sodium–potassium pump moves three sodium ions out of the cell in exchange for two potassium ions entering. This process ultimately addresses the overshoot of the imbalance and the cell returns to the same potential as before, ready for a new signal to create changes, and to propagate that signal itself.

As the chain reaction of this signal approaches the synapse—the axon terminal—the electrochemical potential triggers a reaction leading to the release of neurotransmitters into the synaptic cleft. By binding to and activating receptors in the postsynaptic cleft (the membrane of the cell to which the axon is delivering its signal), the neurotransmitters can affect the downstream behavior of the subsequent cell. There are a great variety of neurotransmitters (over 200 have been identified; Wang et al., 2015) each with distinct roles and/or behaviors. Furthermore, there are a great number of neuronal cell types, some with multiple inputs (dendrites) and/or multiple outputs (axons) that exist in various different parts of the brain. The behavior of each of these cells can vary in terms of the rate of firing that they may elicit, whether or not they are tonic (constantly active), or whether they are excitatory or inhibitory (instigating or halting firing, respectively). The morphological and functional range of neurons are only just beginning to be explored in any great detail, but it is hoped that a catalogued computational understanding will provide a greater understanding of how the brain operates at large (Sunkin et al., 2012). This complexity of molecular and cellular types clearly makes attempts to elucidate the function of the brain a nontrivial matter. Overall, however, the communication between disparate brain regions can be understood as a series of electrochemical signals passed by chain reaction, from one area to another.

In essence, this is how the brain communicates. The series of electrochemical changes are minute on an individual level, but the synchronized activity of thousands of cells causes a large enough change in the electrical potential that can be detected by electroencephalography (EEG). It is of note that this imaging methodology primarily detects activity within the cortex, as that is closest in proximity to the electrode detecting the change in the electrical potential. However, changes deeper within the brain can potentially also be detected, although these are inevitably more difficult to pinpoint and may in fact increase the noisiness of the signal (Whittingstall, Stroink, Gates, Connolly, & Finley, 2003).

While neurons are the principal communicators of the brain, giving rise to all thoughts, feelings, and behaviors, they do not act alone. Around half of the cells within the brain are in fact something else: glia (Azevedo et al., 2009).

3.3 Glia

Originally named after the Greek word for glue, glia have often been considered to be little else than just binding material in the brain. While this is part of their function, it doesn't get across the full truth of the matter. There are in fact several types of glial cells, each with distinct roles in maintaining the functioning of the nervous system. An overview of three of the most widely discussed cell types is given below.

Oligodendrocytes are cells that construct and maintain the myelin sheaths that surround the axon, facilitating rapid conduction of neuronal signals. The process of myelination begins in the first year of life, but this process—of increasing efficient communication between and within brain regions—is not complete until early adulthood (Sowell et al., 2003). The progression of the myelin building begins in the occipital lobe (where visual information is processed) and continues toward the front of the brain. This process has interesting implications for development, as the biological components of adolescent brains, while fully formed in many respects, are still undergoing a process of refinement. It is of particular note that the frontal cortex, as the last area of the brain to be myelinated, has such strong associations with executive functions. Debate continues into the reliability and robustness of associations between the potential behavioral links with biological development (Paus, 2005). In any case, once the oligodendrocytes have completed their role in providing myelin to axons they also provide essential molecules for the continued maintenance of the axon, thus allowing neuronal signaling to continue in a healthy and regular manner.

The brain is often considered as consisting of both gray matter and white matter. The former pertains to areas that are rich in neurons, while the latter is rich in myelin. For example, the corpus callosum, a region of the brain that solely consists of connections between the left and right hemispheres, is myelin rich due to the myelin wrapping neurons that are created and maintained by oligodendrocytes. The cortex is dense with neurons, and in between this area and other subcortical regions is a myelin-rich, white matter area that facilitates rapid communication. Imaging of the white matter can be carried out with a particular application of magnetic resonance imaging (MRI) known as diffusion tensor imaging (DTI). This imaging technique is able to reveal details about the connections between brain areas (Behrens et al., 2003).

Astrocytes are widely dispersed throughout the brain, making contact with various parts of blood vessels, neurons, as well as other astrocytes (Rouach, Glowinski, & Giaume, 2000). They have a wide range of roles and functions, and an interesting morphology that includes receptors and ion channels in a somewhat similar fashion to neurons. A large bulk of astrocyte research has shown their role in maintaining cells, providing needed molecules, and absorbing and recycling material from neurons (Allaman, Bélanger, & Magistretti, 2011). However, research has also shown that in some cases of when absorbing neurotransmitters, signaling can be triggered in astrocytes (Zonta et al., 2002). This can even lead to an activation of neurons, changes in blood flow in the brain, and has even been shown to affect sleeping behavior in mice (Halassa et al., 2009). Future research will likely continue to elucidate the true complexity of function that these cells appear to harbor.

Microglia are the resident immune cells of the brain. Due to impermeability of the blood–brain barrier to white blood cells, microglia substitute the role as singular resident protectors against infectious diseases and foreign material (Graeber & Streit, 2009). The cells can track, ingest, and break apart most harmful material. They are widely distributed throughout the brain.

3.4 Blood Supply to the Brain

All organs of the body require energy to sustain functioning, and the brain is the principal user, accounting for around 20% of total oxygen consumption and 11% of the cardiac output (Siegel, 2011). There are two principal sources of blood to the brain, from the internal carotid artery and vertebral artery, stemming from similar positions in the neck. The blood is carried through the blood–brain barrier, up along the underside of the brain stem. The blood vessels continue to penetrate upward further into the brain, branching off continuously throughout. The internal carotid artery splits into the anterior and middle cerebral artery. In broad terms, the anterior cerebral artery provides blood to the frontal cortex and subcortical regions, while the middle cerebral artery provides blood to the non-posterior cortex. The vertebral artery provides blood to the posterior cortex and generally within the occipital lobe. A restriction of blood to any of these regions (ischemia) can ultimately cause a stroke.

Due to the clear importance of oxygenated blood in both normal functioning and diseased states of the brain, these signals, detected by functional MRI (fMRI), have been particularly influential in the understanding of the brain. MRI detects signals created by hydrogen atoms that are found in abundance within our body (up to 70% of the body is water; Siri, 1956). This is able to create a static map of the brain, showing neuroanatomical structures. In order to examine the functioning of the brain, fMRI can be applied. This approach detects oxygenated blood. As oxygenated blood is required for brain activity, the signal is utilized as a proxy of brain functioning. This signal, while technically an indirect measure of brain activity, has been used in thousands of studies to elucidate the mechanisms of brain function.

4 Conclusions

The brain is a complex organ no matter which way it is examined, from either molecular and cellular perspectives, or functional and behavioral perspectives. Deciphering the processes of function to behavior remains a challenge for researchers working within any brain-based field.

Molecular and cellular approaches have yielded great discoveries about function on a micro-level, but an equivalent and equal translation of these processes into an understanding of function at a macro-level has not occurred. There are a great deal of unknown components and factors that give rise to brain function. This is not to say that attempts at understanding the brain have not allowed researchers to gain a deeper understanding of how our brains, bodies, and behaviors interact, but simply that there is much more to learn.

In the late nineteenth century, phrenology became a popular approach to understanding brain function. This field assumed that brain regions would dictate behavioral outcomes in bizarrely detailed ways, with brain areas purported to be linked to criminality, morality, conscientiousness, etc. The size of particular brain regions was taken as evidence that a person was more likely to engage in certain behaviors. This was obviously discredited with the advent of more thorough investigations into brain function and was subsequently ridiculed. However, there was an element of truth about the phrenological approach—certain brain areas do relate to certain behaviors. It just manifests in a different way than was envisaged (and the relation to size is far from clear). Neuroimaging, in a variety of forms, has allowed a better dissection (sometimes literally) of the relationship between brain and function.

A similar change in understanding of the molecular and cellular aspects of the brain also occurred at the beginning of the twenty-first century, when Ramon y Cajal showed neurons and synapses for the very first time. Before this breakthrough, neuroscientists were essentially limited to crude lesion studies (and the aforementioned phrenological approach), which didn't and couldn't fill in the gaps of knowledge about brain function. A closer inspection of the cellular processes has allowed a better understanding of disease progression, and in accordance with this, disease treatment. This understanding has also pushed forward brain imaging methods themselves, so that clearer ideas regarding brain function can begin to emerge. With this improved insight into the brain, and the molecular and cellular understanding as the backbone of this knowledge, we are now able to draw clearer inferences from brain function and relate them to behaviors.

References

Abbott, S. M., Arnold, J. M., Chang, Q., Miao, H., Ota, N., Cecala, C., … Gillette, M. U. (2013). Signals from the brainstem sleep/wake centers regulate behavioral timing via the circadian clock. *PLoS ONE, 8*(8). https://doi.org/10.1371/journal.pone.0070481.

Adolphs, R., Tranel, D., Damasio, H., & Damasio, A. (1994). Impaired recognition of emotion in facial expressions following bilateral damage to the human amygdala. *Nature, 372*(6507), 669–672. https://doi.org/10.1038/372669a0.

Alexander, G., DeLong, M., & Strick, P. (1986). Parallel organization of functionally segregated circuits linking basal ganglia and cortex. *Annual Review of Neuroscience, 9*(1), 357–381. https://doi.org/10.1146/annurev.neuro.9.1.357.

Alexander, G. E., & Crutcher, M. D. (1990). Functional architecture of basal ganglia circuits: Neural substrates of parallel processing. *Trends in Neurosciences, 13*(7), 266–271. https://doi.org/10.1016/0166-2236(90)90107-l.

Allaman, I., Bélanger, M., & Magistretti, P. J. (2011). Astrocyte–neuron metabolic relationships: For better and for worse. *Trends in Neurosciences, 34*(2), 76–87. https://doi.org/10.1016/j.tins.2010.12.001.

Alvarez, J. A., & Emory, E. (2006). Executive function and the frontal lobes: A meta-analytic review. *Neuropsychology Review, 16*(1), 17–42. https://doi.org/10.1007/s11065-006-9002-x.

Azevedo, F. A., Carvalho, L. R., Grinberg, L. T., Farfel, J. M., Ferretti, R. E., Leite, R. E., … Herculano-Houzel, S. (2009). Equal numbers of neuronal and nonneuronal cells make the human brain an isometrically scaled-up primate brain. *The Journal of Comparative Neurology, 513*(5), 532–541. https://doi.org/10.1002/cne.21974.

Battaglia, F. P., Benchenane, K., Sirota, A., Pennartz, C. M., & Wiener, S. I. (2011). The hippocampus: Hub of brain network communication for memory. *Trends in Cognitive Sciences*. https://doi.org/10.1016/j.tics.2011.05.008.

Baxter, M. G., & Chiba, A. A. (1999). Cognitive functions of the basal forebrain. *Current Opinion in Neurobiology, 9*(2), 178–183. https://doi.org/10.1016/s0959-4388(99)80024-5.

Bechara, A., Damasio, H., Damasio, A. R., & Lee, G. P. (1999). Different contributions of the human amygdala and ventromedial prefrontal cortex to decision-making. *The Journal of Neuroscience, 19*(13), 5473–5481. https://doi.org/10.1523/jneurosci.19-13-05473.1999.

Behrens, T. E., Johansen-Berg, H., Woolrich, M. W., Smith, S. M., Wheeler-Kingshott, C. A., Boulby, P. A., … Matthews, P. M. (2003). Non-invasive mapping of connections between human thalamus and cortex using diffusion imaging. *Nature Neuroscience, 6*(7), 750–757. https://doi.org/10.1038/nn1075.

Buckner, R. (2013). The cerebellum and cognitive function: 25 years of insight from anatomy and neuroimaging. *Neuron, 80*(3), 807–815. https://doi.org/10.1016/j.neuron.2013.10.044.

Burns, R. S., Chiueh, C. C., Markey, S. P., Ebert, M. H., Jacobowitz, D. M., & Kopin, I. J. (1983). A primate model of parkinsonism: Selective destruction of dopaminergic neurons in the pars compacta of the substantia nigra by N-methyl-4-phenyl-1,2,3,6-tetrahydropyridine. *Proceedings of the National Academy of Sciences, 80*(14), 4546–4550. https://doi.org/10.1073/pnas.80.14.4546.

Chau, A., Salazar, A. M., Krueger, F., Cristofori, I., & Grafman, J. (2015). The effect of claustrum lesions on human consciousness and recovery of function. *Consciousness and Cognition, 36*, 256–264. https://doi.org/10.1016/j.concog.2015.06.017.

Crick, F. C., & Koch, C. (2005). What is the function of the claustrum? *Philosophical Transactions of the Royal Society B: Biological Sciences, 360*, 1271–1279.

Defelipe, J., Markram, H., & Rockland, K. S. (2012). The neocortical column. *Frontiers in Neuroanatomy, 6*. https://doi.org/10.3389/fnana.2012.00005.

DeLong, M. R. (1990). Primate models of movement disorders of basal ganglia origin. *Trends in Neurosciences, 13*(7), 281–285. https://doi.org/10.1016/0166-2236(90)90110-v.

Dogan, I., Eickhoff, S. B., Schulz, J. B., Shah, N. J., Laird, A. R., Fox, P. T., & Reetz, K. (2013). Consistent neurodegeneration and its association with clinical progression in Huntingtons disease: A coordinate-based meta-analysis. *Neurodegenerative Diseases, 12*(1), 23–35. https://doi.org/10.1159/000339528.

Dolcos, F., Labar, K. S., & Cabeza, R. (2004). Interaction between the amygdala and the medial temporal lobe memory system predicts better memory for emotional events. *Neuron, 42*(5), 855–863. https://doi.org/10.1016/s0896-6273(04)00289-2.

Eichenbaum, H. (2004). Hippocampus: Cognitive processes and neural representations that underlie declarative memory. *Neuron, 44*, 109–120.

Elbert, T., Flor, H., Birbaumer, N., Knecht, S., Hampson, S., Larbig, W., & Taub, E. (1994). Extensive reorganization of the somatosensory cortex in adult humans after nervous system injury. *NeuroReport, 5*(18), 2593–2597. https://doi.org/10.1097/00001756-199412000-00047.

Feinstein, J. S., Buzza, C., Hurlemann, R., Follmer, R. L., Dahdaleh, N. S., Coryell, W. H., et al. (2013). Fear and panic in humans with bilateral amygdala damage. *Nature Neuroscience, 16*(3), 270–272. https://doi.org/10.1038/nn.3323.

Feldman, D. E., & Brecht, M. (2005). Map plasticity in somatosensory cortex. *Science, 310*(5749), 810–815. https://doi.org/10.1126/science.1115807.

Gallese, V., Fadiga, L., Fogassi, L., & Rizzolatti, G. (1996). Action recognition in the premotor cortex. *Brain, 119*(2), 593–609. https://doi.org/10.1093/brain/119.2.593.

Girardeau, G., Benchenane, K., Wiener, S. I., Buzsáki, G., & Zugaro, M. B. (2009). Selective suppression of hippocampal ripples impairs spatial memory. *Nature Neuroscience, 12*(10), 1222–1223. https://doi.org/10.1038/nn.2384.

Graeber, M. B., & Streit, W. J. (2009). Microglia: Biology and pathology. *Acta Neuropathologica, 119*(1), 89–105. https://doi.org/10.1007/s00401-009-0622-0.

Halassa, M. M., Florian, C., Fellin, T., Munoz, J. R., Lee, S., Abel, T., ... Frank, M. G. (2009). Astrocytic modulation of sleep homeostasis and cognitive consequences of sleep loss. *Neuron, 61*(2), 213–219. https://doi.org/10.1016/j.neuron.2008.11.024.

Hausdorff, J. M., Cudkowicz, M. E., Firtion, R., Wei, J. Y., & Goldberger, A. L. (1998). Gait variability and basal ganglia disorders: Stride-to-stride variations of gait cycle timing in Parkinsons disease and Huntingtons disease. *Movement Disorders, 13*(3), 428–437. https://doi.org/10.1002/mds.870130310.

Hyman, B., Hoesen, G. V., Damasio, A., & Barnes, C. (1984). Alzheimers disease: Cell-specific pathology isolates the hippocampal formation. *Science, 225*(4667), 1168–1170. https://doi.org/10.1126/science.6474172.

Ikeda, K., Kawakami, K., Onimaru, H., Okada, Y., Yokota, S., Koshiya, N., ... Koizumi, H. (2016). The respiratory control mechanisms in the brainstem and spinal cord: Integrative views of the neuroanatomy and neurophysiology. *The Journal of Physiological Sciences, 67*(1), 45–62. https://doi.org/10.1007/s12576-016-0475-y.

Ikemoto, S., & Panksepp, J. (1999). The role of nucleus accumbens dopamine in motivated behavior: A unifying interpretation with special reference to reward-seeking. *Brain Research Reviews, 31*(1), 6–41. https://doi.org/10.1016/s0165-0173(99)00023-5.

Koziol, L. F., Budding, D., Andreasen, N., D'Arrigo, S., Bulgheroni, S., Imamizu, H., ... Yamazaki, T. (2014). Consensus paper: The cerebellums role in movement and cognition. *The Cerebellum, 13*(1), 151–177. https://doi.org/10.1007/s12311-013-0511-x.

Leggio, M. G., Mandolesi, L., Federico, F., Spirito, F., Ricci, B., Gelfo, F., & Petrosini, L. (2005). Environmental enrichment promotes improved spatial abilities and enhanced dendritic growth in the rat. *Behavioural Brain Research, 163*(1), 78–90. https://doi.org/10.1016/j.bbr.2005.04.009.

Livingstone, M., & Hubel, D. (1988). Segregation of form, color, movement, and depth: Anatomy, physiology, and perception. *Science, 240*, 740–749.

Lumpkin, E. A., & Caterina, M. J. (2007). Mechanisms of sensory transduction in the skin. *Nature, 445*, 858–865.

Manto, M., Bower, J. M., Conforto, A. B., Delgado-García, J. M., Guarda, S. N., Gerwig, M., ... Timmann, D. (2011). Consensus paper: Roles of the cerebellum in motor control—The diversity of ideas on cerebellar involvement in movement. *The Cerebellum, 11*(2), 457–487. https://doi.org/10.1007/s12311-011-0331-9.

Martin, J. H., Radzyner, H., & Leonard, M. E. (2012). *Neuroanatomy: Text and atlas*. London: McGraw-Hill.

Massion, J. (1992). Movement, posture and equilibrium: Interaction and coordination. *Progress in Neurobiology, 38*(1), 35–56. https://doi.org/10.1016/0301-0082(92)90034-c.

Middleton, F. A., & Strick, P. L. (2000). Basal ganglia output and cognition: Evidence from anatomical, behavioral, and clinical studies. *Brain and Cognition, 42*(2), 183–200. https://doi.org/10.1006/brcg.1999.1099.

Miller, E. K., & Cohen, J. D. (2001). An integrative theory of prefrontal cortex function. *Annual Review of Neuroscience, 24*(1), 167–202. https://doi.org/10.1146/annurev.neuro.24.1.167.

Mufson, E. J., Ginsberg, S. D., Ikonomovic, M. D., & Dekosky, S. T. (2003). Human cholinergic basal forebrain: Chemoanatomy and neurologic dysfunction. *Journal of Chemical Neuroanatomy, 26*(4), 233–242. https://doi.org/10.1016/s0891-0618(03)00068-1.

Muir, J. L., Page, K. J., Sirinathsinghji, D., Robbins, T. W., & Everitt, B. J. (1993). Excitotoxic lesions of basal forebrain cholinergic neurons: Effects on learning, memory and attention. *Behavioural Brain Research, 57*(2), 123–131. https://doi.org/10.1016/0166-4328(93)90128-d.

Nelson, A. B., & Kreitzer, A. C. (2014). Reassessing models of basal ganglia function and dysfunction. *Annual Review of Neuroscience, 37*(1), 117–135. https://doi.org/10.1146/annurev-neuro-071013-013916.

Ochsner, K. N., Ray, R. D., Cooper, J. C., Robertson, E. R., Chopra, S., Gabrieli, J. D., & Gross, J. J. (2004). For better or for worse: Neural systems supporting the cognitive down- and up-regulation of negative emotion. *NeuroImage, 23*(2), 483–499. https://doi.org/10.1016/j.neuroimage.2004.06.030.

Öngur, D., & Price, J. L. (2000). The organization of networks within the orbital and medial prefrontal cortex of rats, monkeys and humans. *Cerebral Cortex, 10*(3), 206–219. https://doi.org/10.1093/cercor/10.3.206.

Packard, M. G., Cahill, L., & Mcgaugh, J. L. (1994). Amygdala modulation of hippocampal-dependent and caudate nucleus-dependent memory processes. *Proceedings of the National Academy of Sciences, 91*(18), 8477–8481. https://doi.org/10.1073/pnas.91.18.8477.

Paulmann, S., Pell, M. D., & Kotz, S. A. (2008). Functional contributions of the basal ganglia to emotional prosody: Evidence from ERPs. *Brain Research, 1217,* 171–178. https://doi.org/10.1016/j.brainres.2008.04.032.

Pauls, D., Towbin, K., Leckman, J., Zahner, G., & Cohen, D. (1986). Gilles de la Tourette's syndrome and obsessive-compulsive disorder: Evidence supporting a genetic relationship. *Archives of General Psychiatry, 43,* 1180–1182.

Paus, T. (2005). Mapping brain maturation and cognitive development during adolescence. *Trends in Cognitive Sciences, 9*(2), 60–68. https://doi.org/10.1016/j.tics.2004.12.008.

Picard, N., & Strick, L. P. (2003). Activation of the supplementary motor area (SMA) during performance of visually guided movements. *Cerebral Cortex, 13*(9), 977–986. https://doi.org/10.1093/cercor/13.9.977.

Pinto, Y., Haan, E. H., & Lamme, V. A. (2017a). The split-brain phenomenon revisited: A single conscious agent with split perception. *Trends in Cognitive Sciences, 21*(11), 835–851. https://doi.org/10.1016/j.tics.2017.09.003.

Pinto, Y., Neville, D. A., Otten, M., Corballis, P. M., Lamme, V. A., Haan, E. H., ... Fabri, M. (2017b). Split brain: Divided perception but undivided consciousness. *Brain*. https://doi.org/10.1093/brain/aww358.

Pontieri, F. E., Tanda, G., Orzi, F., & Chiara, G. D. (1996). Effects of nicotine on the nucleus accumbens and similarity to those of addictive drugs. *Nature, 382*(6588), 255–257. https://doi.org/10.1038/382255a0.

Ramnani, N. (2006). The primate cortico-cerebellar system: Anatomy and function. *Nature Reviews Neuroscience, 7*(7), 511–522. https://doi.org/10.1038/nrn1953.

Roberts, A. C., Robbins, T. W., & Weiskrantz, L. (2003). *The prefrontal cortex: Executive and cognitive functions*. Oxford: Oxford University Press.

Rouach, N., Glowinski, J., & Giaume, C. (2000). Activity-dependent neuronal control of gap-junctional communication in astrocytes. *The Journal of Cell Biology, 149*(7), 1513–1526. https://doi.org/10.1083/jcb.149.7.1513.

Sakata, H., & Taira, M. (1994). Parietal control of hand action. *Current Opinion in Neurobiology, 4,* 847–856.

Siegel, G. J. (2011). *Basic neurochemistry: Molecular, cellular and medical aspects*. Amsterdam: Elsevier Academic Press.

Siri, W. E. (1956). The gross composition of the body. In *Advances in biological and medical physics* (pp. 239–280). New York: Academic Press.

Sowell, E. R., Peterson, B. S., Thompson, P. M., Welcome, S. E., Henkenius, A. L., & Toga, A. W. (2003). Mapping cortical change across the human life span. *Nature Neuroscience, 6*(3), 309–315. https://doi.org/10.1038/nn1008.

Squire, L. R. (1992). Memory and the hippocampus: A synthesis from findings with rats, monkeys, and humans. *Psychological Review, 99*(2), 195–231. https://doi.org/10.1037//0033-295x.99.2.195.

Squire, L. R. (2009). The legacy of patient H.M. for neuroscience. *Neuron, 61*(1), 6–9. https://doi.org/10.1016/j.neuron.2008.12.023.

Sunkin, S. M., Ng, L., Lau, C., Dolbeare, T., Gilbert, T. L., Thompson, C. L., … Dang, C. (2012). Allen brain atlas: An integrated spatio-temporal portal for exploring the central nervous system. *Nucleic Acids Research, 41*(D1). https://doi.org/10.1093/nar/gks1042.

Torgerson, C. M., Irimia, A., Goh, S. Y., & Horn, J. D. (2014). The DTI connectivity of the human claustrum. *Human Brain Mapping, 36*(3), 827–838. https://doi.org/10.1002/hbm.22667.

Vigneau, M., Beaucousin, V., Hervé, P., Duffau, H., Crivello, F., Houdé, O., … Tzourio-Mazoyer, N. (2006). Meta-analyzing left hemisphere language areas: Phonology, semantics, and sentence processing. *NeuroImage, 30*(4), 1414–1432. https://doi.org/10.1016/j.neuroimage.2005.11.002.

Wang, Y., Wang, M., Yin, S., Jang, R., Wang, J., Xue, Z., & Xu, T. (2015). NeuroPep: A comprehensive resource of neuropeptides. *Database* (Oxford). https://doi.org/10.1093/database/bav038.

Whittingstall, K., Stroink, G., Gates, L., Connolly, J., & Finley, A. (2003). Effects of dipole position, orientation and noise on the accuracy of EEG source localization. *Biomedical Engineering Online, 2,* 14.

Wong, R. O., & Ghosh, A. (2002). Activity-dependent regulation of dendritic growth and patterning. *Nature Reviews Neuroscience, 3*(10), 803–812. https://doi.org/10.1038/nrn941.

Zonta, M., Angulo, M. C., Gobbo, S., Rosengarten, B., Hossmann, K., Pozzan, T., & Carmignoto, G. (2002). Neuron-to-astrocyte signaling is central to the dynamic control of brain microcirculation. *Nature Neuroscience, 6*(1), 43–50. https://doi.org/10.1038/nn980.

Neuroimaging Methods in Neuroergonomics

The EEG Cookbook: A Practical Guide to Neuroergonomics Research

Nathan Sanders, Sanghyun Choo and Chang S. Nam

Abstract Conducting an EEG-based neuroergonomics experiment can be a daunting task for novice researchers. This chapter provides an overview of three aspects of EEG research which we hope will help novice researchers efficiently produce meaningful and replicable results: power analysis, data preprocessing, and reporting. We explain why power analysis and sample size estimation are critical yet often overlooked aspects of experimental research and describe the most common measures of effect size likely to be encountered, Cohen's *d* and eta-squared. We also provide a list of powerful (and free) power analysis tools to facilitate the actual calculations. We also provide step-by-step instructions for data preprocessing with EEGLAB which can be used in preparation for subsequent ERP or connectivity analyses. This includes filtering, artifact removal and correction, independent component analysis, and source localization. Finally, we condense EEG reporting guidelines into a checklist which can be used to ensure that your manuscript draft follows best practices.

1 Introduction

Conducting an electroencephalogram (EEG)-based neuroergonomic experiment can be a daunting task for novice researchers. From conducting a power analysis in the early planning stages to designing experimental manipulations, and from processing data to reporting results, the amount of technical and theoretical knowledge required to produce meaningful results is both broad and deep. The main goal of this chapter is to provide a brief overview of some of the aspects of EEG-based experiments that will serve as an entry point and guide to other researchers who are new to the field.

We first review an underappreciated aspect of the experimental design: power analysis and sample size calculation. While my treatment is not as thorough as some of the sources we have cited, we do enumerate some of the most important

N. Sanders · S. Choo · C. S. Nam (✉)
Edward P. Fitts Department of Industrial & Systems Engineering, North Carolina State University, Raleigh, NC 27695, USA
e-mail: csnam@ncsu.edu

C. S. Nam (ed.), *Neuroergonomics*, Cognitive Science and Technology,
https://doi.org/10.1007/978-3-030-34784-0_3

equations which can be referenced while inputting data into various sample size calculators. Next, we provide a step-by-step "recipe" for preprocessing EEG data in EEGLAB which we think is a little more accessible for the first-time user than the (excellent) tutorial on the EEGLAB wiki. Last, we summarize a set of important reporting guidelines for EEG-based experiments, which can be used as a checklist when writing a manuscript draft.

2 Power Analysis

The power of a study reflects the chance that your experiment will detect a true effect, defined as $1 - \beta$ (the complement of the type-II error rate). It is the probability that you will correctly reject the null hypothesis when it is indeed false. Although the significance of a particular result is interpreted via the p-value, which reflects the chance that your particular observation is a false positive, the p-value by itself does not provide any information about the chance of a false negative. The significance level, power, effect size, and sample size are intimately linked, and generally speaking the power is inversely proportional to the significance level. The less you are willing to accept a false positive, the more likely it will be that you fail to detect a true effect. That is, the less power your experiment will have. In the following section, I will briefly outline why a power analysis should be done, when to do it, and provide an overview of useful formulas and software packages that will help you.

2.1 Why Perform a Power Analysis?

Reproducibility is a cornerstone of the scientific method, but there is an increasing concern that many (or most) published experimental results are false or overinflated (Ioannidis, 2005). This is especially true in the fields of neuroscience and psychology, in which low sample size has been implicated as the primary culprit (Larson & Carbine, 2017). The reason this occurs is twofold. First of all, scientific journals tend not to publish papers which report statistically insignificant results. Second, exploratory experiments tend to have small sample sizes, which makes them underpowered and is simply a part of the discovery process. The result, however, is a selection effect in which only relatively large and unlikely effect sizes (which are random variables) are reported. When other teams attempt to replicate an exploratory study, they start by conducting an a priori power analysis based on an inflated effect size, so they too will create an underpowered study. The result will either be an insignificant (hence contradictory) result which may go unpublished, or by chance they observe an unusually large effect size too and the cycle repeats itself. Both the fixation on p-values and suboptimal power are necessary for this phenomenon to occur (Ioannidis, 2008). To illustrate this effect, Johnson, Payne, Wang, Asher, and Mandal (2017) conducted replications of 100 experimental and correlational studies published in three psychological journals. They found that 97% of the original studies reported significant results ($p < 0.05$), but only 36% of their replications had significant results. Only 47% of the original effect sizes were within the 95% confidence intervals of the replication effect sizes.

The only way to counteract this effect is to conduct a proper a priori power analysis while making a reasonable "factor of safety" correction to published effect sizes, and in turn report all the information that future scientists need to conduct their own power analyses. Unfortunately, the EEG community seems to be rather bad at this. A study by Larson and Carbine (2017) reviewed 100 randomly selected clinical EEG studies and found that none (0%) reported sample size calculations. 77% used repeated measure designs but none of those reported the variances and correlations necessary for future researchers to calculate sample sizes. Only 44% reported an effect size of any kind.

Power analyses are often calculated during the planning stages of an experiment in order to determine the sample size necessary to detect a certain effect. This is called an a priori power analysis, and is one of the four types of statistical power analyses that can be done (the others being post hoc, compromise, and sensitivity analysis) (Faul, Erdfelder, Lang, & Buchner, 2007). In an a priori analysis, sample size is calculated as a function of significance level, power, and effect size. To understand why it is important to do this, consider the situation in which your sample size is too low. Your experiment will be underpowered and may fail to detect an effect, in which case you and your colleagues have wasted time, wasted resources, and needlessly subjected participants to experimentation when there was not a reasonable chance of obtaining a meaningful result. On the other hand, if your sample size is too large, you again waste time and resources by conducting more testing than was needed, and you have again subjected more people to experimentation than was strictly necessary. Both cases are wasteful and unethical (Guo, Logan, Glueck, & Muller, 2013). It is therefore important to at least attempt a power analysis during the planning stages of your experiment in order to provide some rational justification for your sample size.

The second most common type of power analysis is post hoc. In this version, power is computed as a function of significance level, population effect size, and sample size. Situations like this may arise as a result of budgetary or time constraints which may determine the sample size for you. Even though this is termed a post hoc power analysis, it can still be used beforehand to answer questions like "With n observations, you will have a $(1 - \beta)$ chance of detecting an effect of this size. Is that acceptable?" (Lenth, 2001). It may also be used to assess whether or not a published test had a reasonable chance of detecting an effect (Faul et al., 2007), hence its name. It is important to distinguish this test from a retroactive power analysis, in which you use a sample effect size to calculate the "observed power" of a test. The critical difference is that a post hoc test requires an a priori specification of effect size, while a retroactive test assumes that the sampled estimate of effect size is an accurate reflection of the true effect size (Faul et al., 2007). This assumption is almost never true. The use by other scientists of "observed power" to calculate sample sizes for replication studies, especially when the observed power estimates come from pilot studies with small sample sizes, is a primary cause of the replication crisis in science (Ioannidis, 2008).

The last type of analysis we would like to mention is the sensitivity analysis. In this test, you calculate the effect size as a function of significance level, power, and sample size. It can be used to answer the question "What effect size was a study able to detect with a power of $(1 - \beta)$ given its sample size n and significance level α as specified by the authors?" (Faul et al., 2007). Or, if your experiment is in the planning stages and your sample size is constrained, "What effect size is our study powered to detect?" (Lenth, 2001). It will be up to you to determine whether this effect size is of scientific or practical importance.

2.2 *Ingredients of a Power Analysis*

The power of a test is a function of three variables: the significance level α, the sample size n, and the effect size that must be detected (Bezeau & Graves, 2003). Knowing any three of these will allow you to calculate the remaining one. In the case of an a priori analysis, we use power, significance level, and effect size to calculate the critical sample size. Below are the steps of a power analysis (Guo et al., 2013; Lenth, 2001):

- *Step 1. Specify the hypothesis.* In the case of EEG experiments, it is not sufficient to make a general prediction that some measures will differ (Keil et al., 2014). You must be very specific about the (for example) amplitudes, latencies, components, dipole locations, or connectivity metrics which you plan to compare. Your hypothesis will help to determine the type of statistical test you will use.
- *Step 2. Specify the significance level of the test.* This includes any correction factors (e.g., Bonferroni) that you plan to use for multiple comparisons.
- *Step 3. Specify an effect size.* It must be of scientific or practical significance, not merely statistical significance. Consequently, it must be in real units. Do not use standardized effect sizes without considering the means, variances, and correlations which comprise them.
- *Step 4. Specify the power of the test.* This is usually set at 0.80.

With these pieces of information in hand, you are ready to begin. Because specifying significance level and power is relatively straightforward, in the next section, we will focus on the key variable of effect size in the context of a sample size calculation.

2.3 *Effect Size*

Effect size is actually a compound variable composed of a point estimate expressible in real units (e.g., microvolts or milliseconds) and a measure of spread or correlation. An example of one such formula is as follows:

$$\theta = \frac{\mu_1 - \mu_2}{\sigma}. \tag{1}$$

Theta is a dimensionless quantity that relates the difference in population means to the pooled standard deviation. It is important to recognize the distinction between a "standardized" effect size (θ) and an absolute effect size ($\mu_1 - \mu_2$). The absolute effect size is a statement about the scientific significance of an observation; the standardized effect size is not. Even though you will often see various standardized effect sizes reported (and you should report standardized effect sizes yourself) it is important to attempt to calculate these from their component parts, and to provide other researchers with the means and variances they need to do so as well.

There is a plethora of standardized effect sizes which are used for different hypothesis tests. They are all scale free and range upward from zero (Cohen, 1992). In the following sections, we will describe the measures that are used for t-tests and ANOVA, since they are the most common tests you will encounter in EEG research (Larson & Carbine, 2017).

2.3.1 Cohen's *d*

Use Cohen's d_s when estimating sample size for a between-subjects t-test (Lakens, 2013).

$$d_s = \frac{\overline{\overline{X_1}} - \overline{\overline{X_2}}}{s}, \tag{2}$$

where s is the pooled standard deviation:

$$s = \sqrt{\frac{(n_1 - 1)\mathrm{SD}_1^2 + (n_2 - 1)\mathrm{SD}_2^2}{n_1 + n_2 - 2}}. \tag{3}$$

Cohen's d_s can be expressed in terms of the t-value,

$$d_s = t\sqrt{\frac{1}{n_1} + \frac{1}{n_2}}. \tag{4}$$

If only the total sample size is known,

$$d_s \approx \frac{2t}{\sqrt{N}}. \tag{5}$$

Use Cohen's d_z for within-subjects (paired) designs (Lakens, 2013).

$$d_z = \frac{t}{\sqrt{n}}. \tag{6}$$

2.3.2 Classical Eta-Squared

In a single-factor ANOVA, classical eta-squared (the "differentiation ratio") is analogous to r^2 in that it describes the proportion of variation in the dependent variable that can be attributed to a certain predictor, and its value ranges from 0 to 1 (Richardson, 2011):

$$\eta^2 = \frac{SS_{\text{Treatment}}}{SS_{\text{Total}}}. \tag{7}$$

In the case of a single-factor between-subjects ANOVA with k groups and N participants, classical eta-squared can also be calculated from the F-statistic with $(k - 1)$ and $(N - k)$ degrees of freedom (Richardson, 2011):

$$\eta^2 = \frac{F(k-1)}{F(k-1) + (N-k)}. \tag{8}$$

2.3.3 Partial Eta-Squared

Partial eta-squared is "overwhelmingly used" as a measure of effect size in social science research (Richardson, 2011). In the case of multifactor between-subject ANOVAs, partial eta-squared is defined as follows:

$$\eta_p^2 = \frac{SS_{\text{factor A}}}{SS_{\text{factor A}} + SS_{\text{error}}}. \tag{9}$$

It can be calculated separately for each factor in the design, and facilitates effect-size comparison between different studies with the same design. Partial eta-squared can be calculated from an F-statistic with its degrees of freedom:

$$\eta_p^2 = \frac{F \cdot df_{\text{factor A}}}{F \cdot df_{\text{factor A}} + df_{\text{error}}}. \tag{10}$$

In the special case where there are only two groups,

$$\eta_p^2 = \frac{t^2}{t^2 + N - 2}. \tag{11}$$

In terms of Cohen's f^2,

$$\eta_p^2 = \frac{f^2}{1 + f^2}, \tag{12}$$

therefore

$$f^2 = \frac{\eta^2}{(1 - \eta^2)}. \tag{13}$$

There are a few things to look out for when using classical eta-squared and partial eta-squared to compare effect sizes between studies of different designs. For example, consider a situation where you have a single-factor (A) ANOVA and you introduce a new factor (B). There are two possibilities (Richardson, 2011):

(1) Factor B had previously contributed to the within-subject variance but is now controlled, so it decreases the within-subject variance SS_{error} but SS_A remains constant. In this case, classical eta-squared remains constant, but partial eta-squared increases.
(2) Factor B introduces additional variation (perhaps it is a new treatment variable) but does not affect the within-subject variance. In this case, SS_A and SS_{error} remain constant but SS_{total} increases, so classical eta-squared decreases and partial eta-squared remains constant.

It is important to be careful when interpreting eta-squared or using it to compare effect sizes for experiments with different designs. Keep in mind that partial eta-squared cannot be used to compare the effects of within-subject variables with one another or with the effects of between-subject variables (Richardson, 2011). *Generalized* eta-squared was developed for this purpose by Olejnik and Algina (2003). The expression takes a different form depending on the design of the experiment, but essentially it is similar to classical eta-squared but with variation due to observational variables removed from SS_{total} before the proportion is calculated (Richardson, 2011).

2.4 *Canned Effect Sizes*

Jacob Cohen famously computed standardized effect sizes for a large number of psychological research articles as part of a meta-analysis, and went on to suggest a range of effect sizes to serve as rules of thumb for researchers. Table 1 shows a list of suggested effect size compiled from Cohen and Lea (2004), Sawilowsky (2017), and Richardson (2011).

It is recommended to avoid using these canned effects sizes unless absolutely necessary (Kim & Seo, 2013). Because the sample size is completely determined by

Table 1 List of suggested effect size

Effect size	Very small	Small	Medium	Large	Very large	Huge
d	0.01	0.20	0.50	0.80	1.20	2.00
η^2	–	0.01	0.06	0.14	–	–
f^2	–	0.01	0.06	0.16	–	–

significance level, power, and standardized effect size, using a "small," "medium," or "large" effect size is just asking for large, medium, or small sample size. "If only a standardized effect is sought without regard for how this relates to an absolute effect, the sample size calculation is just a pretense." (Lenth, 2001). Instead, you should determine these effect sizes from a literature review.

2.5 *Free Power Analysis Tools*

Because calculating power can be quite complicated for all but the simplest experimental designs, a number of tools have been developed to automate the process. The following are three free programs which can help researchers to calculate sample size and power:

- **pwr**: an R package which provides power analysis functions along the lines of Cohen (1988) (https://cran.r-project.org/package=pwr).
- **G*Power 3**: a powerful stand-alone program which can calculate power for a wide variety of statistical tests and experimental designs (Faul et al., 2007) (http://www.gpower.hhu.de/en.html).
- **GLIMMPSE**: an online tool which provides a step-by-step, user-friendly interface to guide researchers through sample size calculations (Guo et al., 2013) (https://glimmpse.samplesizeshop.org/#/).

3 Data Preprocessing

The following is a data preprocessing pipeline in EEGLAB for single subjects in preparation for performing Granger causality analysis using SIFT (Delorme et al., 2011; Mullen, 2014; Seth, Barrett, & Barnett, 2015). This process was adapted largely from Makoto Miyakoshi's own preprocessing pipeline on the Swartz Center for Computational Neuroscience Wiki and is not an official recommendation.[1] EEGLAB users may also want to check out the PREP pipeline tool, which is a standardized early-stage preprocessing tool developed by Bigdely-Shamlo, Mullen, Kothe, Su, and Robbins (2015). However, for this walkthrough, you will need the following EEGLAB extensions:

- **CleanLine**: adaptively estimates and removes sinusoidal artifacts.
- **clean_rawdata**: cleans continuous data using Artifact Subspace Reconstruction.
- **DIPFIT 2**: source localization of ICA components.
- **Firfilt**: routines for designing linear filters.
- **IC Label**: IC classifier using a neural network trained on hundreds of thousands of ICs.

[1] https://sccn.ucsd.edu/wiki/Makoto's_preprocessing_pipeline.

This walkthrough assumes that you have already loaded your data into EEGLAB and have imported channel locations. The process will vary depending on the software you used to collect the data, so check the EEGLAB single-subject data processing tutorial for detailed information.

Step 1. High-pass Filter: Online analog filters lead to loss of information and should be minimized in favor of digital offline filtering (Keil et al., 2014). Analog low-pass filtering should be limited to what is necessary to prevent aliasing in the analog-to-digital converter, i.e., the Nyquist frequency (Picton et al., 2000). Picton et al. (2000) suggest setting the low-pass cutoff to one-quarter of the sampling rate. In our own setup, we do not use any analog filters and rely exclusively on offline digital filters for low-pass, high-pass, and line-noise filtration.

Widmann et al. (2015) suggest limiting digital high-pass filtering to very low cutoffs (<0.1 Hz) in order to remove baseline drifts, citing Acunzo, MacKenzie, and van Rossum (2012) and Luck (2005). According to Winkler, Debener, Muller, and Tangermann (2015), "As a pre-processing step for [independent component analysis], high-pass filtering between 1 and 2 Hz consistently produced good results in terms of signal-to-noise ratio (SNR), single-trial classification accuracy and the percentage of 'near-dipolar' ICA components."

Low-pass digital filters are optional and are largely cosmetic because high-frequency noise (with the exception of line noise) has very low energy in the EEG spectrum (Widmann et al., 2015). However, in practice, they may cause problems in calculating Granger causality by increasing the model order.

Select *Basic FIR filter* (*new, default*) in the tools menu, then set the lower edge of the passband to 1 Hz and leave the higher edge box blank. Note that *Basic FIR filter* (*new, default*) uses passband edge instead of cutoff frequency. When you specify a passband edge lower than 2 Hz, EEGLAB will adjust the transition band so that it ends at DC (i.e., 0 Hz). The settings are shown in Fig. 1.

Basic FIR filter (*new, default*) is a windowed sin*c* finite impulse response filter. The details of digital filter design are beyond the scope of this chapter, but a key concept is that all filters create some sort of phase delay and this filter will delay all the frequency components of the input signal by the same amount, which means there is no phase distortion, only a group delay which can easily be corrected. Make sure to report all relevant parameters of the filter in your write-up. In this case, that includes the filter type (windowed sin*c* FIR), window type (Hamming), cutoff frequencies (or passband edge here), and filter order (the width of the sliding window, which should be reported by the function during execution).

Step 2. Remove Line Noise with CleanLine: CleanLine is an EEGLAB plug-in developed by Tim Mullen. It uses a sliding window to adaptively estimate the sine wave amplitude to subtract. It does not distort the signal like a notch filter. Select "CleanLine" from the tools menu and run it with default settings.

Step 3. Reject Bad Channels and Correct Continuous Data Using Artifact Subspace Reconstruction: Clean_rawdata() is an EEGLAB plug-in which identifies and adaptively removes high-variance artifacts (Mullen et al., 2015). Select "Clean continuous data using ASR" from the Tools menu.

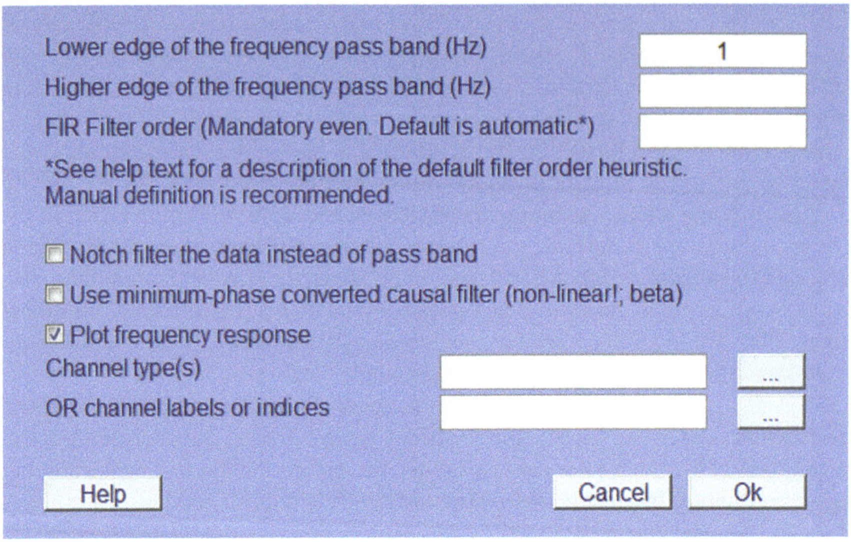

Fig. 1 GUI interface for the basic FIR filter which comes with the Firfilt plug-in. The lower passband edge is set to 1 Hz, and we have opted not to set a low-pass or notch filter

Note that the high-pass filter and line-noise parameters in Fig. 2 are disabled (set to -1) because we have already high-pass filtered the data and removed line noise. The other parameters are set according to the recommended "reasonable" values in the clean_rawdata() help file.

Step 4. Interpolate the Removed Channels: In the next step (after this one), we will re-reference the voltage data to the average scalp potential. If clean_rawdata() has removed a lot of channels, then re-referencing could introduce a large bias, especially if most of the channels were removed from one side of the head. To prevent this, we

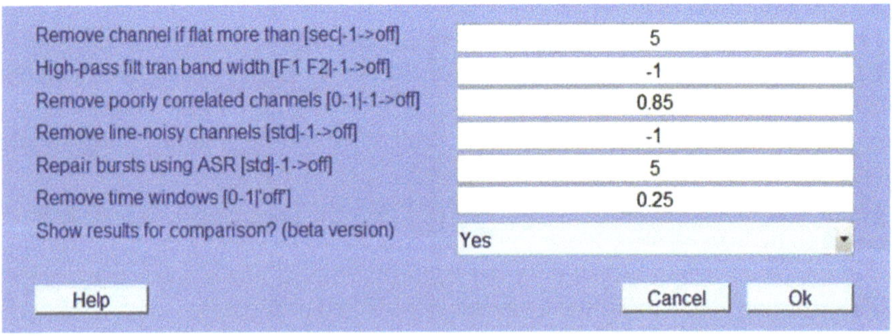

Fig. 2 GUI interface for "Clean continuous data using ASR" implemented by the clean_rawdata plug-in. The recommended settings (as specified in the help file) are used, except for "High-pass filt tran band width" and "Remove line-noisy channels," which are disabled (set to -1)

will interpolate the missing channels first. Note that this causes rank deficiency, so dimension reduction will be necessary before ICA.

Select "Interpolate electrodes" from the Tools menu, then select "Use all channels from other dataset" and choose the index of the original *.set* file (which should contain all the channels, even the ones you removed). It will automatically identify which channels are missing.

Step 5. Re-reference the Data: The choice of reference location is critical when interpreting the results of EEG experiments. This stems from the fact that there is no such thing as electric potential at a single site. Instead, the potential at some point in an electric field can only be understood in reference to another point at which the potential is arbitrarily defined as zero. So what the electrodes on an EEG cap are measuring is the voltage between a particular scalp location and a common reference location, usually the earlobe, mastoids, or someplace else where it is assumed there is no cortical activity. However, noise at the reference electrode (such as EMG) will affect the voltage readings at the active electrodes, so one method to separate this out is to recalculate the reference to the average voltage of the active sites, which is based on the assumption that all potentials across the scalp will sum to zero because they are created by dipoles within the brain, thus the net outflow of the fields produced should be zero. This technique has its advantages and disadvantages, and it is important to choose the right reference scheme for your particular application. Dien (1998) provides an excellent explanation of these, as well as James Luck's *An Introduction to the Event Related Potential Technique*.

One serious disadvantage to the average reference is that ERP waveforms will look different depending on the choice of reference location. Re-referencing the data to average will make it extremely difficult or impossible to compare ERPs across studies with different electrode montages. Thus, if your experiment is investigating an ERP, you should avoid this and instead make sure to use the same reference location as the studies with which you will compare your results, whether it be the earlobe, mastoids, or a particular scalp electrode.

However, re-referencing to average will not significantly change the qualitative nature of scalp topography. Doing so is analogous to changing the reference location of a topographical map of the earth from mean sea level to some arbitrary elevation. The individual elevations and colorations will change, but their difference relative to each other will remain the same. For high-density electrode montages, re-referencing to average may improve source localization based on scalp topography and dipole modeling (Dien, 1998).

In either case, your choice of reference location deserves careful consideration and will need to be justified. In general, if you have a dense montage and are interested in source localization, use the average reference. If you have a sparse montage or are interested in ERPs, choose a reference location that is consistent with previous literature.

In our lab, we are usually interested in the former, so we will re-reference to average before performing independent component analysis and source localization. Select "Re-reference" from the tools menu and then select "Compute average reference."

Step 6. Run ICA: Select "Run ICA" from the tools menu and start ICA using the default settings. This process will produce a set of independent components equal to the rank of the data, which is usually the same as the number of channels. However, rejecting channels will reduce the rank of the data, even if you re-interpolate them. For example, if you have 64 channels and reject 4 of them, ICA should only produce 60 independent components. Likewise, re-referencing the data to average will reduce the rank of the data by 1. "Run ICA" has a rank checker that uses the MATLAB function rank() to ensure that the number of ICs generated equals the rank of the data, even if it is rank deficient as a result of our preprocessing steps.

Step 7. Reject Non-brain Components with IC Label: IC Label provides an automated classifier that has been trained on thousands of labeled components and hundreds of thousands of unlabeled components from SCCN's database. The team is currently asking for volunteers to help classify the unlabeled components, with the goal of creating a classifier that is reliable enough to be used in automated data processing. Their website includes a tutorial on source identification, which can be a good resource for novices in the field looking to gain experience (Fig. 3).

Step 8. Estimate Location of Equivalent Dipoles: Before estimating the location of source dipoles, we need to specify the head model to use. Select "Locate dipoles using DIPFIT 2.x" in the tools menu, and then select "Head and model settings." The result is shown in Fig. 4.

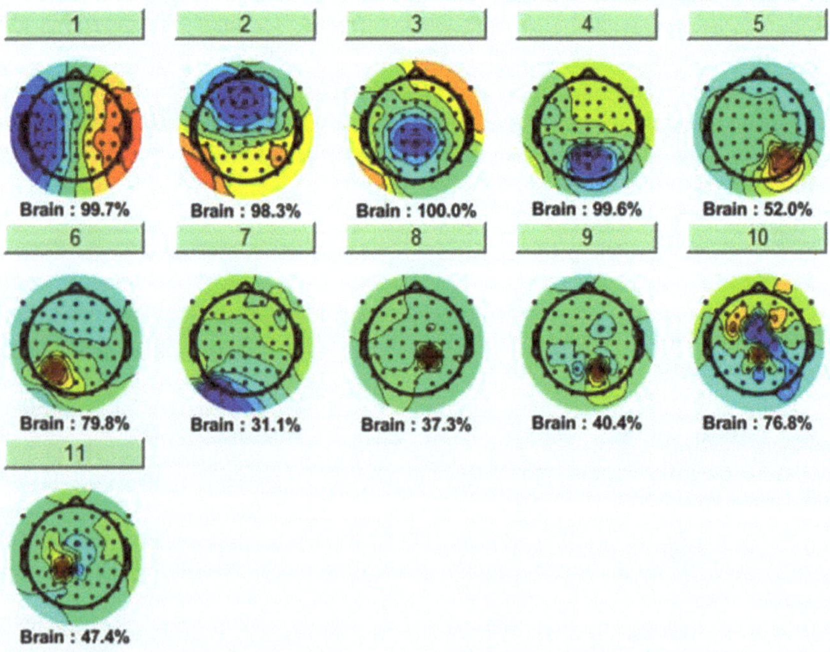

Fig. 3 GUI interface for the IC Label plug-in, shown after removing all sources not identified as "brain"

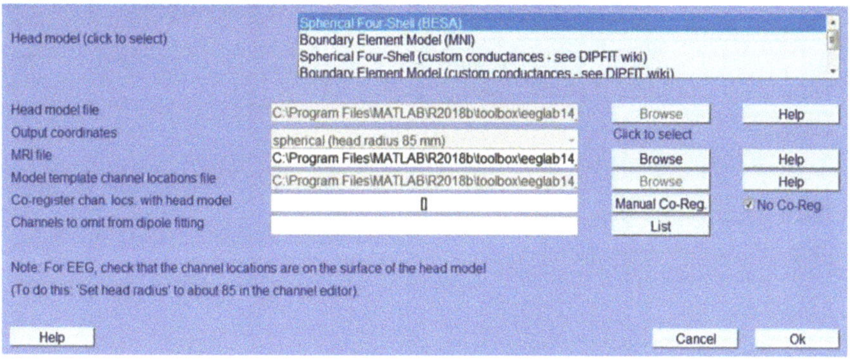

Fig. 4 GUI interface for head and model settings. The head model is BESA and the "No Co-Reg" box is checked

The brain electrical source analysis (BESA) model can be used with no co-registration. Co-registration refers to the alignment of the channel locations with specific points on the surface of the head model. Because we placed our electrodes and labeled them according to the International 10–20 system, the easiest way to align our channels to the head model is to use the standard channel coordinates.

To do this, select "Channel locations" from the edit menu. Click on the *Look up Locs* button and select the appropriate head model, as shown in Fig. 5.

Now we are ready to use DIPFIT 2. Select "Locate dipoles using DIPFIT 2.x," then "Autofit (coarse fit, fine fit & plot)."

The component indices will automatically be filled in according to how many components you rejected in the previous step. In this case, all but 11 were rejected (Fig. 6). Setting the maximum-allowed residual variance to 15% is a good rule of thumb, and we also choose to remove dipoles that are located outside the head.

Plotting the results in summary mode will give you the Talairach coordinates of the dipoles, as shown in Fig. 7. You can use the Yale Bioimage Suite to determine the Brodmann areas that these correspond to. You may find that some sources have been placed in non-physiologically valid (non-cortical) locations. This is a result of the poor spatial resolution of EEG and other biases introduced by the head model and the

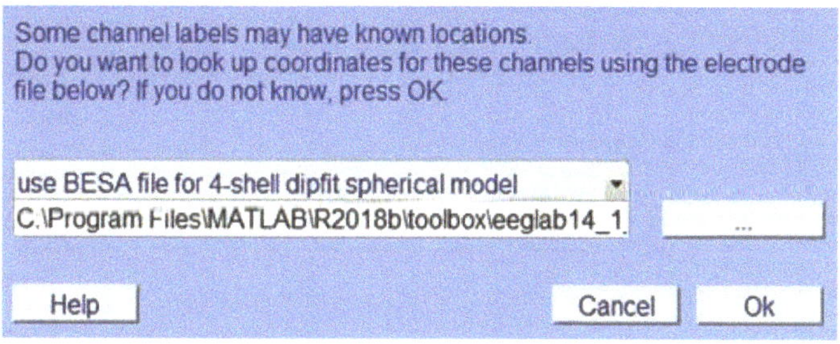

Fig. 5 Make sure the head model here matches the one in the head and model settings

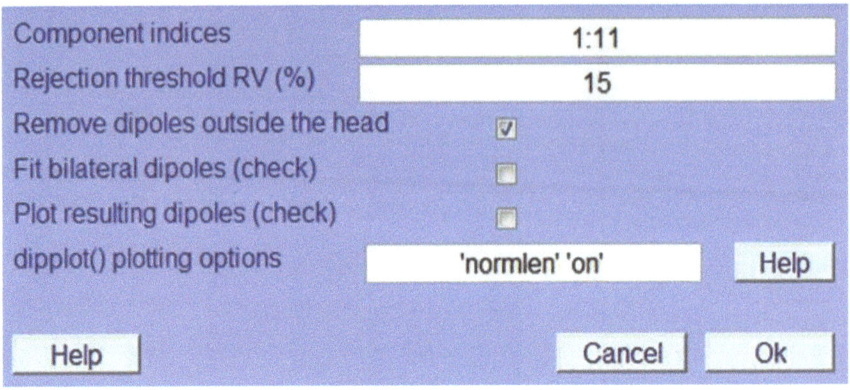

Fig. 6 The GUI options for Autofit. The rejection threshold for residual variance has been set to 15% and we have opted to remove dipoles located outside the head

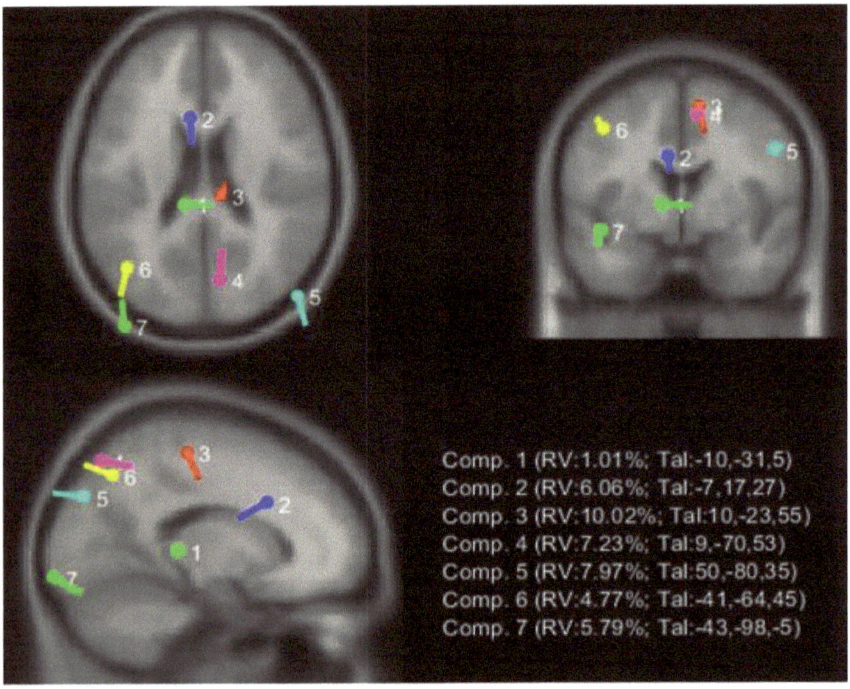

Fig. 7 Plotting dipole locations in summary mode

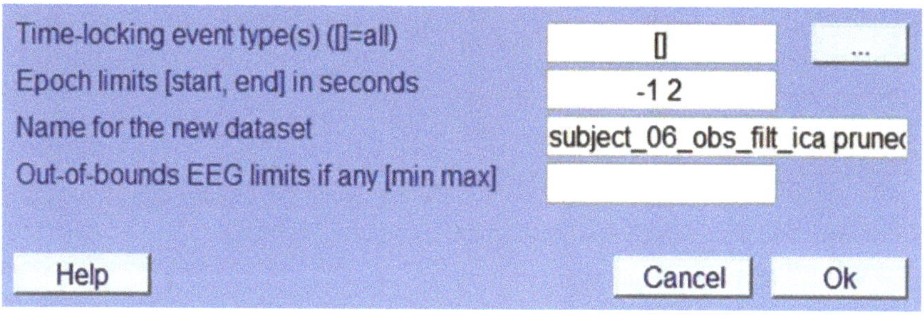

Fig. 8 GUI interface for the extract epochs tool with default values

DIPFIT algorithm. The components can still be used for clustering, but you should be cautious about interpreting their locations. Note that, in this example, dipoles 8–11 have been removed, either because their residual variance was too high or they were outside the head.

Step 9. Extract Epochs and Remove Baseline: Select "Extract epochs" from the tools menu.

In this case, we have selected epochs extending [−1, 2] seconds around all event types (Fig. 8). Epoching can also be done before ICA. Remove the entire 1000 ms baseline period when presented with the next window. This is the pre-stimulus period over which voltage is averaged and then subtracted from the rest of the epoch.

Next Steps: Save your dataset because it will have taken a lot of computational power to get to this point. You are now ready to proceed with the rest of your analysis. You may also create a STUDY to facilitate group analysis.

4 Reporting Guidelines

The following checklist was adapted from Luck (2004), Keil et al. (2014), and Picton et al. (2000) which describe the major considerations in the design, interpretation, and reporting of ERP experiments. It is meant to serve as a quick reference for authors prior to submission for publication.

[1] Study Formulation

- Explain how you came to your ideas and why they are important.
- State hypotheses about specific metrics (e.g., ERP component amplitude, reaction time, etc.) in terms of a positive outcome.
- Describe the instructions given to participants regarding the experimental task and describe the debriefing procedures and outcomes.
- Describe the experimental structure and order of conditions.

[2] Participants

- Report basic descriptive statistics (e.g., number, age, gender).
- Report any exclusion criteria.
- Report how you controlled for medications (including recreational drug and alcohol use).
- Report informed consent.
- Equipment and recording.
- Specify make and model of all equipment (including electrode type and material).
- Specify all settings needed for replication (e.g., gain, resolution, sampling rate, etc.).
- Report electrode locations (e.g., International 10–20 System).
- Specify reference locations and ground location.
- Report electrode scalp impedance or other suitable index of data quality.
- Describe any online filters.

[3] Equipment and Recording

- Specify make and model of all equipment (including electrode type and material).
- Specify all settings needed for replication (e.g., gain, resolution, sampling rate, etc.).
- Report electrode locations (e.g., International 10–20 System).
- Specify reference locations and ground location.
- Report electrode scalp impedance or other suitable index of data quality.
- Describe any online filters.

[4] Stimuli and Responses

- Specify all aspects of the stimuli necessary for replication by others.
- Specify timings (e.g., stimulus durations, interstimulus intervals, etc.).
- Describe how responses were made and how this was controlled.
- Give the number of trials per condition, per session.
- Report the total recording duration and rest periods.
- Give any practice trial information.

[5] Behavioral Responses

- Report response times, accuracy, and other suitable indices of performance available.

[6] Data Preprocessing

- Enumerate all preprocessing steps.
- Describe how time-locking was done.
- Describe digital offline filters, including filter type, cutoff parameters, and whether it was applied to continuous or epoched data
- Specify both online reference and offline reference.
- Report how many channels per subject were rejected and how missing channels were interpolated.
- Specify epoch size and baseline removal period.

- Specify artifact correction/rejection criteria and the proportion of artifacts removed.

[7] Principle Component Analysis and Independent Component Analysis

- Report the algorithm and implementation.
- Specify the type of association matrix (PCA).
- Specify any rotation applied (PCA).
- Describe information used for component clustering (ICA).
- Describe retention/rejection criteria.
- Report the number of components removed or retained per subject.

[8] Source Localization

- Describe the algorithm and implementation.
- Describe the head model.
- Report goodness of fit (e.g., residual variance).

[9] Measurement

- Clearly define the type of measurement (e.g., average amplitude), measurement procedures, and parameters (e.g., latency window).
- Give a priori rationale for measurement parameters.
- Present descriptive statistics before inferential statistics.
- Avoid analytic circularity (do not use a dataset to identify key aspects of the analysis, then use that same dataset for inferential statistics).

[10] Figures

- ERP waveforms for all relevant conditions and comparisons must be shown.
- Line plots must include voltages, polarity convention, sensor location, and have clearly labeled axes.
- Scalp maps must show orientation, sensor locations, use an appropriate spatial resolution, and be clearly labeled.
- Connectivity and time–frequency plots must show physical units, use clearly labeled axes, show the baseline period, and show source locations.

[11] Statistics

- Describe the statistical model and test procedures.
- Report test statistics, *p*-values, sample sizes, and effect sizes.
- Report any corrections for violations of sphericity or other model assumptions.
- Report the adjustment for multiple comparisons (if any).

5 Conclusion

Always at least attempt to calculate the required sample size for your experiment. According to Lenth (2001), the fact that sample size planning is considered at all is a big step forward. Do this by finding one or two papers that describe experimental designs similar to yours and provide detailed information about effect sizes, means, variances, significance levels, and other statistics. If these studies have small sample sizes, consider adding a small factor of safety to your own sample size estimate to account for effect-size inflation. When you are writing your manuscript, focus on providing all the information necessary for another person to replicate your experiment, down to the smallest detail. Be sure to include any effect sizes produced by your analysis software (such as generalized eta-squared, which meta-analysts will appreciate), as well as the means, variances, p-values, and F- or t-values that future researchers will need to perform their own power analyses. Doing so will help to ensure that science is replicable, and that findings can be relied upon.

References

Acunzo, D. J., MacKenzie, G., & van Rossum, M. C. W. (2012). Systematic biases in early ERP and ERF components as a result of high-pass filtering. *Journal of Neuroscience Methods, 209*(1), 212–218. https://doi.org/10.1016/j.jneumeth.2012.06.011.

Bezeau, S., & Graves, R. (2003). Statistical power and effect sizes of clinical neuropsychology research. *Journal of Clinical and Experimental Neuropsychology, 23*(3), 399–406. https://doi.org/10.1076/jcen.23.3.399.1181.

Bigdely-Shamlo, N., Mullen, T., Kothe, C., Su, K.-M., & Robbins, K. A. (2015). The PREP pipeline: Standardized preprocessing for large-scale EEG analysis. *Frontiers in Neuroinformatics, 9*, 1–20. https://doi.org/10.3389/fninf.2015.00016.

Cohen, B. H., & Lea, R. B. (2004). *Essentials of statistics for the social and behavioral sciences*. Hoboken: Wiley. ISBN 0-471-22031-0.

Cohen, J. (1988). *Statistical power analysis for the behavioral sciences*. New York: Routledge. ISBN 978-1-134-74270-7.

Cohen, J. (1992). A power primer. *Psychological Bulletin, 112*(1), 155–159. https://doi.org/10.1038/141613a0.

Delorme, A., Mullen, T., Kothe, C., Akalin Acar, Z., Bigdely-Shamlo, N., Vankov, A., & Makeig, S. (2011). EEGLAB, SIFT, NFT, BCILAB, and ERICA: New tools for advanced EEG processing. *Computational Intelligence and Neuroscience, 2011*, 1–12. https://doi.org/10.1155/2011/130714.

Dien, J. (1998). Issues in the application of the average reference: Review, critiques, and recommendations. *Behavior Research Methods, Instruments, and Computers, 30*(1), 34–43. https://doi.org/10.3758/BF03209414.

Faul, F., Erdfelder, E., Lang, A., & Buchner, A. (2007). G*Power 3: A flexible statistical power analysis program for the social, behavioral, and biomedical sciences. *IEEE International Symposium on Information Theory Proceedings, 39*(2), 175–191. https://doi.org/10.1109/ISIT.2013.6620417.

Guo, Y., Logan, H. L., Glueck, D. H., & Muller, K. E. (2013). Selecting a sample size for studies with repeated measures. *BMC Medical Research Methodology, 13*(1). https://doi.org/10.1186/1471-2288-13-100.

Ioannidis, J. (2008). Why most discovered true associations are inflated. *Epidemiology, 19*(6), 640–648. https://doi.org/10.1097/EDE.ObO.

Ioannidis, J. P. A. (2005). Why most published research findings are false. *PLoS Medicine, 2*(8), 0696–0701. https://doi.org/10.1371/journal.pmed.0020124.

Johnson, V. E., Payne, R. D., Wang, T., Asher, A., & Mandal, S. (2017). On the reproducibility of psychological science. *Journal of the American Statistical Association, 112*(517), 1–10. https://doi.org/10.1080/01621459.2016.1240079.

Keil, A., Debener, S., Gratton, G., Junghöfer, M., Kappenman, E. S., Luck, S. J., … Yee, C. M. (2014). Committee report: Publication guidelines and recommendations for studies using electroencephalography and magnetoencephalography. *Psychophysiology, 51*(1), 1–21. https://doi.org/10.1111/psyp.12147.

Kim, J., & Seo, B. S. (2013). How to calculate sample size and why. *Clinics in Orthopedic Surgery, 5*(3), 235–242. https://doi.org/10.4055/cios.2013.5.3.235.

Lakens, D. (2013). Calculating and reporting effect sizes to facilitate cumulative science: A practical primer for t-tests and ANOVAs. *Frontiers in Psychology, 4*(November), 1–12. https://doi.org/10.3389/fpsyg.2013.00863.

Larson, M. J., & Carbine, K. A. (2017). Sample size calculations in human electrophysiology (EEG and ERP) studies: A systematic review and recommendations for increased rigor. *International Journal of Psychophysiology, 111*, 33–41. https://doi.org/10.1016/j.ijpsycho.2016.06.015.

Lenth, R. V. (2001). Some practical guidelines for effective sample size determination. *The American Statistician, 55*(3), 187–193.

Luck, S. J. (2004). Ten simple rules for designing and interpreting ERP experiments. *Components, 249*, 17–32. Retrieved from http://www.ncbi.nlm.nih.gov/pubmed/7783201.

Luck, S. J. (2005). *An introduction to the event-related potential technique*. Cambridge, MA: MIT Press.

Mullen, T. R. (2014). *The dynamic brain: Modeling neural dynamics and interactions from human electrophysiological recordings* (Order No. 3639187). Available from Dissertations & Theses @ University of California; ProQuest Dissertations & Theses A&I. (1619637939).

Mullen, T. R., Kothe, C. A. E., Chi, Y. M., Ojeda, A., Kerth, T., Makeig, S., … Cauwenberghs, G. (2015). Real-time neuroimaging and cognitive monitoring using wearable dry EEG. *IEEE Transactions on Biomedical Engineering, 62*(11), 2553–2567. https://doi.org/10.1109/TBME.2015.2481482.

Olejnik, S., & Algina, J. (2003). Generalized eta and omega squared statistics: Measures of effect size for some common research designs. *Psychological Methods, 8*(4), 434–447. https://doi.org/10.1037/1082-989X.8.4.434.

Picton, T., Ritter, W., Ruchkin, D., Rugg, M., Taylor, M., Bentin, S., … Miller, G. (2000). Guidelines for using human event-related potentials to study cognition: Recording standards and publication criteria. *Psychophysiology, 37*(2), 127–152. https://doi.org/10.2113/gselements.13.1.29.

Richardson, J. T. E. (2011). Eta squared and partial eta squared as measures of effect size in educational research. *Educational Research Review, 6*(2), 135–147. https://doi.org/10.1016/j.edurev.2010.12.001.

Sawilowsky, S. S. (2017). New effect size rules of thumb. *Journal of Modern Applied Statistical Methods, 8*(2), 597–599. https://doi.org/10.22237/jmasm/1257035100.

Seth, A. K., Barrett, A. B., & Barnett, L. (2015). Granger causality analysis in neuroscience and neuroimaging. *Journal of Neuroscience, 35*(8), 3293–3297. https://doi.org/10.1523/jneurosci.4399-14.2015.

Widmann, A., Schröger, E., & Maess, B. (2015). Digital filter design for electrophysiological data–a practical approach. *Journal of neuroscience methods, 250*, 34–46.

Winkler, I., Debener, S., Muller, K. R., & Tangermann, M. (2015). On the influence of high-pass filtering on ICA-based artifact reduction in EEG-ERP. In *Proceedings of the Annual International Conference of the IEEE Engineering in Medicine and Biology Society, EMBS* (pp. 4101–4105) *2015–November*. https://doi.org/10.1109/EMBC.2015.7319296.

Functional Near-Infrared Spectroscopy (fNIRS) in Neuroergonomics

Liya Thomas and Chang S. Nam

Abstract This chapter focuses on functional near-infrared spectroscopy (fNIRS) and its applications in the field of neuroergonomics. Due to the small, portable, and flexible nature of fNIRS, it has been able to track the neurocognitive behavior of participants during complex and ambulatory activities. This chapter presents a comprehensive summary and review of fNIRS technology with respect to its design, implementation, and measurements. The three different types of fNIRS that exist—continuous wave, frequency domain, and time domain—are further explained through example studies of each tool in action. We also provide detailed technological information on numerous studies that focus on the four main topics that fNIRS has frequently been used for: driving, mental workload, motor function, and decision-making. It then concludes with the directions and potential applications that fNIRS can be used for in the future. The main purpose of this chapter is to show the necessity and value of fNIRS in research and how it is able to provide accurate information in situations that would have otherwise been inaccessible.

1 Introduction

Functional near-infrared spectroscopy or fNIRS is a noninvasive tool used for brain imaging. It has recently gained popularity because it is able to provide the same results as an fMRI, or functional magnetic resonance imagining machine, but is more accommodating in terms of mobility, cost, and accuracy. Moreover, unlike fMRI, fNIRS is wireless and portable. Thus, being less restrictive toward subjects, and resulting in more accurate information in active situations (Piper et al., 2014). It is a valuable and insightful tool in tracking neurocognitive behavior and continues to be present in clinical research. To continue, fNIRS is especially unique in versatility because there are different types of fNIRS: continuous wave (e.g., Boas, Elwell, Ferrari, & Taga, 2014), frequency domain (e.g., Brigadoi et al. 2014), and time

L. Thomas · C. S. Nam (✉)
Edward P. Fitts Department of Industrial & Systems Engineering, North Carolina State University, Raleigh, NC 27695, USA
e-mail: csnam@ncsu.edu

C. S. Nam (ed.), *Neuroergonomics*, Cognitive Science and Technology,
https://doi.org/10.1007/978-3-030-34784-0_4

domain (e.g., Molteni et al., 2012). fNIRS has been frequently used in understanding driving or piloting behavior, mental workload, motor functions, and decision-making.

Figure 1 highlights the advancement of fNIRS throughout time. fNIRS has been used to evaluate a wide array of neurocognitive and physical processes. fNIRS has been used in conjunction with several other neurocognitive tracking devices such as EGG and VR and continues to be present in variety of different topics. It has been used in understanding auditory responses, driving behavior, cognitive tasks, and continues to be a significant piece of technology in understanding degenerative neural diseases such as multiple sclerosis (Stojanovic-Radic, Wylie, Voelbel, Chiaravalloti, & DeLuca, 2015). Throughout time, more accessible and accommodating version of fNIRS has been put out, such as multichannel devices, and then mobile multichannel devices (Piper et al., 2014). It also started to be used in real situations as a mode of tracking the neural behavior of participants and was proven to be a feasible tool to use in complex situations (Gateau, Ayaz, Dehais, & Lotte, 2018).

The remainder of this chapter is organized as follows. Section 2 introduces the structure and instrumentation of fNIRS, while depicting the fundamental principles of fNIRS and a detailed background for how it works. It also provides detailed information on the different types of fNIRS: continuous wave, frequency domain, and time domain. In Sect. 3, we review neuroergonomics studies that used fNIRS, categorized by four topics that fNIRS is frequently used for: driving, mental workload, motor functions, and decision-making. Moreover, important fNIRS structures are further highlighted in the summaries. Finally, Sect. 4 concludes by presenting a list of recommendations for fNIRS as well as the potential research topics that fNIRS can or has been used for. It shows that fNIRS is a valuable tool that needs to be used in the future of neuroergonomics.

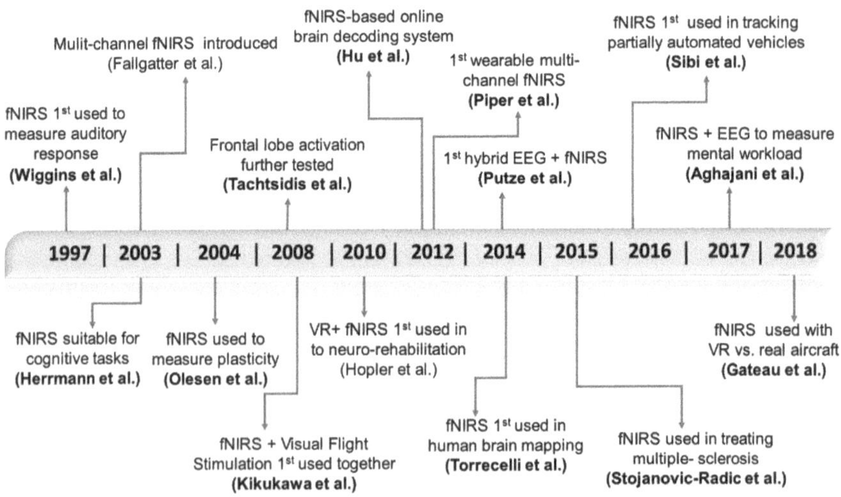

Fig. 1 Timelines of fNIRS technology advancements

2 Design and Implementation of fNIRS

This section elaborates on the key principles and instrumentation that fNIRS requires. It provides the specific tools required and the purpose of each piece in order to achieve the optimal output. Furthermore, it also discusses the connection between fNIRS and the BOLD response, which is the hemodynamic response that results in increased cerebral blood flow in the area of the brain with increased neuronal activity (Causse, Chua, Peysakhovich, Del Campo, & Matton, 2017). Since fNIRS uses the blood oxygen level-dependent or BOLD response as an integral part of understanding and implementing its technology, this relationship is further discussed below (Aghajani, Garbey, & Omurtag, 2017).

2.1 How Does fNIRS Work?

fNIRS is able to track the hemodynamic changes the brain undergoes. Any activity within the brain leads to an increase in oxygen consumption, followed by an increase in cerebral blood flow, which leads to variations in the concentration of deoxygenated (HbR) and oxygenated hemoglobin (HbO) levels, often referred to as the BOLD response (León-Carrión & León-Domínguez, 2012). Figure 2 displays these changes between HbR and HbO as changes in neural activity occur and the response of these signals (Buxton, 2012).

The response begins with a decrease in oxygen supply, as population of neurons begin to activate. In order to compensate for this low oxygen supply, vasoactive agents

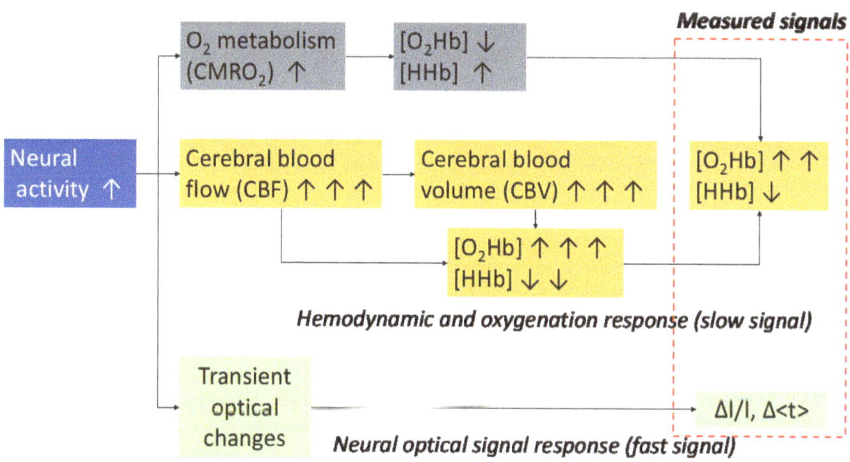

Fig. 2 Overview of the cerebral hemodynamic and oxygenation response when neural activity increased (Buxton, 2012)

alert nearby arterioles to increase oxygen-rich blood flow (Nishiyori, 2016). However, this sudden increase in oxygen-rich blood flow exceeds the normal demand for oxygen, so the concentration of oxygenated hemoglobin increases. In conjunction, as oxygenated hemoglobin increases, there is a decrease in deoxygenated hemoglobin. This process can be considered a form of canonical hemodynamic response function and the difference in deoxygenated and oxygenated hemoglobin is later quantified to determine the changes in neural activity (Curtin & Ayaz, 2018). Oxygenated and deoxygenated hemoglobin have different magnetic properties, in terms of their light absorption spectra and these differences are what fNIRS uses to get measurements (Nishiyori, 2016). By placing sensors in different regions around the head, fNIRS can monitor the hemodynamic activations throughout a subject's brain and deduce the subject's cognitive state through pattern classifications. fNIRS technology is able to monitor a wider region within the brain and can measure a specific location from which neural activity is originating. It is through this ability to measure cerebral functions that functional near-infrared spectroscopy gained its name (Torricelli et al., 2014).

As shown in Fig. 3, there are three different approaches to setting up fNIRS, they include the transmission approach, reflectance approach, and multi-distance approach. These approaches all include some form of an emitter and detector; however, the placement of the emitter and detector shifts between approaches (Wahr, Tremper, Samra, & Delpy, 1996). Additionally, Fig. 3 also displays the separation distances between channels (León-Carrión & León-Domínguez, 2012).

fNIRS is composed of light-emitting diodes or LDs that emit a ray of light, directed to a specific location on the head by an optode. The optodes will be placed on top of the head around the region of interest. The near-infrared light is delivered through the fiber optic cables that go into optode-embedded headgear (Boas et al., 2014).

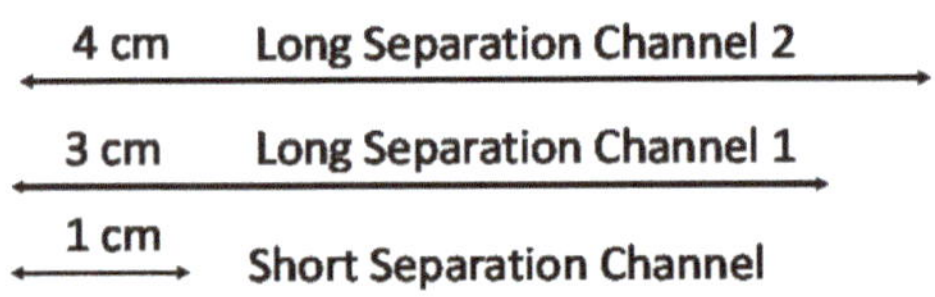

(a) Long and short separations

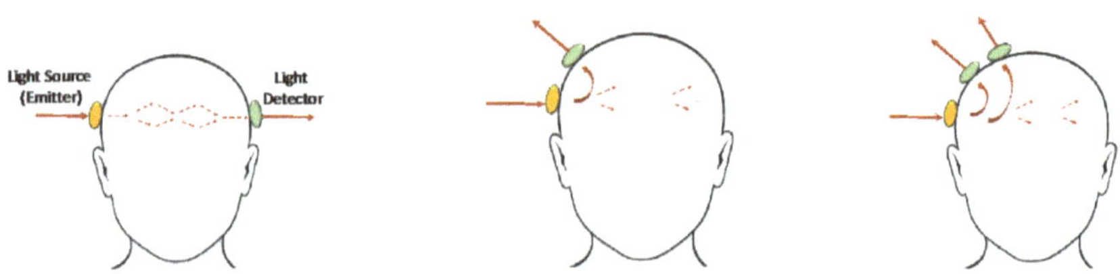

(b) Transmission approach (c) Reflectance approach (d) Multi-distance approach

Fig. 3 Different types of fNIRS setups: **a** transmission approach, **b** reflectance approach, and **c** multi-distance approach (León-Carrión & León-Domínguez, 2012; Wahr et al., 1996)

The headgear will then be placed on the participant's head, allowing for a much more efficient process to position the cap. The near-infrared light is delivered via fiber optic cables that terminate into a specialized headgear. The optodes can be embedded into the headgear before it is placed on the participant's head, allowing for a much quicker and smoother process to precisely position the cap (León-Carrión & León-Domínguez, 2012).

The headgear is one of the most essential and crucial pieces of the technology used to acquire a quality set of data. So, in order to enhance grip, nonslip, rubber-like materials are often used. This material will need to be able to conform to the different shapes of heads, easy to cut, and rigid enough to hold the grommets. The goal of the headgear is to secure the optodes in the desired position on the head and to keep the near-infrared light directed at the scalp (Nishiyori, 2016). Ultimately, the security of optodes will determine how much motion can be tolerated before artifacts are introduced. fNIRS also tends to use multiple channels (León-Carrión & León-Domínguez, 2012). A channel is defined as a path between one emitter and one detector, typically an fNIRS device tends to have several; however, space constraints of a subject must be taken into consideration when arranging the optodes (Scholkmann, Kleiser, et al., 2014; Scholkmann, Metz, & Wolf, 2014).

2.2 *What Does fNIRS Measure?*

In the electromagnetic spectrum, within the light window of 650–1000 mm, biological tissue is transparent. The light that does enter the tissue is mostly absorbed by chromophores or light-absorbing molecules, which include deoxygenated hemoglobin, oxygenated hemoglobin, and water. fNIRS utilizes two different wavelengths: the lower wavelength, which is from 650 to 700 nm, is predominantly absorbed by HbR, while the higher wavelength, which ranges from 800 to 850 mm, is predominantly absorbed by HbO (Curtin & Ayaz, 2018).

Through using the two wavelengths, the total change in the concentration of hemoglobin is able to be calculated. The changes in the concentration of HbO and HbR in blood flow are used to determine the area of interest. Moreover, because light can only travel a few centimeters into the skull and through brain tissues, fNIRS has only been able to measure the cerebral cortex region (Nishiyori, 2016). Furthermore, in order to convert the raw light intensity data into the total hemoglobin concentration, modified Beer–Lambert law (MBLL) is used (Scholkmann et al., 2014; Scholkmann, Kleiser, et al., 2014). The law attributes the loss of light intensity within the tissue as a function of chromophore concentrations (León-Carrión & León-Domínguez, 2012).

2.3 Types of fNIRS

fNIRS can be implemented with respect to three different fNIRS approaches: (1) continuous wave (CW), (2) frequency domain (FD), and (3) time domain (TD) (Torricelli et al., 2014). Table 1 shows comparisons of three different fNIRS techniques. These three different types of fNIRS systems also share some similarities in that they all use an emitter, detector, and a noninvasive method to track cognitive processes (Curtin & Ayaz, 2018). The principles and basics of the three different types of fNIRS are described below, along with their main strengths and weaknesses, as well as example studies that employed each NIRS approach.

2.3.1 Continuous Wave

Continuous wave (CW) fNIRS is a noninvasive method that is composed of light-emitting diodes or LEDs that emit a ray of light, directed to a specific location on the head by an optode (refer to Table 1). CW fNIRS typical modulated at a low frequency (a few kHz) in order to exploit the significant improvements in sensitivity available from phase-locked detection techniques, and of a detection apparatus sensitive to light attenuation changes (e.g., photodiode) (Torricelli et al., 2014). CW devices rely on light scattering through the tissue medium containing chromophores (e.g., light-absorbing substances). Typically, as the ray of light contacts brain tissue, part of it will be absorbed by chromophores. When the light leaves the head, it is collected by

Table 1 Comparisons of the three different fNIRS techniques

	Continuous wave (CW)	Frequency domain (FD)	Time domain (TD)
Schematic representation	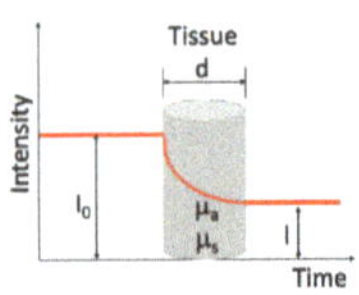	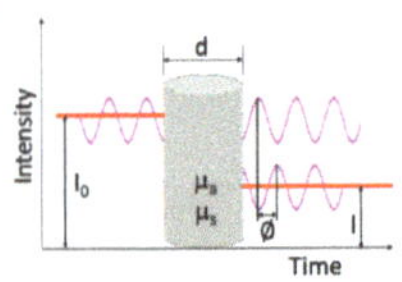	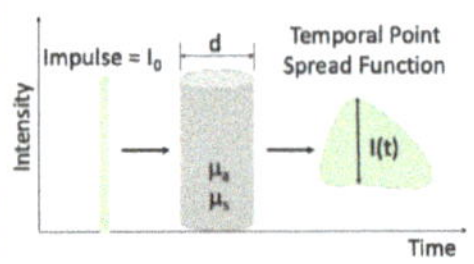
	I_0: incident light signal, I: transmitted light signal, d: thickness of head tissue, μ_a: absorbing coefficient, μ_s: scattering coefficient, φ: phase delay, $I(t)$: temporal point spread function of the transmitted light signal		
Light source	• CW light	• CW light source modulated at 50–500 MHz	• Pulsed laser
Detection and measure	• Changes in transmission • Pulse oximetry	• Phase and amplitude • μ_a and μ_s	• Meantime of flight (TOF) and intensity • μ_a and μ_s
Price and usage	• Inexpensive	• Inexpensive • Commercial versions	• Expensive • Only at lab stage

a photodetector. The optodes will be placed on top of the head around the region of interest (Boas et al., 2014).

The concentration changes within the chromophores—deoxygenated hemoglobin, oxygenated hemoglobin, and water—can be calculated from changes in detected light intensity using modified Beer–Lambert law (Gateau, Durantin, Lancelot, Scannella, & Dehais, 2015).

In addition to the reliance on assumptions made on the scattering of light, another critical shortcoming of the CW is the assumed differential path length (*L*), which represents the estimated distance traveled by the light. This estimate fundamentally limits the accuracy of any interpretation of the output parameters (Davies et al., 2017). For example, a flight stimulation study by Gateau et al. (2015) used the continuous wave fNIRS device. In this study, the optode separation was approximately 25 mm and had two wavelengths of 730 and 850 mm (Gateau et al., 2015). Moreover, the frequency in terms of oxygenation level variations, in comparison to the baseline, was 2 Hz.

2.3.2 Frequency Domain

Frequency-domain (FD) fNIRS is a continuation of the continuous wave fNIRS that is able to measure both light intensity and phase shift, since the intensity of the light emitted from these devices is modulated at a specific frequency, as given in Table 1 (Torricelli et al., 2014).

It is based on amplitude-modulated light sources and on the detection of light amplitude demodulation and phase shift. In an FD fNIRS system, the intensity of the light source is modeled by a sinusoidal wave. The light energy propagates as a spherical wave outward from the source. This *c* scalar wave is often called the diffuse photon density wave and is highly dependent on the optical properties of the tissue (Scholkmann et al., 2014; Scholkmann, Kleiser, et al., 2014). FD is able to provide a more enhanced technique to measure light scattering and allows more consistent measurements from tissue. This is especially valuable when resolution can sometimes be weak as a result of the depth in the brain (Torricelli et al., 2014).

Brigadoi et al. (2014) used a frequency-domain fNIRS device, ISS Imagent, that used 32 laser diodes and four photomultiplier tubes. The source and detector were placed along the region of interest through a probe-placement method. It used approximately 20 channels per wavelength, for each participant and had a sampling frequency of 7.8 Hz. Additionally, there was a motion artifact to consider, the participant's jaw movement whenever there was a vocal response, which in terms of results lead to a large variance in the data.

2.3.3 Time Domain

Time-domain (TD) NIRS relies on the ability to measure the photon distribution of time-of-flight (DTOF) in a diffusive medium (refer to Table 1). Essentially meaning,

TD fNIRS uses a pulsed light source, typically a laser-providing light pulse and a detection apparatus with temporal resolution in a sub-nanosecond scale (Torricelli et al., 2014).

After injecting the light pulse within a turbid medium, the DTOF which is measured at a fixed distance from the injection point is delayed, broadened, and attenuated. Increasing the source–detector distance yields an increased delay and broadening of the DTOF and decreases the number of detected photons. Similar behavior is observed when the scattering increases. Finally, the absorption affects both the signal intensity and the trailing edge of the DTOF, while leaving the temporal position of the DTOF substantially unchanged (Torricelli et al., 2014).

TD is most notably recognized for its effectiveness in its ability to provide thorough and accurate measurements of optical properties like the absorption coefficient and reduced scattering coefficient (Torricelli et al., 2014). However, it also tends to be one of the bulkier pieces of equipment and can be very expensive, so it does require some improvement to meet the standards that CW and FD provide in research (Scholkmann et al., 2014; Scholkmann, Kleiser, et al., 2014). For example, a study that tested frontal brain activation using a working memory task used time-domain NIRS. They operated at wavelengths of 690 and 820 nm, with a repetition rate of 80 MHz. Moreover, the laser heads were connected to multimode graded index fibers that were in various lengths (Molteni et al., 2012).

3 Applications of fNIRS in Neuroergonomics Research

A few common applications of fNIRS are in mental workload, driving, decision-making techniques, and mental health. Additionally, it can be used in patient rehabilitation and military training. fNIRS is able to provide the most effective, inexpensive option for monitoring neural activity in mobile situations. Applications can also range to studies that require long-term monitoring of behaviors or studies that are in realistic or specific settings. Furthermore, it is one of the only techniques that is able to monitor extracellular and intracellular activities (Boas et al., 2014). fNIRS is a tool with a tremendous amount of variability, it is able to yield the same results as an fMRI in a more adaptable manner, making it a unique and powerful tool with great potential (Ayaz et al., 2013).

3.1 Driving

The common research goal among the driving studies involves tracking cognitive changes and cognitive load as the participants are driving. Table 2 shows a summary of the exemplary fNIRS studies in driving research. Several studies also added stress conditions and possible distractions to the drivers. For example, Gateau et al. (2018) asked pilots to recall air traffic control instructions while driving in order to

Table 2 Summary of exemplary fNIRS studies in driving research

Authors	Experimental task	Characteristics of participants	Brain areas measured	No. of channels	fNIRS wavelength (nm)	Sampling rate (SR)/frequency (*F*)	Type of fNIRS
Kikukawa, Kobayashi, and Miyamoto (2008)	Flight simulator	• 4 male pilots • $M_{age} = 42.5$	PFC	2	–	–	CW
Tsunashima & Yanagisawa (2009)	Driving simulator	• 4 male subjects • Age = 20s	PFC	42	–	–	CW
Lin, Maikala, McGorry, and Brunette (2010)	Assembly task	• 20 male subjects • $M_{age} = 36.4$	PFC	2	775, 810, 850, 905	1000 Hz (SR)	CW
Harrison et al. (2014)	Stimulation	• 4 pilots	PFC	16	730, 850	2 Hz (SR)	CW
Oka et al. (2015)	Driving simulator	• 15 adults (7*F*) • $M_{age} = 31.4$	PFC, PMC & M1	48	730, 850	0.1 Hz (*F*)	CW
von Lühmann, Herff, Heger, and Schultz (2015)	Mental arithmetic task	• 12 subjects	FC	4	750, 850	0.5 Hz (*F*)	CW
Gateau et al. (2015)	Flight simulator	• 19 adults (6*F*) • $M_{age} = 27.4$	PFC	16	730, 850	2 Hz (*F*)	CW
Foy, Runham, and Chapman (2016)	Driving simulator	• 32 students • $M_{age} = 21.22$	PFC	16	730, 850	2 Hz (*F*)	CW
Xu et al. (2017)	Driving simulator	• 14 subjects • $M_{age} = 24$	PFC, MC	–	–	10 Hz (SR)	CW
Unni, Ihme, Jipp, and Rieger (2017)	Virtual reality and driving simulator	• 19 subjects (2F) • $M_{age} = 25.2$	FC, PC & TOC	78	730, 850	1.955 Hz (F)	CW

(continued)

Table 2 (continued)

Authors	Experimental task	Characteristics of participants	Brain areas measured	No. of channels	fNIRS wavelength (nm)	Sampling rate (SR)/frequency (*F*)	Type of fNIRS
Liu et al. (2017)	Driving simulator	• 12 male subjects • $M_{age} = 24.5$	PFC	16	–	10 Hz (SR)	CW
Causse, Chua, Peysakhovich, Del Campo, and Matton (2017)	Flight simulator + execute function task	• 26 student pilots (2F) • $M_{age} = 20.6$	PFC	16	730, 850	2 Hz (*F*)	CW
Gateau et al. (2018)	Flight simulator and real flight	• 28 pilots	PFC	16	730, 850	2 Hz (*F*)	CW

Brain regions *PFC* prefrontal cortex, *PMC* premotor cortex, *M1* primary motor cortex, *FC* frontal cortex, *TOC* temporal/occipital cortex, *OC* occipital cortex, *TC* temporal cortex, *VC* visual cortex, *PC* parietal cortex, *FL* frontal lobe, *TL* temporal lobe. **fNIRS devices** *CW* continuous wave, *FD* frequency domain better understand the neurocognitive impact on the drivers. Additionally, through putting stress factors onto the driver they were able to determine if their cognitive performance is able to respond and not potentially pose as a safety threat. This was considered a significant research goal since previous studies have shown that message complexity and length can affect the pilot's ability to execute commands. The main brain area that was studied for the driving domain was the prefrontal cortex, which is responsible for decision-making and planning complex cognitive behavior. However, a few also studied the parietal cortex, a region of the brain that is responsible for sensory information and spatial navigation (Haas & Keel, 2003). The primary motor, premotor, and occipital cortex were areas that were also measured, but only for very specific studies.

The method through which the research goals are being tested is through driving or visual stimulation. Certain studies put strain factors such as mental arithmetic tasks, harsh driving conditions, or memory tasks onto the drivers. Moreover, the study with Liu et al. (2017) involved an actual route, through which the participants were asked to navigate down with 1400-m pathway while monitored by fNIRS instruments and following specific driving rules. The study by Lin (2013) used an assembly situation so it was testing the time it took for participants to complete fastener driving sessions. The participant parameters primarily consisted of young adults, with a few studies involving middle-aged adults. The minimum age starts at 20 and the maximum age is 45. While many studies used pilots, one study did use students to complete the experimental task (Liu et al., 2017). To continue, the gender demographic tends to be skewed slightly toward males; however, majority of studies did have some female participants. However, not all studies were able to provide participant demographics.

The fNIRS technology that is mostly used is continuous wave fNIRS, except for one study by Liu et al. (2017) that used time frequency to track the participants. The specific type of technology tends to vary; however, common fNIRS devices that were used include fNIR100 (Biopac1) and fNIRS OMM-3000. Moreover, in terms of number of channels that are used, there is a large range which includes the minimum number of channels being 2 and a maximum number of 78 channels.

One significant key finding found among all studies is that fNIRS is sensitive to tracking hemodynamic changes, and thus can be used to track cognitive changes and workload effectively. In Oka (2015), it was found that driving on the left side of the road required more brain activity, thus driving on the left side may involve more visual attention. To continue, another study found that driving with adaptive cruise control leads to a less active frontal lobe region (Tsunashima & Yanagisawa, 2009). Furthermore, another study that focused on differences in driving between older and younger drivers actually ended up finding no distinction in age, but rather that males had a higher tendency to overtake than females (Foy et al., 2016). In another study, connection strength among PFC (prefrontal cortex), MA, and VA increased from the resting state to the simple-driving state, whereas the connection strength relatively decreased during the car-following task (Liu et al., 2017). In the same study, it was seen that electrical connectivity can be strengthened by cognitive workload, but can also get overwhelmed by constant cognitive workload that requires many subtasks.

Some limitations of driving research using fNIRS are the applicability of the data due to the limited number of subjects. In the future, it is important to create systems that require an appropriate level of brain activity (Tsunashima & Yanagisawa, 2009). Furthermore, due to the limitations in the ability to recreate the experiment, the replicability and reliability of the data can also be considered as a limitation. Moreover, some uses of the data are the ability to track cognitive workload and determine any possible distractions that pilots and drivers may face that can serve as safety hazards. In the future, this can also expand and help to further understand how to improve car and driving design in order to achieve the optimal performance.

3.2 Mental Workload

The main goal of most studies involved tracking the hemodynamic changes and mental workload in order to better understand the relationships between strenuous tasks and the effect it has on cognitive state of the participants. Table 3 shows a summary of the exemplary fNIRS studies in mental workload research. The primary brain areas that were studied in mental workload mostly involved the prefrontal cortex; however, some studies also targeted the sensorimotor, visual, and occipital regions. The brain regions that are tracked would change depending on the experimental task. All the regions studied are crucial to understanding the effect that mental workload on the brain.

The experimental tasks through which mental workload is tracked involve a wide range of both physical and mental stimulating tasks. Two studies used visual stimulation in order to track hemodynamic changes, while other studies used activities such as solving anagrams or *n*-back tasks. Additionally, one study used navigating through maze software as their experimental task. Essentially any task that is stimulating and can lead to changes in mental workload was utilized as the method.

The participant parameters included mostly young adults; however, a few studies included an older demographic. The minimum age was 18 years and the maximum age was 42 years of age. In terms of gender demographics, there was a normal distribution of females to males. While there were overall more males than females in the participants, it was still close in number with the number of females. However, not all studies were able to provide participant demographics.

The type of fNIRS technology that was used consisted mostly of continuous wave (CW) fNIRS technology, with one study that used frequency-domain fNIRS. The fNIRS device used varied; however, a frequently used fNIRS device used included the Oxymon Mark III. A few of other devices used included NIRO200 NIRS system, ETG-4000 Optical Topography System, and the OxiplexTS from ISS Inc, the frequency-domain fNIRS device. Additionally, the number of channels between the studies did change; however, the smallest number of channels used was 2 and the maximum number of channels used was 52.

A significant key finding was that fNIRS can be used to track the hemodynamic changes associated with mental workload, and this information can be later quantified

Table 3 Summary of fNIRS studies in mental workload research

Authors	Experimental task	Characteristics of participants	Brain areas measured	No. of channels	fNIRS wavelength (nm)	Sampling rate	Type of fNIRS
Tachtsidis, Leung, Devoto, Delpy, and Elwell (2008)	Task stimulation: needed to solve anagrams	• 9 male subjects • $M_{age} = 22.9$	FC	2	–	6 Hz	CW
Lu et al. (2010)	Task session	• 29 subjects (14*F*) • $M_{age} = 21$	Bilateral sensorimotor cortex	52	695, 830	10 Hz	CW
Minati, Kress, Visani, Medford, and Critchley (2011)	Visual stimulation	• 16 subjects (9*F*) • $\mathrm{M}_{age} = 25$	PFC, OC	4	764, 859	–	CW
Bunce et al. (2011)	Warship commander task	• 8 subjects (3*F*) • $\mathrm{M}_{age} = 35$	PFC	16	730, 850	–	CW
Holper, Scholkmann, and Wolf (2012)	PFT task	• 16 subjects (4*F*) • $M_{age} = 28.25$	PFC	4	760, 870	10 Hz	CW
Ayaz et al. (2012)	*n*-back and ATC tasks	• 24 certified professional controllers • Age: 24–55	PFC	32	730, 850	–	CW
Solovey et al. (2012)	Robotic tasks	• 11 subjects (7*F*) • Age: 18–22	Anterior PFC	–	690, 830	6.25 Hz	FD
Shewokis, Ayaz, Curtin, Izzetoglu, and Onaral (2013)	Spatial navigation tasks	• 8 subjects	Inferior frontal gyrus	4	730, 850	2 Hz	CW

(continued)

Table 3 (continued)

Authors	Experimental task	Characteristics of participants	Brain areas measured	No. of channels	fNIRS wavelength (nm)	Sampling rate	Type of fNIRS
Fishburn, Norr, Medvedev, and Vaidya (2014)	*n*-back task	• 16 subjects (10*F*) • $M_{age} = 20.3$	PFC	2	690, 830	41.7 kHz	CW
Herff et al. (2014)	*n*-back task	• 10 subjects (4*F*) • $M_{age} = 22$	PFC	8	765, 856	2 Hz	CW
Balconi and Vanutelli (2017)	Task stimulation	• 28 subjects • $M_{age} = 23.78$	PFC	8	760, 850	6.25 Hz	CW
Aghajani et al. (2017)	*n*-back task	• 17 subjects (1*F*) • $M_{age} = 26.2$	PFC	19	760, 850	8.93 Hz	CW
Li, Liu, Zhu, Zhu, and Huang (2018)	*n*-back task	• 10 drivers (2*F*) • $M_{age} = 25.0$	PFC	16	695, 830	0.1 Hz	CW

Brain regions *PFC* prefrontal cortex, *OC* occipital cortex. **fNIRS devices** *CW* continuous wave, *FD* frequency domain and classified as data to comprehend mental workload. Moreover, brain activation in the prefrontal cortex can be used as a valid indication of mental workload (Ayaz et al., 2012). One study also showed that greater expertise was associated with relatively lower oxygenation (less neural activity) at low to moderate levels of taskload, but higher oxygenation and better performance at high levels of taskload (Bunce et al., 2011). Meanwhile, another indicated that transient systemic blood pressure changes are reflected in brain NIRS measurements (Minati et al., 2011).

The limitation in tracking the mental workload of participants is the ability to directly apply the causation of brain activation to the task that is at hand. Furthermore, due to the limitations in the ability to recreate the experiment, the replicability and reliability of the data can also be considered as a limitation. Finally, the small sample size in each study is also a limitation in fully understanding the potential application. The potential uses of tracking mental workload can help to determine how the brain reacts when facing high mental strain or undergoing emotional distractions or burdens. It can also be used to comprehend behavioral changes in people and can be useful in understanding the effect of neurological conditions on mental workload.

3.3 Motor Functions

The overall goal of majority of the research studies in motor function involved tracking the hemodynamic changes while the participants are in motion or performing an action. Moreover, many of the studies also involved testing a hypothesis or tracking the hemodynamic changes to determine a better mechanism or method to accomplish a goal. Table 4 shows a summary of the exemplary fNIRS studies in motor function research. For example, one study was trying to determine the neural differences between patients with multiple sclerosis and healthy subjects (Stojanovic-Radic et al., 2015). The primary brain region that was being monitored was the frontal lobe, and the prefrontal cortex was also being tracked, and hence other regions like the motor, temporal, and orbitofrontal regions. This was due to the fact some studies were testing how motor function affects language capabilities, and the temporal region functions include language and speech comprehension.

The type of task that was used to track motor function varied between studies. However, they all were some sort of motion or function, for example, speech tasks, *n*-back tasks, finger or head motion, and physical tasks were all methods that were used. Additionally, several studies were testing walking, with some adding changes or constraints, such as talking while walking or increasing the walking difficulty.

The subjects' age demographics ranged from students to middle-aged adults. The minimum age found was 20 and the maximum age was 65. While there was a normal distribution of females and males, there were a few studies that did not have any female participants at all. However, not all studies provided participant demographic information.

Table 4 Summary of fNIRS studies in motor function research

Authors	Experimental task	Characteristics of participants	Brain areas measured	No. of channels	Wavelength (nm)	Frequency (F)/sampling rate (SR)	Type of fNIRS
Wiggins and Prugniel (1997)	Speech (auditory and visual)	• 17 subjects (9F) • $M_{age} = 65.0$	TL	20–24	695, 830	10 Hz (SR)	CW
Fallgatter and Strik (1998)	Performance during WCST	• 10 subjects	FL	2	776.5, 819, 871, 908.7	1 Hz (F)	CW
Herrmann, Ehlis, and Fallgatter (2003)	Verbal fluency test	• 14 subjects	PFC	2	775, 810, 850, 910	2 kHz/2 Hz (SR)	CW
Holper et al. (2010)	Virtual reality stimulator	• 23 subjects (14F) • $M_{age} = 26.0$	Left hemisphere	4	760, 870	100 Hz (SR)	CW
Cui, Bray, and Reiss (2010)	Tracking changes in finger tapping motion	• 10 subjects (6F) • $M_{age} = 26.9$	MC	48	–	10 Hz (SR)	CW
Leamy, Collins, and Ward (2011)	Digital instructions	• 2 subjects • $M_{age} = 32$	MC	7	690, 830	25 Hz (SR)	CW
Scholkmann, Kleiser et al. (2014) and Scholkmann et al. (2014)	Four different tasks	• 2 subjects • Age: 29–31	PFC	3	760, 805, 870	–	CW
Brigadoi et al. (2014)	Tracking task	• 22 subjects (11F) • $M_{age} = 25.54$	FL, PMC	20	690, 830	7.8 Hz (F)	FD

(continued)

Table 4 (continued)

Authors	Experimental task	Characteristics of participants	Brain areas measured	No. of channels	Wavelength (nm)	Frequency (*F*)/sampling rate (SR)	Type of fNIRS
Mirelman et al. (2014)	Walking while dual tasking	• 23 subjects (13*F*) • $M_{age} = 30.9$	FC	6	760, 850	10 Hz (F)	CW
Khan et al. (2018)	Walking and rest tasks	• 9 male subjects • $M_{age} = 30$	MC	12	760, 830	1.81 Hz (SR)	CW
Beurskens, Helmich, Rein, and Bock (2014)	Walking and dual tasks	• 15 young subjects ($M_{age} = 24.5$) • 10 old subjects ($M_{age} = 71.0$)	FC, PFC	14	760, 830	1.81 Hz (SR)	CW
Rupp and Stephane (2009)	Physical tasks	• 15 male subjects • $M_{age} = 25.3 \pm 0.9$	PFC	2	775, 810, 850, 905	2 Hz (F)	CW
Stojanovic-Radic et al. (2015)	*n*-back tasks	• 25 subjects (13MS; 12HC) • 20*F* • $M_{age} = 45.8$ (MS); 31.0 (HC)	OFC	900	760, 830	–	CW
Metzger et al. (2017)	Walking and single tasks	• 12 subjects (8*F*) • $M_{age} = 27.6$	FC, TC, PC	24	787, 827	–	CW

Brain regions *PFC* prefrontal cortex, *PMC* premotor cortex, *FC* frontal cortex, *OC* occipital cortex, *TC* temporal cortex, *VC* visual cortex, *PC* parietal cortex, *FL* frontal lobe, *TL* temporal lobe. **fNIRS devices** *CW* continuous wave, *FD* frequency domain

The fNIRS technology that was used consisted mostly of continuous wave fNIRS, with one study that used frequency-domain fNIRS. The fNIRS device used tended to change; however, the ETG-4000 fNIRS, NIRO300, and DYNOT devices were used by several studies. Moreover, the ISS Imagent was used for the frequency-domain study. To continue, the number of channels used also had a wide distribution which included a minimum number of channels as 2 and a maximum number of channels as 900. However, the study that used 900 channels can be considered an outlier because it exceeds the typical range, and as a result the next highest maximum was 48 channels.

A few significant key themes found in the data include the ability of fNIRS to reliably measure speech-evoked auditory responses. Additionally, several studies show the activation of the frontal lobe for adults occurs during walking and dual-task studies. Moreover, for the studies that included walking and talking, there was increased brain activation in Broca's area, which is responsible for speech (Mirelman et al., 2014). To continue, another study was testing a potential neurocognitive rehabilitation regimen, and it was concluded that the scheme could be used for neurofeedback training and rehabilitation of amputees and paralyzed patients (Khan et al., 2018). Many studies explored the effect of specific treatments on improving motor function in patients with neurologically degenerative disorders. For example, a study was focused on patients with multiple sclerosis, and the results showed that individuals with MS experience a task with the lower cognitive load as more difficult than the healthy control group (Stojanovic-Radic et al., 2015). Other studies were also focused on the effect of motion on neurological processes, one found that while oxy- and deoxy-Hb are generally negatively correlated, head motion causes the correlation to become more positive (Cui et al., 2010).

Some limitations of the studies are that they tend to track different motor functions and as a result there can be a lack of consistency between the experiments. Moreover, the validity of the data can be concerning due to the lack of control in some of the experiments. To continue, however, some uses for these studies can be the ability to understand the neurological effects of diseases that target motor functions. Furthermore, it can be used to test processes that affect these neurological diseases. It can be applied to neurological changes like aging and be used to discover better rehabilitation solutions. In terms of areas for growth, there has to be more relatability and repeatability of the fNIRS response to motor functions as there is currently a lack of attention to detail in the past. Moreover, it is only through this consistency that there can be an effective and wholistic view of motor function in regard to fNIRS that can be established (Torricelli et al., 2014).

3.4 Decision-Making

A few of the main research goals for the decision-making studies involve tracking the significance of certain brain regions in decision-making. Table 5 shows a summary of the exemplary fNIRS studies in decision-making research. Moreover, the method

Table 5 Summary of exemplary fNIRS studies in decision-making research

Authors	Experimental task	Characteristics of participants	Brain areas measured	No. of channels	Wavelength (nm)	Sampling rate (Hz)	Type of fNIRS
Luu and Chau (2008)	Mental evaluation	• 9 subjects (4*F*) • $M_{age} = 23.4$	PFC	48	690, 830	31. 25	FD
Hu, Hong, and Ge (2012)	Visual stimulation	• 8 male subjects • $M_{age} = 29$	PFC	4	760, 830	1.81	CW
Putze et al. (2014)	Memorizing stories	• 12 subjects (6*F*) • $M_{age} = 23.6$	TC, VC	170	690, 830	19.5	FD
McKendrick, Ayaz, Olmstead, and Parasuraman (2014)	Digital memory tasks	• 10 subjects	PFC	16	–	–	CW
de Winkel, Nesti, Ayaz, and Bülthoff (2017)	Motion platform	• 17 subjects (6*F*) • $M_{age} = 24.8$	PFC	–	760, 850	7.8125	CW

Brain regions *PFC* prefrontal cortex, *TC* temporal cortex, *VC* visual cortex. **fNIRS devices** *CW* continuous wave, *FD* frequency domain through which the brain is able to process specific types of decisions, for example, how the brain processes deception. The brain areas that have been measured include mostly the prefrontal cortex, but also the parietal, visual, and temporal cortex.

The types of tasks that have been used to track decision-making involve visual stimulation, task sessions, and one study that actually involved using a motion platform in order to distinguish between decisions (de Winkel et al., 2017). The participant parameters within the study range from students to older adults, the minimum age being 22 and the maximum age being 51 years. Moreover, the gender demographics also have a normal distribution of males to females and every study included female participants.

The fNIRS technology used in the study consists mostly of continuous wave fNIRS but two studies used frequency-domain fNIRS. While the CW devices used tend to vary, the frequency-domain devices used consisted of only one fNIRS device: Imagent from ISS Inc. The channel distribution also had a wide range which included the minimum number of channels of 2 and the maximum number of channels as 64. However, every study used multiple channels.

A significant key finding found in the study includes sensitivity in distinguishing between two separate decision-making processes. Moreover, a continual use of fNIRS is tracking the decision-making processes. Some limitations of using decision-making studies include the changes in constraints and methods, which lead to a lack of consistency in results, and potentially non-reliable information. A possible use of decision-making studies includes the ability to be able to determine what neural changes occur when making a decision and also potentially understanding the neural changes that occur in specific states, such as situations where deception is being used (Hu et al., 2012).

4 Summary and Future Directions

fNIRS is a broad ranging and valuable tool that is able to dissect and understand once complex situations. Due to its nonintrusive, mobile, and minimized nature, it has the potential to be used in a plethora of different areas (Shewokis et al., 2013). Throughout this chapter, the necessity of fNIRS is shown in understanding neurocognitive behavior and providing a deeper meaning to brain. Moreover, a wholistic and critical explanation, use, and application of fNIRS have been thoroughly discussed and evaluated within this chapter.

For continuous wave fNIRS instrumentation, there is a desire to create instruments that include an increased number of light sources and detectors in order to take more advantage of already existing imaging algorithms. These instruments will also offer a higher dynamic range in order to measure at a wider range of distances, which will optimize the depth of these solutions. Frequency-domain devices have seen use in a multitude of different areas, an area to potentially expand upon is in tissue analysis (Spichtig, Hornung, Brown, Haensse, & Wolf, 2009). To continue, frequency domain also requires improving their solid-state devices to a point where phase and intensity can both be measured (Scholkmann et al., 2014; Scholkmann, Kleiser, et al., 2014). Time-domain devices are finding application in areas such as mammography and molecular imaging. The instrument itself tends to be somewhat on the larger side and expensive, and as result it is hopeful that over time it will be able to become more cost-effective and smaller in size (Torricelli et al., 2014). As a whole, there has been discussion that fNIRS can be used in learning and training environments to potentially make the learning process more efficient (Shewokis et al., 2013). Furthermore, the ability to measure neurocognitive behavior without being constrained leads to increased research in areas such as sports. The ability to analyze the brain activity between two or more subjects, also called hyperscanning, is also a discipline that can benefit from fNIRS (Scholkmann et al., 2014; Scholkmann, Kleiser, et al., 2014). Its application is currently focused on understanding the processes that factor into driving, mental workload, decision-making, and motor function; however, as time progresses, the future and application of fNIRS are limitless.

References

Aghajani, H., Garbey, M., & Omurtag, A. (2017). Measuring mental workload with eEG + fNIRS. *Frontiers in Human Neuroscience, 11,* 1–20.

Ayaz, H., Onaral, B., Izzetoglu, K., Shewokis, P. A., McKendrick, R., & Parasuraman, R. (2013). Continuous monitoring of brain dynamics with functional near infrared spectroscopy as a tool for neuroergonomic research: Empirical examples and a technological development. *Frontiers in Human Neuroscience, 7,* 871.

Ayaz, H., Shewokis, P. A., Bunce, S., Izzetoglu, K., Willems, B., & Onaral, B. (2012). Optical brain monitoring for operator training and mental workload assessment. *NeuroImage, 59*(1), 36–47.

Balconi, M., & Vanutelli, M. E. (2017). Brains in competition: Improved cognitive performance and inter-brain coupling by hyperscanning paradigm with functional near-infrared spectroscopy. *Frontiers in Behavioral Neuroscience, 11,* 1–10.

Beurskens, R., Helmich, I., Rein, R., & Bock, O. (2014). Age-related changes in prefrontal activity during walking in dual-task situations: A fNIRS study. *International Journal of Psychophysiology, 92*(3), 122–128.

Boas, D. A., Elwell, C. E., Ferrari, M., & Taga, G. (2014). Twenty years of functional near-infrared spectroscopy: Introduction for the special issue. *NeuroImage, 85,* 1–5.

Brigadoi, S., Ceccherini, L., Cutini, S., Scarpa, F., Scatturin, P., Selb, J., … Cooper, R. J. (2014). Motion artifacts in functional near-infrared spectroscopy: A comparison of motion correction techniques applied to real cognitive data. *NeuroImage, 85,* 181–191.

Bunce, S. C., Izzetoglu, K., Ayaz, H., Shewokis, P., Izzetoglu, M., Pourrezaei, K., & Onaral, B. (2011). *Implementation of fNIRS for monitoring levels of expertise and mental workload. Lecture Notes in Computer Science (Including Subseries Lecture Notes in Artificial Intelligence and Lecture Notes in Bioinformatics), 6780 LNAI* (pp. 13–22).

Buxton, R. B. (2012). Dynamic models of BOLD contrast. *Neuroimage, 62*(2), 953–961.

Causse, M., Chua, Z., Peysakhovich, V., Del Campo, N., & Matton, N. (2017). Mental workload and neural efficiency quantified in the prefrontal cortex using fNIRS. *Scientific Reports, 7*(1), 1–16.

Cui, X., Bray, S., & Reiss, A. L. (2010). Functional near infrared spectroscopy (NIRS) signal improvement based on negative correlation between oxygenated and deoxygenated hemoglobin dynamics. *NeuroImage, 49*(4), 3039–3046.

Curtin, A., & Ayaz, H. (2018). The age of neuroergonomics: Towards ubiquitous and continuous measurement of brain function with fNIRS. *Japanese Psychological Research, 60*(4), 374–386.

Davies, D. J., Clancy, M., Lighter, D., Balanos, G. M., Lucas, S. J. E., Dehghani, H., … & Belli, A. (2017). Frequency-domain vs continuous-wave near-infrared spectroscopy devices: A comparison of clinically viable monitors in controlled hypoxia. *Journal of Clinical Monitoring and Computing, 31*(5), 967–974.

de Winkel, K. N., Nesti, A., Ayaz, H., & Bülthoff, H. H. (2017). Neural correlates of decision making on whole body yaw rotation: An fNIRS study. *Neuroscience Letters, 654,* 56–62.

Fallgatter, A. J., & Strik, W. K. (1998). Frontal brain activation during the Wisconsin Card Sorting Test assessed with two-channel near-infrared spectroscopy. *European Archives of Psychiatry and Clinical Neuroscience, 248*(5), 245–249.

Fishburn, F. A., Norr, M. E., Medvedev, A. V., & Vaidya, C. J. (2014). Sensitivity of fNIRS to cognitive state and load. *Frontiers in Human Neuroscience, 8,* 1–11.

Foy, H. J., Runham, P., & Chapman, P. (2016). Prefrontal cortex activation and young driver behaviour: A fNIRS study. *PLoS ONE, 11*(5), 1–18.

Gateau, T., Ayaz, H., Dehais, F., & Lotte, F. (2018). In silico vs over the clouds: On-the-fly mental state estimation of aircraft pilots, using a functional near infrared spectroscopy based passive-BCI. *Frontiers in Human Neuroscience, 12,* 1–14.

Gateau, T., Durantin, G., Lancelot, F., Scannella, S., & Dehais, F. (2015). Real-time state estimation in a flight simulator using fNIRS. *PLoS ONE, 10*(3), 1–19.

Haas, D., & Keel, C. (2003). An integrative theory of prefrontal cortex function. *Annual Review of Phytopathology, 41*(1), 117–153.

Harrison, J., Izzetoğlu, K., Ayaz, H., Willems, B., Hah, S., Ahlstrom, U., … Onaral, B. (2014). Cognitive workload and learning assessment during the implementation of a next-generation air traffic control technology using functional near-infrared spectroscopy. *IEEE Transactions on Human-Machine Systems, 44*(4), 429–440.

Herff, C., Heger, D., Fortmann, O., Hennrich, J., Putze, F., & Schultz, T. (2014). Mental workload during n-back task—Quantified in the prefrontal cortex using fNIRS. *Frontiers in Human Neuroscience, 7,* 1–9.

Herrmann, M. J., Ehlis, A.-C., & Fallgatter, A. J. (2003). Frontal activation during a verbal-fluency task as measured by near-infrared spectroscopy. *Brain Research Bulletin, 61*(1), 51–56.

Holper, L., Muehlemann, T., Scholkmann, F., Eng, K., Kiper, D., & Wolf, M. (2010). Testing the potential of a virtual reality neurorehabilitation system during performance of observation, imagery and imitation of motor actions recorded by wireless functional near-infrared spectroscopy (fNIRS). *Journal of Neuroengineering and Rehabilitation, 7*(1), 57.

Holper, L., Scholkmann, F., & Wolf, M. (2012). Between-brain connectivity during imitation measured by fNIRS. *NeuroImage, 63*(1), 212–222.

Hu, X. S., Hong, K. S., & Ge, S. S. (2012). fNIRS-based online deception decoding. *Journal of Neural Engineering*, *9*(2).

Khan, R. A., Naseer, N., Qureshi, N. K., Noori, F. M., Nazeer, H., & Khan, M. U. (2018). FNIRS-based neurorobotic interface for gait rehabilitation. *Journal of NeuroEngineering and Rehabilitation, 15*(1), 1–17.

Kikukawa, A., Kobayashi, A., & Miyamoto, Y. (2008). Monitoring of pre-frontal oxygen status in helicopter pilots using near-infrared spectrophotometers. *Dynamic Medicine, 7*(1), 1–10.

Leamy, D. J., Collins, R., & Ward, T. E. (2011). Combining fNIRS and EEG to improve motor cortex activity classification during an imagined movement-based task. *Lecture Notes in Computer Science (Including Subseries Lecture Notes in Artificial Intelligence and Lecture Notes in Bioinformatics), 6780 LNAI* (pp. 177–185).

León-Carrión, J., & León-Domínguez, U. (2012). Functional near-infrared spectroscopy (fNIRS): Principles and neuroscientific applications. In *Neuroimaging-methods*. IntechOpen.

Li, L., Liu, Z., Zhu, H., Zhu, L., & Huang, Y. (2018). Functional near-infrared spectroscopy in the evaluation of urban rail transit drivers' mental workload under simulated driving conditions. *Ergonomics*, *0*(0), 1–14.

Lin, P. Y., Chen, J. J. J., & Lin, S. I. (2013). The cortical control of cycling exercise in stroke patients: An fNIRS study. *Human Brain Mapping, 34*(10), 2381–2390.
Lin, J. H., Maikala, R. V., McGorry, R., & Brunette, C. (2010). NIRS application in evaluating threaded-fastener driving assembly tasks. *International Journal of Industrial Ergonomics, 40*(2), 146–152.
Liu, Z., Zhang, M., Xu, G., Huo, C., Tan, Q., Li, Z., et al. (2017). Effective connectivity analysis of the brain network in drivers during actual driving using near-infrared spectroscopy. *Frontiers in Behavioral Neuroscience, 11,* 1–11.
Lu, C. M., Zhang, Y. J., Biswal, B. B., Zang, Y. F., Peng, D. L., & Zhu, C. Z. (2010). Use of fNIRS to assess resting state functional connectivity. *Journal of Neuroscience Methods, 186*(2), 242–249.
Luu, S., & Chau, T. (2008). Decoding subjective preference from single-trial near-infrared spectroscopy signals. *Journal of Neural Engineering, 6*(1), 16003.
McKendrick, R., Ayaz, H., Olmstead, R., & Parasuraman, R. (2014). Enhancing dual-task performance with verbal and spatial working memory training: Continuous monitoring of cerebral hemodynamics with NIRS. *NeuroImage, 85,* 1014–1026.
Metzger, F. G., Ehlis, A. C., Haeussinger, F. B., Schneeweiss, P., Hudak, J., Fallgatter, A. J., & Schneider, S. (2017). Functional brain imaging of walking while talking—An fNIRS study. *Neuroscience, 343,* 85–93. https://doi.org/10.1016/j.neuroscience.2016.11.032.
Minati, L., Kress, I. U., Visani, E., Medford, N., & Critchley, H. D. (2011). Intra- and extracranial effects of transient blood pressure changes on brain near-infrared spectroscopy (NIRS) measurements. *Journal of Neuroscience Methods, 197*(2), 283–288.
Mirelman, A., Maiden, I., Bernad-Elazari, H., Nieuwhof, F., Reelick, M., Giladi, N., & Hausdorff, J. M. (2014). Increased frontal brain activation during walking while dual tasking: An fNIRS study in healthy young adults. *Journal of NeuroEngineering and Rehabilitation, 11,* 85.
Molteni, E., Contini, D., Caffini, M., Baselli, G., Spinelli, L., Cubeddu, R., … Torricelli, A. (2012). Load-dependent brain activation assessed by time-domain functional near-infrared spectroscopy during a working memory task with graded levels of difficulty. *Journal of Biomedical Optics, 17*(5), 056005.
Nishiyori, R. (2016). fNIRS: An emergent method to document functional cortical activity during infant movements. *Frontiers in Psychology, 7,* 1–11.
Oka, N., Yoshino, K., Yamamoto, K., Takahashi, H., Li, S., Sugimachi, T., … Kato, T. (2015). Greater activity in the frontal cortex on left curves: A vector-based fNIRS study of left and right curve driving. *PLoS ONE, 10*(5), e0127594.
Piper, S. K., Krueger, A., Koch, S. P., Mehnert, J., Habermehl, C., Steinbrink, J., … Schmitz, C. H. (2014). A wearable multi-channel fNIRS system for brain imaging in freely moving subjects. *NeuroImage, 85,* 64–71.
Putze, F., Hesslinger, S., Tse, C. Y., Huang, Y., Herff, C., Guan, C., et al. (2014). Hybrid fNIRS-EEG based classification of auditory and visual perception processes. *Frontiers in Neuroscience, 8,* 373.
Rupp, T., & Stephane, P. (2009). A study on the k-convexity of sequences. *Global Journal of Pure and Applied Mathematics, 11*(1), 311–314.
Scholkmann, F., Kleiser, S., Metz, A. J., Zimmermann, R., Mata Pavia, J., Wolf, U., & Wolf, M. (2014). A review on continuous wave functional near-infrared spectroscopy and imaging instrumentation and methodology. *NeuroImage, 85,* 6–27.
Scholkmann, F., Metz, A. J., & Wolf, M. (2014b). Measuring tissue hemodynamics and oxygenation by continuous-wave functional near-infrared spectroscopy—How robust are the different calculation methods against movement artifacts? *Physiological Measurement, 35*(4), 717–734.
Shewokis, P. A., Ayaz, H., Curtin, A., Izzetoglu, K., & Onaral, B. (2013). Brain in the loop learning using functional near infrared spectroscopy. *Lecture Notes in Computer Science (Including Subseries Lecture Notes in Artificial Intelligence and Lecture Notes in Bioinformatics), 8027 LNAI* (pp. 381–389).

Solovey, E. T., Schermerhorn, P., Scheutz, M., Sassaroli, A., Fantini, S., & Jacob, R. (2012). Brainput: Enhancing interactive systems with streaming fNIRS brain input. In *Proceedings of the 2012 ACM Annual Conference on Human Factors in Computing Systems—CHI '12* (p. 2193).

Spichtig, S., Hornung, R., Brown, D. W., Haensse, D., & Wolf, M. (2009). Multifrequency frequency-domain spectrometer for tissue analysis. *Review of Scientific Instruments, 80*(2).

Stojanovic-Radic, J., Wylie, G., Voelbel, G., Chiaravalloti, N., & DeLuca, J. (2015). Neuroimaging and cognition using functional near infrared spectroscopy (fNIRS) in multiple sclerosis. *Brain Imaging and Behavior, 9*(2), 302–311.

Tachtsidis, I., Leung, T. S., Devoto, L., Delpy, D. T., & Elwell, C. E. (2008). Measurement of frontal lobe functional activation and related systemic effects: A near-infrared spectroscopy investigation. *Advances in Experimental Medicine and Biology, 614,* 397–403.

Torricelli, A., Contini, D., Pifferi, A., Caffini, M., Re, R., Zucchelli, L., & Spinelli, L. (2014). Time domain functional NIRS imaging for human brain mapping. *NeuroImage, 85,* 28–50.

Tsunashima, H., & Yanagisawa, K. (2009). Measurement of brain function of car driver using functional near-infrared spectroscopy (fNIRS). *Computational Intelligence and Neuroscience, 2009.*

Unni, A., Ihme, K., Jipp, M., & Rieger, J. W. (2017). Assessing the driver's current level of working memory load with high density functional near-infrared spectroscopy: A realistic driving simulator study. *Frontiers in Human Neuroscience, 11,* 1–14.

von Lühmann, A., Herff, C., Heger, D., & Schultz, T. (2015). Toward a wireless open source instrument: Functional near-infrared spectroscopy in mobile neuroergonomics and BCI applications. *Frontiers in Human Neuroscience, 9,* 617.

Wahr, J. A., Tremper, K. K., Samra, S., & Delpy, D. T. (1996). Near-infrared spectroscopy: theory and applications. *Journal of Cardiothoracic and Vascular Anesthesia, 10*(3), 406–418.

Wiggins, I., & Prugniel, P. (1997). Kinematical data on early-type galaxies. III. *Astronomy and Astrophysics Supplement Series, 126*(3), 519–523.

Xu, L., Wang, B., Xu, G., Wang, W., Liu, Z., & Li, Z. (2017). Functional connectivity analysis using fNIRS in healthy subjects during prolonged simulated driving. *Neuroscience Letters, 640,* 21–28.

Transcranial Direct Current Stimulation (tDCS): A Beginner's Guide for Neuroergonomists

Jacob Green, Sehyeon Jang, Jinyoung Choi, Sung C. Jun and Chang S. Nam

Abstract This Beginner's Guide is intended to provide the necessary information for neuroergonomists to be able to understand the effects and usages of Transcranial Direct Current Stimulation (tDCS) on the brain. A thorough timeline provides the transformation of tDCS with technological advancements as well as a representation of its benefits throughout human history. It was these benefits that propelled tDCS to still be a modern-day solution to neuroergonomical problems. A large part of this chapter is dedicated to the impacts of tDCS, specifically, its effects on motor and cognitive functions and how the configurations of a session can alter the intended results. By the end of this guide, the reader should have a comprehensive understanding of what tDCS is, how it works, how to accurately design and implement tDCS configurations as well as the impacts it has on cognitive and motor functions of the brain.

1 Introduction

Transcranial Direct Current Stimulation (tDCS) was developed to help patients with brain injuries regain some of their initial conditions. However, over the years it was found to have many more beneficial uses such as increasing cognitive performance (e.g., Fregni, Boggio, Nitsche, Rigonatti, & Pascual-Leone, 2006; Horvath, Forte, & Carter, 2015) and motor functions (e.g., Antal et al., 2004; Inguaggiato, Bolognini, Fiori, & Cioni, 2019). With a growing interest in the unknown potential of tDCS a large effort in various research fields has been launched to better understand the effects tDCS has on the human brain. Technological advances in tDCS such as using

J. Green · C. S. Nam (✉)
Edward P. Fitts Department of Industrial & Systems Engineering, North Carolina State University, Raleigh, NC 27695, USA
e-mail: csnam@ncsu.edu

S. Jang · J. Choi · S. C. Jun
School of Electrical Engineering and Computer Science, Gwangju Institute of Science and Technology, Gwangju 61005, South Korea

C. S. Nam (ed.), *Neuroergonomics*, Cognitive Science and Technology,
https://doi.org/10.1007/978-3-030-34784-0_5

a simple galvanic battery in the eighteenth and nineteenth centuries to microprocessors and microcontroller technologies in the twentieth century have also increased the ability to study its potential benefits and enabled researchers to design tDCS devices with stricter stimulation parameters. While this treatment has been enhanced with modern technology, tDCS has been around for a long time. In fact, the first electricity sources used for electrical stimulation were produced by animal electricity (Sarmiento, San-Juan, & Prasath, 2016). Let's look at a timeline regarding the progression of tDCS technology. Table 1 shows the progression of tDCS from its early beginnings to its now technologically advanced state.

Table 1 Timelines of tDCS technological advancements

[1] **BC**
• **3000 BC**: Ancient Egyptians became aware of electrical properties of Nile catfish
• **400 to 300 BC**: Plato and Aristotle observe and record the torpedo fish's (electric ray) ability to elicit therapeutic effects using electrical discharge
[2] **AD**
• **43 AD**: Roman physician Scribonius Largus, discovers and records pain-relieving properties of electricity after studying a story of a man relieved of gout related pains after stepping on an electric ray. Scribonius would later use electric fish like the torpedo fish for headache treatment
• **143 AD**: Greek physician Claudius Galen realizes torpedo fish had to be alive when applied during treatment, implying the torpedo fish must be actively producing something that causes the pain-relieving phenomenon (electric current!)
[3] **16–19th**
• **Sixteenth-century**: Syrian physician Dawud al-Antaki uses torpedo fish to treat vertigo in his patients
• **1660**: German scientist Otto Von Guericke invents a hand-cranked electrostatic generator, effectively creating the first controllable form of artificial electricity!
• **1745**: Ewald Georg Von Kleist invents the first electric capacitor, dubbed the Leyden Jar
• **1757**: Experimenters and tinkerers like Anton de Haen and Benjamin Franklin uses electrostatic generators and Leyden jars for therapeutic electrification
• **1756**: Italian anatomist Leopoldo Marco Antonio Caldani uses the Guericke electrostatic generator to stimulate muscles in sheep and frogs; first evidence of the importance of electricity in physiology
• **1767**: Middlesex Hospital in England becomes the first hospital to purchase an electrostatic therapy machine
• **1773**: Anatomist John Hunter studies and elucidates the mechanism for the pain relief and numbing caused by the torpedo fish. Discovers torpedo fish produced electricity!
• **1780**: Luigi Galvani and Alessandro Volta induced muscle twitching and movement in animals using an electrical current; more evidence toward the importance of electricity in animal physiology
• **1800**: Volta creates the first battery, which produced direct current (DC). This opened the doors for the first clinical applications of therapeutic direct current stimulation

(continued)

Table 1 (continued)

• **1801**: Galvani's nephew Giovanni Aldini, uses an early form of tDCS to improve the mood of 27-year-old farmer Luigi Lanzarini who presumably suffered from depression "Melancholia". This was one of the first recorded instances of tDCS being used to treat neurological and psychiatric conditions, a major topic of interest in today's medical fields
• **1802**: Hellwag and Jacobi observe the occurrence of phosphenes during tDCS applications
[4] Twentieth century
• **1930–1940**: Electroconvulsive therapy (ECT) quickly gains popularity and overshadows tDCS; persistent inconclusive results, a lack of understanding of operating principles and poor experimental design were several factors leading to the loss of interest in experimental tDCS at that time
• **1960**: tDCS regains its popularity as interest in electro-sleep therapy and electro-anesthesia increases, this leads to an increase in an effort toward understanding how tDCS affects the brain
• **1964–1966**: DJ Albert publishes two major papers illustrating that tDCS could increase or decrease memory retention in rats, depending on the current direction. This shows tDCS' capability for modulating cortical excitability. Around this time, Lippold and Redfearn also reports on anodal current's ability to induce increased alertness, mood and motor activity and cathodal current's ability to induce quietness and apathy; providing further evidence to suggest tDCS' ability to modulate cortical excitability
• **1970**: Advances in pharmaceuticals for psychiatric disorders captures public attention, consequently overshadowing electrical therapy research
• **1998**: Electrical stimulations regain interest due to an effort to develop more effective and targeted treatments with fewer side effects than tradition pharmacological means. New brain imaging techniques and brain stimulation methods also help to reintroduce tDCS
[5] Twenty-first century
• **2000–present**: Progress in microcontroller technology allows engineers to create precision tDCS devices with superior control over stimulation parameters at reduced costs. Research is currently being conducted on tDCS as a means for treating neurological and psychiatric disorders, with applications in depression, schizophrenia, aphasia, addiction, epilepsy, chronic pain, attention and motor rehabilitation. Possible applications of tDCS on healthy individuals are also being explored for cognitive and neurological enhancements. Research is also being conducted to understand tDCS and its interactions with the brain on multiple levels. This is made possible by the advancements of computer processing capabilities and physics simulation software

Note that this summary is based on Sarmiento et al. (2016)

During the BC and AD era, scientists mainly made observations regarding the torpedo fish's ability to produce electrical current and its potential medical applications such as healing headaches. Moving through the sixteenth–nineteenth centuries were the beginnings of tDCS as it is known today. Researchers developed the first electrostatic generator in the mid-sixteenth century. This was further used to conduct controlled experiments studying the beneficial effects of electrical current stimulation. During the twentieth century, tDCS really took off when researchers published papers identifying the potential impacts stimulation on the brain could have on a person's motor and cognitive functional abilities. These findings and modern technology allow twenty-first century researchers to develop precise studies that target specific areas of the brain associated with a selected cognitive or motor function. This chapter's goal is to effectively educate researchers on what tDCS is and how it works. An emphasis is placed on cognitive and motor function effects as they are the reason for such electrical stimulation. In relation to the goal of this guide, it is also worth noting the reason why a beginner's guide is necessary. tDCS can have an intimidating sound when hearing it for the first-time causing people to shy away. However, with a well thought out guide for beginners, tDCS can be better communicated to the public thus increasing its popularity.

This chapter is further broken down into three sections that cover the design and implementation of tDCS, the state-of-the-art of tDCS-based research and lastly future directions of tDCS. Each of these sections contains subsections that go into more detail relative to that section's topic. In Sect. 2, the design and implementation of tDCS will be discussed. This includes how it works, designing parameters, how to use the tDCS system, and safety concerns. Section 3 covers the state-of-the-art of tDCS-based neuroergonomics research. Motor and cognitive functions are the focus of this section going into depth on how each function is actively affected and studied within different research domains. Lastly, Sect. 4 recaps the goals of the guide and provides a list of limitations and future recommendations that will contribute to forthcoming research opportunities.

2 Design and Implementation of tDCS

In order to understand the impacts tDCS has on the human brain one needs to clearly understand the background of its implementation process. Knowing how tDCS works will increase the likely hood of a successful stimulation period compared to an unsuccessful and potentially harmful one. Also, designing parameters such as current intensity and electrode placement is key to achieving stimulation sessions' desired outcome. Without precise parameter configurations stimulation can be unpredictable and harmful. Perhaps the most important topic about tDCS is safety and the concerns of safety that come with it. Anytime a person is exposed to an electrical current there can be harm. In the last part of this section safety concerns and boundaries are discussed on how to keep people safe while receiving stimulation.

2.1 How Does tDCS Work?

tDCS is a non-invasive neuromodulatory technique that delivers a low electric current (e.g., 1–2 mA) to specifically targeted areas of the brain. When applied to healthy adults tDCS has shown increases in cognitive performance on a variety of tasks. Many studies have also shown tDCS can enhance language (e.g., Price, McAdams, Grossman, & Hamilton, 2015) and mathematical ability (e.g., Looi, & Kadosh, 2016), attention span (e.g., Ulam et al., 2015), problem solving (e.g., Zmigrod, Colzato, & Hommel, 2015), memory (e.g., Au et al., 2016) and coordination (e.g., Meesen, Rand, Cuypers, Nitsche, & Nitsche, 2017). Some studies have even included depression (e.g., Meron, Hedger, Garner, & Baldwin, 2015), anxiety (e.g., Ironside, Harmer, Shea, Ebmeier, & Brunoni, 2016), post-traumatic stress disorder or PTSD (e.g., van't Wout-Frank, Shea, Larson, Greenberg, & Philip, 2019) and chronic pain (e.g., Volz, Farmer, & Siegmund, 2016) as potential conditions treatable by tDCS.

tDCS works by applying a positive (anodal) current as seen in Fig. 1B or negative (cathodal) current in Fig. 1C. Current intensity is applied from a current source

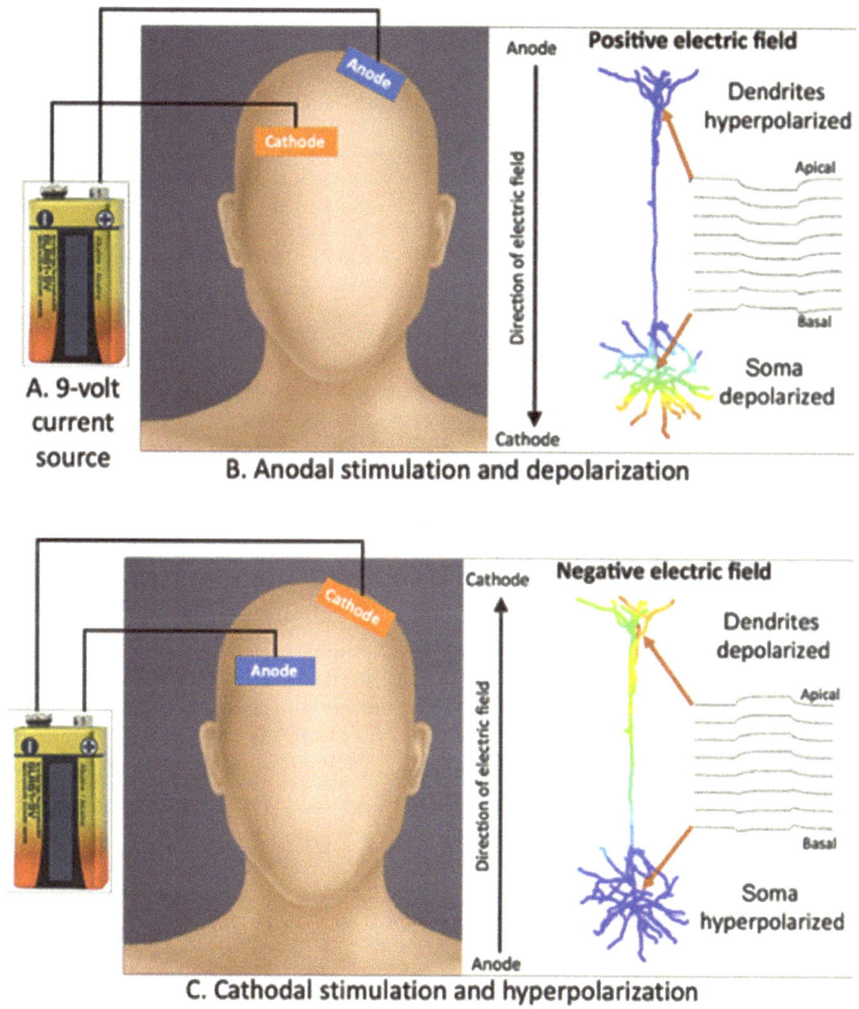

Fig. 1 Effects of anodal (of primary motor cortex or M1) and cathodal (over contralateral supraorbital area) stimulations on membrane polarization

(Fig. 1A) and kept between 1 and 2 mA via electrodes to an area. The current facilitates the depolarization or hyperpolarization of neurons, respectively. The position of anodal and cathodal electrodes determines how current flows and wherein the brain it flows. Basically, the current is not used to trigger an action response from the neurons but rather to drive them closer together or further apart to solicit an appropriate increase or decrease of the effective area. For example, the anodal stimulation depolarizes the neuronal membrane and enhances M1 excitability (Fig. 1B), while the cathodal stimulation hyperpolarizes the neuronal membrane and diminish M1 excitability (Fig. 1C). The normal tDCS devices are small battery-powered devices that usually come with a control panel used to set duration and intensity of the stimulation. As stated earlier, electrodes are placed on the head, held by an elastic strap. Cables connect each electrode to the stimulator which is then turned on sending current from the stimulator to the electrodes thus through the brain. However, research and clinical-grade stimulators have many features such as impedance and current meter that help ensure stimulation is tolerable and reliable.

2.2 *Parameter Configurations*

The region of interest, otherwise known as the predetermined area of the brain associated with the motor or cognitive task being given, is stimulated using an electrode attached over the region of interest, the location of which depends on the hypotheses and task. For example, if the treatment is for aggression then the stimulation should be focused on the prefrontal cortex. Anatomical features such as skull thickness and composition can alter the consistency of current across a subject. With the use of neuro-navigational software, the experimenter can be more accurate when placing electrodes above the defined cortical locations while taking anatomical differences into consideration. The components of a typical tDCS system are illustrated in Fig. 2.

The electrodes should be a distance apart of at least 8 cm when using 35 cm^2 electrodes (Wagner et al., 2007). Additionally, smaller electrode sizes have been correlated with larger shunting effects. The current strength should also be monitored and controlled between 1 and 2 mA. Studies exceeding this interval have shown discomfort in respiratory movements. Electrode sizes are a varying parameter among tDCS while the most commonly used electrodes are sized between 25 and 35 cm^2. The last parameter to be determined is stimulation intensity. Most stimulation durations range between 5 and 30 min with a current intensity between 1 and 2 mA. The main takeaway with stimulation intensity and duration is that it is not necessarily linear and that the relationship between these two variables requires further investigation.

Fig. 2 The components of a typical tDCS system. A Comfortable neoprene head cap based on the 10-10 system with 39 positions available. B 25 cm^2 circular sponge cover. C Carbon rubber electrode. D 10 electrode cable composed of 8-channel, common mode sense or CMS, and driven right leg or DRL. E Neuroelectrics control box. F Saline solution. G Curved syringe for saline solution application. H Power adapter. All components are based on Starstim8 device (Neuroelectrics, Spain)

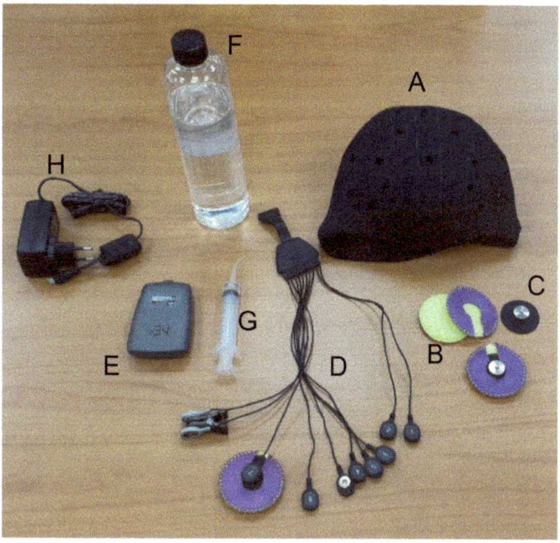

2.3 The Way to Use the tDCS System

Before wearing the head cap, it is advisable to arrange the hair in the areas where the electrodes will be attached so that the sponge pad maintains good contact with the scalp. After the head cap is in place, the target and the return electrodes' real-time impedances are monitored through the software provided (Fig. 3) to evaluate whether the electrodes are in good contact with the scalp. In the case of poor contact or insufficient saline solution, the conductivities become low and result in indicating high impedance levels. In that case, by taking the appropriate methods (rearranging the electrodes' positions, adjusting the head cap's chin strap to achieve closer contact between the sponge and the scalp, or applying more saline solution to the sponge pad), a stable level of impedance can be identified, and then electrical stimulation is initiated.

Physiological saline (0.9% NaCl) is applied commonly to the sponge pad to increase the conductivity between the area in contact with the scalp and the carbon rubber electrode. When excessive saline solution is applied, it may cause some harmful irritation to the skin (Woods et al., 2016). Therefore, it is recommended to use an appropriate amount of saline applied evenly on the sponge pad (approximately 5 ml for a 25 cm^2 area; DaSilva, Volz, Bikson, & Fregni, 2011).

In the software used to set the control box, the channels for the active and the reference electrodes are set, and the channel and electrode are connected accordingly. In addition, configuration parameters, such as stimulus intensity, time, and type, are set appropriately according to the experimental paradigm designed, and then the electrical stimulation is initiated. At the beginning and end of the electrical stimulation, the subject may feel a slight itch when the current increases or decreases gradually to the target intensity (typically approximately 30 s; Bikson et al., 2016). However, the

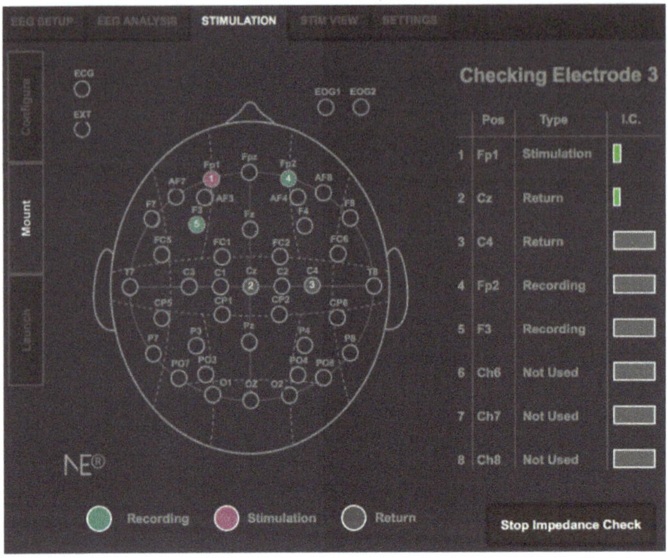

Fig. 3 A screenshot of the tDCS management interface (e.g., Starstim8). Researchers can select or modify stimulation parameters through the software. Stimulation electrode's montage and its impedance can be monitored in real-time

subject will feel no itching while the current remains at the target intensity (Kessler, Turkeltaub, Benson, & Hamilton, 2012). Thus, during the stimulation, the subject scarcely will realize that the electrical stimulation actually is being applied. However, if the impedance is unstable, then the current intensity cannot be maintained consistently, and the subject may feel a slight tingling and become aware of the stimulation (Thair, Holloway, Newport, & Smith, 2017). Therefore, it is important to minimize moving the body and the head during the stimulation to maintain a stable state. The detailed procedure of the tDCS system is illustrated in Fig. 4.

We note that the rubber electrode is perishable and should be replaced when it loses conductivity. For example, the rubber electrode in the equipment above is likely to lose conductivity after approximately 100 h of stimulation (Thair et al., 2017).

2.4 *Safety Concerns and Issues*

Although tDCS has been recognized widely as a safe brain stimulation technique, there are still some potential detrimental effects worthy of mentioning and considering. First, researchers and therapists should pay attention to skin lesions that usually occur under the anodal or cathodal electrodes in patients after treatment with tDCS. Second, in addition to skin lesions, it is also worth mentioning that tDCS can be

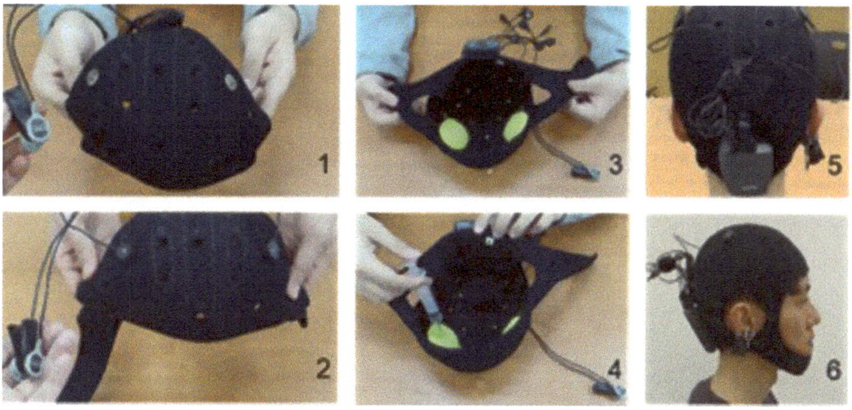

Fig. 4 tDCS system setup procedure. (1) Electrode placement on the stimulation positions. (2) Connection of the channel and the sponge pad electrode. (3) The sponge pad's appearance. (4) Soaking the sponge pad with saline solution. (5) Appearance of back. (6) Appearance of side

associated with a few mild adverse effects, such as mild tingling, mild pain, and transient redness.

In order to ensure that participants undergo safe tDCS stimulation, several procedures are recommended.

- Researchers and therapists should have proper exclusion criteria, such as the presence of metallic implants in the skull or brain and skin diseases (Utz et al., 2010).
- Subjects' skin should be lightly cleaned with a swab, and researchers and therapists should avoid abrading the skin because eroded skin is prone to induce skin lesions. The static impedance of the skin should also be measured, and stimulation should not be carried out unless impedance levels are within limits according to the tDCS device manufacturer.
- Electrode sponges should be soaked with saline solution to minimize skin resistance and ensured that the contact with the skin is sufficiently firm.
- Constant current strength should be provided by the stimulation device. A safety current density limit of 0.029–0.142 mA/cm^2 (Poreisz et al., 2007), as well as a maximum charge density of 40 μC/cm^2 at the stimulating electrode (Agnew, & McCreery, 1987; Yuen et al., 1981), are suggested for a safely delivered tDCS.
- For repeated applications of tDCS, a sufficiently large interval between sessions is suggested in order to avoid undesirable cumulative effects. The interval is determined by the stimulation purpose and procedure. Most importantly, the experimenter or therapist should be well prepared and oversee the whole procedure, especially when examining children and patients.
- All other safety measures suggested by the device manufacturers and/or the lab should be followed.

3 State-of-the-Art of tDCS-Based Neuroergonomics Research

This section covers the findings of up to date neuroergonomic research in the domain of motor and cognitive functions. Each function is discussed in detail about what tDCS does to them and how tDCS can be used in areas of interest relative to the function. Overview, key findings, and limitations subsections are implored for each domain to inform the reader of the many areas of interest tDCS covers.

3.1 Motor Functions

Investigations have applied tDCS to the motor cortex for rehabilitation, to enhance motor skills, or even improve learning. These studies have attempted to determine the motor cortex's involvement in specific tasks and have examined changes in cortical excitement's effect on task performance. Types of tasks that these investigations have addressed range from simple button responses using the index finger to activities of daily living (such as moving small objects) or even speech. Table 2 summarizes exemplary tDCS-based research on motor functions.

Antal et al. (2004) investigated tDCS effects on the extrastriate visual area and the primary motor cortex, both of which may be involved in learning visuo-motor coordination tasks on the part of healthy subjects. They found improved performance in the early phase (first session) of learning the task with anodal tDCS, as well as enhanced excitability in the V5 and M1 areas. Hummel et al. (2005) attempted to improve motor functions in the paretic hand of patients with chronic stroke. They found that Jebsen–Taylor Hand Function Test (JTT) performance with the paretic hand improved significantly in the active stimulation condition. Boggio et al. (2006) also investigated tDCS's effect in improving motor function in healthy subjects' non-dominant hand and found significant enhancement of motor task performance after anodal tDCS of the primary motor cortex. Inguaggiato et al. (2019) evaluated the effect on unilateral gross manual function of a single application of anodal tDCS over the ipsilesional motor cortex in a small group of subjects with unilateral cerebral palsy (UCP). They found immediate improvement in hemiplegics' unimanual gross motor dexterity following the active stimulation. On the other hand, Conley et al. (2015) found no benefit of tDCS over the dominant primary motor cortex on go/no go task performance. They suggested that these discrepant effects may be related in part to differences in task properties. Lastly, Buchwald et al. (2019) performed the first investigation of tDCS effects in speech motor learning. Their results showed that the group that received stimulation during the speech task improved more during practice than did the sham group, indicating that stimulation during the learning task may be particularly helpful during acquisition.

tDCS's effects on motor functions have addressed three principal domains: (1) **Rehabilitation** showing the way tDCS can be used to help patients who have

Table 2 A summary of exemplary tDCS applications in motor functions

Study	Research domain	tDCS type	Research goal	Subject characteristic	Anode and cathode positioning	Current and density	Duration of stimulation
Conley et al. (2015)	Rehabilitation	Anodal	Explore the effectiveness of tDCS varied with timing between stimulation and test	• *N* (52) • Age: adults	• **Anode**: dominant (left) primary motor cortex (M1) • **Cathode**: supraorbital region of the contralateral hemisphere	• 1 mA • 28.6 $\mu A/cm^2$	• tDCS: 20 min
Hummel et al. (2005)			Investigate motor tasks mimicking daily living activities	• *N* (6) • Female (33.3%) • Age: 62.2 ± 7.56 years	• **Anode**: projection of the hand knob area of the primary motor cortex of the affected hemisphere • **Cathode**: contralateral supraorbital region	• 1 mA • 40 $\mu A/cm^2$	• tDCS: 20 min • SHAM: 30 s ramp up
Inguaggiato et al. (2019)			Study the efficacy of tDCS for upper limb motor rehabilitation in congenital or early-acquired stroke	• *N* (8) • Age: 10–28 years	• **Anode**: primary motor cortex (M1) of the ipsilesional hemisphere • **Cathode**: contralateral supraorbital area	• 1.5 mA	• tDCS: 20 min • SHAM: 30 s ramp up

(continued)

Table 2 (continued)

Study	Research domain	tDCS type	Research goal	Subject characteristic	Anode and cathode positioning	Current and density	Duration of stimulation
Buchwald et al. (2018)	Enhancement	Anodal	Test if the timing of tDCS affected performances	• *N* (80) • Female (62.5%) • Age ($M = 23.7$ years)	• **Anode**: left motor cortex. • **Cathode**: right supraorbital area	• 1 mA • 28.6 μA/cm^2	• tDCS: 20 min • SHAM: 30 s ramp up
Boggio et al. (2006)			Assess tDCS and sham over the primary motor cortex	• *N* (8) • Female (100%) • Age: 22–26 years	• **Anode**: C3 or C4 (parietal cortex) • **Cathode**: contralateral supraorbital area	• 1 mA • 28.6 μA/cm^2	• tDCS: 20 min • SHAM: 30 s ramp up
Antal et al. (2004)	Tracking performance	Anodal/cathodal	Follow a dot on a television screen to time responses relative to sight	• *N* (42) • Female (64.3%) • Age: $M = 25.5$ years	• **Anode**: hand area of the left motor cortex or Oz • **Cathode**: right orbital or Cz	• 1 mA • 28.6 μA/cm^2	• tDCS: 10 min

suffered crippling life events, such as stroke (e.g., Conley et al., 2015; Hummel et al., 2005; Inguaggiato et al., 2019); **(2) Enhancement** studies investigating the way tDCS is used to enhance function in a healthy adult's brain, such as increased learning ability (e.g., Boggio et al., 2006; Buchwald et al., 2018); and **(3) Tracking performance** studies using tDCS to determine what parts of the brain are associated with certain tasks (e.g., Antal et al., 2004). These domains play an important role in understanding tDCS' full scope and its effects. One simply cannot operate without the other, and investigating the combination of all three ultimately will provide finite answers to the questions about the direct positive and negative effects tDCS has on the brain when applied to different areas.

3.1.1 Rehabilitation of Motor Functions

Overview: Therapeutic rehabilitation has been one of the most common reasons tDCS has become such a popular research topic. It offers the possibility of increasing one's motor functions after suffering a crippling event. In this paper, there are three individual studies Conley et al. (2015), Hummel et al. (2005) and Inguaggiato et al. (2019) that cover the research domain of rehabilitation. Each of these studies was done with the goal of exploring the effectiveness that tDCS would have on motor functions. Typically, when subjected to certain areas of the brain like the primary motor cortex (M1) which is responsible for generating neural impulses that control the execution of movement and located in the frontal lobe along a bum called the precentral gyrus. The results of these studies shed light on the positive and negative impacts that tDCS has on motor functions and the directions future research studies should focus on.

The application of tDCS has several factors that must be designed prior to implementation in order to achieve an optimal result. Depending on the brain area of interest said factors such as tDCS Type, Subject Type, Anode and Cathode Positioning, Current, Current Density, and Stimulation duration are adjusted to trigger a response in that area. In the case of rehabilitation all studies used Anodal tDCS with the Anode electrode positioned over the primary motor cortex (M1) and the Cathode electrode positioned over the contralateral supraorbital area. A current of 1 mA was applied over 20 min of stimulation with a current density of 28.6 μA/cm^2 (Conley et al., 2015) and 40.0 μA/cm^2 (Hummel et al., 2015). The current density is determined by dividing the current (mA) by the size of the electrodes (cm^2). A Sham test or a fake test designed to resemble the real procedure with a ramp-up of 30 s then turned off was also given (Hummel et al., 2015; Inguaggiato et al., 2019). This test was used to compare the actual tDCS to the Sham test for differences. The ages and types of each study's subjects differed although majority were adults. All three studies incorporated a similar experimental task that involved mimicking daily living activities either before, during or after receiving the tDCS or Sham stimulation.

Key findings: Stimulation had positive effects on response time and movement. Hummel et al. (2005) a double-blind, sham-controlled, crossover study was designed to test the hypothesis that non-invasive stimulation of the motor cortex could improve motor function in the paretic hand of patients with chronic stroke. The Jebsen–Taylor Hand Function Test (JTT), a well-validated test for functional motor assessment that reflects activities of daily living, was used to measure hand functionality. The JTT found that cortical stimulation of motor regions can have beneficial effects on skilled motor functions in the paretic hand of patients suffering from chronic stroke sequelae but not with Sham. Further, they found increased motor cortical excitability using TMS after tDCS intervention sessions and suggest that this interventional strategy in combination with customary rehabilitative treatments may play an adjuvant role in neurorehabilitation (Hummel et al., 2005). Inguaggiato et al. (2019) similarly used a Box and Block Test (BBT), a highly reliable hand dexterity test, to measure the effects of a single session, sham-controlled tDCS session. The aim of this study was to explore the effects anodal tDCS has on motor performance in individuals affected by Unilateral Cerebral Palsy (UCP). The results determined that tDCS had no serious adverse effects either immediately after or in follow-up tests, and they observed immediate improvement in hemiplegics' unimanual gross motor (i.e., hand) dexterity following active stimulation. Moreover, such improvement was maintained for at least 90 min, although it was the result of a single anodal tDCS application. Thus, tDCS has been found to have a positive effect on motor functions overall in rehabilitation.

Utilities and Limitations: Rehabilitation studies also have limitations. First, the damage that a person's brain suffered already is not known completely. This causes uncertainty about potential adverse effects of tDCS in certain areas of the brain, although some studies have found no such obvious effects (Inguaggiato et al., 2019). However, to understand tDCS's influence on rehabilitation fully would require first understanding precisely what requires treatment, and thus far, only positive or neutral feedback has been acquired in this domain. Perhaps research that focuses on adverse effects is needed to improve our understanding of tDCS's effects on rehabilitation overall.

3.1.2 Enhancements of Motor Functions

Overview: Another area of interest is the idea that tDCS can enhance motor function skills of healthy adults. Buchwald et al. (2018) and Boggio et al. (2006) focus on the effects tDCS has on speech and hand motor performance in single session tDCS experiments. There are two types of motor skills that can be enhanced: Gross motor skills and Fine motor skills. Gross motor skills require large muscle groups to perform tasks like walking, balancing, and crawling whereas Fine motor skills require the use of smaller muscle groups to perform smaller movements with the hands, fingers or feet as well as aid in growth of intelligence. For the purpose of this paper, the focus is the Fine motor skills and how they correlate with the brain.

As stated in the Rehabilitation domain, there are factors in every experiment that must be designed accordingly. Enhancement much like rehab has similar factor characteristics. However, since the goal is no longer to fix but rather enhance there is a slight adjustment in the strategy. Anodal tDCS is used with a current of 1 mA and current density of 28.6 μA/cm^2. A single session of either tDCS for 20 min or Sham with a 30-s ramp-up is implemented. The Anode electrode is positioned over the C3 or C4 located on top of the primary somatosensory cortex which is responsible for processing touch and sensation as well as keeping track of the location of your body parts (Boggio et al., 2006) and over the left motor cortex (Buchwald et al., 2018). The Cathode electrode is positioned over the contralateral supraorbital area for both. Subjects consist of healthy adults around the age of 22–26 years. Experimental goals differed for the two studies: *Buchwald* focused on the enhancement of motor learning whereas Boggio et al. (2006) focused on the enhancement of hand motor performance.

Key findings: Buchwald et al. (2018) was a single session tDCS experiment that explored the unexplored domain of speech motor learning and the role that timing played in its impacts. tDCS was giving in two-time phrases: before and during the experiment. These differences found that individuals who received active stimulation prior to the task maintained increased short-term accuracy relative to those who received sham stimulation or active stimulation during a task. Moreover, the group that received tDCS immediately before the task exhibited outstanding error correction during acquisition compared to the other groups. They suggested that these properties may derive from the sufficient amount of current that was delivered during the 20 min of stimulation before training. Supporting the claim that tDCS can improve speech motor learning and is considerably more effective when applied immediately before activity. Similarly, Boggio et al. (2006) conducted one session of both sham and active anodal stimulation of the non-dominant or dominant primary motor cortex. The Jebsen–Taylor Hand Function Test (JTT) was used to assess motor function differences. The study found that anodal tDCS of the primary motor cortex enhanced healthy subjects' motor performance with the non-dominant hand, an improvement of 9.41% compared to the shams 1.3% improvement (Boggio et al., 2006). However, stimulation had no effect on performance with the dominant hand, suggesting that weak cortical plasticity caused by disuse or impairment may be one of the factors in which there is room for improvement. Both studies showed that when tDCS is applied with the right performance factors, fine motor skills were enhanced.

Utilities and Limitations: A major concern that arises when enhancing the brain's function is the potential adverse effects after stimulation. While it is known now that tDCS can improve our ability to learn and function, it is not known how long the effects last or the effects the stimulation may have on other parts of the brain. There also are limits, in that the brain can tolerate only a certain current (0.1–2.0 mA) and duration without being harmed. This means that there is a limit on the degree to which tDCS can enhance function and crossing the safety threshold offers another set of issues that have yet to be addressed. Boggio et al. (2006) suggested using an NMDA-antagonist that can block training-induced cortical plasticity to help investigate tDCS's mechanisms of action on motor function. Moreover, these studies, which were performed with healthy subjects, provide preliminary evidence that tDCS may enhance performance in learning motor tasks under specific stimulation conditions.

3.1.3 Tracking Performance

Overview: Tracking performance of the human brain is when tDCS is applied over certain areas and studies are conducted to see how certain parts of the brain perform in response. This can be used to determine the assignments each part of the brain has as well as map the effects on each part of the brain when tDCS is applied in certain areas. Antal et al. (2004) performed a study to determine if the human middle temporal and primary motor cortex were involved in the learning of visuo-motor coordination tasks.

Tracking performance measures are set up the same way enhancement experiments are which is to design them around an area of interest. The difference is that with tracking, there is an experimental hypothesis that a certain response will come from the targeted area, but the response can be negative or positive rather than trying to solicit a particular response in enhancement. tDCS type of Anodal and Cathodal are used with a current of 1 mA and a current density of 28.6 $\mu A/cm^2$ for a single session of 10 min. Anodal electrode positioning over the hand area of the left motor cortex (Oz) and Cathodal electrode positioning over the right orbital cortex (Cz). The 42 subjects consisted of 64.3% female and a mean age of 25.5 years. The experiment was to follow a white dot on a Sony monitor to a specific stopping point and hit a button signaling the stop.

Key findings: Antal et al. (2004) was a single session anodal and cathodal stimulation study to determine if the middle temporal visual (MT or V5) and primary motor cortex (M1) was involved with visuo-motor coordination tasks. The reason behind this study was to make a credible statement that tDCS can help map the human brain and its performance areas/levels. The results of 10 min of anodal stimulation demonstrated that anodal tDCS on the left V5 and M1 areas improved tracking performance significantly in the early learning phase. Cathodal stimulation had no significant effect. However, tDCS effects did not persist for longer than 5 min, which can be expected as a potential period to modify the visuo-motor learning process externally. Moreover, tDCS' beneficial effect on the V5 and M1 areas can be explained, in that V5 and M1 are involved in the early phase of learning visuo-motor coordination. Again, it is worth noting that this study is an example of how tDCS can be used to track performance measures of the human brain. This domain can lay a foundation on which future studies in other domains may be conducted. For example, a future study involving the improvement of visual tasks can now be performed since the V5 and M1 were found to play a direct role in the human brain's ability to perceive images.

Utilities and Limitations: Tracking performance can be a very beneficial tool in mapping the areas that tDCS affects. Unfortunately, little research is available to support these types of studies. If more research were to be conducted in this domain, focusing on the effects of each tDCS setup is important. Comparing the findings to other similar studies would also be advised to explore variation in experimental design concepts. Moreover, more sophisticated investigations also are required to obtain a clear understanding of tDCS effects. For example, Antal et al. (2004) mentioned that one possible explanation for enhanced performance may be increased neuronal sensitivity enhanced cortical excitability causes rather than cortical coupling that is involved in visuo-motor learning.

3.2 *Cognitive Functions*

Having a general understanding of cognitive skills is necessary to perform procedures that could affect them. These skills are the core skills the brain uses together in order to take in information, process it and store it. tDCS has been used to investigate cognitive functions in three main domains: **(1) Rehabilitation** studies regarding the effects tDCS has on cognitive issues such as depression, learning disorders, and perception; **(2) Tracking Performance** focusing on the overall performance that tDCS has on cognitive functions; and **(3) Enhancement** research on the beneficial effects tDCS can have on enhancing a healthy adults' cognitive skills.

All three of these domains represent different fields within cognitive effects that produce results that can be combined to form a conclusion. Table 3 summarizes exemplary tDCS-based research on cognitive functions.

3.2.1 Rehabilitation of Cognitive Function

Overview: In the above table only one paper falls into the Rehabilitation domain type (Fregni et al., 2006) with the goal of evaluating the cognitive adverse effects of tDCS in severely depressed patients. In modern-day, mental issues have never been a bigger deal. More and more people are showing signs of depression and with that comes a rise in demand for a solution. Thus, there has been an increase in the research area of tDCS on cognitive functions and if it can be used in rehab to help reduce a patient's depression.

The same factors are included for Cognitive functions as motor functions but are adjusted to target a different part of the brain. Anodal tDCS with the current of 1 mA and current density of 28.6 μA/cm^2 was applied for 20 min or a sham ramp-up of 20 s then turned off. This was repeated five times of five consecutive days. The Anodal electrode was positioned over the frontal cortex (F3) and the Cathodal electrode over the contralateral supraorbital area. The experiment consisted of 18 patients of which 61.1% were female with a mean age of 46.4 years and the goal was to investigate the cognitive effects of 5 days of anodal stimulation.

Key Findings: Recent studies have shown that using different stimulation parameters and different electrode sizes/locations can cause shifts in neuronal membrane excitability. These shifts might be helpful for stroke recovery and depression treatment. Fregni et al. (2006) conducted a double-blind, randomized, controlled study to evaluate the cognitive effects tDCS has on patients with major depression. Five days of repeated sessions of active tDCS were found to improve one aspect of cognitive performance-working memory while the sham stimulation had no effect. No correlation was found between mood and cognitive improvement. Fregni et al. (2005) also

Table 3 A summary of exemplary tDCS applications in cognitive functions

Study	Research domain	tDCS type	Research goal	Subject characteristic	Anode and cathode positioning	Current and density	Duration of stimulation
Fregni, Boggio, Nitsche, Rigonatti, and Pascual-Leone (2006)	Rehabilitation	Anodal	Investigate the cognitive effects of 5 days of anodal stimulation	• *N* (18) • Female (61.1%) • Age: 46.4 ± 9.4 years	• **Anode**: F3 (frontal cortex) • **Cathode**: contralateral supraorbital area	• 1 mA • 28.6 μA/cm^2	• tDCS: 20 min • SHAM: 5 s
Horvath et al. (2015)	Tracking performance	Anodal/cathodal	Put together research findings that tDCS has on healthy adults	• *N* (59)	• Cortical regions relevant to the outcome measure of interest	• 1 mA • 28.6 μA/cm^2	• tDCS: < 7 min
Fregni et al. (2005)	Enhancement	Anodal	Investigate the effects of anodal stimulation of the DLPFC on working memory	• *N* (15) • Female (73.3%) • Age: 19–22 years	• **Anode**: F3 (frontal cortex) • **Cathode**: contralateral supraorbital area	• 1 mA • 28.6 μA/cm^2	• tDCS: 10 min • SHAM: 5 s

(continued)

Table 3 (continued)

Study	Research domain	tDCS type	Research goal	Subject characteristic	Anode and cathode positioning	Current and density	Duration of stimulation
Ohn et al. (2008)			Investigate the effects of anodal stimulation of the DLPFC on working memory	• (15) • Female (66.7%) • Age: 26.5 ± 3.5 years	• **Anode**: F3 (frontal cortex) • **Cathode**: contralateral supraorbital area	• 1 mA • 40.0 $\mu A/cm^2$	• tDCS: 30 min • SHAM: 5 s
Clark et al. (2012)			Identified threat-related objects concealed in naturalistic virtual surroundings	• *N* (96)	• **Anode**: right inferior frontal cortex (F10) or right parietal cortex • **Cathode**: contralateral upper arm	• 0.1–2.0 mA • 9–181 $\mu A/cm^2$	• tDCS: 30 min

conducted a single session of anodal, cathodal, and sham stimulation study aimed to determine if anodal tDCS influenced working memory. Anodal stimulation of the left prefrontal area was found to improve working memory further clarifying the effect tDCS has on cognitive functions. Cathodal and sham stimulations showed no improved effect compared to that of anodal stimulation. However, the study showed no correlation between mood improvement and cognitive improvement. Both studies confirm that stimulation can, in fact, result in improved working memory, further crediting tDCS effects on cognitive functions. Future studies would need to consider a different series of montages that may influence different but important results.

Utilities and Limitations: Limitations in using tDCS in rehabilitation consists of a few factors: (1) The severity of a patient's cognitive issues can differ from another's; (2) To diagnose and treat a patient for one specific disorder means to have a full understanding of the brain area affected which is extremely difficult to pinpoint; and (3) There were a few patients in this study that reported no effect whereas other did. There is also a limited amount of safety data available to researchers. This restricts the ability to conduct larger clinical trials that might show a greater distinction of repeated sessions of tDCS results. But until more safety data has been published, researches must stay within the current safety limits. Further research would need to be done to determine if tDCS has a positive impact on everyone or specific types of people.

3.2.2 Tracking Performance

Overview: Tracking performance is used to keep up with tDCS and its effects in either different domains or with a particular one. Horvath et al. (2015) collects papers of single session tDCS data on healthy adults and their cognitive outcome measures to address the paradigms utilized and variable results reported in the literature. Completing a necessary performance measure on the effectiveness tDCS has on cognitive functions.

Factors in tracking performance consist of collecting data and reporting on that data. Similar papers were collected and summarized. Anodal and Cathodal tDCS was used at a current of 1 mA and a current density of 28.6 $\mu A/cm^2$ over a stimulation period of less than 7 min. Anodal and Cathodal electrode positioning varied based on the cortical regions relevant to the outcome measure of interest. A total of 59 studies were collected and analyzed. The experimental task was to collect every single session tDCS data on healthy adults and their cognitive outcome measure reported by at least two different research groups.

Key Findings: Like most studies, there are negative results among positive ones. Horvath et al. (2015) was not a study but rather a quantitative review of cognitive data. They collected 59 papers on single session tDCS on healthy adults. Within their review, they found no evidence that single session tDCS has a reliable effect on cognition in healthy adult populations. It is important to note, however, that these findings may be due to state dependency effects which, with elucidation, can be controlled for and leveraged. In addition, these findings do not preclude the possibility that tDCS influences different populations (juvenile, elderly and infirm), when utilized multiple-times over several days or weeks, or on behavioral tasks. Nor does this preclude the possibility that tDCS could be effective if utilized in a novel fashion (hi-definition tDCS, spinal tDCS, pulsed current tDCS, etc.). Many studies have shown a positive improvement after implementing anodal tDCS. A wide range of parameter configurations and parts play a large role in the results of an experiment. It is important to note possible flaws in experimental design or tool failure when analyzing the results of a study. Nonetheless, tracking the performance tDCS has on cognitive functions is pivotal to advancing future research and it starts with identifying possible defects in the results before determining there is no effect at all.

Utilities and Limitations: Many studies have suggested that flaws in design have led to inaccuracies in results which could possibly lead to the conclusion Horvath et al. (2015) came to. A major limitation of this analysis is the lack of comparable research available in the current tDCS literature. Of the 50 cognitive outcome measures replicated between two different research groups included in this paper, 35 include only two or three papers. Another limitation is the populations utilized in the studies. These studies focused on the effects of healthy adults so it remains to be seen whether or not tDCS can influence these measures in different health conditions.

3.2.3 Cognitive Enhancement

Overview: Much like the enhancement of motor functions, tDCS can be used to enhance Cognitive functions/skills. Cognitive skills are skills your brain uses to think, read, learn, remember, reason, and pay attention. Enhancing these skills would allow people to operate at a higher capacity, think of it like adding memory to a computer. Fregni et al. (2005), Ohn et al. (2008) and Clark et al. (2012) all fall into this domain type and whether tDCS can directly increase the working memory of healthy adults.

The enhancement of cognitive functions is mainly due to the positioning of the electrodes and slight adjustments to other factors. Anodal tDCS was applied at a current of 1 mA and a current density of 28.6 μA/cm^2 for 10 min of active stimulation or sham with a ramp-up of 5 s (Fregni et al., 2005). Ohn et al. (2008) and Clark et al. (2012) applied current of 1 mA and 0.1–2.0 mA with current density of 40.0 μA/cm^2 and 9.0–181.8 μA/cm^2, respectively, over 30 min of active stimulation or sham with a 5 s ramp up. Anodal electrode positioning over the Frontal Cortex (F3) or Right inferior frontal cortex (F10) and the Cathodal electrode positioning over the contralateral supraorbital area except Clark et al. (2012) over the contralateral upper arm. All studies consisted of healthy adults of similar ages and male-to-female ratios. The experimental tasks were to test the effect tDCS has on working memory by having the subject perform cognitive tasks.

Key Findings: Fregni et al. (2005) results indicate that anodal stimulation of the left prefrontal area increases the accuracy of the task performance when compared with sham stimulation of the same area. Furthermore, our control experiment showed that the effect of the anodal stimulation of the left dorsolateral prefrontal cortex (DLPFC) was relatively focal and depends on the polarity of stimulation. This accuracy enhancement during active stimulation cannot be accounted for by slowed responses, as response times were not changed by stimulation. These results, in a sequential-letter-matching task, showed that left prefrontal anodal stimulation leads to an enhancement of working memory performance.

Similarly, Ohn et al. (2008) results indicated that anodal tDCS over the left DLPFC enhanced verbal working memory in healthy participants in a time-dependent manner. The accuracy of verbal working memory tasks increased after 10 min of tDCS application, and this effect was further enhanced by 30 min of stimulation. The accuracies at 30 min of stimulation were significantly different between anodal and sham tDCS. Furthermore, this memory-enhancing effect was maintained at 30 min after discontinuation of tDCS. Error rates, reaction times, concentration, and fatigue did not change significantly during or after intervention.

Lastly, Clark et al. (2012) used neuroimaging to identify the placement for tDCS electrodes, and tDCS was then used to verify the contribution of these brain regions to this task through increased performance and rate of learning. The study found that tDCS over the right inferior frontal and right parietal cortex greatly improved learning and performance.

Utilities and Limitations: Much like the limitations of the enhancement of motor functions, the same ones apply to the cognitive side. There is a limit that the brain can take and that's a current (0.1–2.0 mA) and duration without being hurt. This means there is a limit on the amount tDCS can enhance. In Ohn et al. (2008) study there was a concern about the possible influence of the exciting motor cortex in the stimulated hemisphere. In response to this, participants were instructed to perform the tasks with their left hands while the left hemisphere was being stimulated. This response ensured that reaction times were not biased because only one hemisphere was excited.

4 Summary and Future Directions

To reiterate, the goal of this chapter is to layout a foundation for researchers including neuroergonomists to understand the conceptual and physical processes of tDCS. Reading this guide will provide them with the tools they need to be successful in starting a tDCS study design. They will know how tDCS works, how to configure parameters, operate the tDCS system, and understand the safety concerns and precautions of electrical stimulation. Furthermore, the researchers are taught about tDCS applications on motor and cognitive functions, what each function does, the effect stimulation has on it, why it is important and limitations on research. Section 2 provides the necessary information to understand what tDCS is and Sect. 3 provides real-world applications and feedback. Together, the researchers should have a good understanding of what tDCS is and how to approach future research.

A large amount of information is still needed to continuously move forward with tDCS research. Anytime there is a process being implemented on a human there comes a demand for safety regulations and known facts before moving forward.

The delay in receiving needed information before moving on is considered a major limitation and is one among many in tDCS research. Refer to the list below for some more limitations:

- **Long-term effects**: tDCS studies typically stimulate for short periods of time in order to study the effect stimulation has on a subject. However, this limits the ability to study the long-term effects stimulation has over longer periods of time.
- **Safety concerns**: While outlines have been made to protect the subject during stimulation there is still a concern about the long-term effects tDCS has on the human brain. Until future research is published acknowledging this concern then research will be limited.
- **Sample size**: The size of a study can influence the results. The bigger the sample size the better the results will be due to less variability. The issue here is that specific studies may not have access to large sample sizes and thus rely on small groups that meet limited criteria. This reduces the credibility of the study's findings and calls for future studies to contain increased sample sizes.

References

Agnew, W. F., & McCreery, D. B. (1987). Considerations for safety in the use of extracranial stimulation for motor evoked potentials. *Neurosurgery, 20*(1), 143–147.

Antal, A., Nitsche, M. A., Kincses, T. Z., Kruse, W., Hoffmann, K. P., & Paulus, W. (2004). Facilitation of visuo-motor learning by transcranial direct current stimulation of the motor and extrastriate visual areas in humans. *European Journal of Neuroscience, 19*(10), 2888–2892.

Au, J., Katz, B., Buschkuehl, M., Bunarjo, K., Senger, T., Zabel, C., ... Jonides, J. (2016). Enhancing working memory training with transcranial direct current stimulation. *Journal of Cognitive Neuroscience, 28*(9), 1419–1432.

Bikson, M., Grossman, P., Thomas, C., Zannou, A. L., Jiang, J., Adnan, T., ... Brunoni, A. R. (2016). Safety of transcranial direct current stimulation: Evidence-based update 2016. *Brain Stimulation, 9*(5), 641–661.

Boggio, P. S., Castro, L. O., Savagim, E. A., Braite, R., Cruz, V. C., Rocha, R. R., ... Fregni, F. (2006). Enhancement of non-dominant hand motor function by anodal transcranial direct current stimulation. *Neuroscience Letters, 404*(1–2), 232–236.

Buchwald, A., Calhoun, H., Rimikis, S., Lowe, M. S., Wellner, R., & Edwards, D. J. (2019). Using tDCS to facilitate motor learning in speech production: The role of timing. *Cortex, 111*, 274–285.

Buchwald, A., Falconer, C., Rykman-Peltz, A., Cortes, M., Pascual-Leone, A., Thickbroom, G. W., ... Chang, J. (2018). Robotic arm rehabilitation in chronic stroke patients with aphasia may promote speech and language recovery (but effect is not enhanced by supplementary tDCS). *Frontiers in Neurology, 9*, 853.

Clark, V. P., Coffman, B. A., Mayer, A. R., Weisend, M. P., Lane, T. D., Calhoun, V. D., ... Wassermann, E. M. (2012). TDCS guided using fMRI significantly accelerates learning to identify concealed objects. *Neuroimage, 59*(1), 117–128.

Conley, A. C., Marquez, J., Parsons, M. W., Fulham, W. R., Heathcote, A., & Karayanidis, F. (2015). Anodal tDCS over the motor cortex on prepared and unprepared responses in young adults. *PloS ONE, 10*(5), e0124509.

DaSilva, A. F., Volz, M. S., Bikson, M., & Fregni, F. (2011). Electrode positioning and montage in transcranial direct current stimulation. *Journal of Visualized Experiments: JoVE, 51*, e2744.

Fregni, F., Boggio, P. S., Nitsche, M., Bermpohl, F., Antal, A., Feredoes, E., … Pascual-Leone, A. (2005). Anodal transcranial direct current stimulation of prefrontal cortex enhances working memory. *Experimental Brain Research, 166*(1), 23–30.

Fregni, F., Boggio, P. S., Nitsche, M. A., Rigonatti, S. P., & Pascual-Leone, A. (2006). Cognitive effects of repeated sessions of transcranial direct current stimulation in patients with depression. *Depression and Anxiety, 23*(8), 482–484.

Horvath, J. C., Forte, J. D., & Carter, O. (2015). Quantitative review finds no evidence of cognitive effects in healthy populations from single-session transcranial direct current stimulation (tDCS). *Brain Stimulation, 8*(3), 535–550.

Hummel, F., Celnik, P., Giraux, P., Floel, A., Wu, W. H., Gerloff, C., & Cohen, L. G. (2005). Effects of non-invasive cortical stimulation on skilled motor function in chronic stroke. *Brain, 128*(3), 490–499.

Inguaggiato, E., Bolognini, N., Fiori, S., & Cioni, G. (2019). Transcranial direct current stimulation (tDCS) in unilateral cerebral palsy: A pilot study of motor effect. *Neural Plasticity, 2019,* 1–10.

Ironside, M., Harmer, C., O'Shea, J., Ebmeier, K., & Brunoni, A. R. (2016). *An exploration of the mechanisms of action of transcranial direct current stimulation (tDCS) in mood and anxiety disorder research* (Doctoral dissertation, University of Oxford).

Kessler, S. K., Turkeltaub, P. E., Benson, J. G., & Hamilton, R. H. (2012). Differences in the experience of active and sham transcranial direct current stimulation. *Brain Stimulation, 5*(2), 155–162.

Looi, C. Y., & Kadosh, R. C. (2016). Brain stimulation, mathematical, and numerical training: Contribution of core and noncore skills. In *Progress in brain research* (Vol. 227, pp. 353–388). Elsevier.

Meesen, R. L., Rand, M. K., Cuypers, K., Nitsche, M., & Jamil, A. (2017). Enhancing bimanual motor coordination in healthy young and older adults using transcranial direct current stimulation.

Meron, D., Hedger, N., Garner, M., & Baldwin, D. S. (2015). Transcranial direct current stimulation (tDCS) in the treatment of depression: Systematic review and meta-analysis of efficacy and tolerability. *Neuroscience & Biobehavioral Reviews, 57,* 46–62.

Ohn, S. H., Park, C. I., Yoo, W. K., Ko, M. H., Choi, K. P., Kim, G. M., … Kim, Y. H. (2008). Time-dependent effect of transcranial direct current stimulation on the enhancement of working memory. *Neuroreport, 19*(1), 43–47.

Poreisz, C., Boros, K., Antal, A., & Paulus, W. (2007). Safety aspects of transcranial direct current stimulation concerning healthy subjects and patients. *Brain research bulletin, 72*(4–6), 208–214.

Price, A. R., McAdams, H., Grossman, M., & Hamilton, R. H. (2015). A meta-analysis of transcranial direct current stimulation studies examining the reliability of effects on language measures. *Brain Stimulation, 8*(6), 1093–1100.

Sarmiento, C. I., San-Juan, D., & Prasath, V. B. S. (2016). Letter to the editor: Brief history of transcranial direct current stimulation (tDCS): From electric fishes to microcontrollers. *Psychological Medicine, 46*(15), 3259–3261.

Thair, H., Holloway, A. L., Newport, R., & Smith, A. D. (2017). Transcranial direct current stimulation (tDCS): A beginner's guide for design and implementation. *Frontiers in Neuroscience, 11,* 641.

Ulam, F., Shelton, C., Richards, L., Davis, L., Hunter, B., Fregni, F., & Higgins, K. (2015). Cumulative effects of transcranial direct current stimulation on EEG oscillations and attention/working memory during subacute neurorehabilitation of traumatic brain injury. *Clinical Neurophysiology, 126*(3), 486–496.

Utz, K. S., Dimova, V., Oppenländer, K., & Kerkhoff, G. (2010). Electrified minds: Transcranial direct current stimulation (tDCS) and galvanic vestibular stimulation (GVS) as methods of non-invasive brain stimulation in neuropsychology—a review of current data and future implications. *Neuropsychologia, 48*(10), 2789–2810.

van't Wout-Frank, M., Shea, M. T., Larson, V. C., Greenberg, B. D., & Philip, N. S. (2019). Combined transcranial direct current stimulation with virtual reality exposure for posttraumatic stress disorder: Feasibility and pilot results. *Brain Stimulation, 12*(1), 41–43.

Volz, M. S., Farmer, A., & Siegmund, B. (2016). Reduction of chronic abdominal pain in patients with inflammatory bowel disease through transcranial direct current stimulation: A randomized controlled trial. *Pain, 157*(2), 429–437.
Wagner, T., Fregni, F., Fecteau, S., Grodzinsky, A., Zahn, M., & Pascual-Leone, A. (2007). Transcranial direct current stimulation: a computer-based human model study. *Neuroimage, 35*(3), 1113–1124.
Woods, A. J., Antal, A., Bikson, M., Boggio, P. S., Brunoni, A. R., Celnik, P., … Knotkova, H. (2016). A technical guide to tDCS and related non-invasive brain stimulation tools. *Clinical Neurophysiology, 127*(2), 1031–1048.
Yuen, T. G., Agnew, W. F., Bullara, L. A., Jacques, S., & McCreery, D. B. (1981). Histological evaluation of neural damage from electrical stimulation: Considerations for the selection of parameters for clinical application. *Neurosurgery, 9*(3), 292–299.
Zmigrod, S., Colzato, L. S., & Hommel, B. (2015). Stimulating creativity: Modulation of convergent and divergent thinking by transcranial direct current stimulation (tDCS). *Creativity Research Journal, 27*(4), 353–360.

Computational Approaches to Neuroergonomics

Adaptive Control of Thought-Rational (ACT-R): Applying a Cognitive Architecture to Neuroergonomics

Nayoung Kim and Chang S. Nam

Abstract For neuroergonomists who wish to apply Adaptive Control of Thought-Rational (ACT-R) to investigate the human mind and its structure including learning, performance, and problem-solving skills, this chapter aims at providing an overview of ACT-R with an emphasis on its modules, buffers, and sub-symbolic levels. ACT-R is a high-level computational simulation of human cognitive processing and one of the cognition theories that seek to predict human performance in real-world settings. A group of previous studies on behavioral- and neural-based cognitive modeling of human cognition using ACT-R will also be discussed. Finally, this chapter presents future directions of ACT-R for neuroergonomics research.

1 Introduction

In 1983, Anderson published his seminal work, *The Architecture of Cognition* (Anderson, 1983). Cognitive architecture refers to both a unified theory of cognition—the outline of the structure of the various parts of the mind—and computational implementation of the theory with specified rules and associative networks.

As shown in Fig. 1, ACT-R has a series of cognitive models in the ACT family; it succeeds ACT, ACTE, and ACT* (Whitehill, 2013). With ACT*, Anderson endeavored to describe the mechanisms through which memory recall and cognition take place by describing how goals are broken down into subgoals using "production rules." Later, in 1990, Anderson took a step back from the existing ACT* and investigated how cognitive processes such as memory and problem-solving could be cast as an optimal solution to the tasks that humans commonly encounter. With his new-found optimization approach, Anderson returned in 1993 to the ACT production rule-based framework and revised ACT* to create ACT-R. The "R" (rational) implies that the human mind is behaving "rationally," in the sense that she/he wishes

N. Kim · C. S. Nam (✉)
Edward P. Fitts Department of Industrial & Systems Engineering, North Carolina State University, Raleigh, NC 27695, USA
e-mail: csnam@ncsu.edu

C. S. Nam (ed.), *Neuroergonomics*, Cognitive Science and Technology,
https://doi.org/10.1007/978-3-030-34784-0_6

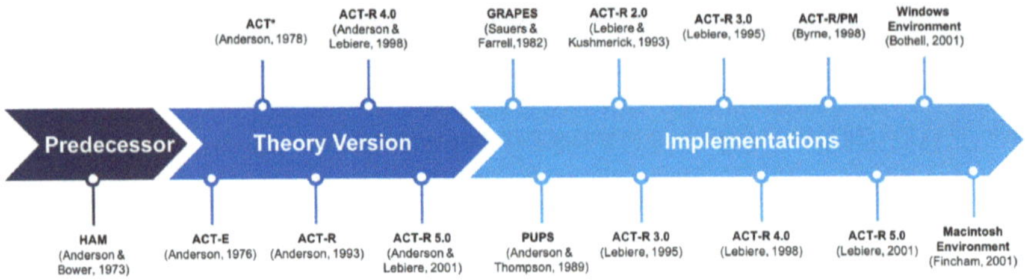

Fig. 1 Historical main events of ACT-R in human cognition and behavior

to solve problems to maximize reward and minimize cost. See Whitehill (2013) for more details.

The symbiotic relationship between cognitive modeling and cognitive neuroscience results in palpable progress towards the shared goal of better understanding the functional architecture of human cognition. Specifically, cognitive architectures are frameworks that can be used to develop computational models of human cognitive processes (Langley, Laird, & Rogers, 2009; Smart, Scutt, Sycara, & Shadbolt, 2016; Taatgen & Anderson, 2010). Various cognitive architectures have been proposed and investigated, such as SOAR or State, Operator, And Result (Laird, 2012), Adaptive Control of Thought-Rational (ACT-R) (Anderson et al., 2004; Anderson, 2007), Connectionist Learning with Adaptive Rule Induction On-line (CLARION) (Sun, 2006, 2007), etc. Cognitive architectures can be used to develop various intelligent systems and agents (e.g., cognitive robots), while helping us understand human cognition in a specific task environment.

ACT-R is a high-level computational simulation of human cognitive processing, which is a useful tool for unifying multiple disciplines (see Fig. 2) by incorporating physical, neurophysiological, behavioral, and cognitive data into one computational model. In this chapter, we choose ACT-R as a solid theoretical framework with which neuroergonomists can better understand the underlying physical and cognitive processes of the operator's cognition and performance in complex workplaces.

The remainder of this chapter is organized as follows. Section 2 briefly overviews ACT-R with following the most important components; modules, buffers and subsymbolic levels in this architecture. In Sect. 3, we reviewed previous studies on (1) behavioral and (2) neural-based cognitive modeling using ACT-R. Finally, Sect. 4 concludes by presenting a list of recommendations for future directions of ACT-R for neuroergonomics research.

2 Overview of ACT-R

ACT-R is a rule-based system that has been widely used by cognitive scientists to model various aspects of human cognitive performance. It is also one of the few cognitive architectures that have an explicit link to research in the neurocognitive

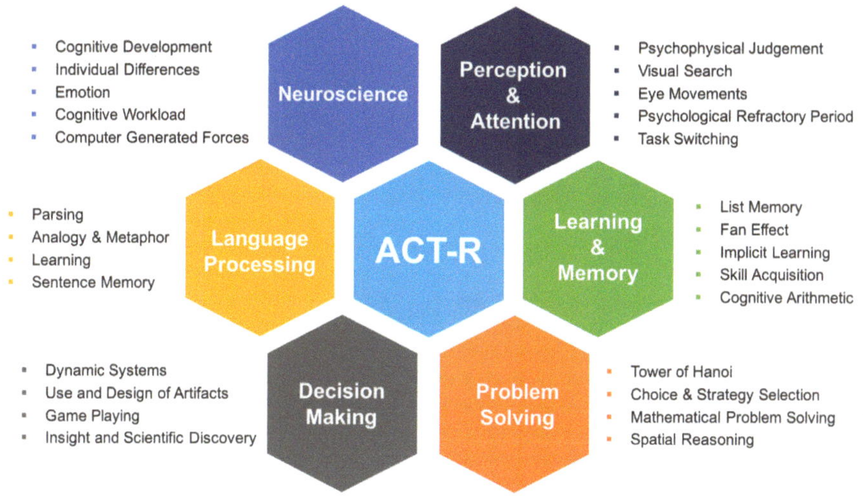

Fig. 2 ACT-R applications

domain: the structural elements of the core ACT-R architecture (i.e., its modules and buffers) map onto different regions of the human brain (Anderson, 2007). This enables cognitive modelers to predict the activity of different brain regions at specific junctures in a cognitive task (Anderson, 2007). In this section, three major components of ACT-R (modules, buffers, and sub-symbolic levels) were presented.

2.1 *ACT-R Modules*

ACT-R is a formalized, integrated cognitive architecture that combines the *Spreading Activation Memory* theory with a production system to model the high-level of cognitive tasks. Like many successful architectures, ACT-R is a modular theory of mind that aims to provide an integrated account of human cognition. It treats the mind as being composed of distinct modules that exist for particular functions. The ACT-R modules represent the functions of the brain as well as how these functions are mapped to different parts of the brain.

ACT-R incorporates both declarative knowledge (e.g., addition facts) and procedural knowledge (e.g., rules for solving multi-column addition) into a production system where procedural rules act on declarative chunks (Anderson et al., 2004). In ACT-R, the structures of declarative knowledge are called chunks and held in the Declarative module, whereas those of procedural knowledge are called rules and held in the Procedural module. The rules also have access to other modules, including the Visual module for perception, the Manual module for action, the Imaginal module for storing visual problem representation, and the Goal module for keeping track of current intentions. These modules are linked to specific areas of the brain:

Manual in the motor cortex (BA 3/4), Imaginal in the parietal cortex (BA 39/40), Declarative in the dorsolateral prefrontal cortex or DLPFC (BA 45/46), Goal in the anterior cingulate cortex or ACC (BA 24/32), Visual in the fusiform gyrus (BA 37), and Procedural in the caudate of the basal ganglia (Matessa, 2008).

2.2 ACT-R Buffers

Each module of ACT-R has its own buffer that can store only one chunk of information extracted from the corresponding module. In ACT-R, the condition statements of all production rules are compared with the current contents of buffers every 50 ms (Anderson, 2007).

When there are one or more matching productions, the matching one or the one with the highest expected value among them, respectively, is selected to fire, and the action statement of the selected rule then makes each module exchange chunks of information or creates new chunks in the particular modules (Taatgen & Lee, 2003). At this time, only one chunk can be processed through each buffer at a time, while all modules can be executed simultaneously. For example, the declarative module cannot retrieve the additional memory chunk until it has completed the previous retrieval. In contrast, the manual module can perform the physical movement, while the visual module looks at a new object. Moreover, there is a limit to the number of objects and a time limit in the visual and retrieval buffer. These are related to the working memory limitation of human cognitive processing (Anderson et al., 2004).

2.3 Sub-symbolic Level in ACT-R

ACT-R also has a sub-symbolic level in which continuously varying quantities are processed in parallel to produce much of the qualitative structure of human cognition. These sub-symbolic quantities participate in neural-like activation processes that determine the speed and success of access to chunks in declarative memory as well as the conflict resolution among production rules. The core of ACT-R memory modeling can be summarized in the following equations (Anderson et al., 2004). The activation $A(t)$ of a certain chunk is defined as the summation of $B(t)$, the base activation of the chunk at time, and the total strength of the related chunks, defined as "the elements in the current goal chunk and the elements currently being processed in the perceptual field."

$$A(t) = B(t) + \sum_{j} w_j s_{ij} + \varepsilon \tag{1}$$

where each $w_j \in [0, 1]$ reflects the salience or validity of [related] element j.

The $s_{ij} \in [-\infty j + \infty]$ are "the strengths of association to i from elements j in the current context." ε: The noise value as described in the last unit.

The base activation $B(t)$ is increased whenever it is "used," either through practice ("learning events") or when matched to a production rule.

$$B(t) = \ln \sum_{k} t_k^{-d} + B \tag{2}$$

If we make a retrieval request and there is a matching chunk, that chunk will only be retrieved if it exceeds the retrieval activation threshold, τ. The probability of this happening depends on the expected activation, A_i, and the amount of noise in the system which is controlled by the parameter s:

$$\text{recall probability}_i = \frac{1}{1 + e^{\frac{\tau - A_i}{s}}} \tag{3}$$

Inspection of that formula shows that, as A_i tends higher, the probability of recall approaches 1, whereas, as t tends higher, the probability decreases. In fact, when $\tau = A_i$, the probability of recall is 0.5. The s parameter controls the sensitivity of recall to changes in activation. If s is close to 0, the transition from near 0% recall to near 100% will be abrupt, whereas when s is larger, the transition will be a slow sigmoidal curve.

The activation of a chunk also determines how quickly it can be retrieved. When a retrieval request is made the time it takes until the chunk that is retrieved is available in the retrieval buffer is given by this equation:

$$\text{Time} = Fe^{-A} \tag{4}$$

A: The activation of the chunk which is retrieved.
F: The latency factor parameter.

If no chunk matches the retrieval request, or no chunk has an activation which is greater than the retrieval threshold then a retrieval failure will occur (Taatgen, Lebiere, & Anderson, 2006). These equations provide an integrated foundation for theory development in memory encoding, storage, and retrieval. For more details, visit http://act-r.psy.cmu.edu/.

3 Cognitive Modeling Using ACT-R

Computational cognitive models such as ACT-R are formal theories of cognition that provide predictions of human performance in cognitive tasks such as multitasking (Salvucci, Kushleyeva, & Lee, 2004). These models generally include simulated representations of human cognitive processes such as memory, attention, visual and motor processing, problem-solving, learning, and other related phenomena. Important insights about the cognitive mechanisms behind multitasking and how brain networks are affected by changes in internal cognitive strategies, external interface properties, and task demands arise by comparing simulated human cognitive data from the ACT-R architecture to real human data.

To understand cognitive models using ACT-R, this section was categorized by two different approaches; behavioral and neural-based ACT-R modeling. Those previous well-defined structures and researches allow us to enhance our understanding about the interactions of operator's behavioral and neural systems.

3.1 Behavioral Data-Based ACT-R Modeling

Lebiere, Anderson, and Bothell (2002) proposed a workload prediction method based on ACT-R to predict the workload in air traffic control simulations. In the latter method, the workload of an ACT-R model was simply assumed as the scaled ratio between the times spent on critical unit tasks to the total time of the task. Because he considered only time as the resource related to mental workload, however, it was difficult to explain the low mental workload of time-consuming but minimally resource-demanding tasks. Gray, Schoelles, and Sims (2005) suggested cognitive metrics profiling, also based on ACT-R, as a theory-based prediction of transient changes in the workload demanded by dynamic task environments, but they did not introduce the detailed procedure or mathematical representation to predict the mental workload over time. Table 1 presents a summary of exemplary ACT-R-based cognitive modeling.

Retrieval by Accumulating Evidence in an Architecture (RACE/A) is an extension of the ACT-R that models the process of retrieving memory (van Maanen, & Hedderik van Rijn, 2012). In RACE/A, retrieval of a chunk from declarative memory is a process in which evidence is accumulated for the likelihood that a chunk will be needed, like sequential sampling models of cognitive behavior (Ratcliff, 1978; Usher & McClelland, 2001). The dynamics of the competition between memory chunks govern the retrieval time from declarative memory. Because RACE/A dynamically updates the activation values of chunks, it can explain the interference effects observed in semantic memory retrieval better than existing ACT-R models.

3.2 Neural Data-Based ACT-R Modeling

The ACT-R architecture has widely been used to develop neuro-computational models. In terms of neuroergonomics research, it's critical to explain how they are connected to other brain regions and understand their functionality. Several mappings between ACT-R modules and brain regions exist (Anderson, 2007; Anderson et al.,

2008; Nijboer, Borst, Van Rijn, & Taatgen, 2016). One of the current research areas with ACT-R is utilizing the buffers to track the activity of their associated modules and then comparing that activity to data from neuroimaging studies (fMRI, MEG, or EEG) in order to find correlations between regions of the brain and particular buffer/module activity in ACT-R models (see Table 1).

Originally, the existence of a connectionist implementation of ACT-R was sufficient proof of the neural plausibility of the architecture (Lebiere & Anderson, 1993).

Table 1 Summary of exemplary ACT-R-based cognitive modeling studies

Author and year	Task	Key finding	Modeling approach	
Lebiere et al. (2001)	Air traffic controller	Modeling mental workload when low mental workload time-consuming	Behavioral data-based	• No parameter revised • Attempted to predict mental workload during ATC task
Gray et al. (2005)	Blocks-world task	Attempted to replace ACT-R's section mechanism		• Predicted driver's mental workload • Working within ACT-R's expected value equation • Replacing the binary success/failure scheme
van Maanen and Hedderik van Rijn (2012)	Picture-word interference task	Developed ***RACE/A*** controlled by accumulated likelihood that a chunk will be needed		• Drift rate manipulation, based on sequential sampling models of cognitive behavior • Use of accumulated probability of retrieval
Van Vugt and Groningen (2012)	Attentional blink task	First attempt to link EEG data and ACT-R module activation	Neural data-based	• ERP peak point was matched with ACT-R module activation
Nijboer et al. (2016)	2-back and tracking task	Created cognitive models of a dual-tasking paradigm by using ACT-R and fMRI data		• Revised parameter by comparison of human data and fMRI data • Revised retrieval threshold and Latency-scaling factor
Kim et al. (2018)	AF-MATB	ACT-R modules' activation timing to explain multitasking human behavior		• Brain effective connectivity measures were matched to ACT-R modules' activation • Predicted causality between modules by EEG data

But with the increased focus on mapping functionality to specific brain regions of interest (ROIs), the question of how the ROIs implement their respective modules' functionality has risen. This is especially so for modules that are part of the basic information processing circuit (Anderson et al., 2008), which function independently from specific inputs or outputs (Anderson et al., 2007). Explaining how they are connected to other brain regions is a crucial part of understanding their functionality.

When brain regions are active, their metabolic demands increase (Anderson, 2007). As a result, the oxygen flow in the blood towards these regions increases. Research using fMRI has been able to develop an understanding of the roles of various brain regions and assign the cognitive functionality of those regions to ACT-R modules and buffers (Cassenti, Kerick, & McDowell, 2011). Despite the strong correlation between fMRI data and ACT-R, it is unknown whether ACT-R also holds a strong correlation with EEG data. Recently, interest in this area has piqued with researchers, who have been outlining methods to study the neural correlates of ACT-R in electrophysiological data and have been demonstrating how different ACT-R modules can be associated with observable EEG data (Cassenti et al., 2011; Van Vugt & Groningen, 2012; Van Vugt, 2014).

Kim, Kim, Yun, and Nam (2018) analyzed EEG-based brain networks and linked them to module activations of ACT-R data. They found causal relationships between brain regions associated with ACT-R modules' activation during multitasking. Based on these studies, future neuroergonomics research may continue to update and improve the previous neural-based ACT-R architectures.

4 Future Directions of ACT-R for Neuroergonomics Research

Neuroergonomics aims at integrating cognitive neuroscience methods into the practice of human factors (Parasuraman, Christensen, & Grafton, 2012). A potential application of neurocognitive architectures might include modelling cognitive training program for increasing specific cognitive abilities (McKendrick & Parasuraman, 2014) or adaptive aiding (Wilson & Russell, 2007).

One of the important goals in neuroergonomics is the accurate understanding and prediction of the behaviors and decision-making processes of humans at work. It is especially demanding to achieve this for real dynamic settings, characterized by constant changes. Individual differences in decision-making and behavior make this even more challenging. The area of human–machine interaction looks at a series of decisions and multifactor situations that are influenced by corresponding feedback. Cognitive modeling provides us with a method to understand and explain how such dynamic decisions are made. Human performance modeling (HPM) is a relatively new research area in human factors and ergonomics and it provides a new set of methods (e.g., mathematical models and computer simulation models) to quantify and predict human performance (e.g., task completion time, work time, motor performance, results of decision-making, errors, and cognitive and physical workload), so that human–machine systems' productivity and safety can be maximized. The one of neurocognitive architecture, ACT-R can quantify human performance based on its fundamental cognitive and physiological mechanisms and theories, unify experimental findings using neurophysiological techniques, in existing empirical studies, and can be embedded and used in intelligent systems evaluation and design.

In the same sense future research in neuroergonomics and neuroscience will go hand in hand and computational neuroscience based on systematic cognitive architecture can reach a higher theoretical and methodological level by further informing and enriching each other through complementary and synergistic efforts. The practical value of computational neuroergonomics will be best demonstrated when it produces specific and practical benefits to the design and improvement of human–machine system.

Cognitive architectures are increasingly employed as blueprints for building intelligent agents to be endowed with various perceptive and cognitive abilities. The neurocognitive architecture will emulate the putative functional aspects of various salient brain sub-systems via a learning memory modeling approach. It can be a strong framework of self-organizing connectionist learning to induce high-level symbolic knowledge autonomously, and support for meta-cognitive functions to pave a way to expand the neuroergonomics field. Finally, neurocognitive architectures would allow us to understand in many real-world situations, from mundane everyday tasks to the most complex, demanding work environments.

References

Anderson, J. R. (1983). *Cognitive science series. The architecture of cognition.* Hillsdale, NJ, US.

Anderson, J. R. (2007). *How can the human mind occur in the physical universe?* Oxford University Press.

Anderson, J. R., Bothell, D., Byrne, M. D., Douglass, S., Lebiere, C., & Qin, Y. (2004). An integrated theory of the mind, *111*(4), 1036–1060. https://doi.org/10.1037/0033-295X.111.4.1036.

Anderson, J. R., Carter, C. S., Fincham, J. M., Qin, Y., Ravizza, S. M., & Rosenberg-lee, M. (2008). Using fMRI to test models of complex cognition. *Cog, 32*(8), 1323–1348. https://doi.org/10.1080/03640210802451588.

Anderson, J. R., Qin, Y., Jung, K. J., & Carter, C. S. (2007). Information-processing modules and their relative modality specificity. *Cognitive psychology, 54*(3), 185–217.

Cassenti, D. N., Kerick, S. E., & McDowell, K. (2011). Observing and modeling cognitive events through event-related potentials and ACT-R. *Cognitive Systems Research, 12*(1), 56–65. https://doi.org/10.1016/j.cogsys.2010.01.002.

Gray, W. D., Schoelles, M. J., & Sims, C. R. (2005). Adapting to the task environment : Explorations in expected value. Action editor : Christian Schunn, *6*, 27–40. https://doi.org/10.1016/j.cogsys.2004.09.004.

Kim, N., Kim, W., Yun, M. H., & Nam, C. S. (2018). *Behavioral and neural correlates of hysteresis effects during multitasking, 2010–2012*. https://doi.org/10.1177/1541931218621003.

Laird, J. E. (2012). *The SOAR cognitive architecture*. MIT press.

Langley, P., Laird, J. E., & Rogers, S. (2009). Cognitive architectures: Research issues and challenges. *Coginitive Systems Research, 10*(2), 141–160. https://doi.org/10.1016/j.cogsys.2006.07.004.

Lebiere, C., & Anderson, R. A. (1993). A connectionist implementation of the ACT-R production system. In *The Fifteenth Annual Conference of the Cognitive Science Society* (pp. 635–640).

Lebiere, C., Anderson, J. R., & Bothell, D. (2002). *Multi-tasking and cognitive workload in an ACT-R model of a simplified air traffic control task.* Paper presented at the 10th Conference on Computer Generated Forces and Behavioral Representation, Norfolk, VA.

Lebiere, C., Anderson, J. R., Bothell, D., Lebiere, C., & Anderson, J. R. (2001). Multi-tasking and cognitive workload in an ACT- R model of a simplified air traffic control task.

Matessa, M. (2008). An ACT-R representation of information processing in autism ACT-R. In *Proceedings of the Thirtieth Annual Conference of the Cognitive Science Society* (pp. 2168–2173).

McKendrick, R., & Parasuraman, R. (2014). Using functional near infrared spectroscopy (fNIRS) to evaluate the neurocognitive effects of transient events: Design matrix mixed effects analysis. In *Proceedings of the Human Factors and Ergonomics Society Annual Meeting* (pp. 235–239). Los Angeles, CA: Sage Publications.

Nijboer, M., Borst, J., Van Rijn, H., & Taatgen, N. (2016). Contrasting single and multi-component working-memory systems in dual tasking. *Cognitive Psychology, 86,* 1–26. https://doi.org/10.1016/j.cogpsych.2016.01.003.

Parasuraman, R., Christensen, J., & Grafton, S. (2012). Neuroergonomics: The brain in action and at work. *Neuroimage, 59*(1), 1.

Ratcliff, R. (1978). A theory of memory retrieval. *Psychological Review, 85*(2), 59.

Salvucci, D. D., Kushleyeva, Y., & Lee, F. J. (2004). Toward an ACT-R general executive for human multitasking. In *ICCM* (pp. 267–272).

Smart, P. R., Scutt, T., Sycara, K., & Shadbolt, N. R. (2016). Integrating ACT-R cognitive models with the unity game engine (pp. 74–86).

Sun, R. (2006). The CLARION cognitive architecture: Extending cognitive modeling to social simulation. In *Cognition and Multi-Agent Interaction* (pp. 79–99).

Sun, R. (2007). Cognitive social simulation incorporating cognitive architectures. *IEEE Intelligent Systems, 22*(5).

Taatgen, N., & Anderson, R. (2010). The past, present, and future of cognitive architectures. *Topics in Cognitive Science, 2*(4), 693–704. https://doi.org/10.1111/j.1756-8765.2009.01063.x.

Taatgen, N., Lebiere, C., & Anderson, J. (2006). Modeling paradigms in ACT-R. In *Cognition and Multi-agent Interaction: From Cognitive Modeling to Social Simulation* (pp. 29–52).

Taatgen, N. A., & Lee, F. J. (2003). Production compilation: A simple mechanism to model complex skill acquisition. *Human Factors, 45*(1), 61–76.

Usher, M., & McClelland, J. L. (2001). The time course of perceptual choice: The leaky, competing accumulator model. *Psychological Review, 108*(3), 550.

van Maanen, L., & Hedderik van Rijn, N. T. (2012). RACE/A: An architectural account of the interactions between learning, task control, and retrieval dynamics. *Cognitive Science, 36,* 62–101. https://doi.org/10.1111/j.1551-6709.2011.01213.x.

Van Vugt, M. K. (2014). NeuroImage cognitive architectures as a tool for investigating the role of oscillatory power and coherence in cognition. *NeuroImage, 85,* 685–693. https://doi.org/10.1016/j.neuroimage.2013.09.076.

Van Vugt, M. K. & Groningen, A. G. (2012). Relating ACT-R buffer activation to EEG activity during an attentional blink task. In N. Rußwinkel, U. Drewitz, & H. van Rijn (Eds.), (p. 218).

Whitehill, J. (2013). *Understanding ACT-R—An outsider's perspective.* arXiv [Preprint]. arXiv: 1306.0125.

Wilson, G. F., & Russell, C. A. (2007). Performance enhancement in an uninhabited air vehicle task using psychophysiologically determined adaptive aiding. *Human Factors, 49*(6), 1005–1018.

Deep Learning Techniques in Neuroergonomics

Sanghyun Choo and Chang S. Nam

Abstract There is increasing interest in using deep learning (DL) for neuroergonomics research that investigates the human brain in relation to behavioral performance in natural environments and everyday settings. But a better understanding of how to design and implement DL techniques is still needed for neuroergonomists. Written for novice neuroergonomists as well as experienced investigators, this chapter presents the history of advancements in DL, its concepts, and applications of DL in neuroergonomics research. In addition to artificial neural network (ANN) which is a basic model for DL, this chapter introduces popular DL models such as the multilayer perceptron (MLP), deep belief network (DBN), convolutional neural network (CNN), and recurrent neural networks (RNN). DL-based neuroergonomics research on four main research areas (i.e., mental workload, motor imagery, driving safety, and emotion recognition) will then be reviewed. Insights into how to model and apply DL techniques will be helpful for neuroergonomics researchers, in particular those who are not familiar with DL, but want to predict and classify brain states under various contexts.

1 Introduction

Deep learning (DL) is a subset of machine learning (ML), which aims to achieve specific tasks (e.g., prediction and classification). DL consists of multiple layers with various activation functions for feature extraction and transformation. DL techniques have validated their high performance in a variety of areas, such as image (e.g., Krizhevsky, Sutskever, & Hinton, 2012), video (e.g., Karpathy et al., 2014), voice recognition (e.g., Deng, Hinton, & Kingsbury, 2013), text mining (e.g., Lai, Xu, Liu, & Zhao, 2015), and natural language processing (e.g., Young, Hazarika, Poria, & Cambria, 2018).

S. Choo · C. S. Nam (✉)
Edward P. Fitts Department of Industrial & Systems Engineering, North Carolina State University, Raleigh, NC 27695, USA
e-mail: csnam@ncsu.edu

C. S. Nam (ed.), *Neuroergonomics*, Cognitive Science and Technology,
https://doi.org/10.1007/978-3-030-34784-0_7

There are several reasons for achieving this high performance. First, feature selection to learn data can be minimized in DL since it has an ability to select the features automatically. Feature is extracted automatically by DL structures in comparison with conventional ML techniques. This is a very useful characteristic since the existing ML methods have a limitation extracting features subjectively for training dataset. Second, it is a good fit for high dimension and large-scale datasets. DL can learn datasets efficiently due to its deep layer and diverse activation functions in each layer. Third, it is very adaptable for various domains (e.g., computer vision, signal processing, and data analytics). Given the successful stories in the various fields and DL's advantages, it has been applied into numerous neuroergonomics fields (e.g., mental workload, motor imagery, driving safety, and emotion recognition) with its high performance, especially classification accuracy. Neuroergonomics uses neural measurements to be applied in the applications. Especially, in the case of EEG, it needs signal processing to extract features and classify human cognition based on the signals. DL can extract the features and classify the cognition effectively due to its model property. From the high classification accuracy by DL, it will be more helpful for the neuroergonomics domains by minimizing their model errors.

The goal of this chapter is to provide neuroergonomics researchers who are novice in DL with the history of advancements, its concepts, and applications of DL in EEG-based neuroergonomics studies. Knowing the DL's concepts is difficult for the novice readers due to its complexity, even though it has been used in neuroergonomics. This chapter will play a role to guide neuroergonomics researchers to know basic concepts of the DL and its successful application domains. The remainder of this chapter is organized as follows: Sects. 2 and 3 provide DL history and methods including artificial neural network (ANN), deep belief network (DBN), convolutional neural network (CNN), and recurrent neural networks (RNN). Section 4 explains applications of DL-based neuroergonomics. Finally, conclusions and future directions are summarized.

2 Advancements of Deep Learning Techniques

Neuroergonomics is the study combining neuroscience and ergonomics (Parasuraman, 2003; Parasuraman & Rizzo, 2007). This study aims at integrating neuroscience and human factors to study brain function and human performance in natural environments (Parasuraman, 2003). This multidisciplinary field has been extensively studied in a variety of domains such as adaptive automation (Aricò et al., 2016), rehabilitation (Meinel, Castaño-Candamil, Reis, & Tangermann, 2016), and driving safety (Lees, Cosman, Lee, Rizzo, & Fricke, 2010). Neuroimaging has played a key role in the study of neuroergonomics to explain relationships between brain structures, functions, and human performance. Neuroimaging methods for neuroergonomics fall into two categories: first, direct measurements of neural signal (e.g., EEG and event-related potentials (ERPs), and second, indirect measurements of neural activity (e.g., fMRI, fNIRS, and PET (Mehta & Parasuraman, 2013). However, each method has its own respective advantages and disadvantages. EEG data have significant high temporal resolution but have low spatial resolution. Conversely, fMRI and PET data have low temporal resolution but have significant high temporal resolution. Since the main utility of DL is to predict and classify, EEG is more appropriate for DL due to its huge dataset based on time resolution compared with other neuroimaging methods.

There are several main reasons for the popularity of DL. First, the dramatically high computation power by hardware, especially GPU unit, became available. Second, improved regularization methods (e.g., dropout, batch normalization, and new activation functions) were developed. Third, useful DL tools (e.g., TensorFlow, PyTorch, and Theano) were made. These hardware and software have been made high performance of DL possible. The biggest issues in DL were gradient vanishing problem and computation for learning. The gradient vanishing problem is that as the model has a deeper layer, learning could not be propagated because of the characteristic of activation function like sigmoid function, which has the range (0,1). Also, there are a large amount of computations to train DL models since it has a lot of layers, nodes, and free parameters. However, these issues have been solved by improved techniques (e.g., pretraining by DBN, dropout, and ReLU function) and hardware (e.g., improved CPU and GPU).

There have been a lot of studies dedicated to developing DL models. McCulloch and Pitts (1943) created the initial neural network (NN) model trying to mimic human brain processes. Their model is called McCulloch–Pitts neurons or MCP model, which has contributed the development of artificial neural network (ANN). Since MCP model, there have been many contributions to develop the ANN, which is basis of DL techniques (Voulodimos, Doulamis, Bebis, & Stathaki, 2018). Figure 1 shows the important milestones in DL advancements. Especially, LeNet (Lecun et al., 1990), which is a kind of convolutional neural network (CNN), and long short-term memory (LSTM) (Hochreiter & Urgen Schmidhuber, 1997), which is a kind of recurrent neural network (RNN), led to DL era (Voulodimos et al., 2018).

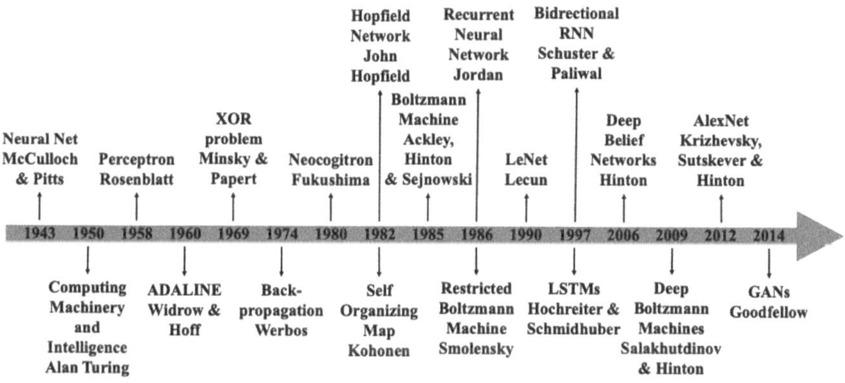

Fig. 1 Timelines of deep learning advancements

Before the deep belief network (DBN) (Hinton, Osindero, & Teh, 2006) was introduced, there were limitations to learning the DL models (e.g., gradient vanishing problem). DBN provided the breakthroughs to learn deep models by using pretraining based on unsupervised learning from restricted Boltzmann machine (RBM) (Smolensky, 1986). Since the DBN model, many methods to learn deep model effectively and efficiently have been developed (e.g., pretraining methods, drop out, and diverse activation functions). Recently, generative adversarial network (GAN), which is a kind of generative model, has been spotlighted due to its performance of data reconstruction.

ML techniques have been applied in neuroergonomics. These ML techniques include Gaussian mixture model (GMM) (Rosipal et al., 2007a, 2007b), NN (Baldwin & Penaranda, 2012), linear discriminant analysis (LDA) (Naseer, Noori, Qureshi, & Hong, 2016), and support vector machine (SVM) (Durantin, Scannella, Gateau, Delorme, & Dehais, 2016). However, these ML techniques have limitations. First, they would still use subjective biomarkers in selecting features for classification (Plis et al., 2014). This can cause ineffective feature extraction and low performance of neuroergonomics applications. Second, complex large-scale data could not be handled by traditional ML techniques due to its algorithm structures (Razzak, Naz, & Zaib, 2017). Signal data from brain are very complicated (e.g., multichannel EEG signals). It is also hard for us to extract meaningful feature extraction by hand in brain data because of its complexity. However, DL methods can be a good tool to solve these problems. As aforementioned before, DL can extract feature automatically and handle large-scale data well.

Now, we need to focus on DL methods for neuroergonomics since its advantages can increase performance of brain-based applications. Applying DL to neuroergonomics requires a huge dataset. So, we will focus on EEG-based neuroergonomics application since EEG can have the huge dataset due to its high temporal property than fMRI, fNIRS, and PET datasets. Utilizing deep learning is important in neuroergonomics area. This area is affected significantly by the model's accuracy in predicting and classifying mental events (e.g., mental workload, mental fatigue, and motor imagery). Therefore, DL would be able to help the performance of the mental states' classification improve based on its abilities (e.g., expression of complex data, extraction of meaningful features from brain data). This chapter aims to critically review the state of the art of DL while also discussing its applications to neuroergonomics. This is because the classification performance of existing ML techniques in neuroergonomics fields is not enough to apply real life. In addition, this chapter will also identify significant outstanding research problems and promising directions in order to reach neuroadaptive system, which is a final goal of neuroergonomics. There are several main techniques for DL (ANN, DBN, CNN, RNN, and GAN). However, the GAN model has little validation in its effectiveness in neuroergonomics fields since it is somewhat newer version of DL compared to others. Therefore, in this chapter, we will focus on other DL models for neuroergonomics except for the GAN.

3 Deep Learning Families

This section presents a basic DL model, ANN, with several types of activation function as well as popular DL models with learning methods of them such as multilayer perceptron (MLP), DBN, CNN, and RNN with its properties and structures.

3.1 Artificial Neural Network (ANN)

The ANN is a ML technique that is inspired by neural network of the human brain (van Gerven & Bohte, 2017). This model is a basic model for DL. A basic ANN is composed of single-layer perceptron (SLP) (see Fig. 2a) and multilayer perceptron (MLP) (see Fig. 2b).

Input nodes in SLP are connected to a summation node. The summation node has a weighted sum of input nodes and bias. An activation function is applied into the summation value. Then, the final value after the activation function becomes the output. A summation formula for multiple outputs is as follows:

$$u_j = \sum_{i=1}^{n} x_i w_i + b_j. \tag{1}$$

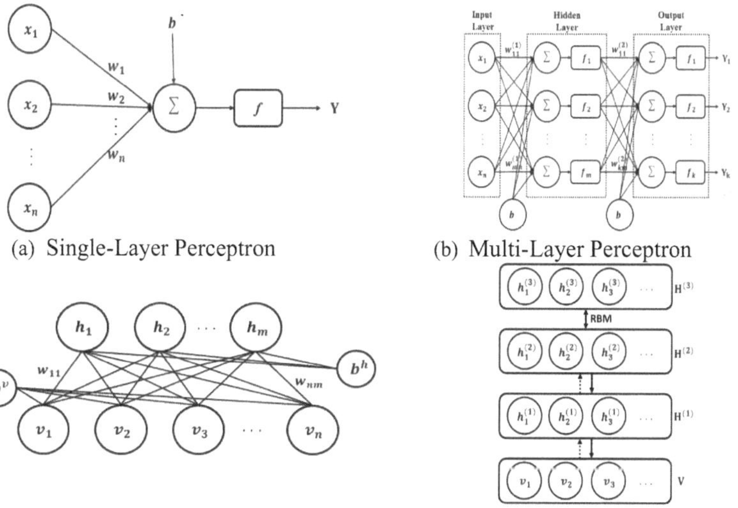

(a) Single-Layer Perceptron (b) Multi-Layer Perceptron

(c) Restricted Boltzmann Machine (d) Deep Belief Network

Fig. 2 Deep learning families

The final output is as follows:

$$z_j = f(u_j). \tag{2}$$

The way to learn the weight parameter is by minimizing the difference between target value and output value. This means that error between the target and output is decreased by each learning state. The weight parameter updating rule is as follows:

$$w_{ji}^{t+1} = w_{ji}^{t} + \eta x_i(t_j - z_j), \tag{3}$$

where w_{ji} are the weights connecting input and summation nodes, η is the learning rate, x_i is the input node's value, t_j is the target value, and z_j is the final output value.

Activation Function: Although many types of activation functions have used in neural network, sigmoid function, hyperbolic function, and Rectified Linear Unit (ReLU) function are often used for constructing networks.

Sigmoid also known as logistic function are as follows:

$$f(x) = \frac{1}{1 + e^{-x}}. \tag{4}$$

This function takes the whole real number as a domain and values between 0 and 1 as the range. Instead of the sigmoid function, a hyperbolic function like the sigmoid function can also be used.

$$f(x) = \frac{e^{2x} - 1}{e^{2x} + 1}. \tag{5}$$

Although the hyperbolic function has different range (−1, 1) compared to the sigmoid function, its characteristics are similar. Both the sigmoid and the hyperbolic functions have constant values in maximal or minimal input values, but have a changing nature between remaining input values. This changing nature is motivated by a nature of nerve cell.

ReLU function is an activation function to train DL efficiently (Glorot, Bordes, & Bengio, 2011). The ReLU function is as follows:

$$f(x) = \max(0, x). \tag{6}$$

The ReLU function always has positive values regardless of input values. It can reduce the computation of DL due to its linear property.

3.2 Multilayer Perceptron (MLP)

MLP was made to overcome a problem of SLP, which was not able to train complex data. Especially, SLP could not solve XOR problem, which is a simple linear pattern. However, MLP was able to overcome the limitation of SLP by adding hidden layers with a backpropagation (BP) algorithm (Werbos, 1974). MLP includes input layer, hidden layer, and output layer with biases, summation nodes, and activation functions. The information of input nodes flows to the right (feedforward) and finally has become processed values.

The way nodes in the hidden layer receive input values is as follows:

$$u_j^{(1)} = \sum_{i=1}^{n} x_i w_{ji}^{(1)} + b_j. \tag{7}$$

The nodes in hidden layers are explained by the weighted sum of the input values with bias. The results from the hidden nodes pass activation function for the next input values about output layer. The following formula provides the output of hidden nodes:

$$z_j^{(1)} = f\left(u_j^{(1)}\right). \tag{8}$$

The output of hidden nodes becomes input values of output layer. The input values of output layer are as follows:

$$u_l^{(2)} = \sum_{i=1}^{n} z_j^{(1)} w_{lj}^{(1)} + b_l. \tag{9}$$

The final output values are derived by an activation function processing the input values from hidden layer.

$$z_l^{(2)} = f\left(u_l^{(2)}\right). \tag{10}$$

MLP is learned by a BP algorithm. The BP algorithm is a learning method used to minimize error between target values and output values. The error (E) of MLP is as follows:

$$E = \frac{1}{2} \sum_{l=1}^{k} t_l - z_l^{(2)}. \tag{11}$$

In order to minimize E, a gradient descent method can be used for weights parameter based on E. The parameter updating rule for the weights is induced by differential of E including sigmoid function as an activation function.

$$w_{ji}^{t+1} = w_{ji}^{t} + \Delta w_{ji}, \tag{12}$$

$$\Delta w_{ji} = \eta \delta_j x_i, \tag{13}$$

where δ_j are local gradients.

3.3 *Deep Belief Network (DBN)*

DBN (see Fig. 2d) is a stochastic architecture of DL based on RBM. The RBM is used to pretrain DBN to solve the gradient vanishing issue.

3.3.1 Restricted Boltzmann Machine (RBM)

An RBM (see Fig. 2c) is a type of a stochastic neural network that only includes visible and hidden nodes unlike the existing Boltzmann machine (BM), which has full connections between visible layers and hidden layers as well as their own connections in each layer. The RBM is proposed by Smolensky in 1986.

The energy function of RBM is defined as

$$E(v,h) = -\sum_{i=1}^{V}\sum_{j=1}^{H} v_i h_j w_{ij} - \sum_{i=1}^{V} v_i b_i^v - \sum_{j=1}^{H} h_j b_j^h, \tag{14}$$

where v_i is the binary data of visible node i, h_j is the binary data of hidden node j, w_{ij} are the weights between visible node i and hidden node j, and b_i^v and b_j^h are biases of visible and hidden nodes, respectively.

The conditional probabilities $p(h)$ and $p(v)$ can be derived by the energy function (14)

$$p(v|h) = \frac{1}{1 + e^{-\left(\sum_{j=1}^{H} h_j w_{kj} + b_k^v\right)}}, \tag{15}$$

$$p(h|v) = \frac{1}{1 + e^{-\left(\sum_{i=1}^{V} v_i w_{ik} + b_k^h\right)}}. \tag{16}$$

Each parameter updating rule of RBM based on maximum likelihood estimation (MLE) is as follows:

$$\Delta w_{ij} = \epsilon_w\left(E_{data}\left[v_i h_j\right] - E_{model}\left[v_i h_j\right]\right), \tag{17}$$

$$\Delta b_i^v = \epsilon_{b^v}(E_{data}[v_i] - E_{model}[v_i]), \tag{18}$$

$$\Delta b_j^h = \epsilon_{b^h}\left(E_{data}\left[h_j\right] - E_{model}\left[h_j\right]\right), \tag{19}$$

where ϵ is the learning rate in each parameter, E_{data} is the expected value underlying distribution of the model with the original input data, and E_{model} is the expected value underlying distribution of the model by the contrastive divergence (CD) algorithm (Hinton, 2002). In order to approximate E_{model} by the CD algorithm, CD-1 can be used since it is effective enough to sample the model (Hinton, 2010).

The joint distribution of DBN between visible node v and the l hidden layers h^k is as follows:

$$P\left(x, h^1, \ldots, h^l\right) = \left(\prod_{k=0}^{l-2} P\left(h^k | h^{k+1}\right)\right) P\left(h^{l-1}, h^l\right), \tag{20}$$

where $v = h^0$, $P\left(h^k | h^{k+1}\right)$ is a conditional probability of the visible nodes by the hidden nodes of RBM with $k + 1$ layer, and $P\left(h^{l-1}, h^l\right)$ is the joint distribution in the top layer RBM.

3.3.2 Learning Method

DBN can be learned by a greedy layer-wise unsupervised training (Hinton et al., 2006). The processes of DBN learning include the following stages:

1. Pretraining: The nodes from two layers (V, H^1) on the bottom are trained using the method of RBM learning. Then, the values of hidden nodes in H^1 are computed by the weight parameter from the learning result. The computed hidden values become input values for the next learning. The input values and hidden nodes of H^2 are trained using the RBM learning again. These sequences are being kept by stacking the layers until it reaches the top layer.
2. Fine-tuning: After pretraining by layer-wise learning, it can be trained by back-propagation algorithm in order to be used in classification problems. The training process is same with MLP with the initial parameters by pretraining.

The main difference of DBN and deep neural network (DNN) is that DBN includes a pretraining session for getting reasonable initial parameter values but DNN is trained without any pretraining process. Also, DBN can be used for an unsupervised model unlike DNN.

3.4 Convolutional Neural Network (CNN)

Convolutional neural network (CNN) is a type of DL for image processing with special layers (convolution layers, pooling layers, and fully connected layers) (LeCun, Bottou, Bengio, & Haffner, 1998). Figure 3a shows the example of CNN structure.

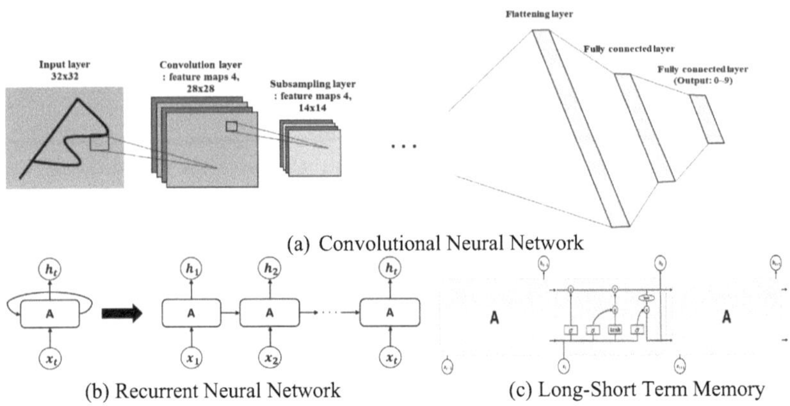

Fig. 3 Schematic representation of CNN, RNN, and LSTM

CNN has an ability to extract features effectively from input data using several layers, which include functions such as the extraction of meaningful features and the size reduction of the features.

Convolution Layer: The convolution layer generates a variety of feature maps of input using convolution kernels. The whole image is convolved by the kernels. The advantage of convolution is that it reduces complexity of images because of weight sharing by convolution (Lecun, Bengio, & Hinton, 2015).

Pooling Layer: The pooling Layer reduces spatial resolution of the feature maps. This layer plays a role of dimension reduction, which can decrease the complexity of images. There are operators for the pooling layers. Usually, average pooling and max pooling are used for subsampling. The effects of these pooling methods are that it makes the feature maps insensitive to shift and distortion of them. Besides these pooling methods, there are other pooling methods (e.g., stochastic pooling (Wu & Gu, 2015) and deformation constrained pooling (def-pooling) (Ouyang et al., 2015)).

Fully Connected Layer: After the convolution and pooling stages, a fully connected layer is performed. All neurons in fully connected layers are connected to all activated neurons of the previous layer. The fully connected layer transforms 2D feature maps into 1D feature maps. Then, these feature maps are connected to nodes of categories for classification.

3.5 Recurrent Neural Networks (RNN)

RNN (Fig. 3b) is utilized to train sequential data (e.g., time series, text, and video). RNNs include recurrent states in a hidden layer to handle sequential data. The left side of the RNN structure can be unfolded like right side of Fig. 3b. The RNN can be considered as multiple copies of the same network, each passing implied data to its next layer. So, it can handle time-series data due to its structures. The RNNs can be trained by backpropagation through time (BPTT) (Werbos, 1990). BPTT is similar to the BP algorithm for MLP. BPTT can also be considered as a BP algorithm expanding in time step (Werbos, 1990).

However, there is a limitation of general RNN model. It is not effective for learning long-term dependencies because of the gradient vanishing problem (Graves, 2013). This means that the RNNs cannot be learned about training data when it has high sequential or long-term data. In order to solve this problem, long short-term memory (LSTM) networks were introduced by Hochreiter and Urgen Schmidhuber (1997). LSTM (see Fig. 3c) is a special type of RNNs designed to overcome long-term dependency problem.

LSTM has the structure adding cell states in the hidden layer of RNN. Also, the LSTM includes the operations, which play a role as a gate to decide information flows. The following equations show the formula of LSTM cells:

$$f_t = \sigma\left(I_f x_t + W_f h_{t-1} + b_f\right), \tag{21}$$

$$i_t = \sigma(I_i x_t + W_i h_{t-1} + b_i), \tag{22}$$

$$o_t = \sigma(I_o x_t + W_o h_{t-1} + b_o), \tag{23}$$

$$g_t = \tanh(I_c x_t + W_c h_{t-1} + b_c), \tag{24}$$

$$c_t = f_t \odot c_{t-1} + i_t \odot g_t, \tag{25}$$

$$h_t = o_t \odot \tanh(c_t), \tag{26}$$

where f_t is a forget gate, i_t is an update gate, o_t is an output gate, g_t is a cell candidate, c_t is a cell output, h_t is a hidden output, I and W are weights for input and hidden nodes, respectively, b is a bias, and $\odot$ is Hadamard product.

To train the LSTMs, a BPTT algorithm based on a framework of Graves and Schmidhuber (2005) can be used. Graves (2013) demonstrated the effect of the LSTM in the gradient vanishing problem. Given the formula of LSTM cells and the BPTT algorithm, the LSTM can overcome the limitation of conventional RNN (gradient vanishing in long term).

4 Applications

There are several paradigms for neuroergonomics such as mental workload, visual attention, motor control, human–automation interaction, adaptive automation, and driving safety. Even though the ability of DL methods has been proven in a lot of fields (e.g., speech recognition, image recognition), there is still a lack of research on

DL-based neuroergonomics. However, mental workload, motor imagery, and driving fatigue among areas of neuroergonomics have been studied a lot in terms of applying the DL methods to neuroergonomics compared to other fields of neuroergonomics. Therefore, we will review previous DL-based neuroergonomics research on these four research topics.

4.1 Mental Workload

Mental workload refers to the amount of working memory processing a task in the restricted time (Horat et al., 2016). There are several reasons why mental workload is important in neuroergonomics. First, high cognitive load of an individual can decrease learning ability and cause a state of distraction (Sweller, Van Merrienboer, & Paas, 1998). Second, it can restrict the ability of an individual in a specific cognitive task since there is a limit to the amount of cognitive workload for the individual (Bashivan, Rish, Yeasin, & Codella, 2016). Therefore, the assessment of mental workload can play a key role in preventing accidents in a variety of areas, especially unmanned aerial vehicle (UAV) (Zhang & Li, 2017).

Also, there have been a lot of studies to predict mental workload, especially using ML methods, such as SVM (Aghajani, Garbey, & Omurtag, 2017), LDA (Cinaz, Arnrich, La Marca, & Tröster, 2013), and hidden Markov model (HMM) (Solhjoo, Nasrabadi, Reza, & Golpayegani, 2005). These studies have shown good performances in estimating mental workload based on EEG data. However, there are several limitations for the existing estimation methods. First, selecting appropriate features for the models takes lots of time. Second, these models cannot handle high-dimensional input data efficiently like EEG. In order to overcome these problems, the assessment of mental workload based on DL has been studied (Bashivan & Bidelman, 2015; Bashivan et al., 2016; Zhang & Li, 2017; Hefron, Borghetti, Kabban, Christensen, & Estepp, 2018; Kuanar, Athitsos, Pradhan, Mishra, & Rao, 2018). From the deep learning-based studies, it is shown that DL was higher performance in estimating cognitive workload than the traditional ML techniques since the DL model can extract appropriate features and learn high-dimensional EEG data.

Bashivan and Bidelman (2015) demonstrated high prediction performance of DBN in terms of four different levels of cognitive load compared with SVM during a memory task. They found that mean spectral power (MSP) can be a better feature for DBN than wavelet entropy (WE) in prediction of the cognitive load. Bashivan, Rish, Yeasin, and Codella (2016) proposed a novel approach for learning EEG data based on topology-preserving multispectral images using combined DL (CNN+RNN). They used the spatial, spectral, and temporal features of EEG as image dataset to train their proposed DL model. It showed significant improvements of classification accuracy in mental workload domain with the comparison of several conventional machine learning models (e.g., SVM, random forest (RF), and logistic regression (LR)). Zhang and Li (2017) suggested a RBM-based model for mental workload classification. They reduced EEG channels using RBM with trial-and-error and entropy-based pruning methods. They showed that the reduced features with DBN increased the accuracy of mental workload classification.

Hefron et al. (2018) studied DL's validation in terms of cognitive state assessment in a non-stimulus-locked task environment with cross-participant. They showed that increased length of temporal sequence of EEG data was not enough to solve dissimilarities between individuals to improve classification accuracy. Also, they demonstrated the convolutional-recurrent model based on residual, bidirectional layers, and multipath subnetworks could increase the classification accuracy of mental workload using statistical analysis. Kuanar, Athitsos, Pradhan, Mishra, and Rao (2018) proposed a deep RNN to solve noise and inter-subject differences of EEG data in the classification of cognitive load. They used transformed EEG time series data including spatial information to learn the RNN. This EEG feature was less sensitive to variations. Also, the proposed deep RNN with the spectral, spatial, and temporal features showed high classification accuracy of four different levels of cognitive load. Table 1 shows overall classification performance of mental workload according to classifiers of authors.

Table 1 Results of classification accuracy of mental workload according to classifiers

(%)	Bashivan and Bidelman (2015)	Bashivan et al. (2016)	Zhang and Li (2017)	Hefron et al. (2018)	Kuanar et al. (2018)
DBN	89.5	85.0	96.1	–	–
CNN	–	88.7	–	71.1	–
RNN (LSTM)	–	–	–	86.8	–
Combined DL	–	91.1	–	–	92.4
ANN	–	–	64.8	86.2	–
SVM	89.1	84.7	–	–	85.0
LDA	–	–	82.6	–	–
LR	–	84.7	–	–	85.6
RF	–	87.4	–	–	87.8
NB	–	–	57.5	–	–

**DBN* Deep belief network; *CNN* Convolutional neural network; *RNN* Recurrent neural network; *LSTM* Long short-term memory; *DL* Deep learning; *ANN* Artificial neural network; *SVM* Support vector machine; *LDA* Linear discriminant analysis; *LR* Linear regression; *RF* Random forest; *NB* Naïve Bayes

4.2 Motor Imagery

Motor imagery (MI) is one of commonly used BCI models. MI can be defined as a cognitive process for stimulating an action by human body. It has been widely used in diverse areas (e.g., rehabilitation, sport training) since many studies have shown it is effective in enhancing physical performance (Daly & Huggins, 2016; Lebon, Collet, & Guillot, 2010). However, MI in BCI aims at governing hardware based on only cognitive task without any individual's movement (Lee & Choi, 2018). In order to become a good model for MI BCI, classification accuracy of MI is the most important part. Many classification methods based on ML has been applied to handle MI classification. These studies used the following methods to classify MI tasks: logistic regression (LR) (Li, Wu, & Yang, 2011), LDA (Djemal, Bazyed, Belwafi, Gannouni, & Kaaniche, 2016), and SVM (Ma et al., 2016). However, they were not enough to handle highly nonstationary EEG data since they had a limitation to extract meaningful information from the high-dimensional data. Given the results of studies using DL techniques for MI classification, DL can be a good basis for MI classification (Chu et al., 2018; Kumar, Sharma, Mamun, & Tsunoda, 2016; Lee & Choi, 2018; Sakhavi & Guan, 2017; Uktveris & Jusas, 2017; Yang, Sakhavi, Ang, & Guan, 2015; Zhang, Li, & Wang, 2017a; Zhang, Yan, & Gong, 2017b; Zhou, Meng, Gao, Ma, & Zhang, 2018). These results were that DL can improve the accuracy of MI classification with effective feature selection or special features (e.g., augmented common spatial pattern (ACSP) and topology-preserving multispectral images of EEG) and have better classification performance of MI than traditional ML methods.

Yang et al. (2015) investigated a CNN-based classification method in terms of multi-class motor imagery. They proposed a frequency complementary feature map selection (FCMS) with constraints of dependency among frequency bands. They demonstrated that FCMS-based CNN had higher performance about motor imagery classification than filter-bank CSP (FBCSP).

Kumar, Sharma, Mamun, and Tsunoda (2016) presented a DL approach for motor imagery classification. They used CSP feature-based DNN to classify motor imagery dataset. They showed the effectiveness of DNN in the classification of motor imagery with 10×10-fold cross-validation errors. Sakhavi and Guan (2017) used FBCSP-based CNN to explore the possibility of knowledge transfer using motor imagery dataset. They trained CNN model using multiple subjects and fine-tuning the model using a small amount of data from new subjects. They showed that the classification accuracy was increased even though the small amount of datasets were used for training the model. Uktveris and Jusas (2017) proposed CNN using fast Fourier transform (FFT)-based image feature map to classify four-class motor imagery. They showed that the simple FFT image map is enough to reach the state-of-the-art classification accuracy of motor imagery with the CNN performance of the dataset. Zhang et al. (2017a, 2017b) proposed CNN model for motor imagery classification using time–frequency images by short-time Fourier transformation (STFT). They used three types of activation functions to learn the model: the rectified Linear Unit (ReLU), exponential linear unit (ELU), and scaled exponential linear unit (SELU).

Table 2 Results of classification accuracy of motor imagery according to classifiers

(%)	Yang et al. (2015)	Sakhavi and Guan (2017)	Uktveris and Jusas (2017)	Zhang et al. (2017a, 2017b)	Lee and Choi (2018)	Zhou et al. (2018)	Chu et al. (2018)
DBN	–	–	–	–	–	–	70.7
CNN	69.3	69.7	68.0	92.7	78.9	–	–
RNN (LSTM)	–	–	–	–	–	91.4	–
SVM	–	–	–	83.7	–	85.7	69.9

**DBN* Deep belief network; *CNN* Convolutional neural network; *RNN* Recurrent neural network; *LSTM* Long short-term memory; *SVM* Support vector machine

They demonstrated the effectiveness of their models with the comparison of existing method and showed SELU function can be better activation function than others. Lee and Choi (2018) proposed a novel approach for classification of motor imagery using continuous wavelet transform (CWT). They used two types of wavelet transform (e.g., Morlet and Bump). The results showed that Morlet-based CWT can be a better feature for CNN than Bump-based feature in terms of motor imagery classification. Zhou et al. (2018) suggested an RNN (LSTM)-based classification method of motor imagery data using combined features of discrete wavelet decomposition (DWD) and Hilbert transformation (HT). These features include both amplitude and frequency modulation of the EEG. The results showed that the proposed method can have higher classification accuracy in terms of motor imagery data. Chu et al. (2018) proposed a novel classification method of incomplete motor imagery data using DBN with Lomb–Scargle periodogram (LSP)-based power spectrum density (PSD) features. The LSP can estimate robust PSD including incomplete EEG data. They showed that the proposed method had higher classification accuracy than other comparison methods (FTT-based models and Welch-based models). Table 2 shows overall classification performance of motor imagery according to classifiers of authors.

4.3 Driving Safety

Driving safety is one of the core applications of neuroergonomics. Especially, mental fatigue in driving safety like drowsiness is a main factor of traffic accidents. Drowsiness in driving consists of two main categories. First category is the behavior detection (e.g., the deviation of vehicle, the steering wheel movement) (Hung, Wang, Prasad, & Lin, 2017). Second category is the physiological detection (e.g., respiratory, heart rate, and brain wave) (Borghini, Astolfi, Vecchiato, Mattia, & Babiloni, 2014). Also, there have been many studies associated with detection of the mental fatigue or drowsiness in driving using EEG data (Balkin, Horrey, Graeber, Czeisler, & Dinges, 2011; Borghini et al., 2014; Hung et al., 2017). In order to detect these cognitive states based on EEG data in driving, many studies have been used diverse ML techniques including SVM (Wang, Jung, & Lin, 2015), ANN (Johnson et al., 2011), and GMM (Rosipal et al., 2007a, 2007b). Although these studies showed good performances of the detection in terms of driving fatigue, DL-based detection methods of the driving fatigue have been tried a lot due to its advantages (e.g., appropriateness in high-dimensional data and effective feature extraction). Many studies have demonstrated that DL techniques can be a good tool for detecting mental fatigue in driving (Li, Jiang, & Su, 2016; Hung et al., 2017; Chai et al., 2017; Zeng et al., 2018; Guarda, López, Moura, & Ramos, 2018; Hernández, Mozos, Ferrández, & Antelis, 2018). These studies showed that DL methods had higher accuracy of driving fatigue detection than conventional ML methods. Also, the DL methods can extract meaningful features significantly for training classification models and learn high-dimensional data like EEG.

Li et al. (2016) proposed detection method of mental fatigues including alert, slight fatigue and severe fatigue which is important issue in driving safety using EEG-based DBN model. They used 21 features including spectral powers and frequency bands to learn the model. The results showed that DBN-based detection method of mental fatigue had better classification performance than existing ML methods (e.g., RF, SVM, etc.). Hung et al. (2017) proposed 3D CNN-based drowsiness detection in the status of driving using limited weight sharing and channel-wise convolution. The proposed model can extract various frequency bands and the spatial information of EEG. Also, this method showed higher performance of the classification in terms of fatigue level during driving than NN and support vector regression (SVR). Chai et al. (2017) suggested classification method of driving fatigue using sparse-DBN based on autoregressive (AR) features. The sparse-DBN can prevent overfitting by controlling hidden nodes' activation with a regularization term. They demonstrated that the proposed model can classify driving fatigue better than ANN and Bayesian neural networks (BNN). Zeng et al. (2018) proposed prediction methods in terms of mental states of drivers using CNN and CNN with residual learning based on EEG. They tested the proposed model with intra- and inter-subject data. The results showed that both models had higher classification performance about the mental states than conventional LSTM and SVM. Guarda et al. (2018) presented DL-based detection method of drowsiness using EEG. They used "ULg Multimodality Drowsiness Database" and EEG spectrograms as features to learn the model. They showed that the conventional CNN outperformed existing machine learnings (e.g., RF, NN, and SVM) in terms of drowsiness classification. Hernández et al. (2018) proposed CNN-based detection method of emergency braking from driver's EEG. They utilized noise removed EEG to train the model. In order to show the effectiveness of the CNN model, they compared the model with SVM. The results showed that CNN has much higher performance about the detection than SVM model. Table 3 shows overall classification performance of mental fatigue according to classifiers of authors.

Table 3 Results of classification accuracy of mental fatigue according to classifiers

(%)	Li et al. (2016)	Hung et al. (2017)	Chai et al. (2017)	Zeng et al. (2018)	Guarda et al. (2018)	Hernández et al. (2018)
DBN	98.9	–	90.7	–	–	–
CNN	–	63.2	–	91.8	86.7	71.8
RNN (LSTM)	–	–	–	85.1	–	–
Proposed DL model	–	–	94.1	92.7	–	–
ANN	–	62.2	79.9	–	74.5	–
SVM	96.7	59.7	–	88.1	73.0	71.1
RF	89.5	–	–	–	–	–
NB	67.9	–	–	–	–	–
KNN	80.4	–	–	–	–	–
Boosting	67.7	–	–	–	–	–
Bagging	88.1	–	–	–	–	–
DT	82.2	–	–	–	–	–
BNN	–	–	84.2	–	–	–

**DBN* Deep belief network; *CNN* Convolutional neural network; *RNN* Recurrent neural network; *LSTM* Long short-term memory; *DL* Deep learning; *ANN* Artificial neural network; *SVM* Support vector machine; *RF* Random forest; *NB* Naïve Bayes; *KNN* K-nearest neighbors; *DT* Decision tree; *BNN* Bayesian neural network

4.4 Emotion Recognition

Emotion recognition play a key role in human life and also is able to give a lot of opportunities improving human–computer interaction like an affective interaction for disabled (Sohaib, Qureshi, Hagelbäck, Hilborn, & Jerčić, 2013). There have been several approaches for emotion recognition (e.g., biosignals (EEG; EMG) (Sharma & Gedeon, 2012; Soleymani, Asghari-Esfeden, Fu, & Pantic, 2016) and neuroimaging methods (fMRI; PET) (Phan, Wager, Taylor, & Liberzon, 2002; Hattingh et al., 2013)). In this study, we focus on emotion recognition based on EEG data since EEG data is appropriate for applying DL techniques due to property of high dimension and the amount of training dataset. Studies associated with emotion recognition based on EEG have been investigated using ML techniques such as Bayesian network (BNT) (Friedman, Geiger, & Goldszmidt, 1997), ANN (Lahane & Sangaiah, 2015), K-nearest neighbors (KNN) (Murugappan, Ramachandran, & Sazali, 2010), and SVM (Sohaib et al., 2013). Although these methods are applied to emotion recognition successfully, their performance of classification is not enough to classify emotions because of subject dependency in terms of emotion (Gao, Lee, & Mehmood, 2015). Recently, DL techniques have been applied to emotion recognition based on

EEG data (Alhagry, Aly, & El-Khoribi, 2017; Gao et al., 2015; Li et al., 2017; Soleymani et al., 2016; Tripathi, Acharya, Ranti, Mittal, & Bhattacharya, 2017; Wang, Zhong, Peng, & Liu, 2018; Zheng & Lu, 2015). These studies demonstrated that the DL techniques (e.g., CNN, DBN, and RNN) can derive higher classification accuracy in terms of emotion recognition in the EEG environment than the conventional ML methods.

Gao et al. (2015) proposed a method of emotion recognition using DNN with raw EEG data. They used a pretraining method with RBMs and a fine-tuning method with BP algorithm to learn the model. The results showed that the proposed framework achieved better recognition accuracy than traditional machine learning methods (e.g., KNN, SVM, and ANN). Zheng and Lu (2015) suggested DBN-based emotion recognition using multichannel EEG data. They extracted EEG features using DE, PSD, differential asymmetry (DASM), and rational asymmetry (RASM). They tested the proposed model with SVM according to the features and frequency bands. The results showed that the recognition model had the highest performance in all frequency bands when DE was used.

Alhagry et al. (2017) proposed a raw EEG-based LSTM model for emotion recognition. The proposed model showed higher average accuracy of emotion recognition in comparison with the conventional emotion recognition methods. Li, Huang, Zhou, and Zhong (2017) suggested the hybrid neural networks, which combined CNN and LSTM, to recognize emotion based on EEG features including spatial characteristics, frequency domain, and temporal characteristics. They used the open-source dataset DEAP (a dataset for emotion analysis using EEG, physiological, and video signals) to validate their model. The results showed that their hybrid DL outperformed the conventional machine learnings (e.g., KNN, RF, and SVM) and the DLs (CNN and RNN). Tripathi et al. (2017) explored two DL models (DNN and CNN) in terms of emotion recognition based on DEAP dataset. They tested the DL models with various combinations of hyperparameters (e.g., the number of hidden layers, learning rate, dropout, and mini batch) to classify the emotion data. The results showed that CNN classified the emotions better than DNN. Wang et al. (2018) proposed a data augmentation method to increase sample dataset of EEG in the domain of emotion recognition for DL training. They used Gaussian noise as the data augmentation method and the differential entropy (DE) as the feature extraction for CNN training. They validated the method of the augmented sample data with the classification accuracy of the CNN to emotion recognition. Table 4 shows overall classification performance of emotion recognition according to classifiers of authors.

5 Conclusions and Future Directions

This chapter introduced overall concepts of DL especially with EEG-based neuroergonomics domain. Neuroergonomics is significantly associated with mental states (e.g., mental workload, mental fatigue, emotion, etc.). High classification accuracy of the mental states is necessary to be applied in neuroergonomics since the study

Table 4 Results of classification accuracy of emotion recognition according to classifiers

(%)	Gao et al. (2015)	Zheng and Lu (2015)	Alhagry et al. (2017)	Li et al. (2017)	Tripathi et al. (2017)	Wang et al. (2018)
DNN	–	–	–	–	57.1	–
DBN	68.4	86.1	–	–	–	–
CNN	–	–	–	–	62.2	–
RNN (LSTM)	–	–	86.4	–	–	–
Combined DL	–	–	–	60.7	–	–
Proposed DL model	–	–	–	67.5	–	–
ANN	60.0	–	–	–	–	–
SVM	60.8	84.0	–	–	–	74.2
LR	–	82.7	–	–	–	–
RF	–	–	–	42.7	–	–
KNN	51.3	72.6	–	59.5	–	–
PCA+SVM	–	–	–	–	–	58.1
LeNet	–	–	–	–	–	49.6
ResNet	–	–	–	–	–	34.2

**DNN* Deep neural network; *DBN* Deep belief network; *CNN* Convolutional neural network; *RNN* Recurrent neural network; *LSTM* Long short-term memory; *DL* Deep learning; *ANN* Artificial neural network; *SVM* Support vector machine; *LR* Linear regression; *RF* Random forest; *KNN* K-nearest neighbors; *PCA* Principle component analysis; *ResNet* Residual neural network

has tried to reveal the cause of human behavior by neural measurements. Also, the definitive goal of neuroergonomics is neuroadaptive system which helps humans to be convenient using brain measurements so that the classification performance by the measurements can have an effect on the system's performance. Given the reasons why the classification accuracy of the mental states by the cognitive measurements is important, DL can be a solution to improve the accuracy since the performance of DL has been validated in not only image classification, voice recognition, and signal processing but also neuroergonomics fields (e.g., mental workload, motor imagery, mental fatigue, and emotion recognition).

There are several advantages of DL techniques. First, they have a good fit for training high complex data like EEG data due to their models' structures including deep layers, a lot of nodes, and activation functions. Second, they can extract meaningful features from the complex data without any prior information of this. Actually, effectiveness for feature extraction of DL has not been proved mathematically yet. However, many researches have demonstrated their ability of feature extraction experimentally. Third, they are versatile in a variety of areas (e.g., image classification, signal processing, and speech recognition). This means that DL methods can be applied to neuroergonomics fields appropriately since neuroergonomics usually use neuroimages and brain signals for its application. However, there are still limitations for the DL techniques. First, they require huge dataset for training. It means that the deep model cannot be a good model in small or simple dataset. Second, in order to get results from DL, it takes a lot of time for computation. But it is alleviated by development of CPU, GPU, and learning methods including dropout and ReLU function. Finally, although many studies have studied regularization problem (overfitting) of DL, it has not been solved completely yet.

Given the perspective of neuroergonomics domain and the limitations of the DL techniques, we suggest as follows:

- Neuroergonomics domain using DL is needed a lot since there are still the lack of studies of DL-based neuroergonomics.
- For effective training of DL in neuroergonomics, appropriate features as input in DL are needed. Although DL techniques can extract features automatically from selected feature datasets, their performance will be improved if they can focus on specific features for model training due to the decrease of computation for unnecessary information.
- Development of efficient algorithms to learn DL are needed in order to be applied to neuroadaptive systems. Computation time for DL should be reduced for real-time system since neuroadaptive system requires fast response in real time.
- Free database should be constructed for huge EEG dataset depending on application domains. DL requires enormous dataset to train the model. So, if we can store the huge dataset according to specific neuroergonomics fields, we can apply adaptive learning to implement real-time neuroadaptive system so that it would contribute to reducing learning time and improving classification performance of the model.

References

Aghajani, H., Garbey, M., & Omurtag, A. (2017). Measuring mental workload with EEG+fNIRS. *Frontiers in Human Neuroscience, 11,* 1–20.

Alhagry, S., Aly, A., & El-Khoribi, R. A. (2017). Emotion recognition based on EEG using LSTM recurrent neural network. *International Journal of Advanced Computer Science and Applications, 8*(10), 355–358.

Aricò, P., Borghini, G., Di Flumeri, G., Colosimo, A., Bonelli, S., Golfetti, A., & Babiloni, F. (2016). Adaptive automation triggered by EEG-based mental workload index: A passive brain-computer interface application in realistic air traffic control environment. *Frontiers in Human Neuroscience, 10*, 1–13.

Baldwin, C. L., & Penaranda, B. N. (2012). Adaptive training using an artificial neural network and EEG metrics for within- and cross-task workload classification. *NeuroImage, 59,* 48–56.

Balkin, T. J., Horrey, W. J., Graeber, R. C., Czeisler, C. A., & Dinges, D. F. (2011). The challenges and opportunities of technological approaches to fatigue management. *Accident Analysis and Prevention, 43,* 565–572.

Bashivan, P., & Bidelman, G. M. (2015). Single trial prediction of normal and excessive cognitive load through EEG feature fusion. In *Proceedings of IEEE Signal Processing in Medicine and Biology Symposium* (pp. 1–5).
Bashivan, P., Rish, I., Yeasin, M., & Codella, N. (2016). Learning representations from EEG with deep recurrent-convolutional neural networks. In *International Conference on Learning Representations (ICLR)*. arXiv:1511.06448.
Borghini, G., Astolfi, L., Vecchiato, G., Mattia, D., & Babiloni, F. (2014). Measuring neurophysiological signals in aircraft pilots and car drivers for the assessment of mental workload, fatigue and drowsiness. *Neuroscience and Biobehavioral Reviews, 44,* 58–75.
Chai, R., Ling, S. H., San, P. P., Naik, G. R., Nguyen, T. N., Tran, Y., & Nguyen, H. T. (2017). Improving EEG-based driver fatigue classification using sparse-deep belief networks. *Frontiers in Neuroscience, 11.*
Chu, Y., Zhao, X., Zou, Y., Xu, W., Han, J., & Zhao, Y. (2018). A decoding scheme for incomplete motor imagery EEG with deep belief network. *Frontiers in Neuroscience, 12,* 1–17.
Cinaz, B., Arnrich, B., La Marca, R., & Tröster, G. (2013). Monitoring of mental workload levels during an everyday life office-work scenario. *Personal and Ubiquitous Computing, 17,* 229–239.
Daly, J. J., & Huggins, J. E. (2016). Brain-computer interface: Current and emerging rehabilitation applications. *Archives of Physical Medicine and Rehabilitation, 96*(30), S1–S7.
Deng, L., Hinton, G., & Kingsbury, B. (2013). New types of deep neural network learning for speech recognition and related applications: An overview. In *ICASSP, IEEE International Conference on Acoustics, Speech and Signal Processing—Proceedings* (pp. 8599–8603).
Djemal, R., Bazyed, A. G., Belwafi, K., Gannouni, S., & Kaaniche, W. (2016). Three-class EEG-based motor imagery classification using phase-space reconstruction technique. *Brain Sciences, 6*(36).
Durantin, G., Scannella, S., Gateau, T., Delorme, A., & Dehais, F. (2016). Processing functional near infrared spectroscopy signal with a Kalman filter to assess working memory during simulated flight. *Frontiers in Human Neuroscience, 9,* 1–9.
Friedman, N., Geiger, D., & Goldszmidt, M. (1997). Bayesian network classifier. *Machine Learning, 29,* 131–163.
Gao, Y., Lee, H. J., & Mehmood, R. M. (2015). Deep learning of EEG signals for emotion recognition. In *2015 IEEE International Conference on Multimedia and Expo Workshops, ICMEW* (pp. 1–5).
Glorot, X., Bordes, A., & Bengio, Y. (2011). Deep sparse rectifier neural networks. In *AISTATS* (Vol. 15, pp. 315–323).
Graves, A. (2013). *Generating sequences with recurrent neural networks* (pp. 1–43). arXiv:1308.0850.
Graves, A., & Schmidhuber, J. (2005). Framewise phoneme classification with bidirectional LSTM and other neural network architectures. *Neural Networks, 18*(5–6), 602–610.
Guarda, L., López, E., Moura, M., & Ramos, M. (2018). Drowsiness detection using electroencephalography signals : A deep learning based method. In *14th PSAM International Conference on Probabilistic Safety Assessment and Management.*
Hattingh, C. J., Ipser, J., Tromp, S. A., Syal, S., Lochner, C., Brooks, S. J., & Stein, D. J. (2013). Functional magnetic resonance imaging during emotion recognition in social anxiety disorder: An activation likelihood meta-analysis. *Frontiers in Human Neuroscience, 6,* 1–7.
Hefron, R., Borghetti, B., Kabban, C. S., Christensen, J., & Estepp, J. (2018). Cross-participant EEG-based assessment of cognitive workload using multi path convolutional recurrent neural networks. *Sensors, 18*(1339).
Hernández, L. G., Mozos, O. M., Ferrández, J. M., & Antelis, J. M. (2018). EEG-based detection of braking intention under different car driving conditions. *Frontiers in Neuroinformatics, 12,* 1–14.
Hinton, G. (2002). Training products of experts by minimizing contrastive divergence. *Neural Computation, 14*(8), 1711–1800.
Hinton, G. (2010). A practical guide to training restricted boltzmann machines. *Momentum, 9*(1), 926.

Hinton, G. E., Osindero, S., & Teh, Y. W. (2006). A fast learning algorithm for deep belief nets. *Neural Computation, 18*(7), 1527–1554.

Hochreiter, S., & Urgen Schmidhuber, J. (1997). Long short-term memory. *Neural Computation, 9*(8), 1735–1780.

Horat, S. K., Herrmann, F. R., Favre, G., Terzis, J., Debatisse, D., Merlo, M. C. G., & Missonnier, P. (2016). Assessment of mental workload: A new electrophysiological method based on intra-block averaging of ERP amplitudes. *Neuropsychologia, 82*, 11–17.

Hung, Y. C., Wang, Y. K., Prasad, M., & Lin, C. T. (2017). Brain dynamic states analysis based on 3D convolutional neural network. In *2017 IEEE International Conference on Systems, Man, and Cybernetics* (pp. 222–227).

Johnson, R. R., Popovic, D. P., Olmstead, R. E., Stikic, M., Levendowski, D. J., & Berka, C. (2011). Drowsiness/alertness algorithm development and validation using synchronized EEG and cognitive performance to individualize a generalized model. *Biological Psychology, 87*, 241–250.

Karpathy, A., Toderici, G., Shetty, S., Leung, T., Sukthankar, R., & Li, F. F. (2014). Large-scale video classification with convolutional neural networks. In *Proceedings of the IEEE Computer Society Conference on Computer Vision and Pattern Recognition* (pp. 1725–1732).

Krizhevsky, A., Sutskever, I., & Hinton, G. E. (2012). ImageNet classification with deep convolutional neural networks. In *Advances in Neural Information Processing Systems* (pp. 1–9).

Kuanar, S., Athitsos, V., Pradhan, N., Mishra, A., & Rao, K. R. (2018). Cognitive analysis of working memory load from Eeg, by a deep recurrent neural network. In *ICASSP, IEEE International Conference on Acoustics, Speech and Signal Processing—Proceedings* (pp. 2576–2580).

Kumar, S., Sharma, A., Mamun, K., & Tsunoda, T. (2016). A deep learning approach for motor imagery EEG signal classification. In *Proceedings of APWC CSE* (pp. 34–39).

Lahane, P., & Sangaiah, A. K. (2015). An approach to eeg based emotion recognition and classification using kernel density estimation. *Procedia Computer Science, 48*, 574–581.

Lai, S., Xu, L., Liu, K., & Zhao, J. (2015). Recurrent convolutional neural networks for text classification. In *Proceedings of the Conference of the Association for the Advancement of Artificial Intelligence (AAAI)*.

Lebon, F., Collet, C., & Guillot, A. (2010). Benefits of motor imagery training on muscle strength. *The Journal of Strength and Conditioning Research, 24*, 1680–1687.

Lecun, Y., Bengio, Y., & Hinton, G. (2015). Deep learning. *Nature, 521*, 436–444.

LeCun, Y., Bottou, L., Bengio, Y., & Haffner, P. (1998). Gradient-based learning applied to document recognition. In *Proceedings of the IEEE* (Vol. 86, no. 11, pp. 2278–2324).

Lecun, Y., Henderson, J., Le Cun, Y., Denker, J. S., Henderson, D., Howard, R. E., & Jackel, L. D. (1990). Handwritten digit recognition with a back-propagation network. *Advances in Neural Information Processing Systems, 2*, 396–404.

Lee, H. K., & Choi, Y. S. (2018). A convolution neural networks scheme for classification of motor imagery EEG based on wavelet time-frequency image. In *International Conference on Information Networking (ICOIN)* (pp. 906–909).

Lees, M. N., Cosman, J. D., Lee, J. D., Rizzo, M., & Fricke, N. (2010). Translating cognitive neuroscience to the driver's operational environment: A neuroergonomics approach. *American Journal of Psychology, 123*(4), 391–411.

Li, Y., Huang, J., Zhou, H., & Zhong, N. (2017). Human emotion recognition with electroencephalographic multidimensional features by hybrid deep neural networks. *Applied Sciences, 7*, 1060.

Li, P., Jiang, W., & Su, F. (2016). Single-channel EEG-based mental fatigue detection based on deep belief network. In *Proceedings of the Annual International Conference of the IEEE Engineering in Medicine and Biology Society, EMBS, 2016–October* (pp. 367–370).

Li, Y., Wu, J., & Yang, J. (2011). Developing a logistic regression model with cross-correlation for motor imagery signal recognition. In *2011 IEEE/ICME International Conference on Complex Medical Engineering* (pp. 502–507).

Ma, Y., Ding, X., She, Q., Luo, Z., Potter, T., & Zhang, Y. (2016). Classification of motor imagery EEG signals with support vector machines and particle swarm optimization. In *Computational and Mathematical Methods in Medicine* (pp. 1–8).

McCulloch, W. S., & Pitts, W. H. (1943). A logical calculus of the ideas immanent in nervous activity. *Bulletin of Mathematical Biophysics, 5,* 115–133.

Mehta, R. K., & Parasuraman, R. (2013). Neuroergonomics: A review of applications to physical and cognitive work. *Frontiers in Human Neuroscience, 7,* 1–10.

Meinel, A., Castaño-Candamil, S., Reis, J., & Tangermann, M. (2016). Pre-trial EEG-based single-trial motor performance prediction to enhance neuroergonomics for a hand force task. *Frontiers in Human Neuroscience, 10,* 1–17.

Murugappan, M., Ramachandran, N., & Sazali, Y. (2010). Classification of human emotion from EEG using discrete wavelet transform. *Journal of Biomedical Science and Engineering, 3,* 390–396.

Naseer, N., Noori, F. M., Qureshi, N. K., & Hong, K.-S. (2016). Determining optimal feature-combination for LDA classification of functional near-infrared spectroscopy signals in brain-computer interface application. *Frontiers in Human Neuroscience, 10,* 1–10.

Ouyang, W., Wang, X., Zeng, X., Qiu, S., Luo, P., Tian, Y., & Tang, X. (2015). DeepID-Net: Deformable deep convolutional neural networks for object detection. In *Proceedings of the IEEE Computer Society Conference on Computer Vision and Pattern Recognition.*

Parasuraman, R. (2003). Neuroergonomics: Research and practice. *Theoretical Issues in Ergonomics Science, 4*(1–2), 5–20.

Parasuraman, R., & Rizzo, M. (2007). *Neuroergonomics: The brain at work.* Oxford; New York: Oxford University Press.

Phan, K. L., Wager, T., Taylor, S. F., & Liberzon, I. (2002). Functional neuroanatomy of emotion: A meta-analysis of emotion activation studies in PET and fMRI. *NeuroImage, 16,* 331–348.

Plis, S. M., Hjelm, D. R., Slakhutdinov, R., Allen, E. A., Bockholt, H. J., Long, J. D., … Calhoun, V. D. (2014). Deep learning for neuroimaging: A validation study. *Frontiers in Neuroscience, 8,* 1–11.

Razzak, M. I., Naz, S., & Zaib, A. (2017). *Deep learning for medical image processing: Overview, challenges and the future.* arXiv:1704.06825.

Rosipal, R., Peters, B., Göran Kecklund, T. Å., Gruber, G., Woertz, M., Anderer, P., & Dorffner, G. (2007a). EEG-based drivers' drowsiness monitoring using a hierarchical gaussian mixture model. In *Foundations of Augmented Cognition* (pp. 294–303).

Rosipal, R., Peters, B., Kecklund, G., Åkerstedt, T., Gruber, G., Woertz, M., & Dorffner, G. (2007b). EEG-based drivers' drowsiness monitoring using a hierarchical gaussian mixture model. In *Proceedings of the HCII2007—Augmented Cognition* (pp. 294–303).

Sakhavi, S., & Guan, C. (2017). Convolutional neural network-based transfer learning and knowledge distillation using multi-subject data in motor imagery BCI. In *8th International IEEE EMBS Conference on Neural Engineering* (pp. 588–591).

Sharma, N., & Gedeon, T. (2012). Objective measures, sensors and computational techniques for stress recognition and classification: A survey. *Computer Methods and Programs in Biomedicine, 108,* 1287–1301.

Smolensky, P. (1986). *Information processing in dynamical systems: Foundations of harmony theory.* MA, USA: MIT Press Cambridge.

Sohaib, A. T., Qureshi, S., Hagelbäck, J., Hilborn, O., & Jerčić, P. (2013). Evaluating classifiers for emotion recognition using EEG. In *Foundations of Augmented Cognition* (pp. 492–501). Berlin, Heidelberg: Springer.

Soleymani, M., Asghari-Esfeden, S., Fu, Y., & Pantic, M. (2016). Analysis of EEG signals and facial expressions for continuous emotion detection. *IEEE Transactions on Affective Computing, 7*(1), 17–28.

Solhjoo, S., Nasrabadi, A. M., Reza, M., & Golpayegani, H. (2005). Classification of chaotic signals using Hmm classifiers: Eeg-based mental task classification. In *Proceedings of European Signal Processing Conference.*

Sweller, J., Van Merrienboer, J. J. G., & Paas, F. G. W. C. (1998). Cognitive Architecture and Instructional Design. *Educational Psychology Review, 10*(3), 251–296.
Tripathi, S., Acharya, S., Ranti, S., Mittal, S., & Bhattacharya, S. (2017). Using deep and convolutional neural networks for accurate emotion classification on DEAP dataset. In *Proceedings of the Twenty-Ninth AAAI Conference on Innovative Applications* (pp. 4746–4752).
Uktveris, T., & Jusas, V. (2017). Application of convolutional neural networks to four-class motor imagery classification problem. *Information Technology and Control, 46*(2), 260–273.
van Gerven, M., & Bohte, S. (2017). Editorial: Artificial neural networks as models of neural information processing. *Frontiers in Computational Neuroscience, 11*, 1–2.
Voulodimos, A., Doulamis, N., Bebis, G., & Stathaki, T. (2018). Recent developments in deep learning for engineering applications. In *Computational Intelligence and Neuroscience* (pp. 1–2).
Wang, Y. K., Jung, T. P., & Lin, C. T. (2015). EEG-based attention tracking during distracted driving. *IEEE Transactions on Neural Systems and Rehabilitation Engineering, 23*(6), 1085–1094.
Wang, F., Zhong, S., J. Peng, J. J., & Liu, Y. (2018). Data augmentation for EEG-based emotion recognition with deep convolutional neural networks. In *International Conference on Multi-media Modeling (MMM)* (pp. 82–93). Springer.
Werbos, P. J. (1974). *Beyond regression: New tools for prediction and analysis in the behavioral sciences*. Ph.D. thesis. Harvard University.
Werbos, P. J. (1990). Backpropagation through time: What it does and how to do it. *Proceedings of IEEE, 78*(10), 1550–1560.
Wu, H., & Gu, X. (2015). Towards dropout training for convolutional neural networks. *Neural Networks*.
Yang, H., Sakhavi, S., Ang, K. K., & Guan, C. (2015). On the use of convolutional neural networks and augmented CSP features for multi-class motor imagery of EEG signals classification. In *Proceedings of the Annual International Conference of the IEEE Engineering in Medicine and Biology Society, EMBS* (pp. 2620–2623).
Young, T., Hazarika, D., Poria, S., & Cambria, E. (2018). Recent trends in deep learning based natural language processing. *IEEE Computational Intelligence Magazine, 13*(3), 55–75.
Zeng, H., Yang, C., Kong, G. D., Qin, F., Zhang, J., & Kong, W. (2018). EEG classification of driver mental states by deep learning. *Cognitive Neurodynamics, 12*(6), 597–606.
Zhang, J., & Li, S. (2017). A deep learning scheme for mental workload classification based on restricted Boltzmann machines. *Cognition, Technology & Work, 19*(4), 607–631.
Zhang, J., Li, S., & Wang, R. (2017a). Pattern recognition of momentary mental workload based on multi-channel electrophysiological data and ensemble convolutional neural networks. *Frontiers in Neuroscience, 11*, 1–16.
Zhang, J., Yan, C., & Gong, X. (2017b). Deep convolutional neural network for decoding motor imagery based brain computer interface. In *2017 IEEE International Conference on Signal Processing, Communications and Computing, ICSPCC 2017* (pp. 1–5).
Zheng, W. L., & Lu, B. L. (2015). Investigating critical frequency bands and channels for EEG-based emotion recognition with deep neural networks. *IEEE Transactions on Autonomous Mental Development, 7*(3), 162–175.
Zhou, J., Meng, M., Gao, Y., Ma, Y., & Zhang, Q. (2018). Classification of motor imagery EEG using wavelet envelope analysis and LSTM networks. In *Proceedings of the 30th Chinese Control and Decision Conference, CCDC 2018* (pp. 5600–5605).

Dynamic Causal Modeling (DCM) for EEG Approach to Neuroergonomics

Jiali Huang and Chang S. Nam

Abstract To study the underlying neural mechanisms of human cognitive and physical activities, various neuroimaging techniques such as fMRI, fNIRS, and EEG have been used. Dynamic causal modeling (DCM) for EEG method has gained a considerable amount of attention in recent years for exploring the directional causal relationship between brain areas. Studies using the DCM for EEG method have investigated contemporary neuroergonomics research topics such as mechanisms behind the event-related potential (ERP) components, mismatch negativity (MMN), abnormal coupling of neural regions in neurologically disordered patients, and causal connections supporting motor tasks. This chapter is a good guide for both novice and experienced neuroergonomics researchers as it introduces the motivations behind DCM for EEG and its developmental timelines, as well as explains fundamentals of the DCM for EEG including the neural mass models, state and observer's equation, and the model selection processes. This chapter also summarizes recent studies that used the DCM for EEG method in the past 10 years in terms of the tasks, preprocessing steps, model construction, and selection. Lastly, recommendations for the future DCM-based neuroergonomics research are presented.

1 Introduction

Due to the nature of brain, it is not easy to directly observe and analyze connectivity among brain areas. Brain connectivity is operationally defined as the estimation of the relation between brain areas, which is categorized into a pattern of anatomical links (i.e., *anatomical connectivity*), of statistical dependencies (i.e., *functional connectivity*) or of causal interactions (i.e., *effective connectivity*) between distinct brain regions (Friston, 2011; Peterson & Fling, 2018). In particular, recent studies have developed the structural equation modeling (SEM), the granger causality

J. Huang · C. S. Nam (✉)
Edward P. Fitts Department of Industrial & Systems Engineering, North Carolina State University, Raleigh, NC 27695-7906, USA
e-mail: csnam@ncsu.edu

C. S. Nam (ed.), *Neuroergonomics*, Cognitive Science and Technology,
https://doi.org/10.1007/978-3-030-34784-0_8

Table 1 Comparisons of granger causality and dynamic causal modeling (Friston, 2009)

	Granger causality	Dynamic causal modeling
Model basis	• Phenomenological model of temporal dependencies among the data itself • No reference to how those dependencies were caused	• Explicit forward or generative model of how observed data was caused • Models invoke hidden neuronal and biophysical states that generate data
Model complexity	• Simple and generic models	• Complicated and domain-specific models
Number of models and objectives	• Only two models: one with and one without a particular functional connection • Objective: infer the dependency	• Multiple competing models • Compare and inference on the biophysical parameters of the model
Meaning of causality	• In a colloquial fashion • Does not mean the data from one part of the brain "causes" data in another part	• In a control theory sense • With this model, activity in one brain area causes dynamics in another
Basis of the method	• Temporal precedence • Assumes the data reflect states that cause each other	• Based on the differential equations of motion, where activity in one brain region changes activity in another

(GC), and the dynamic causal modeling (DCM) methods to assess effective connectivity of the brain (see Table 1). Unlike SEM and GC, DCM constructs a reasonably realistic neuronal model of interacting cortical regions and a supplementary forward model of how neuronal activity is transformed into a measured response (Stephan et al., 2010).

Since the concept of DCM was first introduced in 2003, it has been recognized as an effective approach to study the causal relationships connecting regions of the brain (see Fig. 1). It was first developed for fMRI data using a set of neuronal state equations and hemodynamic state equations. DCM for M/EEG was then introduced in 2006, and it primarily focused on the evoked responses. In 2008, DCM for induced response was introduced and followed by stochastic DCM in 2009. DCM for EEG was used to study mismatch negativity, psychological diseases, physical impairment due to brain damage, and much more. Recent developments in DCM are focused on addressing limitations in current method. To provide a more robust statistical inference technique, Markov chain Monte Carlo was used instead of inversions of DCMs. To lift the limitation on the number of ROI used in DCM, the regression method was introduced in 2017.

The remainder of this chapter is organized as follows. Section 2 introduces the fundamental structure and methods of DCM for EEG. In Sect. 3, we review recent investigations in neuroergonomics topics using DCM for EEG in the past ten years. Finally,

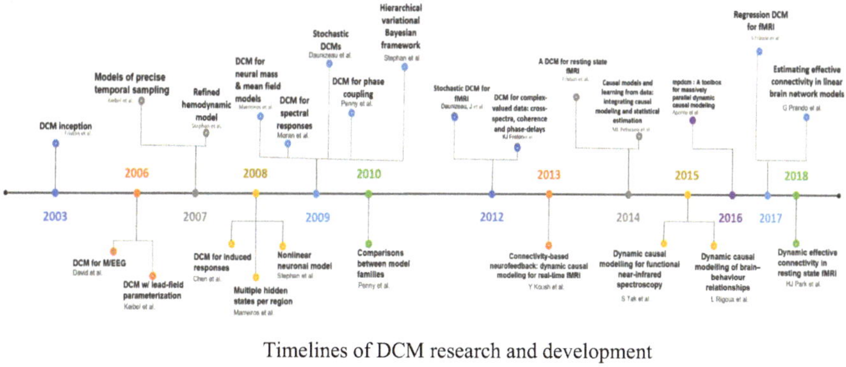

Fig. 1 Timelines of DCM research and development

Sect. 4 concludes by presenting a list of recommendations for future development of using DCM for EEG for neuroergonomics study.

2 Fundamentals of DCM for EEG

DCM was first introduced as a method to infer effective connectivity using fMRI data. The DCM for EEG was later introduced in 2006. This chapter focuses on DCM for EEG. We first introduce the models DCM for EEG is based on, and then present a description of the neuronal and observer's equations. Finally, we present the Bayesian model selection method for choosing the winning model.

2.1 Neuronal Model and Electromagnetic Model

The dynamic causal modeling (DCM) uses a dynamic input-state-output system with multiple inputs and outputs. Experimental manipulation can be viewed as modulatory input to the system and can evoke cortical responses and change causal architectures among the region of interest. The state variables cover both the neuronal activities and other neurophysiological or biophysical variables needed to form the outputs (Kiebel et al., 2006). The outputs in DCM for EEG models are depolarizations of a neural mass of pyramidal cells.

A hierarchical cortical model is used here as a DCM embodies directed extrinsic connections among a number of sources, with each being based on Jansen's model (Jansen & Rit, 1995). Jansen's model includes three interconnected neural populations: excitatory pyramidal cells in supragranular cortical layers as an output receiving end, a group of excitatory spiny stellate cells in granular layer IV, and a group of inhibitory interneurons in infragranular cortical layers. These three populations lie in the order of inhibitory, excitatory (spiny stellate), and excitatory from the skull. The connections between these three populations are as follows: ***(i) bottom-up or forward connections*** that originate in the infragranular layers and terminate in the granular layer; ***(ii) top-down or backward connections*** that connect infragranular to agranular layers; and ***(iii) lateral connections*** that originate in infragranular layers and target all layers (Knösche, 2013).

The responding dendritic signal of the pyramidal subpopulation of the source is detected remotely on the scalp surface in an EEG. The spatial expression of each area is caused by one equivalent current dipole (ECD). The head model for the dipoles is based on four concentric spheres, each with homogeneous and isotropic conductivity. The four concentric spheres are approximately the brain, skull, cerebrospinal fluid (CSF), and scalp. The lead field of each ECD is a function of three location and three orientation or moment parameters (Kiebel et al., 2006). Electromagnetic forward model is used to decide how much each subpopulation's signal contributes to the signal measured.

2.2 State Equation and Observer Equation

The neuronal model for a DCM can be expressed as the following equation:

$$\dot{x} = f(x, u, \theta). \tag{1}$$

In this equation, x denotes the hidden neural state of each cortical sources, u represents the experimental inputs, and θ is the parameters of the state. Equation (1) is a first-order differential equation derived from the behaviors of three neuronal subpopulations : excitatory subpopulation in the granular layers, inhibitory subpopulation in the supragranular layer, and deep pyramidal cells in the infragranular layer. The integration of the differential equations can be expressed as a convolution of the experimental input (David & Friston, 2003).

The convolution converts presynaptic voltage into postsynaptic voltage using the firing rate and excitatory postsynaptic potentials (ESP)/inhibitory postsynaptic potentials (IPSP) slope. The convolution kernel is given by

$$p(t) = \frac{H}{\tau} t \exp(-\frac{t}{\tau}) \quad t \geq 0. \tag{2}$$

The parameters H tune the maximum amplitude of PSPs, and τ is the lumped representation of the sum of the rate constants of passive membrane and other spatially distributed delays in the dendritic tree (David & Friston, 2003). An operator S converts the potential into a firing rate, which is the input to other subpopulations. This transformation is assumed to be instantaneous and is described by the sigmoid function:

$$S_k(x) = \frac{c_k^1 e_0}{1 + \exp(r(x_0 - c_k^2 x))}. \tag{3}$$

For the kth subpopulation, c_k^1, e_0, r, and x_0 are parameters that determine its shape (e.g., voltage sensitivity) (David & Friston, 2003). The strengths of the intrinsic connections are controlled by internal coupling constants $\gamma_{1,2,3,4}$. For each subpopulation, the changes in connectivity are different in the sense of its effect. Connections are expressed in forward, backward, and lateral connections and are modeled by coupling gains A^F, A^B, and A^L:

$$\dot{x}_i = \frac{H}{\tau}((A^{F/B/L} + \gamma_j I)S(x_0) + Cu) - \frac{2x_i}{\tau} - \frac{x_j}{\tau^2}. \tag{4}$$

Equation (4) describes the output of i and is modulated by input Cu, intrinsic connection j, and extrinsic forward/backward/lateral connection $S(x_0)$. The response of the system is stated as

$$h = g(x, \theta). \tag{5}$$

The system's response is a function of neuronal state and parameters of the system. For EEG signals:

$$h = L(\theta^L)x_0. \tag{6}$$

Using a conventional linear electromagnetic forward model, Eq. (6) links the measured data on scalp to neuronal state data. In this application, assuming each source as one ECD, the head model for the dipoles is based on four concentric spheres: the brain, skull, cerebrospinal fluid, and scalp. Lead field $L(\theta^L)$ was parameterized in terms of the radii and conductivities of each of the four layers of head model (Kiebel, Garrido, Moran, Chen, & Friston, 2009).

The observation model for the vectorized data y is

$$y = vec(h(\theta) + X\theta^X) + \varepsilon. \tag{7}$$

2.3 Bayesian Inference

For a given model m, parameter estimation corresponds to approximating the moments of the posterior distribution. The likelihood $p(y|\theta, m)$, the prior $p(\theta|m)$, and the posterior are given by Bayes rule:

$$p(\theta|y, m) = \frac{p(y|\theta, m)p(\theta|m)}{p(y|m)}. \tag{8}$$

Then, model evidence $p(y|m) = \int p(y|\theta, m)p(\theta|m)\partial\theta$ is used to decide how difficult is to achieve a proper fit. The estimation procedure employed in DCM is described in Friston et al. (2002) as Bayesian inference, which proceeds by using the conditional or posterior density estimated by the EM algorithm (Kiebel et al., 2006). The most likely model is the one with the largest log-evidence. To compare between two competing models, the Bayes factor is used to determine the winning model. The Bayes factor is simply the ratio of the two model evidences. For model i and j, the Bayes factor B_{ij} can be expressed as $\ln B_{ij} = \ln p(y|m = i) - \ln p(y|m = j)$. By convention, a model can be said to be the best among a selection of other models, with strong evidence, if its log-model evidence exceeds all other log-model evidences by at least 3 (Ashburner et al., 2013).

When applied to multiple subjects' data, fixed effect metrics such as the group Bayes factor (GBF) and random effect methods are introduced to accommodate different characteristics of the data. Fixed effect methods do not account for group heterogeneity of outliers (Stephan, Penny, Daunizeau, Moran, & Friston, 2009), and it assumes the optimal model for each subject in the group is the same. GBF is calculated by multiplying all single Bayes factors together and represents the evidence for one model relative to the evidence for another model. On the other hand, random effect calculates positive evidence ratio (PER) by counting the ratio of how many subjects showed positive evidence for one model relative to another. The results are expected probability and exceedance probability. They represent the expected fraction of subjects whose data was generated by the specific model and the probability the specific model is more likely than all others, respectively.

3 Applications of DCM for EEG

Because of the high temporal resolution achieved by EEG, in recent years more and more researches have paid attention to DCM for EEG. DCM for EEG was used to test out the neuronal connection underlying auditory and visual mismatch negativity among normal populations as well as schizophrenic patients, Alzheimer patients, and sleep-deprived people. DCM is mainly used to compare DCMs that differ in terms of which inputs affect the system, where these inputs enter, whether mechanisms are linear or nonlinear, which anatomical connections exist, or which priors are best (Acs & Greenlee, 2008; Chen, Henson, Stephan, Kilner, & Friston, 2009; Garrido et al., 2008; Stephan, Weiskopf, Drysdale, Robinson, & Friston, 2007; Stephan et al., 2009). Like approximation of the model evidence such as the Akaike and Bayesian information criteria, or free energy and the ensuing Bayes factor, it is not applied to models that are fitted to different data, and any measure of model goodness is relative, not absolute (Stephan et al., 2010).

Due to the poor spatial resolution of EEG, most former research based their source reconstruction results on previous fMRI studies or applied an fMRI-EEG combination paradigm. During the literature search, we also discovered that the most studied EEG responses were event-related potentials. Some studies concerning induced responses have surfaced in recent years and presented valuable results. One advantageous feature of DCM is that it does not rely on subject group size to obtain a statistically significant result. As shown in the table, the number of participants in these studies ranged from 2 subjects to 84 subjects. This broadened range allowed for a wide usage for future DCM studies.

In this section, we go through recent DCM for EEG studies and summarize the topics discussed in those papers based on the models constructed and selected. Parameters used in preprocessing are also presented in the summary. Data processing and DCM inversion were performed in SPM (https://www.fil.ion.ucl.ac.uk/spm/).

3.1 Cognitive Neuroergonomics

In this section, we describe some of recent investigations into cognitive neuroergonomics topics including mismatch negativity, studies using visual stimulus, and abnormal coupling of neural regions in neurological disordered patients.

3.1.1 Mismatch Negativity (MMN)

For cognitive neuroergonomics, many DCM studies investigated Mismatch Negativity during auditory oddball tasks. Mismatch negativity (MMN) is one of the ERP components elicited by any discriminable violation in the acoustic regularity (Garrido et al., 2008). Garrido, Kilner, Kiebel, Stephan, and Friston (2007), Garrido et al. (2008, 2009) constructed DCMs using A1, STG, and IFG as network nodes differentiated each model based on directions of the connections. The results showed that the MMN rests on plastic change in both extrinsic (between-source) and intrinsic (within-source) connections (Garrido et al., 2007).

Using these results, Garrido et al. (2007) explored the neural network underlying the MMN, the mechanism eliciting MMN, and its implication on brain plasticity. Investigating the connectivity of the connection, he concluded the strength is related to the amplitude of MMN. Garrido et al. approach was innovative but simple, yet it served as the basis for future studies concerning auditory MMN. Following a similar route, Silchenko, Adamchic, Hauptmann, and Tass (2013) continued the study among patients treated with acoustic CR Neuromodulation and added A2, PCC, PAC, DLPFC, OF, and ACC to build models. They tested three models using delta, alpha, and gamma band EEG signals and found the tinnitus-related neuronal connections to determine whether the therapy reversed the abnormal pattern. In this case, DCM was used to compare group difference in connectivity patterns by contrasting the known pattern to group data. Also exploring connection patterns of neuronal disordered patients, Ranlund et al. (2016) and Díez et al. (2017) compared neuronal reactions of people with psychosis and their relatives.

Cooray, Garrido, Brismar, and Hyllienmark (2016) used DCM for EEG to study the evolution of the MMN network for the maturation of adolescents. Based on previous studies of MMN using DCM, they found the network is different among young adolescences and adults. This study showed the possibility of using DCM to study brain plasticity.

3.1.2 Visual Processing

Modulatory effects of extrinsic connections elicited by visual stimulus were investigated in Fogelson, Litvak, Peled, Fernandez-del-Olmo, and Friston (2014). They used PC, IT, V5, and V1 as the nodes. To investigate changes in backward and intrinsic (recurrent) connections, the models were narrowed down to these two categories. Comparing ERPs of the observed signal and predicted signal, the author concluded the effect of the connectivity is apparent in the observed diagram.

Legon et al. (2016) combined EEG data with fMRI recordings to identify whole-brain networks and specific neuronal synaptic mediators within the prefrontal cortex with visual stimulus. This provided a model that is both spatially and temporally accurate. The network identified in fMRI-DCM provided a basis for the further exploration of presynaptic modulation within a specific time frame.

Sato et al. (2017) took advantage of the accurate recordings of intracranial EEG to investigate differences in effective connectivity when subjects were exposed to mosaic pictures or pictures of faces. The results showed that specific connections were present when observing a picture of faces. Li et al. (2018) compared DCM results for visual stimulus tasks between healthy subjects and schizophrenia patients, and found that different brain connectivity patterns were observed after the same visual task. Sumner et al. (2018a, 2018b) presented results resembling Fogelson's research in that they all focused on the visual stimuli paradigm.

3.1.3 Studies of Patients with Neuronal Deficiency

Studies focusing on epilepsy patients expanded the realm of DCM to finding the causal mechanism of neuropathology. Cooray et al. (2015) constructed network models based on the different neuron subpopulations from distinct cortical layers: excitatory granular cells, superficial and deep pyramidal cells, and inhibitory interneurons. This is distinctively different from the paper we mentioned before because it focused on the neuronal basis of all brain regions used as network nodes. Among the eight models, the author found that distinctive changes in excitatory–inhibitory balance were necessary to explain observed seizure activity and that these changes were conserved over seizures. This points out a possible path to investigate the regional effective connectivity.

Li et al. (2018) investigated different connectivity patterns of P300 tasks for healthy subjects and patients with schizophrenia. Comparing the winning models for the two groups, they discovered a flow from ACC to IPS was replaced by the directed flow from DLPFC to IPS in schizophrenia patients. Comparing connections among psychotic patients and healthy subjects, Ranlund et al. (2016) and Díez et al. (2017) also provided valuable insights using DCM for EEG (Table 2).

3.2 *Physical Neuroergonomics*

For physical neuroergonomics, fewer studies were conducted compared to that for cognitive neuroergonomics. Herz et al. (2012) explored the potential of using DCM to characterize context-specific changes in coupling within functional brain networks. By experimenting on 13 healthy subjects and their index finger extension–flexion movement and right arm isometric contraction, four models were built within LPM, M1, and SMA. Coupling parameters were inferred for the two tasks and distinct connectivity patterns revealed not only the frequency but also the spatial distribution of oscillatory coupling. He then expanded the study in dopamine-deficient patients and compared connection models with that of healthy subjects.

Similar studies like Auksztulewicz and Blankenburg (2013), Liu et al. (2017) and Kim, Park, Lee, Im, and Kim (2018) all investigated the underlying connectivity within healthy subjects. Research in Dopamine-deficient patients (Herz et al., 2013, 2014) and in stroke patients (Larsen et al., 2018) expanded on the research within healthy subjects. Studies on physical neuroergonomics generally choose the motor cortex area and the prefrontal cortex as their region of interest, and DCM results show directional connection within these areas. Investigations among healthy control and neurological disordered patients usually showed different connection patterns when conducting the same task. Future studies could explore the default network for physical actions and serve as a basis for further investigation into neuronal disorder and its connection with abnormal physical movement (Table 3).

4 Recommendations for Future DCM-Based Neuroergonomics Research

Effective connectivity reveals the directional relationship within ROIs and can explain the cause of neuroergonomics related phenomena. We introduced the DCM for EEG method to infer connectivity patterns underlying neuroergonomics topics. Compared with GCM and other methods studying effective connectivity, the DCM method's strength lies in that it is a spatiotemporal source reconstruction algorithm, which uses additional temporal constraints given by neural mass dynamics and long-range effective connectivity (Litvak et al., 2011). Combined with the high temporal resolution brought by EEG, DCM for EEG provides a glimpse of light into the unknown domain of the human brain. It employs an explicit forward or generative model of how observed data is caused and provides the most likely parameters for any given model, as well as the model evidence itself (Friston, 2009).

Table 2 Summary of exemplary EEG DCM-based cognitive neuroergonomics studies

	Mental task	Preprocessing	Subjects	ffx/rfx[a]	Model construction[b]
Garrido et al. (2007)	Auditory oddball	• **Epoch**—100–400 ms • **Filtering** 0.5–40 Hz • **Downsampled** 200 Hz • **Re-referenced** to bilateral earlobes	• 13 subjects (5F) • Age: 24–35	FFX	• F/B/FB connections between A1 (primary auditory cortex), STG (superior temporal gyrus), and IFG (inferior frontal gyrus)
Garrido et al. (2008)	Auditory oddball	• **Epoch**—100–400 ms • **Filtering** 0.5–40 Hz • **Downsampled** 200 Hz • **Robust averaging**	• 12 subjects (4F) • Age: 24–34	FFX	• F/B/Intrinsic connections between A1, STG and IFG
Garrido et al. (2009)	Auditory oddball	• **Epoch**—100–400 ms • **Filtering** 0.5–40 Hz • **Re-referenced** to the nose • **Robust averaging**	• 10 subjects (4F) • Age: 24–34	FFX	• F/B/Bilateral connections between A1, STG, and IFG
Brown and Friston (2012)	Visual changing sensory perception stimuli	• **Epoch**—100–400 ms • **Filtering** 0.5–45 Hz • **Downsampled** 200 Hz • **Referenced** to average	• 18 subjects (9F) • Age: 20–56	FFX	• Bidirectional connections between SOrbG (superior occipital gyrus), SOG (superior orbital gyrus), IPC (inferior parietal cortex), and IOG (inferior occipital gyrus)
Babajani-Feremi, Gumenyuk, Roth, Drake, and Soltanian-Zadeh (2012)	Auditory oddball	• **Epoch**—100–500 ms • **Filtering** 0.5–30 Hz • **Referenced** to average	• 8 normal; age: 24–36 • 9 habitual short sleepers; age: 25–45	FFX and RFX	• F/B/Intrinsic connections between A1, STG and IFG
Dietz, Roepstorff, and Garrido (2012)	Auditory location oddball		• 12 subjects • Age: 21–35	FFX	• Bilateral connections between A1, STG and IFG and IPL

(continued)

Table 2 (continued)

	Mental task	Preprocessing	Subjects	ffx/rfx[a]	Model construction[b]
Parkinson, Korzyukov, Larson, Litvak, and Robin (2013)	Self feedback auditory oddball	• **Epoch**—1–200 ms • **Referenced** to average	• 10 subjects (8F) • Age: 20.6 ± 1.8	RFX	• L/R/Bilateral connections between IFG, PMC (premotor cortex), and STG
Brown and Friston (2013)	Visual stimuli	• **Epoch**—50–400 ms • **Downsampled** 200 Hz • **Filtering** 2–32 Hz • **Robust averaging**	• 14 subjects (6F) • Age: 20–54	FFX	• Extrinsic excitatory modulation among PC (superior parietal sources), V2 (visual sources), V3 (bilateral dorsal), and V5 (occipitotemporal-parietal junction)
Silchenko et al. (2013)	Matching auditory stimuli	• **Referenced** to Cz • **Filtering** 0.8–130 Hz • **Artifact corrected**	• 28 patients treated with acoustic CR neuromodulation	FFX	• Connections among A1, A2 (primary auditory cortex), PCC (posterior cingulum), PAC (parietal cortex), DLPFC (dorsal lateral prefrontal cortex), OF (orbitofrontal cortex), and ACC (anterior cingulum)
Fogelson et al. (2014)	Predictable and unpredictable visual targets	• **Epoch**—150–1150 ms • **Bad trial exclusion** > 75 μV	• 25 normal; Age: 33.1 ± 2.1 • 25 schizophrenia patients; Age: 33.7 ± 2.7	FFX	• Forward connections between PC, V5, and V1 (primary visual cortex)
Shayegh, Amirfattahi, and Ansari-asl (2014)	Temporal lobe epilepsy occurrence	–	• 11 patients suffering from temporal lobe epilepsy		• Multisource/multichannel (MSMC) hippocampal model

(continued)

Table 2 (continued)

	Mental task	Preprocessing	Subjects	ffx/rfx[a]	Model construction[b]
Dietz, Friston, Mattingley, Roepstorff, and Garrido (2014)	Audio-spatial stimuli	• **Epoch**—100–400 ms • **Filtering** 0.5–30 Hz • **Downsampled** 250 Hz • **Referenced** to average • **Baseline corrected** • **Bad trial exclusion** > 80 μV	• 12 subjects (7F) • Age: 20–35	RFX	• Connections within A1, STG, IFG, and IPC
Hauser et al. (2014)	Probabilistic reversal learning task	• **Epoch**—100–700 ms • **Low-pass filter** 30 Hz • **Downsampled** 256 Hz • **Referenced** to average	• 15 subjects (9F) • Age: 25.7 ± 2.7	RFX	• Connections among vmPFC (ventral medial prefrontal cortex), DLPFC, amygdala, striatum, and dACC (anterior cingulum)
Cooray et al. (2015)	Seizure	• **Filtering** 0.5–50 Hz • **Referenced** to Fz	• 2 anti-NMDA-R encephalitis patients • Age: 19–31	FFX	• F/B connections among inhibitory, excitatory, deep pyramidal neuromas population
Legon et al. (2016)	Visual stimuli	• **Epoch**—400–1500 ms • **Filtering** 1–60 Hz • **Downsampled** 250 Hz • **Artifact** manually checked and disused	• 24 older; age: 65–78 • 30 younger subjects; age: 18–30	FFX	• F/B/FB connections among IFG, VAA (visual association area), and HPC (hippocampus)
Cooray et al. (2016)	Auditory oddball	• **Epoch**—100–500 ms • **Filtering** 0.5–30 Hz • **Downsampled** 250 Hz • **Referenced** to average	• 52 adolescences; age: 10–18 • 26 adults; age: 20–35	RFX	• Bidirectional/bilateral connections among A1, STG, and IFG

(continued)

Table 2 (continued)

	Mental task	Preprocessing	Subjects	ffx/rfx[a]	Model construction[b]
Ranlund et al. (2016)	Auditory oddball	• **Epoch**—100–300 ms • **Filtering** 0.5–70 Hz • **Downsampled** 200 Hz • **Referenced** to average • **Bad trial exclusion** > 70 μV	• 24 psychotic patients; age: 23–54 • 25 unaffected relatives; age: 16–62 • 35 healthy controls; age: 19–69	FFX	• Intrinsic connections among A1, STG, and IFG
Sato et al. (2017)	Face recognition	• **Epoch**—100–2000 ms • **Filtering** 0.5–120 Hz • **Downsampled** 250 Hz	• Patients with pharmacologically intractable focal epilepsy • Age: 34.5 ± 7.9	RFX	• Intrinsic connections among IOG and amygdala
Díez et al. (2017)	Auditory oddball	• **Epoch**—100–300 ms • **Filtering** 0.5–70 Hz • **Downsampled** 200 Hz	• 24 psychotic patients • 24 unaffected relatives • 25 healthy controls	–	• F/B connections among A1, STG, and IFG
Kibleur et al. (2017)	Emotional Stroop task with faces	• **Filtering** 0.6–30 Hz • **Downsampled** 250 Hz	• Unipolar TRD patients • Age: 52 ± 4	FFX	• Connections among LOC (lateral occipital cortex), MFC (medial frontal cortex), OFC (occipital frontal cortex), FG (fusiform gyrus), TP (temporal), and V1
Adebimpe, Bourel-Ponchel, and Ballois (2018)	Epilepsy with centrotemporal spikes	• **Epoch**—0–400 ms • **Filtering** 1–70 Hz • **Referenced** to average	• 12 BCECTS patients (5F) • Age: 9.38 ± 2.39	FFX	• F/B/FB connections among PFC (prefrontal cortex), rC, rTP (temporal), and rTPJ (temporal parietal junction)
Oestreich, Whitford, Garrido, and Garrido (2018)	Talk, passive listen, and cued listen conditions	• **Epoch**—100–0 ms • **Filtering** 0.5–30 Hz • **Referenced** to average	• 75 subjects (47F) • Age: 18–44	RFX	• Null/arcuate direct/indirect pathway among A1, SMA (supplementary motor area), Wernicke's area, and Broca's area

(continued)

Table 2 (continued)

	Mental task	Preprocessing	Subjects	ffx/rfx[a]	Model construction[b]
Tsai et al. (2018)	Steady state	• **Filtering** 0.5–80 Hz • **Downsampled** 250 Hz • **Artifacts rejection** based on ICA	• 14 patients (6F); age: 38.57 ± 13.91 • 14 healthy control (8F); age: 38.71 ± 11.78	–	• Connections among STG (superior and transverse temporal gyrus), MOG (middle occipital gyrus), SFG (superior frontal gyrus), and SPC (superior parietal cortex)
Penny, Iglesias-Fuster, Quiroz, Lopera, and Bobes (2018)	Paired visual stimuli	• **Epoch**—100–900 ms • **Filtering** 0.5–30 Hz • **Downsampled** 200 Hz • **Referenced** to average	• 15 NonCs • 13 PreCs	FFX	• Lateral connections among MTL (middle temporal cortex), IT (inferotemporal), and MOG (middle occipital gyrus)
Li et al. (2018)	Visual oddball	• **Epoch**—300–500 ms • **Filtering** 0.5–45 Hz • **Referenced** to average	• 24 schizophrenia patients (14F) • Age: 33.63 ± 8.36	FFX	• F/B connections among DLPFC, ACC, IPS (inferior parietal sulcus), and cuneus
Sumner et al. (2018a, 2018b)	Roving MMN task and the visual LTP[c] task	• **Epoch**—1000–4000 ms • **Filtering** 0.1–50 Hz • **Downsampled** 512 Hz	• 20 female subjects • Age: 21–23	FFX	• F/B/FB connections among A1 STG and IFG
Rachael Sumner et al. (2018)	Visual stimuli	• **Epoch**—0.5–1 ms • **Referenced** to average	• 20 female subjects • Age: 21–23	–	• Connections within superficial pyramidal, spiny stellate, and deep pyramidal cells

[a] *FFX* fixed effect analysis; *RFX* random effect analysis
[b] *F* forward connection; *B* backward connection; *FB* forward–backward connection
[c] *LTP* long-term plasticity

Table 3 Summary of exemplary EEG DCM-based physical neuroergonomics studies

	Physical task	Preprocessing	Participants	FFX/RFX[a]	Model construction[b]
Herz et al. (2012)	Index finger extension–flexion movement Right arm isometric contraction	• **Epoch**—100–2000 ms • **Filtering** 0.5–48 Hz • **Downsampled** 200 Hz	• 13 subjects (6F) • Age: 18–26	RFX	• Extrinsic/intrinsic connections within LPM (lateral premotor cortex), M1 (primary motor cortex), and SMA (supplementary motor area)
Auksztulewicz and Blankenburg (2013)	Electrical constant current pulse stimuli	• **Epoch**—100–400 ms • **Filtering** 48–52 Hz • **Downsampled** 512 Hz	• 20 subjects (9F) • Age: 22–33	RFX	• Connections between PMC (primary motor cortex), SII (secondary somatosensory cortex), and CSI (contralateral somatosensory cortex)
Herz et al. (2013)	Repetitive extension–flexion movements of right index finger	• **Epoch**—100–2000 ms • **Filtering** 0.5–48 Hz • **Downsampled** 200 Hz	• 13 healthy; Age: 58 ± 9.9 • 11 PD patients; age: 64 ± 7.2	RFX	• Bidirectional connection among PFC (prefrontal cortex), SMA, IPM (inferior parietal), and M1
Herz et al. (2014)	Repetitive movements of the right index finger	• **Epoch**—100–2000 ms • **Filtering** 0.5–48 Hz • **Downsampled** 200 Hz	• 12 healthy; age: 60.5 ± 9.4 • 11 patients; age: 63.8 ± 7.2	RFX	• Connections between left SMA, IPM, and M1
Loehrer et al. (2016)	Button press in a sequence	• **Filtering** 1–48 Hz • **Downsampled** 250 Hz	• 28 young; age: 25 ± 2.2 • 32 elderly; age: 60.1 ± 7.9	RFX	• Connections among PFC, SMA, IPM, and M1

(continued)

Table 3 (continued)

	Physical task	Preprocessing	Participants	FFX/RFX[a]	Model construction[b]
Liu et al. (2017)	Auditory cue followed by motor movement	• **Epoch**—200–800 ms • **Filtering** 0.01–100 Hz • **Referenced** to FCz • **Artifact rejected** ±50 μV	• 10 subjects (2F) • Age: 20–27		• F/B connections among PFC, SMA, IPM, and M1
Kim et al. (2018)	Motor imagery and execution	• **Epoch**—1000–4000 ms • **Filtering** 0.1–50 Hz • **Downsampled** 512 Hz	• 20 subjects (10F) • Age: 25.7 ± 3.1	FFX	• F/B connections between PFC, SMA, IPM, and M1
Larsen et al. (2018)	Grooved pegboard test	• **Epoch**—250–2250 ms • **Filtering** 5–48 Hz • **Downsampled** 256 Hz	• 15 patients after stroke (2F) • Age: 31–86	FFX	• Bilateral connections among PFC, SMA, IPM, and M1

[a] *FFX* fixed effect analysis; *RFX* random effect analysis
[b] *F* forward connection; *B* backward connection; *FB* forward–backward connection

DCM for EEG, however, still has room for exploration and improvement. Here are our recommendations for future DCM-based neuroergonomics research:

1. The poor spatial resolution limits the ROI identification and model construction in DCM for EEG. To address this limitation, DCM for other neuroimaging methods such as fMRI-EEG, fNIRS, are under study. We recommend future studies to combine high spatial resolution results with the high temporal resolution of EEG. We believe the simultaneous neuroimage recording would provide us with more comprehensive insight into the neural mechanisms underlying cognitive as well as physical tasks.
2. The DCM for EEG method is widely used to study evoked responses. Yet to understand the dynamic changes occurring during neuroergonomics tasks, steady-state effective connectivity should also be included in the picture. For this purpose, DCM for induced and steady-state responses should also be further investigated in the future. This would allow the further exploration of neuronal connections underlying static tasks.
3. So far, studies using DCM for EEG are limited to relatively simple experiment paradigms. To use DCM to its full potential, experiments that are more practical should be conducted and the interpretation of the DCM results should be in the context of the experiments.

For neuroergonomics study, the use of DCM could expand our current understanding of the cognitive processes such as working memory, motor imagery, image recognition, and so on. It could help develop the "default mode" network for healthy subjects and in turn help identify causes for other neuronal diseases. DCM for EEG is a relatively new method, yet a rapidly developing one. We look forward to seeing much more accomplished with the help of this method.

References

Acs, F., & Greenlee, M. W. (2008). Connectivity modulation of early visual processing areas during covert and overt tracking tasks. *NeuroImage, 41,* 380–388.

Adebimpe, A., Bourel-Ponchel, E., & Wallois, F. (2018). Identifying neural drivers of benign childhood epilepsy with centrotemporal spikes. *Neuroimage-Clinical, 17*(December), 739–750.

Ashburner, J., Barnes, G., Chen, C., Daunizeau, J., Moran, R., Henson, R., ... Phillips, C. (2013). SPM12 manual the FIL methods group (and honorary members). *Functional Imaging Laboratory*, 475–1.

Auksztulewicz, R., & Blankenburg, F. (2013). Subjective rating of weak tactile stimuli is parametrically encoded in event-related potentials. *Journal of Neuroscience, 33*(29), 11878–11887.

Babajani-Feremi, A., Gumenyuk, V., Roth, T., Drake, C. L., & Soltanian-Zadeh, H. (2012). Connectivity analysis of novelty process in habitual short sleepers. *NeuroImage, 63*(3), 1001–1010.

Brown, H. R., & Friston, K. J. (2012). Dynamic causal modelling of precision and synaptic gain in visual perception—An EEG study. *Neuroimage, 63*(1), 223–231.

Brown, H. R., & Friston, K. J. (2013). The functional anatomy of attention: A DCM study. *Frontiers in Human Neuroscience, 7.*

Chen, C. C., Henson, R. N., Stephan, K. E., Kilner, J. M., & Friston, K. J. (2009). Forward and backward connections in the brain: A DCM study of functional asymmetries. *NeuroImage, 45*(2), 453–462.

Cooray, G. K., Garrido, M. I., Brismar, T., & Hyllienmark, L. (2016). The maturation of mismatch negativity networks in normal adolescence. *Clinical Neurophysiology, 127*(1), 520–529.

Cooray, G. K., Sengupta, B., Douglas, P., Englund, M., Wickstrom, R., & Friston, K. (2015). Characterising seizures in anti-NMDA-receptor encephalitis with dynamic causal modelling. *Neuroimage, 118*, 508–519.

David, O., & Friston, K. J. (2003). A neural mass model for MEG/EEG: Coupling and neuronal dynamics. *NeuroImage, 20*(3), 1743–1755.

Dietz, M. J., Friston, K. J., Mattingley, J. B., Roepstorff, A., & Garrido, M. I. (2014). Effective connectivity reveals right-hemisphere dominance in audiospatial perception: Implications for models of spatial neglect. *Journal of Neuroscience, 34*(14), 5003–5011.

Dietz, M., Roepstorff, A., & Garrido, M. (2012). *Functional asymmetry in the brain's response to left and right auditory space: A DCM study (1999).*

Díez, Á., Ranlund, S., Pinotsis, D., Calafato, S., Shaikh, M., Hall, M.-H. H., … Bramon, E. (2017). Abnormal frontoparietal synaptic gain mediating the P300 in patients with psychotic disorder and their unaffected relatives. *Human Brain Mapping, 38*(6), 3262–3276.

Fogelson, N., Litvak, V., Peled, A., Fernandez-del-Olmo, M., & Friston, K. (2014). The functional anatomy of schizophrenia: A dynamic causal modeling study of predictive coding. *Schizophrenia Research, 158*(1–3), 204–212.

Friston, K. (2009). Causal modelling and brain connectivity in functional magnetic resonance imaging. *PLoS Biology, 7*(2), 0220–0225.

Friston, K. J. (2011). Functional and effective connectivity: A review. *Brain Connectivity, 1*(1), 13–36.

Friston, K. J., Glaser, D. E., Henson, R. N. A., Kiebel, S., Phillips, C., & Ashburner, J. (2002). Classical and Bayesian inference in neuroimaging : Applications, *512*, 484–512.

Garrido, M. I., Friston, K. J., Kiebel, S. J., Stephan, K. E., Baldeweg, T., & Kilner, J. M. (2008). The functional anatomy of the MMN: A DCM study of the roving paradigm. *NeuroImage, 42*(2), 936–944.

Garrido, M. I., Kilner, J. M., Kiebel, S. J., Stephan, K. E., Baldeweg, T., & Friston, K. J. (2009). Repetition suppression and plasticity in the human brain. *Neuroimage, 48*(1), 269–279.

Garrido, M. I., Kilner, J. M., Kiebel, S. J., Stephan, K. E., & Friston, K. J. (2007). Dynamic causal modelling of evoked potentials: A reproducibility study. *Neuroimage, 36*(3), 571–580.

Hauser, T. U., Iannaccone, R., Staempfli, P., Drechsler, R., Brandeis, D., Walitza, S., & Brem, S. (2014). The feedback-related negativity (FRN) revisited: New insights into the localization, meaning and network organization. *Neuroimage, 84*, 159–168.

Herz, D. M., Christensen, M. S., Reck, C., Florin, E., Barbe, M. T., Stahlhut, C., … Timmermann, L. (2012). Task-specific modulation of effective connectivity during two simple unimanual motor tasks: A 122-channel EEG study. *Neuroimage, 59*(4), 3187–3193.

Herz, D. M., Florin, E., Christensen, M. S., Reck, C., Barbe, M. T., Tscheuschler, M. K., … Timmermann, L. (2014). Dopamine replacement modulates oscillatory coupling between premotor and motor cortical areas in Parkinson's disease. *Cerebral Cortex, 24*(11), 2873–2883.

Herz, D. M., Siebner, H. R., Hulme, O. J., Florin, E., Christensen, M. S., & Timmermann, L. (2013). Levodopa reinstates connectivity from prefrontal to premotor cortex during externally paced movement in Parkinson's disease. *Neuroimage, 90,* 15–23.

Jansen, B. H., & Rit, V. G. (1995). Electroencephalogram and visual evoked potential generation in a mathematical model of coupled cortical columns. *Biological Cybernetics, 73*(4), 357–366.

Kibleur, A., Polosan, M., Favre, P., Rudrauf, D., Bougerol, T., Chabardes, S., & David, O. (2017). Stimulation of subgenual cingulate area decreases limbic top-down effect on ventral visual stream: A DBS-EEG pilot study. *Neuroimage, 146*, 544–553.

Kiebel, S., David, O., Friston, K. J., Kiebel, S. J., David, O., & Friston, K. J. (2006). Dynamic causal modelling of evoked responses in EEG/MEG with lead field parameterization. *Neuroimage, 30*(4), 1273–1284.

Kiebel, S. J., Garrido, M. I., Moran, R., Chen, C.-C., & Friston, K. J. (2009). Dynamic causal modeling for EEG and MEG. *Human Brain Mapping, 30*(6), 1866–1876.

Kim, Y. K., Park, E., Lee, A., Im, C. H., & Kim, Y. H. (2018). Changes in network connectivity during motor imagery and execution. *PLoS ONE, 13*(1), 1–18.

Knösche, T. (2013). Jansen-Rit model. *Encyclopedia of Computational Neuroscience, 2*(1), 1–5.

Larsen, L. H., Zibrandtsen, I. C., Wienecke, T., Kjaer, T. W., Langberg, H., Nielsen, J. B., & Christensen, M. S. (2018). Modulation of task-related cortical connectivity in the acute and subacute phase after stroke. *European Journal of Neuroscience, 47*(8), 1024–1032.

Legon, W., Punzell, S., Dowlati, E., Adams, S. E., Stiles, A. B., & Moran, R. J. (2016). Altered prefrontal excitation/inhibition balance and prefrontal output: Markers of aging in human memory networks. *Cerebral Cortex, 26*(11), 4315–4326.

Li, F., Wang, J., Jiang, Y., Si, Y., Peng, W., Song, L., … Xu, P. (2018). Top-down disconnectivity in schizophrenia during P300 tasks. *Frontiers in Computational Neuroscience, 12*.

Litvak, V., Mattout, J. J., Kiebel, S., Phillips, C., Henson, R., Kilner, J., … Friston, K. (2011). EEG and MEG data analysis in SPM8. In *Computational Intelligence and Neuroscience, 2011* (June 2016).

Liu, T., Li, F., Jiang, Y., Zhang, T., Wang, F., Gong, D., … Qiu, K. (2017). Cortical dynamic causality network for auditory-motor tasks. *IEEE Transactions of Neural Systems and Rehabilitation Engineering, 25*(8), 1092–1099.

Loehrer, P. A., Nettersheim, F. S., Jung, F., Weber, I., Huber, C., Dembek, T. A., … Timmermann, L. (2016). Ageing changes effective connectivity of motor networks during bimanual finger coordination. *Neuroimage, 143*, 325–342.

Oestreich, L. K. L., Whitford, T. J., Garrido, M. I., & Garrido I. M. (2018). Prediction of speech sounds is facilitated by a functional fronto-temporal network. *Frontiers in Neural Circuits, 12*(May), 1–10.

Parkinson, A. L., Korzyukov, O., Larson, C. R., Litvak, V., & Robin, D. A. (2013). Modulation of effective connectivity during vocalization with perturbed auditory feedback. *Neuropsychologia, 51*(8), 1471–1480.

Penny, W., Iglesias-Fuster, J., Quiroz, Y. T., Lopera, F. J., & Bobes, M. A. (2018). Dynamic causal modeling of preclinical autosomal-dominant Alzheimer's disease. *Journal of Alzheimers Disease, 65*(3), 697–711.

Peterson, D. S., & Fling, B. W. (2018). How changes in brain activity and connectivity are associated with motor performance in people with MS. *NeuroImage: Clinical, 17*(June 2017), 153–162.

Ranlund, S., Adams, R. A., Díez, Á., Constante, M., Dutt, A., Hall, M.-H. H., … Bramon, E. (2016). Impaired prefrontal synaptic gain in people with psychosis and their relatives during the mismatch negativity. *Human Brain Mapping, 37*(1), 351–365.

Sato, W., Kochiyama, T., Uono, S., Matsuda, K., Usui, K., Usui, N., … Toichi, M. (2017). Bidirectional electric communication between the inferior occipital gyrus and the amygdala during face processing. *Human Brain Mapping, 38*(9), 4511–4524.

Shayegh, F., Amirfattahi, R., & Ansari-asl, K. (2014). Hippocampal effective synchronization values are not pre-seizure indicator without considering the state of the onset channels. *Network: Computation in Neural Systems* (July).

Silchenko, A. N., Adamchic, I., Hauptmann, C., & Tass, P. A. (2013). Impact of acoustic coordinated reset neuromodulation on effective connectivity in a neural network of phantom sound NeuroImage Impact of acoustic coordinated reset neuromodulation on effective connectivity in a neural network of phantom sound. *NeuroImage, 77*(March).

Stephan, K. E., Penny, W. D., Daunizeau, J., Moran, R. J., & Friston, K. J. (2009). Bayesian model selection for group studies. *Neuroimage, 46*(4), 1004–1017.

Stephan, K. E., Penny, W. D., Moran, R. J., den Ouden, H. E. M. M., Daunizeau, J., & Friston, K. J. (2010). Ten simple rules for dynamic causal modeling. *Neuroimage, 49*(4), 3099–3109.

Stephan, K. E., Weiskopf, N., Drysdale, P. M., Robinson, P. A., & Friston, K. J. (2007). Comparing hemodynamic models with DCM. *NeuroImage, 38,* 387–401.

Sumner, R. L., Mcmillan, R. L., Shaw, A. D., Singh, K. D., Sundram, F., & Muthukumaraswamy, S. D. (2018a). Peak visual gamma frequency is modified across the healthy menstrual cycle. *Human Brain Mapping* (March), 3187–3202.

Sumner, R. L., Spriggs, M. J., Mcmillan, R. L., Sundram, F., Kirk, I. J., & Muthukumaraswamy, S. D. (2018b). Neural plasticity is modified over the human menstrual cycle: Combined insight from sensory evoked potential LTP and repetition suppression. *Neurobiology of Learning and Memory, 155*(August), 422–434.

Tsai, M.-C., Cai, Y.-X., Wang, C.-D., Zheng, Y.-Q., Ou, J.-L., & Chen, Y.-H. (2018). Tinnitus abnormal brain region detection based on dynamic causal modeling and exponential ranking. *Biomed Research International.*

Neuroergonomics Assessments of Cognitive and Physical Performance

Physical Activity and Sedentary Behavior Influences on Executive Function in Daily Living

Brett Baker and Darla Castelli

Abstract Today, work and learning environments are obesogenic, as individuals of all ages are seated for more than 6 h a day. This is despite the emerging evidence that increased cerebral blood flow may enhance executive function, immediately following a brief bout of physical activity. This chapter will provide an overview of the direct and indirect effects of physical activity from both acute and chronic perspectives. In addition, brain imaging evidence will be described. Specifically, one subsection would overview the effects of sedentary (i.e., sitting) and active behavior (i.e., standing and walking) on O_2 uptake and neural activation on brain function and structure.

1 Introduction: Physical Activity and Public Health

Over the last few decades, physical inactivity has been increasingly recognized as a global health problem. Since the industrial revolution, the development of new technologies has enabled people to reduce the amount of physical labor needed to accomplish many tasks in their daily lives (Hallal et al., 2012). The use of many of these technologies has been driven by the goal of increased individual worker productivity and reduced physical hardships caused by jobs entailing heavy labor. Fewer jobs today require physical work as the United States (US) has changed from an industrial to a service-based economy, with increased labor-saving technology, increased self-sufficiency in the home and reduced a percentage of individuals walking and cycling (World Health Organization [WHO], 2011). Although the technological revolution has been of great benefit to many populations throughout the world, it has come at a significant cost in terms of the contribution of physical inactivity to the worldwide epidemic of noncommunicable diseases (WHO, 2010). In 2009, physical inactivity was identified as the fourth leading risk factor for noncommunicable diseases and accounted for more than 3 million preventable deaths (WHO, 2009).

B. Baker · D. Castelli (✉)
The University of Texas at Austin, Austin, TX, USA
e-mail: dcastelli@utexas.edu

C. S. Nam (ed.), *Neuroergonomics*, Cognitive Science and Technology,
https://doi.org/10.1007/978-3-030-34784-0_9

The rise in physical inactivity has led to an increased prevalence of obesogenic environments for all age groups, thus increasing the likelihood of cardiovascular disease and cognitive impairment risk factors manifesting. Physical activity, a behavior requiring energy expenditure and use of large skeletal muscles (e.g., gardening, walking, running; ACSM, 2017), is one of the most modifiable disease risk markers for the prevention of cardiovascular and brain diseases. Yet, over the last 50 years, the rate of physical activity participation has declined exponentially (Fox & Hillsdon, 2007), with the overall average reduction in energy expenditure being roughly estimated to be 250–500 kcal/day (Fox & Hillsdon, 2007). This decrease in energy expenditure can cause reductions in insulin sensitivity and chronically high levels of circulating glucose and insulin in the bloodstream, contributing to cardiovascular risk factors and cognitive impairment. To combat the adverse health effects of physical inactivity, the Centers for Disease Control and Prevention (CDC) created the physical activity guidelines. First implemented in 2008, and revised in 2018, these guidelines describe what type, intensity, and amount of physical activity are believed to produce health-enhancing effects. The guidelines are commonly utilized to delineate someone who is "physically active" from someone who is not. For optimal cardiovascular health benefits, the CDC currently states that adults over the age of 18 should do at least 150 min a week of moderate-intensity or 75 min a week of vigorous-intensity aerobic physical activity, or an equivalent combination of moderate- and vigorous-intensity aerobic activity. However, the CDC currently lists the percentage of Americans meeting the recommended physical activity guidelines at 23%. Knowing that far too few people meet the physical activity guidelines is disturbing considering the known, empirical benefits of physical activity participation on cognitive functioning. Research suggests that one session of physical activity (i.e., acute) can increase cerebral blood flow, attention, and arousal as well as upregulate endorphin release. Chronic physical activity (i.e., regular or routine participation in physical activity, like running each day) can lead to cerebrovascular remodeling (Szostak & Laurant, 2011; Voss, Carr, Clark, & Weng, 2014) which includes angiogenesis (Voss et al., 2014), arteriogenesis (Voss et al., 2014) and neurogenesis (Redila & Christie, 2006; Schinder & Poo, 2000; Van der Borght et al., 2009; Van Praag, Kempermann, & Gage, 1999). As will be discussed further in detail later, physically active individuals significantly outperform their age-matched physically inactive counterparts during executive function tests of inhibition, cognitive flexibility, and working memory.

What is becoming an intriguing line of research is the physical activity paradigm (Fig. 1). Once synonymous with physical inactivity, sedentary behavior, or the very low expenditure of energy ($\leq$1.5 metabolic equivalents (METs) such as sitting and lying down) is being increasingly recognized as its own state on the physical activity continuum (Tudor-Locke, Craig, Thyfault, & Spence, 2013). Emerging evidence supports the notion that sedentary behavior is independent of exercise and other moderate-to-vigorous intense physical activities, suggesting that one can be highly active yet also be sedentary for most of one's waking hours (Craft et al., 2012). For example, an individual may go for a morning 3-mile run, but then be seated on the job for eight hours during the day. An individual can be physically active and sedentary, as seen in the above example, and someone can also be physically

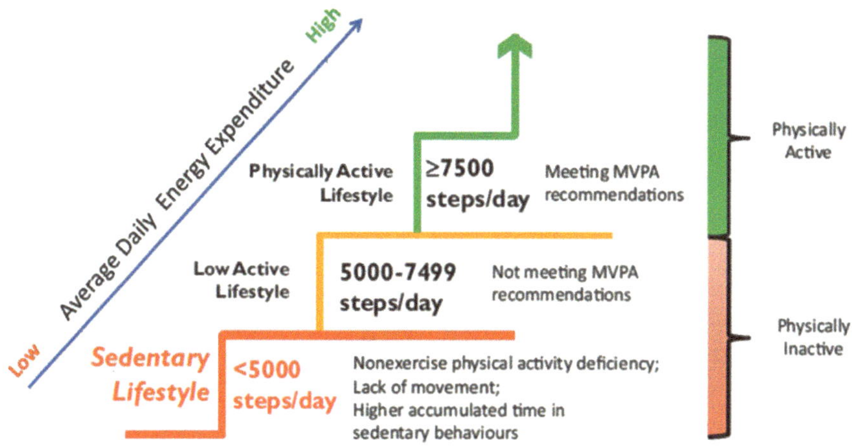

Fig. 1 Step-defined sedentary lifestyle index for adults, MVPA, moderate-to-vigorous physical activity. Applied physiology, nutrition, and metabolism by Canadian Science Publishing. Reproduced with permission of ©Canadian Science Publishing in the format Republish in a book via Copyright Clearance Center

inactive and sedentary. The delineation between someone who is physically active and someone who is physically inactive is whether or not they meet the physical activity guidelines.

Furthermore, sedentary behavior has been observed as a risk factor for several chronic diseases independent of physical activity (Duvivier et al., 2013; Hamilton, Hamilton, & Zderic, 2007; Healy, Matthews, Dunstan, Winkler, & Owen, 2011; Schmidt et al., 2002). Sedentary behavior has been shown to suppress lipoprotein lipase; suppression of lipoprotein lipase results in ineffective triglyceride metabolism and increases in visceral fat (Hamilton et al., 2007). In other words, without energy expenditure, it will be stored in the body as adipose tissue or fat. This empirical evidence suggests that sedentary behavior is associated with less efficient glucose metabolism (e.g., greater 2-h glucose plasma after glucose tolerance challenge), reduced insulin sensitivity, and insulin resistance (Duvivier et al., 2013; Healy et al., 2011). The evidence is also available coupling obesity-related inflammation to cognitive dysfunction. Inflammation has been shown to predict a higher risk of dementia (Schmidt et al., 2002) and many inflammatory biomarkers have been associated with reduced total brain volume (Jefferson et al., 2007) and cognition (Athilingam et al., 2013; Yaffe et al., 2003) in cross-sectional studies. Sedentary behavior also has shown to suppress neurogenesis and neurotrophic growth factors (Jefferson et al., 2007). Even though physical activity can be a protective factor against diseases of the brain, sedentary behavior may counteract the positive impact of physical activity on brain health if sedentary behavior and physical activity share overlapping molecular pathways (Ahlskog, Geda, Graff-Radford, & Petersen, 2011; Voss et al., 2014). In short, both acute and chronic physical activity participation has metabolic, physiologic, and cognitive benefits, whereas sedentary behaviors not only do not have benefits but likely also may have deleterious effects.

The purposes of this chapter are as follows: (1) to review the effects of both physical activity and physical fitness on cognition and executive functioning, (2) to summarize the influences of sedentary behavior on cognition, (3) to examine limitations and measurement issues with measuring cognition during activity, (4) and to propose future directions and research strategies to optimally determine the influences of physical activity and sedentary behavior on brain health.

2 Physical Activity and Executive Function

Much research has determined that both acute and chronic bouts of physical activity affect cognition. The amount of research conducted among children and older adults has grown exponentially in the last 10 years as the primary age group of interests. Studies investigating an acute bout of physical activity on young adults, as a distinct portion of the life span, are minimal as it is well established that executive functioning peaks in young adulthood (Åberg et al., 2009; Hansen, Johnsen, Sollers, Stenvik, & Thayer, 2004; Kamijo & Takeda, 2010; Themanson & Hillman, 2006; Themanson, Pontifex, & Hillman, 2008). Because of this, the cognitive effects of an acute bout of physical activity may be minimal in younger adults. Studies that do employ young adults are typically using them as a reference group for cognitive changes across the life span when comparing to an older adult group. However, in general, the findings from several reviews suggests that there is a positive relationship between acute physical activity and cognitive functioning (Brisswalter, Collardeau, & René, 2002; McMorris & Graydon, 2000; Tomporowski, 2003a, 2003b). A recent meta-analytic review by Lambourne and Tomporowski (2010) concluded that acute exercise in young adults improved cognition tested after exercise ($d = 0.20$). Moreover, a systematic review commissioned by the American College of Sports Medicine (ACSM) that examined 73 studies, further confirmed that the effects of physical activity need to continue to be explored, specifically by physical activity type, amount, frequency, and timing, despite the global support for the positive benefits (Donnelly et al., 2016). The question is what do we specifically know and do not know? And what are the best methods for quantifying the relationships between and among the variables of interest?

2.1 Physical Activity, Task Switching, and Response Inhibition

When measuring cognition, most studies utilize executive function tests. Executive function is a group of higher order cognitive processes that include inhibition, attention, planning, mental flexibility, and working memory (Chan, Shum, Toulopoulou,

& Chen, 2008). Executive functions are essential as they peak in young adulthood and deteriorate with age. Common executive functions include task switching, inhibition, and working memory. Task switching and inhibition are discussed here while working memory is discussed in the next subsection. We chose to include these executive functions as they are heavily researched and are very relevant cognitive functions for everyday life. Individuals are constantly having to inhibit actions and behaviors, switch back and forth between several tasks, and store and update information for performance on a future task.

One executive function is cognitive flexibility which is measured as task switching. Similarly, if not synonymous with mental flexibility, task switching requires individuals to respond to the same stimuli but with different sets of rules (Guiney & Machado, 2013). An example is as follows: if you see a number that is blue, say whether the number is higher or less than 25; however, if you see a number that is yellow, say whether the number is odd or even. There are non-switch trials (e.g., congruent with one's expectancy) and switch trials (e.g., incongruent with one's expectancy) that facilitate examination of neural substrates of cognitive control in the prefrontal cortex. The non-switch trials are when one follows the original rules as described above. The switch trials are when the rules are modified from the original request response. The difference in the reaction times between the non-switch and the switch trials is called the switching costs. Switching costs describe the time it takes for someone to mentally switch between rules (Guiney & Machado, 2013); smaller switching costs describe more efficient executive functioning (Banich, 2009; Monsell, 2003). Thus, researching the effects of physical activity, whether acute or chronic, rely heavily on the speed of information processing and the corresponding behavior response (e.g., button click, item selection). Further, most research requires a visual or audio prompt and a motor response. We include this detail in our description because the type of behavioral response is perceived as both a strength and a limitation within the existing body of literature.

Cross-sectional studies comparing both young and older adults have revealed that switching costs are not associated with self-reported physical activity volume per week (Hillman, Kramer, Belopolsky, & Smith, 2006; Themanson, Hillman, & Curtin, 2006). However, event-related potential data from Hillman et al. (2006) exhibited shorter P3 latencies among more active participants during task switching when compared to age-matched sedentary counterparts (Hillman et al., 2006). In a study using functional near-infrared spectroscopy (fNIR), they identified age differences in the behavior response tasks, suggesting that older adults were significantly slower than in young adults (Vasta et al., 2018). Brain oxygen–hemoglobin concentration (HbO_2) was also significantly different during task switching tasks by age. Further, it was suggested that fNIRS were a viable measure of O_2 uptake in the brain when examining the effects of task switching. These findings are possibly the result of the sensitivity of the physical activity measure, timing of the cognitive testing about the physical activity, and the type of executive function targeted. To date, there are no known studies that have examined task switching after physical activity participation using fNIR technology.

Response inhibition is another executive function and is commonly measured by either the Stroop task or the Eriksen flanker task. Response inhibition is an executive function that requires participants to suppress prepotent responses in regard to particular stimuli. For the Stroop task, participants are required to indicate the color of ink that a word appears in Guiney and Machado (2013). The Stroop task has both noninterference and interference trials. Noninterference trials, for example, would be when the word "red" is written in red ink. Interference trials, for example, would be when the word "red" is written in black ink. The interference trials of the Stroop task are considered a successful measure of executive functioning because individuals must inhibit themselves from calling out the color of the word and selectively attend to the ink color of the word (Miyake et al., 2000).

The flanker task asks participants to specify the direction of a centralized stimulus while ignoring stimuli located adjacent to the central stimulus (these peripheral stimuli are known as "flankers") (Guiney & Machado, 2013). One of the more commonly used flanker approaches is to present a participant with five arrowheads and asking them to specify the direction of the central arrowhead by pressing either the left or right shift key. There are both congruent and incongruent flanker trials. During the congruent trials, the "flankers" are pointed in the same direction as the central stimulus (e.g., > > > > >); during the incongruent trials, the "flankers" are pointed in the opposite direction of the central stimulus (e.g., > > < > >). Incongruent trials are typically associated with slower reaction times and less accurate responses due to the distracting flankers in the periphery. The difference in both reaction time and accuracy between the congruent and incongruent trials is known as the flanker effect (Guiney & Machado, 2013); with smaller flanker effects reflecting more efficient executive functioning (Callejas, Lupiáñez, & Tudela, 2004).

The results of cross-sectional studies utilizing older adults showed that higher levels of aerobic fitness were associated with more substantial activation in the prefrontal cortex (using fMRI), greater accuracy, and lower amounts of interference (faster reaction times in the interference trials) during the Stroop task (Prakash et al., 2011). A cross-sectional study utilizing older adults demonstrated that those with high aerobic fitness scores had lower flanker effects than those with lower aerobic fitness scores (Colombe et al., 2004). In addition, fMRI data highlighted that during the flanker task, older adults with high aerobic fitness scores had more extensive activation in brain regions associated with regulating selective attention (middle and superior frontal gyri). Switching costs, flanker effects, and Stroop interference are all known to increase with age (Zhu, Zacks, & Slade, 2010). Although there are no known applications of using fNIR technologies during gross motor, higher intensity physical activity, several studies have observed an increase in prefrontal cortex activation during the Stroop task after chronic and acute bouts of physical activity; specifically, during the interference trials when compared to the noninterference trials (Ehlis, Herrmann, Wagener & Fallgatter, 2005; Zysset, Schroeter, Neumann, & von Cramon, 2007). Because of physiological and motion artifacts generated by the subject, deployment of fNIR-based technologies into the wild has been questioned. In 2015, a systematic review investigating imagined walking and real-world walking revealed that gait increases brain activity in prefrontal, premotor, and motor cortices as determined by fNIR, EEG, and PET techniques (Hamacher, Hamacher, Rehfeld, Hökelmann, & Schega, 2015; Hamacher, Herold, Wiegel, Hamacher, & Schega, 2015). Typically, motor areas of the brain experience high activation during walking (Hanakawa et al., 1999). Findings from experiments examining the effects of dual tasks (e.g., imaging while walking and talking) have shown that activation in the prefrontal cortex increases when the walking condition includes a cognitive task (Holtzer et al., 2011; Ohsugi, Ohgi, Shigemori, & Schneider, 2013).

Conversely, no significant differences in brain activity were detected in walking and talking versus standing and talking conditions (Beurskens, Helmich, Rein, & Bock, 2014). The findings in these studies could have been explained by confounding variables such as the difficulty of the task, the type of imaging utilized, and the walking speed. Generally, using fNIRs to examine brain function before, during, and after exercise is feasible if the motion and physiological artifacts are measured and controlled (Giles et al., 2014).

2.2 *Physical Activity and Working Memory*

Working memory is an executive function that requires holding information and updating that information as needed to respond accurately to a subsequent task (Baddeley & Hitch, 1974). A frequently used working memory test is the n-back task. The n-back task has several variations, including the 1-back, 2-back, and 3-back tests. Individuals are shown a sequence of letters, with each letter being displayed individually. For example, in the 2-back paradigm, individuals are asked whether the currently displayed letter matches the letter that was displayed two letters ago (Cohen et al., 1994). The n-back task is a good measure of working memory as it requires participants to continually update information and hold onto that information to respond correctly (Smith & Jonides, 1997). To date, only a few studies have used fNIR technology to determine the effects of an acute bout of physical activity on cognitive function. Acute submaximal exercise increased HbO_2 about cognitive workload (Bediz et al., 2016). Participants completed a 2-back test before and after a submaximal Wingate anaerobic cycling test which lasted 30 s. HbO_2 levels post exercise were significantly higher as were 2-back performance scores over pretest values. Further, peak power measures were correlated with oxy-, deoxy-, and total Hb measures.

Cross-sectional studies in older adults (Voelcker-Rehage, Godde, & Staudinger, 2010) and young adults (Hansen et al., 2004) have shown that aerobic fitness is a significant predictor of response accuracy during the n-back task. Also, not measured based off of physical activity, studies have shown that increasing the n-back demand (e.g., from 1 to 2 to 3 back) significantly increases prefrontal cortex activity (Molteni et al., 2012).

Another working memory test is the number span test. In the forward span test, participants are presented with numbers one at a time and are then asked to correctly recall the entire sequence (Guiney & Machado, 2013). In the backward span test, participants are asked to correctly recall the sequence in the reverse order. In both the forward span and the backward span tests, the length of the presented numbers is increased until the responses become incorrect. A cross-sectional study in older adults determined that higher levels of physical activity were associated with more extended sequences of numbers during the backward span test (Weuve et al., 2004).

3 Physical Fitness and Executive Function

While physical activity and sedentarism are behaviors, physical fitness is considered to be a stable trait. To achieve a state of health and well-being where one can carry out tasks without fatigue, considered to be physical fitness, one must participate in the minimum recommendations outlined in the 2018 Physical Activity Guidelines, of 150 min of moderate-to-vigorous engagement per week, thus reaping the health-enhancing benefits of habitual participation. Globally, physically fit individuals exhibit more efficient cognitive responses than unfit individuals and, because cerebral blood flow declines with age, sedentary adults.

3.1 Physical Fitness and Response Inhibition

As described earlier, response inhibition is one's ability to suppress an expected response. Using multichannel fNIRs and Stroop test measures, the effects of an acute moderate-intensity bout of cycling were examined among 20 healthy adults (Yanagisawa et al., 2010). Findings suggest that enhanced activation of the prefrontal cortex resulted from participation in the physical activity produced improved interference score. Cardiorespiratory protective factors emerge from the achievement and maintenance of physical fitness, most robustly among older adults. Evidence of such effects were discovered when high-fit women demonstrated significantly higher HbO_2 than low-fit women when performing controlled counting (Albinet, Mandrick, Bernard, Perrey, & Blain, 2014). In another study, higher fitness has also been linked to performance on the Stroop test, when high-fit women were faster than low-fit women (Dupuy et al., 2015).

4 Sedentary Behaviors and Cognition

Mentioned earlier, much evidence is starting to converge on the idea that sedentary behavior is a distinct activity separate from physical inactivity. Being physically inactive is described as someone who does not meet the physical activity guidelines that were outlined earlier, whereas sedentary behavior is any task or activity performed while in a sitting or reclining posture. Because of this definition, everyone is susceptible to prolonged bouts of sedentary behavior; one who is highly active can still be sedentary for most of their day (Craft et al., 2012). Prolonged sitting and total sitting time are both adversely associated with a higher risk of chronic diseases (Hamilton et al., 2007), metabolic health (Healy et al., 2008) and premature mortality (Dunstan et al., 2012). Furthermore, it has been shown that all-cause mortality is increased by 15% for individuals who sit for more than 8 h per day and by 40% for those who sit for more than 11 h per day when compared to more physically active people (Van der Ploeg, Chey, Korda, Banks, & Bauman, 2012). Even though physical activity can be a neuroprotective factor against diseases of the brain (Ahlskog et al., 2011), sedentary behavior may counteract the positive impact of physical activity on brain health if sedentary behavior and physical activity share overlapping molecular pathways (Voss et al., 2014).

It is not yet well known how sedentary behavior affects cognition or brain health, as well as controlled experimental studies, are lacking in human research. What we do know of the relationship between sedentary behavior and cognition comes from cross-sectional studies and experimental studies involving rodents. It is hypothesized that sedentary behavior influences cognition through similar mechanisms as physical activity. This comes from existing evidence that sedentary behavior influences other physiological mechanisms through the same molecular processes as physical activity; which we will turn to now.

It is possible that physical activity and sedentary behavior have overlapping mechanisms on the brain structure as they both affect one's energy balance (Stranahan & Mattson, 2008, 2011); sedentary behavior promotes insulin resistance and fat deposits while physical activity decreases insulin resistance and fat deposits (Lees and Booth, 2004). The absence of muscle contractions due to sedentary behavior can suppress lipoprotein lipase (LPL) (Hamilton et al., 2007) which is an essential enzyme for regulating triglyceride metabolism. This loss of LPL activity will increase fat accumulation due to poor triglyceride metabolism (Voss et al., 2014) which will then increase leptin release. Leptin is a crucial hormone for appetite regulation (Shizgal & Hyman, 2013) through its binding sites in the hypothalamus. In addition to the hypothalamus, binding sites for leptin also exist in the hippocampus (Håkansson, Brown, Ghilardi, Skoda, & Meister, 1998). Highlighted by Moult et al. (2010) and Shanley, Irving, and Harvey (2001), leptin can influence hippocampal synapse plasticity through its role in regulating glutamate activity (Moult et al., 2010) and long-term potentiation (Shanley et al., 2001). Therefore, insufficient fat metabolism and increased levels of leptin can cause built up leptin resistance, decreased levels of leptin binding in the hippocampus, and alleviation of synapse plasticity in the hippocampus (Knight, Hannan, Greenberg, & Friedman, 2010).

In addition to sedentary behavior-induced reductions in LPL activity, there is compelling evidence that sedentary behavior is associated with less efficient glucose clearance and increased insulin resistance (Duvivier et al., 2013; Healy et al., 2011). This is important as physical activity is known to decrease insulin resistance and fasting blood glucose levels. Insulin receptors are also found in the hippocampus, and it is presumed that insulin resistance can occur in the brain. Like leptin, insulin also affects hippocampal synapse plasticity (Messier & Teutenberg, 2005). These data are supported by the decreases seen in memory, learning, and hippocampal atrophy in diabetic patients with insulin resistance (Craft, 2009; Williamson, McNeilly, & Sutherland, 2012).

Chronic participation in physical activity has been shown to increase neurogenesis in the hippocampus, a structure vital for learning and memory (Van Praag et al., 1999). Also, removing rodents' running wheels returned hippocampal neurogenesis to baseline measures within a day (Van der Borght et al., 2009). Chronic participation in physical activity can lead to cerebrovascular remodeling including angiogenesis (increases in the number of capillaries) and arteriogenesis (increases in existing arterial wall diameters) (Voss et al., 2014). Van der Borght et al. (2009) demonstrated cerebrovascular remodeling after just one day of physical activity in mice. However, the increase in angiogenesis and neurogenesis in the hippocampus returned to baseline after just one day of forced sedentariness (Van der Borght et al., 2009). Again, this shows intriguing evidence that the molecular effects of sedentary behavior and physical activity occur through similar mechanisms.

5 Mediators and Moderators of Executive Function

As mentioned earlier, an acute bout of physical activity seems to increase cognitive performance immediately following activity. This has been shown for executive functions such as working memory, response inhibition, and task switching in both young and older adults. However, there are some factors that can moderate the relationship between physical activity and cognition that we need to be measured or controlled. We will discuss the effects of physical activity and fitness on cognitive performance.

Investigation of how physical activity intensity influences cognition typically utilizes the inverted-U hypothesis, which suggests that the highest cognitive benefits will come from moderate-intensity exercise and that lower cognitive benefits will be seen following both bouts of low and high-intensity exercise (Fig. 2). A meta-analytic review of 79 studies performed by Chang, Labban, Gapin, and Etnier (2012) determined that very light, light, and moderate-intensity physical activity positively and significantly influenced cognition. Findings from their review also determined that physical activity that was hard, very hard, or maximal did not significantly influence cognitive performance. Studies have confirmed the inverted-U hypothesis by examining the effects of progressive cycling on cognitive performance. Salmela and Ndoye (1986) found that spatial reaction time was fastest during cycling rates corresponding to moderate-intensity activity as measured by heart rate, as compared to slower and faster pedaling rates. At higher rates of cycling, they reported a lengthening of reaction time. Brisswalter, Durand, Delignieres, and Legros (1995) examined how reaction time was influenced by seven different pedaling rates with fixed power output. Like Salmela and Ndoye, an inverted-U-shaped function was observed. Midrange pedaling rates corresponded with the fastest reaction times while the highest pedaling rate corresponded with the longest reaction times.

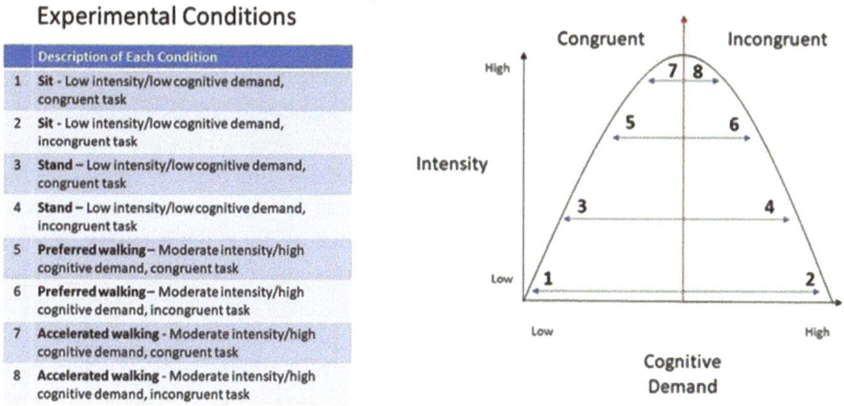
Experimental Conditions

	Description of Each Condition
1	**Sit** - Low intensity/low cognitive demand, congruent task
2	**Sit** - Low intensity/low cognitive demand, incongruent task
3	**Stand** – Low intensity/low cognitive demand, congruent task
4	**Stand** – Low intensity/low cognitive demand, incongruent task
5	**Preferred walking** – Moderate intensity/high cognitive demand, congruent task
6	**Preferred walking** – Moderate intensity/high cognitive demand, incongruent task
7	**Accelerated walking** - Moderate intensity/high cognitive demand, congruent task
8	**Accelerated walking** - Moderate intensity/high cognitive demand, incongruent task

Fig. 2 Inverted-U hypothesis applied to experiments examining the dual-task effects of physical activity intensity and cognitive demand

The most common mechanism for explaining enhanced cognitive performance following an acute bout of physical activity is that of arousal. Acute physical activity can promote increased cerebral blood flow, increased endorphin release, and increased attention; all of which promote an enhancement in arousal levels. Most studies that investigate the impact of intense or maximal exercise on cognitive performance are under the assumption that intense exercise will produce a fatigued state in individuals. This fatigue state can lead to a lack of arousal and of attentional resources necessary to optimize cognitive functioning.

When the effects of physical activity are examined at the molecular and cellular levels, it was discovered that physical activity enhances learning and memory as well as facilitates angiogenesis, the development of new blood vessels (Stillman, Cohen, Lehman, & Erickson, 2016). Cross-sectional research had demonstrated that the association between physical fitness and memory had implicated the hippocampal region as the mechanism facilitating the response. Hippocampal regions fluctuate by stage of development and as such could mediate the effects of physical activity and fitness on memory. Depending on study design, the prefrontal cortex could also be implicated as a brain region mediating cognitive performance (Weinstein et al., 2012).

6 Exercise and Measurement Issues

Physical activity participation places stress on the body producing physiological and metabolic responses as well as a motion artifact. Yet, to maintain health and quality of life, we must move. The neuroergonomic investigation of neural activation has merit before, during, and after exercise, but there are limitations. First, physiological effects of increased energy expenditure must be quantified. Heart rate monitors, blood pressure sensors, and measures of heart rate variability are common ways to measure energy expenditure. Remember, energy expenditure is relative to body weight, type of physical activity, and level of physical fitness. Second, the motion artifacts must be precisely measured using nine-axis accelerometer. Such a measure can account for extraneous movement during the task. Third, determining the psychological state of the participant is also recommended. To overcome the limitations associated with existing technologies that measure neural activation, such as the requires stillness for EEG and MRI data collection methods, collecting CBF and total blood volume using fNIRS is plausible when potential limitations are addressed through addition direct measures.

Robust evaluation of the effects of physical activity on executive function considers the frequency, duration, intensity, and type of physical activity. Scholars must determine where individuals fall on the Physical Activity Continuum (Tremblay, Colley, Saunders, Healy, & Owen, 2010) when neural imaging is being conducted. As mentioned earlier, because someone can exceed the physical activity guideline by participating in moderate-to-vigorous physical activity each day, and yet remain seated for up to 8 h while at work, there is a conundrum about how to comprehensively investigate the effects of physical activity and fitness on neurologic function and cognitive performance.

Findings presented in this chapter also vary by the type of executive function and the tool used to assess said function. Some variation is expected, but for the field of neuroeconomics to fully maximize the potential of quantifying the effects of physical activity and fitness, standard measures and dual-task conditions must be established and agreed upon by a community of scholars.

7 Future Directions

There are several emerging areas of interest that may assist the shaping of standardized protocols and help to enumerate the adverse effects of prolonged sedentary behavior and risk for memory loss in later life. fNIR technologies may be positioned to measure cognitive health risk in relation to health protective factors such as education, environment, and access to health care. Given the increased prevalence in physical inactivity, reduction in the risk for disease may already be dependent on neuroscience to mechanistically determine how to stave off the effects of dementia and other memory and cognitive health issues in later life. Two examples of futuristic foci of research are the use of virtual reality and expanded examination of sedentary behavior.

7.1 Virtual Reality

Using virtual reality, a computer-generated environment that can be interacted with to complete a task or engage a simulated has been used in the treatment of neurologic disorders and to examine behavior in a controlled setting that could be reproduced. Using virtual reality allows participants to experience highly realistic and interactive environments that extend well-beyond simply watching movies or visually interpreting statically presented stimuli. Virtual reality has been a particularly useful tool in multiple disciplines and training programs such as the military. Rustic virtual reality units such as Wii gaming stations to elaborate simulation chambers, already employ measurement of reaction time and accurate as part of game play. Immersive, virtual reality activates brain regions similar to what would be activated in the real-world situation (Campbell et al., 2009). Further, virtual reality could be used for the creation of authentic environments, replicable conditions, or rehabilitation.

One study using fNIR technologies and virtual reality discovered that there was enhanced parietal stimulation during completion of visuomotor tasks, which appeared to be a reliable measure of hemodynamic response (Seraglia et al., 2011). This study was not without limitations as the helmet could not be adjusted in size the procedures did not account for head movement and eye tracking. fNIRs and virtual reality were found to be an effect diagnostic tool when measuring prefrontal cortex activation during a balancing task (Moro et al., 2014). Sixteen healthy males were asked to maintain equilibrium while experiencing perturbations. Given the ability of the researchers to control the environment using VR the researcher were able to isolate the HbO_2 responses as the stimuli were experienced. The capacity of presenting identical real-life circumstance to different individuals can be used for research findings as well as specified training programs that could enhance neuroplasticity through neurorehabilitation (Holper et al., 2010).

In general, whether the VR is a set of goggles or an enclosed chamber, the potential for examining to how humans respond to real-world scenarios has limitless implications for research and development. Whether it is designing driverless vehicles or an application to identify our health risk, VR in combination with fNIRs can identify hemodynamic responses in real time has the potential to revolutionized how we training military personnel, new drivers, and aid in the recovery from traumatic brain injury. Although there is some positive emerging research this future direction needs to focus on the operationalization and standardization of research designs and experimental protocols.

7.2 Sedentary Behavior

As has been discussed, sedentary behavior is negatively associated with several chronic diseases and it may influence cognitive processes through similar physiological mechanisms as physical activity. With research on sedentary behavior starting to garner more interest, there are several areas in which current literature is lacking.

It is well established that physical activity participation leads to cardioprotective benefits, many of which stem from changes in the vasculature. One such example is increased endothelial function. Physical activity increases shear stress, nitric oxide production (Green, Maiorana, O'driscoll, & Taylor, 2004) and smooth muscle relaxation. Chronically, this can lead to reductions in blood pressure and increases in the ability of arteries to buffer pulsatile stress. In contrast, prolonged sitting causes reductions in shear stress (Newcomer, Sauder, Kuipers, Laughlin, & Ray, 2008; Padilla, Sheldon, Sitar, & Newcomer, 2009) which leads to decreases in nitric oxide bioavailability and attenuated endothelial function (Thosar, Johnson, Johnston, & Wallace, 2012). Additionally, prolonged sitting leads to calf blood pooling with increases in blood pressure and arterial resistance (Shvartz, Reibold, White, & Gaume, 1982; Shvartz, Gaume, White, & Reibold 1983; Pekarski, 2004). More research needs to be done that includes vascular health as a contributor to the effects of daily physical activity and sedentary behavior on brain health.

Studies utilizing EEG have shown that both higher fit children and adults have a higher amplitude of the P3 component and shorter P3 latency when compared to their lower fit counterparts (Dustman et al., 1990; Hillman et al., 2009). P3 amplitude is a measure of the attentional resources one can allocate to a given task while P3 latency is a measure of reaction time, or processing speed (Polich & Kok, 1995). Additionally, cross-sectional studies have revealed positive associations between cardiorespiratory fitness and white matter integrity (Johnson, Kim, Clasey, Bailey, & Gold, 2012) and cardiorespiratory fitness and hippocampal volume (Chaddock et al., 2010). Furthermore, a 12-month aerobic exercise intervention in older adults demonstrated that increases in cardiorespiratory fitness are associated with increased white matter integrity (Voss et al., 2013). Even with the plethora of studies documenting the effects of physical activity on brain volume, to date, no studies have investigated the effects of sedentary behavior on brain volume (Voss et al., 2014). Future studies have specifically examined the association between sedentary behavior and functional brain outcomes.

Lastly, the physical activity guidelines are used as recommendations for how active individuals should be to optimize health. However, there are currently no guidelines for sedentary behavior. Future studies are needed to (1) determine how long sitting needs to be before it is qualified as "prolonged" and (2) to determine the minimum dosage of sedentary behavior necessary before adverse metabolic and cognitive effects begin. In essence, it should be determined if there is a dose–response relationship between sitting time and cognitive health, as has been shown for physical activity and cardio-metabolic outcomes.

7.3 *Physical Activity and Cognitive Reserve*

Habitual physical activity likely also increases cognitive reserve, particularly when accumulated in early life. Cognitive reserve is a concept generally used to describe how individuals' brains can withstand deterioration as they age generally, and how they can withstand the effects of Alzheimer's disease and memory loss. Cognitive reserve suggests that the brain actively attempts to cope with brain damage by utilizing cognitive processes or compensatory mechanisms (Stern, 2002). Although used interchangeably, there is a discrete difference between cognitive reserve and brain reserve. Unlike cognitive reserve, brain reserve is the concept that larger brains, either derived from brain size or neuronal count, can sustain more pathology before clinical issues emerge (Katzman, 1993). In a prospective and retrospective study of 1826 subjects, it was determined that regular early life physical activity was related to better cognitive function and memory in later life (Reas et al., 2019), as one example of a cross-sectional study supporting the potential existence of cognitive reserve.

The hypothesized model focuses more on how tasks are processed as opposed to the anatomic differences (e.g., larger brain volume, more synapses, etc.). Cognitive reserve is often estimated using proxy variables for lifetime exposures and cognitive activity; years of education, physical activity participation, number of intellectually stimulating leisure activities, the degree of functional complexity, and socioeconomic status are all commonly used to create an estimate of CR (Stern, 2002). The concept of cognitive reserve explains why many studies have demonstrated that higher levels of educational and occupational attainment, as well as more time spent in leisure time physical activity, are good predictors of which individuals can sustain greater brain damage before demonstrating functional decline. Studies have demonstrated that those with greater cognitive reserve, as determined by greater time spent in leisure time physical activity, higher educational attainment, and greater occupational attainment, have greater neural efficiency and capacity (Stern et al., 2003). Neural efficiency describes the amount of neural resources needed to accomplish a task whereas neural capacity refers to how much brain activity in a particular area can be increased. Individuals with greater cognitive reserve have greater efficiency in the fact that they can perform the same, if not better on a particular task while recruiting fewer neural resources to do so (Habeck et al., 2005). Further, individuals with higher cognitive reserve have greater neural capacity in that they can increase brain activity in a particular region during a task more so than individuals with lower cognitive reserve (Holtzer et al., 2009; Stern et al., 2003). These factors can be seen as protective mechanisms for compensating for brain pathology. Because of the efficiency with which high cognitive reserve individuals process tasks, they can withstand more brain pathology than can an individual with lower cognitive reserve and perform better on that task.

Again, engagement in leisure time physical activity is a proxy variable that is used to estimate the amount of cognitive reserve one has. The greater the physical activity level of an individual, the greater their brain health is predicted to be. However, as will be discussed, there is mounting support that physical activity and sedentary behavior share molecular mechanisms of action. Because of this, more research is warranted to determine how sedentary behavior influences the cognitive reserve theory. The question remains, could fNIR technology be used to quant- and qualify the potentiation of cognitive reserve. We acknowledge this research is in its infancy and that hemodynamics would only be a contributing factor to the clustered effects of other health indicators. Presently, we lack consensus on how cognitive reserve is defined, operationalized and measured (Harrison et al., 2015), but given the advancement of technologies such as fNIRs, the idea is worth mentioning as a future direction of research.

8 Implications of Using FNIRs to Study the Effects of Human Movement

In sum, there are physical and cognitive benefits that result from single sessions and regular engagement in physical activity. The study of hemodynamic response in the brain regions provides measures of cerebral blood flow and deoxy- and oxygenation. This is a real-time, noninvasive device that is economical. The use of light sources and detectors identify the change that brain tissue undergoes. Such technology has endless applications. Since physical inactivity is the world's 4th ranked cause for mortality, the benefits of physical activity and adverse effects of sedentary behavior need to be examined. This chapter calls for a standardization in protocols and accountability of mediators, moderators, and confounding variables.

References

Åberg, M. A. I., Pedersen, N. L., Torén, K., Svartengren, M., Bäckstrand, B., Johnsson, T., et al. (2009). Cardiovascular fitness is associated with cognition in young adulthood. *Proceedings of the National Academy of Sciences of the United States of America, 106*(49), 20906–20911.

Ahlskog, J. E., Geda, Y. E., Graff-Radford, N. R., & Petersen, R. C. (2011). Physical exercise as a preventive or disease-modifying treatment of dementia and brain aging. *MayoClinic Proceedings, 86*(9), 876–884.

Albinet, C. T., Mandrick, K., Bernard, P. L., Perrey, S., & Blain, H. (2014). Improved cerebral oxygenation response and executive performance as a function of cardiorespiratory fitness in older women: A fNIRS study. *Frontiers in Aging Neuroscience, 6,* 272.

Athilingam, P., Moynihan, J., Chen, L., D'Aoust, R., Groer, M., & Kip, K. (2013). Elevated levels of interleukin 6 and C-reactive protein associated with cognitive impairment in heart failure. *Congestive Heart Failure, 19*(2), 92–98.

Baddeley, A. D., & Hitch, G. (1974). Working memory. *Psychology of Learning and Motivation, 8,* 47–89.

Banich, M. T. (2009). Executive function: The search for an integrated account. *Current Directions in Psychological Science, 18*(2), 89–94.

Bediz, C. S., Oniz, A., Guducu, C., Ural Demirci, E., Ogut, H., Gunay, E., et al. (2016). Acute supramaximal exercise increases the brain oxygenation in relation to cognitive workload. *Frontiers in Human Neuroscience, 10,* 174.

Beurskens, R., Helmich, I., Rein, R., & Bock, O. (2014). Age-related changes in prefrontal activity during walking in dual-task situations: A fNIRS study. *International Journal of Psychophysiology, 92*(3), 122–128.

Brisswalter, J., Collardeau, M., & René, A. (2002). Effects of acute physical exercise characteristics on cognitive performance. *Sports Medicine, 32*(9), 555–566.

Brisswalter, J., Durand, M., Delignieres, D., & Legros, P. (1995). Optimal and non-optimal demand in a dual-task of pedaling and simple reaction time: Effects on energy expenditure and cognitive performance. *Journal of Human Movement Studies, 29*(1), 15–34.

Callejas, A., Lupiáñez, J., & Tudela, P. (2004). The three attentional networks: On their independence and interactions. *Brain and Cognition, 54*(3), 225–227.

Campbell, Z., Zakzanis, K. K., Jovanovski, D., Joordens, S., Mraz, R., & Graham, S. J. (2009). Utilizing virtual reality to improve the ecological validity of clinical neuropsychology: An FMRI case study elucidating the neural basis of planning by comparing the Tower of London with a three-dimensional navigation task. *Applied Neuropsychology, 16*(4), 295–306.

Chaddock, L., Erickson, K. I., Prakash, R. S., Kim, J. S., Voss, M. W., VanPatter, M., et al. (2010). A neuroimaging investigation of the association between aerobic fitness, hippocampal volume, and memory performance in preadolescent children. *Brain Research, 1358,* 172–183.

Chan, R. C. K., Shum, D., Toulopoulou, T., & Chen, E. Y. H. (2008). Assessment of executive functions: Review of instruments and identification of critical issues. *Archives of Clinical Neuropsychology, 23*(2), 201–216.

Chang, Y. K., Labban, J. D., Gapin, J. I., & Etnier, J. L. (2012). The effects of acute exercise on cognitive performance: A meta-analysis. *Brain Research, 1453,* 87–101.

Cohen, J. D., Forman, S. D., Braver, T. S., Casey, B. J., Servan-Schreiber, D., & Noll, D. C. (1994). Activation of the prefrontal cortex in a nonspatial working memory task with functional MRI. *Human Brain Mapping, 1*(4), 293–304.

Colcombe, S. J., Kramer, A. F., Erickson, K. I., Scalf, P., McAuley, E., Cohen, N. J., et al. (2004). Cardiovascular fitness, cortical plasticity, and aging. *Proceedings of the National Academy of Sciences, 101*(9), 3316–3321.

Craft, S. (2009). The role of metabolic disorders in Alzheimer disease and vascular dementia: Two roads converged. *Archives of Neurology, 66*(3), 300–305.

Craft, L. L., Zderic, T. W., Gapstur, S. M., Vaniterson, E. H., Thomas, D. M., Siddique, J., et al. (2012). Evidence that women meeting physical activity guidelines do not sit less: An observational inclinometry study. *International Journal of Behavioral Nutrition and Physical Activity, 9*(1), 122.

Donnelly, J. E., Hillman, C. H., Castelli, D., Etnier, J. L., Lee, S., Tomporowski, P., et al. (2016). Physical activity, fitness, cognitive function, and academic achievement in children: A systematic review. *Medicine and Science in Sports and Exercise, 48*(6), 969–1225.

Dunstan, D. W., Kingwell, B. A., Larsen, R., Healy, G. N., Cerin, E., Hamilton, M. T., et al. (2012). Breaking up prolonged sitting reduces postprandial glucose and insulin responses. *Diabetes Care, 35*(5), 976–983.

Dupuy, O., Gauthier, C. J., Fraser, S. A., Desjardins-Crèpeau, L., Desjardins, M., Mekary, S., et al. (2015). Higher levels of cardiovascular fitness are associated with better executive function and prefrontal oxygenation in younger and older women. *Frontiers in Human Neuroscience, 9,* 66.

Dustman, R. E., Emmerson, R. Y., Ruhling, R. O., Shearer, D. E., Steinhaus, L. A., Johnson, S. C., et al. (1990). Age and fitness effects on EEG, ERPs, visual sensitivity, and cognition. *Neurobiology of Aging, 11*(3), 193–200.

Duvivier, B. M. F. M., Schaper, N. C., Bremers, M. A., van Crombrugge, G., Menheere, P. P. C. A., Kars, M., & Savelberg, H. H. C. M. (2013). Minimal intensity physical activity (Standing and walking) of longer duration improves insulin action and plasma lipids more than shorter periods of moderate to vigorous exercise (cycling) in sedentary subjects when energy expenditure is comparable. *PLOS One, 8*(2), e55542.

Ehlis, A. C., Herrmann, M. J., Wagener, A., & Fallgatter, A. J. (2005). Multi-channel near-infrared spectroscopy detects specific inferior-frontal activation during incongruent Stroop trials. *Biological Psychology, 69*(3), 315–331.

Fox, K. R., & Hillsdon, M. (2007). Physical activity and obesity. *Obesity Reviews, 8,* 115–121.

Giles, G. E., Brunyé, T. T., Eddy, M. D., Mahoney, C. R., Gagnon, S. A., Taylor, H. A., et al. (2014). Acute exercise increases oxygenated and deoxygenated hemoglobin in the prefrontal cortex. *NeuroReport, 25*(16), 1320–1325.

Green, D. J., Maiorana, A., O'driscoll, G., & Taylor, R. (2004). Effect of exercise training on endothelium-derived nitric oxide function in humans. *The Journal of Physiology, 561*(1), 1–25.

Guiney, H., & Machado, L. (2013). Benefits of regular aerobic exercise for executive functioning in healthy populations. *Psychonomic Bulletin & Review, 20*(1), 73–86.

Habeck, C., Rakitin, B. C., Moeller, J., Scarmeas, N., Zarahn, E., Brown, T., et al. (2005). An event-related fMRI study of the neural networks underlying the encoding, maintenance, and retrieval phase in a delayed-match-to-sample task. *Cognitive Brain Research, 23*(2), 207–220.

Håkansson, M. L., Brown, H., Ghilardi, N., Skoda, R. C., & Meister, B. (1998). Leptin receptor immunoreactivity in chemically defined target neurons of the hypothalamus. *Journal of Neuroscience, 18*(1), 559–572.

Hallal, P. C., Andersen, L. B., Bull, F. C., Guthold, R., Haskell, W., Ekelund, U., & Lancet Physical Activity Series Working Group. (2012). Global physical activity levels: Surveillance progress, pitfalls, and prospects. *The Lancet, 380*(9838), 247–257.

Hamacher, D., Hamacher, D., Rehfeld, K., Hökelmann, A., & Schega, L. (2015). The effect of a six-month dancing program on motor-cognitive dual-task performance in older adults. *Journal of Aging and Physical Activity, 23*(4), 647–652.

Hamacher, D., Herold, F., Wiegel, P., Hamacher, D., & Schega, L. (2015). Brain activity during walking: A systematic review. *Neuroscience & Biobehavioral Reviews, 57*, 310–327.

Hamilton, M. T., Hamilton, D. G., & Zderic, T. W. (2007). Role of low energy expenditure and sitting in obesity, metabolic syndrome, type 2 diabetes, and cardiovascular disease. *Diabetes, 56*(11), 2655–2667.

Hanakawa, T., Katsumi, Y., Fukuyama, H., Honda, M., Hayashi, T., Kimura, J., et al. (1999). Mechanisms underlying gait disturbance in Parkinson's disease: A single photon emission computed tomography study. *Brain, 122*(7), 1271–1282.

Hansen, A. L., Johnsen, B. H., Sollers, J. J., Stenvik, K., & Thayer, J. F. (2004). Heart rate variability and its relation to prefrontal cognitive function: The effects of training and detraining. *European Journal of Applied Physiology, 93*(3), 263–272.

Harrison, S. L., Sajjad, A., Bramer, W. M., Ikram, M. A., Tiemeier, H., & Stephan, B. C. (2015). Exploring strategies to operationalize cognitive reserve: A systematic review of reviews. *Journal of Clinical and Experimental Neuropsychology, 37*(3), 253–264.

Healy, G. N., Matthews, C. E., Dunstan, D. W., Winkler, E. A. H., & Owen, N. (2011). Sedentary time and cardio-metabolic biomarkers in US adults: NHANES 2003–06. *European Heart Journal, 32*(5), 590–597.

Healy, G. N., Wijndaele, K., Dunstan, D. W., Shaw, J. E., Salmon, J., Zimmet, P. Z., et al. (2008). Objectively measured sedentary time, physical activity, and metabolic risk: The Australian Diabetes, Obesity and Lifestyle Study (AusDiab). *Diabetes Care, 31*(2), 369–371.

Hillman, C. H., Kramer, A. F., Belopolsky, A. V., & Smith, D. P. (2006). A cross-sectional examination of age and physical activity on performance and event-related brain potentials in a task switching paradigm. *International Journal of Psychophysiology, 59*(1), 30–39.

Hillman, C. H., Pontifex, M. B., Raine, L. B., Castelli, D. M., Hall, E. E., & Kramer, A. F. (2009). The effect of acute treadmill walking on cognitive control and academic achievement in preadolescent children. *Neuroscience, 159*(3), 1044–1054.

Holper, L., Muehlemann, T., Scholkmann, F., Eng, K., Kiper, D., & Wolf, M. (2010). Testing the potential of a virtual reality neurorehabilitation system during performance of observation imagery and imitation of motor actions recorded by wireless functional near-infrared spectroscopy (fNIRS). *Journal of Neuroengineering and Rehabilitation, 7*(1), 57.

Holtzer, R., Mahoney, J. R., Izzetoglu, M., Izzetoglu, K., Onaral, B., & Verghese, J. (2011). fNIRS study of walking and walking while talking in young and old individuals. *Journals of Gerontology Series A: Biomedical Sciences and Medical Sciences, 66*(8), 879–887.

Holtzer, R., Rakitin, B. C., Steffener, J., Flynn, J., Kumar, A., & Stern, Y. (2009). Age effects on load-dependent brain activations in working memory for novel material. *Brain Research, 1249*, 148–161.

Jefferson, A. L., Massaro, J. M., Wolf, P. A., Seshadri, S., Au, R., Vasan, R. S., et al. (2007). Inflammatory biomarkers are associated with total brain volume: The Framingham Heart Study. *Neurology, 68*(13), 1032–1038.

Johnson, N. F., Kim, C., Clasey, J. L., Bailey, A., & Gold, B. T. (2012). Cardiorespiratory fitness is positively correlated with cerebral white matter integrity in healthy seniors. *Neuroimage, 59*(2), 1514–1523.

Kamijo, K., & Takeda, Y. (2010). Regular physical activity improves executive function during task switching in young adults. *International Journal of Psychophysiology, 75*(3), 304–311.

Katzman, R. (1993). Education and the prevalence of dementia and Alzheimer's disease. *Neurology, 43*(1), 13–20.

Knight, Z. A., Hannan, K. S., Greenberg, M. L., & Friedman, J. M. (2010). Hyperleptinemia is required for the development of leptin resistance. *PLoS ONE, 5*(6), e11376.

Lambourne, K., & Tomporowski, P. (2010). The effect of exercise-induced arousal on cognitive task performance: A meta-regression analysis. *Brain Research, 1341,* 12–24.

Lees, S. J., & Booth, F. W. (2004). Sedentary death syndrome. *Canadian Journal of Applied Physiology, 29*(4), 447–460.

McMorris, T., & Graydon, J. (2000). The effect of incremental exercise on cognitive performance. *International Journal of Sport Psychology, 31*(1), 66–81.

Messier, C., & Teutenberg, K. (2005). The role of insulin, insulin growth factor, and insulin-degrading enzyme in brain aging and Alzheimer's disease. *Neural Plasticity, 12*(4), 311–328.

Miyake, A., Friedman, N. P., Emerson, M. J., Witzki, A. H., Howerter, A., & Wager, T. D. (2000). The unity and diversity of executive functions and their contributions to complex "frontal lobe" tasks. *Cognitive Psychology, 41*(1), 49–100.

Molteni, E., Baselli, G., Cerutti, S., Bianchi, A. M., Contini, D., Caffini, M., et al. (2012). Load-dependent brain activation assessed by time-domain functional near-infrared spectroscopy during a working memory task with graded levels of difficulty. *Journal of Biomedical Optics, 17*(5), 056005.

Monsell, S. (2003). Task switching. *Trends in Cognitive Sciences, 7*(3), 134–140.

Moro, S. B., Bisconti, S., Muthalib, M., Spezialetti, M., Cutini, S., Ferrari, M., et al. (2014). A semi-immersive virtual reality incremental swing balance task activates prefrontal cortex: A functional near-infrared spectroscopy study. *Neuroimage, 85,* 451–460.

Moult, P. R., Cross, A., Santos, S. D., Carvalho, A. L., Lindsay, Y., Connolly, C. N., et al. (2010). Leptin regulates AMPA receptor trafficking via PTEN inhibition. *Journal of Neuroscience, 30*(11), 4088–4101.

Newcomer, S. C., Sauder, C. L., Kuipers, N. T., Laughlin, M. H., & Ray, C. A. (2008). Effects of posture on shear rates in human brachial and superficial femoral arteries. *American Journal of Physiology-Heart and Circulatory Physiology, 294*(4), H1833–H1839.

Ohsugi, H., Ohgi, S., Shigemori, K., & Schneider, E. B. (2013). Differences in dual-task performance and prefrontal cortex activation between younger and older adults. *BMC Neuroscience, 14*(1), 10.

Padilla, J., Sheldon, R. D., Sitar, D. M., & Newcomer, S. C. (2009). Impact of acute exposure to increased hydrostatic pressure and reduced shear rate on conduit artery endothelial function: A limb-specific response. *American Journal of Physiology-Heart and Circulatory Physiology, 297*(3), H1103–H1108.

Pekarski, S. E. (2004). A gravitational hypothesis of essential hypertension as a natural adaptation to increased gravitational stress caused by regular, prolonged sitting typical of modern life. *Medical Science Monitor, 10*(6), HY27–HY32.

Polich, J., & Kok, A. (1995). Cognitive and biological determinants of P300: An integrative review. *Biological Psychology, 41*(2), 103–146.

Prakash, R. S., Voss, M. W., Erickson, K. I., Lewis, J. M., Chaddock, L., Malkowski, E., et al. (2011). Cardiorespiratory fitness and attentional control in the aging brain. *Frontiers in Human Neuroscience, 4,* 229.

Reas, E. T., Laughlin, G. A., Bergstrom, J., Kritz-Silverstein, D., Richard, E. L., Barrett-Connor, E., et al. (2019). Lifetime physical activity and late-life cognitive function: The Rancho Bernardo study. *Age and Ageing, 48*(2), 241–246.

Redila, V. A., & Christie, B. R. (2006). Exercise-induced changes in dendritic structure and complexity in the adult hippocampal dentate gyrus. *Neuroscience, 137*(4), 1299–1307.

Salmela, J. H., & Ndoye, O. D. (1986). Cognitive distortions during progressive exercise. *Perceptual and Motor Skills, 63*(3), 1067–1072.

Schinder, A. F., & Poo, M. M. (2000). The neurotrophin hypothesis for synaptic plasticity. *Trends in Neurosciences, 23*(12), 639–645.

Schmidt, R., Schmidt, H., Curb, J. D., Masaki, K., White, L. R., & Launer, L. J. (2002). Early inflammation and dementia: A 25-year follow-up of the Honolulu-Asia aging study. *Annals of Neurology, 52*(2), 168–174.

Seraglia, B., Gamberini, L., Priftis, K., Scatturin, P., Martinelli, M., & Cutini, S. (2011). An exploratory fNIRS study with immersive virtual reality: A new method for technical implementation. *Frontiers in Human Neuroscience, 5,* 176.

Shanley, L. J., Irving, A. J., & Harvey, J. (2001). Leptin enhances NMDA receptor function and modulates hippocampal synaptic plasticity. *Journal of Neuroscience, 21*(24), RC186–RC186.

Shizgal, P. B., & Hyman, S. E. (2013). Homeostasis, motivation, and addictive states. *Principles of Neural Science, 5,* 1095–1115.

Shvartz, E., Gaume, J. G., White, R. T., & Reibold, R. C. (1983). Hemodynamic responses during prolonged sitting. *Journal of Applied Physiology, 54*(6), 1673–1680.

Shvartz, E., Reibold, R. C., White, R. T., & Gaume, J. G. (1982). Hemodynamic responses in orthostasis following 5 hours of sitting. *Aviation, Space and Environmental Medicine, 53*(3), 226–231.

Smith, E. E., & Jonides, J. (1997). Working memory: A view from neuroimaging. *Cognitive Psychology, 33*(1), 5–42.

Stern, Y. (2002). What is cognitive reserve? Theory and research application of the reserve concept. *Journal of the International Neuropsychological Society, 8*(03), 448–460.

Stern, Y., Zarahn, E., Hilton, H. J., Flynn, J., DeLaPaz, R., & Rakitin, B. (2003). Exploring the neural basis of cognitive reserve. *Journal of Clinical and Experimental Neuropsychology, 25*(5), 691–701.

Stillman, C. M., Cohen, J., Lehman, M. E., & Erickson, K. I. (2016). Mediators of physical activity on neurocognitive function: A review at multiple levels of analysis. *Frontiers in Human Neuroscience, 10,* 626.

Stranahan, A. M., & Mattson, M. P. (2008). Impact of energy intake and expenditure on neuronal plasticity. *NeuroMolecular Medicine, 10*(4), 209–218.

Stranahan, A. M., & Mattson, M. P. (2011). Bidirectional metabolic regulation of neurocognitive function. *Neurobiology of Learning and Memory, 96*(4), 507–516.

Szostak, J., & Laurant, P. (2011). The forgotten face of regular physical exercise: A 'natural' anti-atherogenic activity. *Clinical Science, 121*(3), 91–106.

Themanson, J. R., & Hillman, C. H. (2006). Cardiorespiratory fitness and acute aerobic exercise effects on neuroelectric and behavioral measures of action monitoring. *Neuroscience, 141*(2), 757–767.

Themanson, J. R., Hillman, C. H., & Curtin, J. J. (2006). Age and physical activity influences on action monitoring during task switching. *Neurobiology of Aging, 27*(9), 1335–1345.

Themanson, J. R., Pontifex, M. B., & Hillman, C. H. (2008). Fitness and action monitoring: Evidence for improved cognitive flexibility in young adults. *Neuroscience, 157*(2), 319–328.

Thosar, S. S., Johnson, B. D., Johnston, J. D., & Wallace, J. P. (2012). Sitting and endothelial dysfunction: The role of shear stress. *Medical Science Monitor: International Medical Journal of Experimental and Clinical Research, 18*(12), RA173–RA180.

Tomporowski, P. D. (2003a). Cognitive and behavioral responses to acute exercise in youths: A review. *Pediatric Exercise Science, 15*(4), 348–359.

Tomporowski, P. D. (2003b). Effects of acute bouts of exercise on cognition. *Acta Psychologica, 112*(3), 297–324.

Tremblay, M. S., Colley, R. C., Saunders, T. J., Healy, G. N., & Owen, N. (2010). Physiological and health implications of a sedentary lifestyle. *Applied Physiology, Nutrition and Metabolism, 35*(6), 725–740.

Tudor-Locke, C., Craig, C. I., Thyfault, J. P., & Spence, J. C. (2013). Step-defined sedentary lifestyle index: <5000 steps/day. *Applied Physiology, Nutrition and Metabolism, 38,* 100–114. https://doi.org/10.1139/apnm-2012-0235.

Van der Borght, K., Kóbor-Nyakas, D. É., Klauke, K., Eggen, B. J., Nyakas, C., Van der Zee, E. A., et al. (2009). Physical exercise leads to rapid adaptations in hippocampal vasculature: Temporal dynamics and relationship to cell proliferation and neurogenesis. *Hippocampus, 19*(10), 928–936.

Van der Ploeg, H. P., Chey, T., Korda, R. J., Banks, E., & Bauman, A. (2012). Sitting time and all-cause mortality risk in 222–497 Australian adults. *Archives of Internal Medicine, 172*(6), 494–500.

Van Praag, H., Kempermann, G., & Gage, F. H. (1999). Running increases cell proliferation and neurogenesis in the adult mouse dentate gyrus. *Nature Neuroscience, 2*(3), 266.

Vasta, R., Cutini, S., Cerasa, A., Gramigna, V., Olivadese, G., Arabia, G., & Quattrone, A. (2018). Physiological aging influence on brain hemodynamic activity during task-switching: A fNIRS study. *Frontiers in Aging Neuroscience, 9,* 433.

Voelcker-Rehage, C., Godde, B., & Staudinger, U. M. (2010). Physical and motor fitness are both related to cognition in old age. *European Journal of Neuroscience, 31*(1), 167–176.

Voss, M. W., Carr, L. J., Clark, R., & Weng, T. (2014). Revenge of the "sit" II: Does lifestyle impact neuronal and cognitive health through distinct mechanisms associated with sedentary behavior and physical activity? *Mental Health and Physical Activity, 7*(1), 9–24.

Voss, M. W., Heo, S., Prakash, R. S., Erickson, K. I., Alves, H., Chaddock, L., et al. (2013). The influence of aerobic fitness on cerebral white matter integrity and cognitive function in older adults: Results of a one-year exercise intervention. *Human Brain Mapping, 34*(11), 2972–2985.

Weinstein, A. M., Voss, M. W., Prakash, R. S., Chaddock, L., Szabo, A., White, S. M., et al. (2012). The association between aerobic fitness and executive function is mediated by prefrontal cortex volume. *Brain Behavior Immunology, 26*(5), 811–819.

Weuve, J., Kang, J. H., Manson, J. E., Breteler, M. M. B., Ware, J. H., & Grodstein, F. (2004). Physical activity, including walking, and cognitive function in older women. *JAMA: The Journal of the American Medical Association, 292*(12), 1454–1461.

WHO. (2009). *Global health risks: Mortality and burden of disease attributable to selected major risks*. Geneva: World Health Organization.

WHO. (2010). *World health statistics 2010.* World Health Organization.

WHO. (2011). *Global status report on noncommunicable diseases 2010.* Geneva: World Health Organization.

Williamson, R., McNeilly, A., & Sutherland, C. (2012). Insulin resistance in the brain: An old-age or new-age problem? *Biochemical Pharmacology, 84*(6), 737–745.

Yaffe, K., Lindquist, K., Penninx, B. W., Simonsick, E. M., Pahor, M., Kritchevsky, S., et al. (2003). Inflammatory markers and cognition in well-functioning African American and white elders. *Neurology, 61*(1), 76–80.

Yanagisawa, H., Dan, I., Tsuzuki, D., Kato, M., Okamoto, M., Kyutoku, Y., et al. (2010). Acute moderate exercise elicits increased dorsolateral prefrontal activation and improves cognitive performance with Stroop test. *Neuroimage, 50*(4), 1702–1710.

Zhu, D. C., Zacks, R. T., & Slade, J. M. (2010). Brain activation during interference resolution in young and older adults: An fMRI study. *NeuroImage, 50*(2), 810–817.

Zysset, S., Schroeter, M. L., Neumann, J., & von Cramon, D. Y. (2007). Stroop interference, hemodynamic response and aging: An event-related fMRI study. *Neurobiology of Aging, 28*(6), 937–946.

Neuroergonomics and Its Relation to Psychophysiology

Ji-Eun Kim and Tae-Ho Lee

Abstract Psychophysiology measures physiological activities originating from the autonomic nervous system, including eye movement, heart rate, electromyography, and skin conductance resistance, while operators conduct behavioral tasks. Researchers have distinguished psychophysiology from neuroergonomics, the latter of which focuses on neural activities in brain functions using brain imaging techniques. Psychophysiology has served as a supplementary measure for representing indicators of the neural activity levels involved in physical and cognitive work. In affective ergonomics, however, the brain's subcortical pathways play a key role, making it difficult to depend solely on the brain imaging techniques used in industrial task settings to quantify and predict human performance. In this chapter, the authors review and compare the strengths and weaknesses of psychophysiology metrics and their applications to cognitive/affective engineering.

1 The Definition and Scope of Neuroergonomics

Neuroergonomics has emerged as an interdisciplinary field of scholarly interest in recent years. Neuergonomics is located at the intersection of neuroscience and ergonomics; thus, it is defined as "the study of brain and behavior at work" (Parasuraman, 2003). Neuroergonomics is distinct from neuroscience, which focuses on the study of brain structure and function, in that neuroergonomics uses task settings that take into consideration real-world environments. Yet neuroergonomics does not use exactly the same terminology as the human factors and ergonomics (HFE) field, which examines the human use of technology in diverse work settings. HFE does not always involve the investigation of neural systems, whereas neuroergonomics does.

J.-E. Kim (✉)
University of Washington, Seattle, USA
e-mail: jikim@uw.edu

T.-H. Lee
Virginia Tech, Blacksburg, USA

C. S. Nam (ed.), *Neuroergonomics*, Cognitive Science and Technology,
https://doi.org/10.1007/978-3-030-34784-0_10

The goals of neuroergonomics are both fundamental and practical. Neuroergonomics researchers aim not only to understand the neural activities related to human performance in real-world task settings, but also to design work systems in more efficient, effective, and safer ways (Parasuraman, 2003). To reach these goals, several techniques have been used and reviewed in the field of neuroergonomics. Neuroergonomic techniques capture changes in the hemodynamic or electromagnetic activities of the brain. Techniques measuring changes in hemodynamic responses include functional magnetic resonance imaging (fMRI), which detects the ratio of oxy- to deoxy-hemoglobin (HbO_2 and Hb, respectively) to capture the local blood flow of the brain (Gazzaniga, Ivry, & Mangun, 2008). Positron emission tomography (PET) also provides information about cerebral blood flow using radioactive tracers, and near-infrared spectroscopy (NIR) detects the changes in both oxy- and deoxy-hemoglobin using the near-infrared spectrum (Son, Guhe, Gray, Yazici, & Schoelles, 2005). Another stream of neuroergonomic methods depends on the electrical and magnetic dynamics of brain activities. Electroencephalography (EEG) measures changes in voltage responses that originate in populations of neurons between overall electrodes and reference electrodes. If repeated trials of specific tasks are given with an interest in specific electrodes, we call it event-related potential (ERP), since averaged repeated trials were used. Similarly, magnetoencephalography (MEG) records changes in the magnetic field given repeated trials of specific events.

2 Challenges in Neuroergononic Methods

Each technique used in neuroergonomics has its well-known advantages and disadvantages in terms of its spatial and temporal resolution for measuring brain signals. Brain imaging techniques that depend on hemodynamic changes have low temporal resolution, while techniques that capture electromagnetic changes show low spatial resolution. For example, MRI, fMRI, and PET are known to have poor temporal resolution compared to EEG, ERP, and MEG. On the other hand, EEG and ERP have constraints that mean the localization of an area of the brain matters. Since the EEG and ERP measure voltage changes from millions of neurons combined on the scalp of the brain, these techniques are limited in their ability to measure the surface area of the brain, such as the visual and motor cortex.

In addition to the constraints posed by temporal and spatial resolution, the techniques' portability, sensitivity of movement, and cost are very significant criteria since the devices are used in task settings that are usually located outside of laboratories (Mehta & Parasuraman, 2013). With respect to the portability, it is almost impossible to move the equipment for MRI, fMRI, and PET out of the laboratory. This is because these techniques require that participants are placed in enormous MRI or PET scanners. Furthermore, these hemodynamic brain imaging techniques are very sensitive to participants' movements, meaning that subjects are required to stay completely still during the brain scan. Although it is true that the scanner allows simple motor responses such as button-pressing, participants in the scanner cannot move their body or head. Finally, fMRI is fairly expensive, with the primary costs stemming from the required personnel and amortization of the machine. Neuroimaging studies currently cost USD 400–900 for a single participant, and an fMRI machine can reach USD 3 million.

3 Methods of Psychophysiology

There are currently no cost-effective neuroergonomic methodologies that have high spatial and temporal resolution, high portability, and low sensitivity of movement (Parasuraman, 2011). Researchers have thus used other methods, including psychophysiological measurements, that merit greater attention. Noninvasive psychophysiological measures provide the data that can facilitate an understanding of the relationship between physiological changes and cognitive and emotional states (Andreassi, 2010; Dirican & Göktürk, 2011). There are different types of physiological measures including eye movements, pupil diameter, skin conductance response (SCR), heart rate (HR), and electromyography (EMG), in addition to EEG and ERP.

We note that most psychophysiological responses are modulated and controlled by the autonomic nervous system (ANS) (Andreassi, 2010). The ANS is an involuntary system that controls the functioning of the glands and organs of the body, including the modulation of pupil dilation, sweat glands, and heart rate. The ANS is subdivided into two complementary systems, the sympathetic nervous system (SNS) and the parasympathetic nervous system (PNS). The SNS generates energy and arousal for activity, and the PNS conserves energy for the resting state. The SNS responds to stressful situations that trigger the fight-or-flight response by increasing a person's heart rate, blood pressure, and pupil dilation (Costanzo, 2018). Along with the ANS, the somatic nervous system that controls muscular activities makes up the peripheral nervous system, which includes nervous tissues outside of the brain and spinal cord. The peripheral nervous system and the central nervous system, the latter of which consists of the brain and spinal cord, comprise the human nervous system. In general, the peripheral nervous system includes many psychophysiological recordings of interest, while the central nervous system includes many neuroergonomic techniques of interest.

Table 1 summarizes each psychophysiological recording's measures and features in terms of temporal resolution, portability, and cost. Up-to-date camera-based eye trackers capture eye movements to provide information on the location and duration of observers' sight. Engineers and scientists are interested in saccadic eye movements, which refer to the voluntary movements from one fixation point to another. A fixation is defined as a moment in which the eye dwells on a particular location for more than 100–400 ms (Rayner, 1998). Eye trackers can also measure variations in pupil diameter, called pupil dilation, which is controlled by the SNS and PNS. As camera technology has advanced, the temporal interval between the time an image is captured by the camera and the time the gaze data is available has become as short as

Table 1 Relative strengths of various psychophysiological recordings

Psychophysiological recordings	Measures	Nervous system	Temporal resolution	Portability	Cost
Eye movements	Fixation, saccadic movement	Central System	High	Moderate to high	Moderate to low
Pupil dilation	Pupil size	ANS (SNS, PNS)	High	Moderate to high	Moderate to low
SCR	Sweat gland activity	ANS (SNS)	Low	High	Low
ECG	Heart rate	ANS (SNS, PNS)	High	High	Low
EMG	Muscle activity	Somatic System	High	Moderate to high	Moderate to low

25 ms, which enables real-time data recordings. Researchers can record eye movement and pupil dilation while participants observe computer monitors equipped with eye trackers (monitor-mounted type) or wear glasses-type eye trackers at a much lower cost than that of brain imaging technology such as fMRI.

Skin conductance response (SCR), also called galvanic skin response (GSR) or electrodermal activity (EDA), measures tonic and phasic changes in sweat gland activity, which is mainly controlled by the SNS. Monopolar or bipolar silver–silver chloride electrodes are generally placed on the palmar surface of the fingers or hands where the eccrine sweat glands are concentrated (Stern et al. 2001). One disadvantage of SCR is its low temporal resolution; the time lag between the onset of a stimulus and SCR initiation ranges from 1.0 to 3.0 s. Recently developed SCR recording equipment reads skin conductance by imposing a constant voltage (Stern et al., 2001), while advances in wireless technology have enabled the development of portable SCR equipment that weighs fewer than 500 g and costs less than USD 10,000.

The measurement of the electrical potential of the heart is called electrocardiography (ECG or EKG, the latter of which is derived from the German spelling). ECG captures changes in a person's heart rate during the contractions of the heart that are controlled by the SNS and PNS. Electrodes placed on the surface of the skin read the electrical potentials. Using the most prominent R wave that is generated during ventricular depolarization, heart rate (HR) can be measured as the number of beats per fixed time frame, usually per minute. The delay between a person's real-time heart rate and the information displayed on an HR monitoring systems varies depending on the type of device, but it has been reported to be very low, with an average of about 90 ms (Liang & Balasingham, 2007). Wearable heart rate monitoring devices for the purposes of self-tracking have constituted a growing market in recent years, and the costs of ECG and wearable heart rate monitoring equipment continue to decrease in the marketplace.

Electromyography (EMG) measures electrical potentials linked to muscle contractions controlled by the somatic nervous system (Andreassi, 2010; Marek, Karwowski, & Rice, 2010). Electrodes are attached to the surface of the skin where muscles of interest are located. Some clinical researchers use needle electrodes. In general, the frequency of EMG signals ranges from 1–1000 Hz, while the sample frequency of recent wearable EMG equipment recording muscular activity is over 1000 Hz and available at a low cost. For high-quality recording for clinical purposes, there are commercial EMG systems that provide multi-channel and high sampling frequency; these, however, typically cost over USD 50,000 (Supuk, Skelin, & Cic, 2014).

In summary, psychophysiological measures have advantages in terms of temporal resolution, portability, and cost compared to neuroimaging techniques. Since most physiological recordings of interest are controlled by the peripheral nervous system, psychophysiology measurements can have better temporal resolutions in the milliseconds range. That is, psychophysiology such as eye movement, pupil dilation, or ECG can efficiently address some temporal-related questions that neuroimaging cannot. Moreover, in line with the recent development of wearable sensors, wearable eye trackers, SCR, ECG, and EMG sensors are available at affordable prices compared to neuroimaging techniques. The increased portability and lower cost of physiological measurement devices suggest they can be helpful in many applied research settings by having the participants whose physiological measurements are being collected be free to walk around or redirect their head and body positions, such as driving simulations or workplace simulations.

4 The Advantages of Psychophysiology Compared to Neuroimaging for Measuring Arousal

Arousal, which refers to the physiological and psychological state of alertness or wakefulness (Thayer, 1989) is crucial in modulating human behavior and performance (Mather & Sutherland, 2011). Both human and animal studies have consistently demonstrated that arousal is related to the neural activity of regions located alongside the subcortical pathway. This pathway is located beneath the cerebral cortex and receives input from the peripheral sense organs and receptors (Morgane, Galler, & Mokler, 2005). Indeed, using neuroimaging approaches such as fMRI, researchers have demonstrated that the neural changes in subcortical regions including the brainstem, amygdala, hippocampus, thalamus, striatum, and orbitofrontal/medial prefrontal cortex (Fig. 1) can serve as neural biomarkers of an individual's arousal level and help in predicting various aspects of human performance (Clewett, Huang, Velasco, Lee, & Mather, 2018; Greening, Lee, & Mather, 2016; Lee, Sakaki, Cheng, Velasco, & Mather, 2014; Lee et al., 2018; Mather et al., 2017).

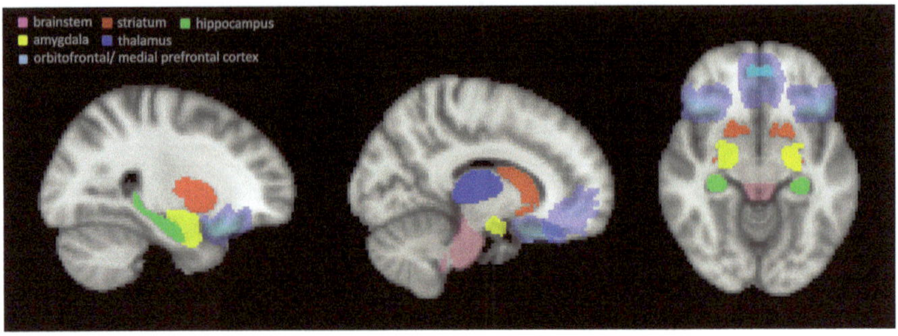

Fig. 1 Subcortical regions associated with arousal changes in the brain

In experimental and industrial ergonomic settings, however, it is not easy to utilize neuroimaging techniques to quantify an individual's arousal level for the prediction of human performance because the neuroimaging techniques are vulnerable to the slightest head and body movements. In applications of experimental ergonomics, participants may be asked to walk around or reorient their heads or bodies to respond to simulated environments such as driving or working. Thus, it may not be optimal for researchers in human factors engineering to use the human neuroimaging approach as their main tool for measuring changes in arousal in experimental situations.

In this context, it is worth noting that psychophysiological measurements such as SCR, ECG, and eye movement have provided reliable metrics of arousal changes in previous studies investigating how, and to what extent, arousal predicts various aspects of human cognitive performance (Lee, Itti, & Mather, 2012; Lee et al., 2014; Lee, Greening, & Mather, 2015). For example, Lee et al. (2014) tested whether an arousing cue associated with the electric stimulation of fingers could enhance perceptual sensitivity compared to a non-arousing cue. In this study, the researchers measured participants' SCR-level changes and found that increased SCR, as an index of heightened arousal, subsequently made low-contrast and low spatial frequency stimuli more easily perceptible. In addition, Lee and his colleagues demonstrated arousal-induced memory enhancement (Lee et al., 2015). In this study, participants were asked to remember neutral objects (i.e., learning phase) on the screen while hearing an emotionally arousing or non-arousing tone. They were then asked to recall what they had seen (i.e., test phase) during the learning phase. As a result, the researchers found that hearing a highly arousing tone elevated participants' SCR levels during the learning phase and subsequently led participants to better remember what they had seen.

HR also reliably reflects a participant's arousal level changes such that visually arousing stimuli significantly increase the R-R interval variability of HR compared to non-arousing visual stimuli (Choi et al., 2017). Recognizing the convenience and reliability of HR changes for measuring arousal, researchers have developed various classification algorithms to identify arousal level changes based on the variability of HR. For instance, Yu and Chen (2015) developed an emotion recognition system for a range of emotions including neutral, happy, stressful, and sad emotions based on individuals' HR pattern changes. To do this, they first categorized the four emotional states based on four HR features, namely, time domain, frequency domain, Poincare plot, and differential features. They then trained the support vector machine with these HR features. As a result, the classifier was able to successfully classify the four emotion states based on the new HR dataset with a high recognition rate of around 90%, indicating that HR patterns can effectively reflect different emotional states.

Another example of how psychophysiology can be used to measure arousal level associated with human performance is eye movement and pupil dilations. Kim and Nembhard (2019) conducted a study in which they experimented with manipulating time pressure. The participants had to complete a task within a certain time frame, and the researchers predicted that increased arousal level due to time pressure would induce more task errors. They found that time pressure-induced arousal reduced task performance level (i.e., increased the number of erroneous responses), and this arousal effect was successfully indexed by eye movement such that participants showed an increased number of fixations in eye movement. Moreover, it is well documented that pupil dilation significantly increases as a function of arousal level. Pupil dilation has been shown to occur under fearful and stressful conditions manipulated by electrical shock as a form of negative feedback for participants (Andreassi, 2010; Polt, 1970).

Interestingly, fMRI studies indicate that psychophysiological measures reliably detect arousal changes in humans. For example, pupil size, SCR, and HR changes are directly linked to the neural activity of subcortical regions such as the locus coeruleus (Clewett et al., 2018; Lee et al., 2014, 2018; Mather et al., 2017), which is known as a major arousal modulating system in the brain (Aston-Jones & Cohen, 2005; Mather, Clewett, Sakaki, & Harley, 2016) as well as the amygdala, the region that triggers noradrenergic-sympathetic arousal changes (de Voogd, Fernández, & Hermans, 2016; Laine, Spitler, Mosher, & Gothard, 2009; Williams et al., 2004; Wood, Ver Hoef, & Knight, 2014). In particular, Clewett et al. (2018) have shown that the neural activity of the locus coeruleus is modulated as a function of pupil size. When they used pupil size changes as parametric regressors in a general-linear model to estimate the brain activities measured by fMRI, they found that the neural activity of the locus coeruleus was associated with physiological arousal increments, such that increased pupil size induced a greater augmented locus coeruleus response in the brain. That is, pupil size changes were linked to the emotional states of the brain. Given that the psychophysiological responses have been found to have such a robust and valid relationship with behavioral and neural changes of arousal, physiological measures can provide a useful index for assessing human arousal changes in affective ergonomic research, thus allowing researchers to move beyond the neuroimaging approach.

5 Applications to Product Design and Evaluation

The reliable measurement of arousal changes using psychophysiological techniques can be applied to industrial settings, including user experience evaluations and designs (Dirican & Göktürk, 2011). These days, many researchers have adopted physiological measures to quantify users' subjective and emotional changes. Affective engineering is one such area in which researchers investigate the interaction between users' emotional changes and product design (Nagamachi, 2008). For example, facial EMG has been used to quantify emotional changes while assessing several website prototypes (Benedek & Hazlett, 2005; Wise, Alhabash, & Park, 2010). Prototypes that showed increased zygomatic muscle activation in the cheek area of the face corresponded to the prototype that was top-ranked on the self-reported desirability questionnaire. Meanwhile, SCR has been used to provide reliable measurements of users' stress and frustration levels while playing video games (Ambinder, 2011). In addition to current research that uses neuroergonomic and physiological metrics to evaluate individuals' arousal levels, future research should consider using physiological metrics as inputs for industrial systems. This will allow researchers to design personalized systems that meet the needs of diverse groups of people.

References

Ambinder, M. (2011, March). Biofeedback in gameplay: How valve measures physiology to enhance gaming experience. In *Game Developers Conference* (Vol. 2011).

Andreassi, J. L. (2010). *Psychophysiology: Human behavior and physiological response*. Psychology Press.

Aston-Jones, G., & Cohen, J. D. (2005). An integrative theory of locus coeruleus-norepinephrine function: Adaptive gain and optimal performance. *Annual Review of Neuroscience, 28,* 403–450.

Benedek, J., & Hazlett, R. (2005). Incorporating facial emg emotion measures as feedback in the software design process. In *Proceedings of Human Computer Interaction Consortium.*

Choi, K. H., Kim, J., Kwon, O. S., Kim, M. J., Ryu, Y. H., & Park, J. E. (2017). Is heart rate variability (HRV) an adequate tool for evaluating human emotions?—A focus on the use of the International Affective Picture System (IAPS). *Psychiatry Research, 251,* 192–196.

Clewett, D., Huang, R., Velasco, R., Lee, T.-H., & Mather, M. (2018). Locus coeruleus activity strengthens prioritized memories under arousal. *Journal of Neuroscience*, 2017–2097.

Costanzo, L. S. (2018). *Physiology* (6th ed.). Philadelphia, PA: Elsevier, Web.

de Voogd, L. D., Fernández, G., & Hermans, E. J. (2016). Disentangling the roles of arousal and amygdala activation in emotional declarative memory. *Social Cognitive and Affective Neuroscience, 11*(9), 1471–1480.

Dirican, A. C., & Göktürk, M. (2011). Psychophysiological measures of human cognitive states applied in human computer interaction. *Procedia Computer Science, 3,* 1361–1367.

Gazzaniga, M. S., Ivry, R. B., & Mangun, G. R. (2008). *Cognitive neuroscience: The biology of the mind* (5th ed.). W. W. Norton & Company.

Greening, S., Lee, T. H., & Mather, M. (2016). Individual differences in anticipatory somatosensory cortex activity for shock is positively related with trait anxiety and multisensory integration. *Brain Sciences, 6*(1), 2.

Kim, J. E., & Nembhard, D. A. (2019). Eye movement as a mediator of the relationships among time pressure, feedback, and learning performance. *International Journal of Industrial Ergonomics, 70,* 116–123.

Laine, C. M., Spitler, K. M., Mosher, C. P., & Gothard, K. M. (2009). Behavioral triggers of skin conductance responses and their neural correlates in the primate amygdala. *Journal of Neurophysiology, 101*(4), 1749–1754.

Lee, T. H., Greening, S. G., & Mather, M. (2015). Encoding of goal-relevant stimuli is strengthened by emotional arousal in memory. *Frontiers in Psychology, 6,* 1173.

Lee, T.-H., Greening, S. G., Ueno, T., Clewett, D., Ponzio, A., Sakaki, M., & Mather, M. (2018). Arousal increases neural gain via the locus coeruleus–noradrenaline system in younger adults but not in older adults. *Nature Human Behaviour, 2*(5), 356.

Lee, T. H., Itti, L., & Mather, M. (2012). Evidence for arousal-biased competition in perceptual learning. *Frontiers in Psychology, 3,* 241.

Lee, T.-H., Sakaki, M., Cheng, R., Velasco, R., & Mather, M. (2014). Emotional arousal amplifies the effects of biased competition in the brain. *Social Cognitive and Affective Neuroscience, 9*(12), 2067–2077.

Liang, X., & Balasingham, I. (2007, May). Performance analysis of the IEEE 802.15. 4 based ECG monitoring network. In *Proceedings of the 7th IASTED International Conferences on Wireless and Optical Communications* (pp. 99–104).

Marek, T., Karwowski, W., & Rice, V. (Eds.). (2010). *Advances in understanding human performance: Neuroergonomics, human factors design, and special populations*. CRC Press.

Mather, M., Clewett, D., Sakaki, M., & Harley, C. W. (2016). GANEing traction: The broad applicability of NE hotspots to diverse cognitive and arousal phenomena. *Behavioral and Brain Sciences, 39.*

Mather, M., & Sutherland, M. R. (2011). Arousal-biased competition in perception and memory. *Perspectives on Psychological Science, 6*(2), 114–133.

Mather, M., Yoo, H. J., Clewett, D. V., Lee, T.-H., Greening, S. G., Ponzio, A., … Thayer, J. F. (2017). Higher locus coeruleus MRI contrast is associated with lower parasympathetic influence over heart rate variability. *Neuroimage, 150,* 329–335.

Mehta, R. K., & Parasuraman, R. (2013). Neuroergonomics: A review of applications to physical and cognitive work. *Frontiers in Human Neuroscience, 7,* 889.

Morgane, P. J., Galler, J. R., & Mokler, D. J. (2005). A review of systems and networks of the limbic forebrain/limbic midbrain. *Progress in Neurobiology, 75*(2), 143–160.

Nagamachi, M. (2008). Perspectives and the new trend of Kansei/affective engineering. *The TQM Journal, 20*(4), 290–298.

Parasuraman, R. (2003). Neuroergonomics: Research and practice. *Theoretical Issues in Ergonomics Science, 4*(1–2), 5–20.

Parasuraman, R. (2011). Neuroergonomics: Brain, cognition, and performance at work. *Current Directions in Psychological Science, 20*(3), 181–186.

Polt, J. M. (1970). Effect of threat of shock on pupillary response in a problem-solving situation. *Perceptual and Motor Skills, 31*(2), 587–593.

Rayner, K. (1998). Eye movements in reading and information processing: 20 years of research. *Psychological bulletin, 124*(3), 372.

Son, I. Y., Guhe, M., Gray, W. D., Yazici, B., & Schoelles, M. J. (2005, May). Human performance assessment using fNIR. In *Biomonitoring for Physiological and Cognitive Performance During Military Operations* (Vol. 5797, pp. 158–170). International Society for Optics and Photonics.

Stern, R. M., Stern, R. M., Ray, W. J., & Quigley, K. S. (2001). *Psychophysiological recording*. USA: Oxford University Press.

Supuk, T., Skelin, A., & Cic, M. (2014). Design, development and testing of a low-cost sEMG system and its use in recording muscle activity in human gait. *Sensors, 14*(5), 8235–8258.

Thayer, R. (1989). *The psychobiology of mood and arousal*. Oxford: Oxford University Press.

Williams, L. M., Das, P., Harris, A. W., Liddell, B. B., Brammer, M. J., Olivieri, G., … Peduto, A. (2004). Dysregulation of arousal and amygdala-prefrontal systems in paranoid schizophrenia. *American Journal of Psychiatry, 161*(3), 480–489.
Wise, K., Alhabash, S., & Park, H. (2010). Emotional responses during social information seeking on facebook. *Cyberpsychology, Behavior, and Social Networking, 13*(5), 555–562.
Wood, K. H., Ver Hoef, L. W., & Knight, D. C. (2014). The amygdala mediates the emotional modulation of threat-elicited skin conductance response. *Emotion, 14*(4), 693.
Yu, S. N., & Chen, S. F. (2015, August). Emotion state identification based on heart rate variability and genetic algorithm. In *2015 37th Annual International Conference of the IEEE Engineering in Medicine and Biology Society (EMBC)* (pp. 538–541). IEEE.

"Hello Computer, How Am I Feeling?", Case Studies of Neural Technology to Measure Emotions

Ian Daly and Duncan Williams

Abstract Emotion is a core part of the human experience. Many artistic and creative applications attempt to produce particular emotional experiences, for example, films, games, music, dance, and other visual arts. However, while emotional states are ubiquitous, they are also complex, proving difficult to describe to others by conventional psychometric means (e.g., traditional self-report mechanisms). Neural technology offers the potential to circumvent these difficulties by allowing the creation of a real-time, objective, metric of felt emotion to assist in emotionally driven experience design across a range of disciplines. This chapter discusses how neural technology based on the processing of the electroencephalogram may be used to measure human emotions in natural environments. We also present a set of case studies of applications that use neural technology to measure emotions. We are particularly interested in the use of neural technology to inform applications which can respond to the felt experience of the individual. We describe two case studies focused on driving scenarios and brain–computer music interfacing. The chapter concludes with a discussion of the challenges inherent in developing neural technology to measure emotion and a set of suggestions for future research directions in developing applications that use neural technology as an objective measure of emotion.

1 Introduction

The experience of emotion is a core part of our humanity; it is the driver behind much of our behavior, and forms the bedrock of the motivations behind many of our actions throughout our lives. Yet, despite the overwhelming importance of emotion

I. Daly (✉)
Brain-Computer Interfacing and Neural Engineering Lab, Department of Computer Science and Electronic Engineering, University of Essex, Colchester CO4 3SQ, UK
e-mail: i.daly@essex.ac.uk

D. Williams
Digital Creativity Labs (Computer Science Department), University of York, York YO10 5DD, UK

C. S. Nam (ed.), *Neuroergonomics*, Cognitive Science and Technology,
https://doi.org/10.1007/978-3-030-34784-0_11

in defining who we are and what we do, we still do not fully understand the neural or physiological bases of emotion (Darwin, 1998; Reeve, 2014).

An emotion may, broadly speaking, be defined as an action, or a pattern of multiple actions that occur in response to one or more external stimuli or that arise spontaneously in an individual's (Barrett, Mesquita, Ochsner, & Gross, 2007). Emotions may arise either as a result of environmental stimuli or from the body (for example, by memories or in response to physical pain or discomfort) (Mier et al., 2010).

In order to scientifically understand emotions, it is first necessary to develop a quantifiable framework by which emotion may be measured. Efforts to do so in the field of psychology have resulted in the development of multidimensional frameworks for describing emotions within a variety of contexts as well as quantifiable labeling systems to allow the systematic description of emotions.

The scientific study of emotion first seeks to identify a system by which different emotions (affective states) may be categorized. A large number of different systems have been proposed to achieve this goal, including category labeling systems and multidimensional systems spanning different numbers of dimensions.

The multidimensional schemes developed to categorize emotions include the valence–arousal circumplex model of affect (Russell, 1980), the three-dimensional Schimmack and Grob model (2000), and other multidimensional models such as the three-dimensional pleasure-arousal-dominance model (Mehrabian, 1996). Alternatively category labeling systems, such as the Geneva Emotional Music Scale (GEMS) (Zentner, Grandjean, & Scherer, 2008) or the Ortony, Clore, and Collins category model (1988), may be used to categorize affect via a set of discrete labels.

Multidimensional categorization schemes map affective states to a continuous space. For example, the valence–arousal circumplex model maps emotions onto a continuous two-dimensional plane and specific affective states can be placed at any location. This approach has the advantage of allowing fine-grained measurement and differentiation of affective states, but may not be the most suitable or accurate categorization.

Indeed it has been argued that, despite the widespread use of the two-dimensional valence–arousal circumplex, emotions are not accurately represented using just two dimensions; and dimensionless category labeling schemes, such as GEMS, are more accurate (Fontaine, Scherer, Roesch, & Ellsworth, 2007).

Arguably, the most widely used emotion categorization system is the two-dimensional valence–arousal circumplex (Russell, 1980). This framework classifies all emotions along two axes, valence (the pleasantness or unpleasantness of the emotion) and arousal (the level of excitement of the emotion). Thus, for example, "fear" may be placed on the two-dimensional plane of the circumplex at a position with low (negative, or "unpleasant") valence and high (positive, or "exciting") arousal while "excitement" may be placed on the circumplex at a position of high valence (very pleasant) and high arousal (very exciting).

The valence–arousal circumplex is illustrated in Fig. 1.

This two-dimensional framework has been extended by others to include a third dimension, tension, which is defined as the level of strain or stress a particular emotion involves (Schimmack & Grob, 2000). Indeed, some research suggests this

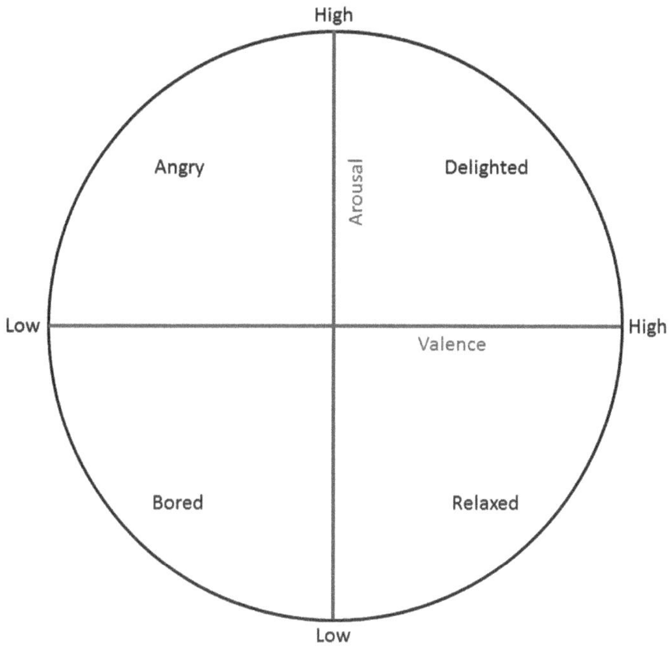

Fig. 1 The valence–arousal circumplex

three-dimensional framework may provide a more complete and accurate description of emotional responses (Daly et al., 2014). However, others suggest that the linear two- or three-dimensional relationships suggested by these frameworks are misleading and in fact emotion is a highly nonlinear, multidimensional experience that is not well described by two- or three-dimensional Cartesian geometry (Fontaine et al., 2007).

Efforts to quantify and describe emotions in a multidimensional space have made use of labels to describe emotional responses in a discipline-specific framework. For example, the Geneva Emotional Music Scale (GEMS) attempts to describe emotional responses to a piece of music via a set of selected adjectives, which an individual is asked to select from in order to describe their current emotional experience. However, such systems are not as readily amenable to quantification or translation between languages or cultures. Thus, for the present, the two- or three-dimensional quantifiable frameworks, such as the valence–arousal circumplex, remain widely used.

Although emotions are ubiquitous in life and analogous to vision or hearing in their effect on our moment-by-moment perception of the world, they, historically, have received far less interest from the scientific community (Kassam, Markey, Cherkassky, Loewenstein, & Just, 2013). Consequently, the neural basis of emotions is far less well understood than the neural basis of our other senses (Kassam et al., 2013).

Meta-analysis of neuroimaging studies investigating the neural basis of emotion has revealed a network of brain regions that are involved in emotional responses (Koelsch, 2010). This network includes, among other brain areas, the amygdala, hippocampal formation, right ventral striatum, pre-supplementary motor cortex, left caudate nucleus, and the cingulate cortex (Koelsch, 2014).

Exploration of this network via techniques such as functional magnetic resonance imaging (fMRI) (which measures slow changes in concentrations of oxyhemoglobin throughout the brain) or electroencephalography (EEG) (which measures fast changes in electro-potential in the cortex) has revealed particular patterns of neural activity to relate to specific felt emotions. For example, changes in a participant's felt valence (how pleasant or unpleasant they find a stimulus or situation), also referred to as "approach–withdrawal" behavior (Hunter & Schellenberg, 2010), involves changes in activity in the paralimbic system and amygdala, among others (Blood & Zatorre, 2001). These changes in neural activity are also reflected in the EEG, for example, in the prefrontal asymmetry (Coan & Allen, 2004).

However, there remains much to discover about how our brains and our bodies work together to generate the experience of emotions.

A better understanding of the basis of emotions can lead to numerous benefits. The most valuable of these, arguably, relate to medical interventions for individuals living with mood disorders such as depression or anxiety (Hou et al., 2017). However, more accurate understanding of the basis of emotions can also lead to numerous other applications, including within the field of neuroergonomics to design systems that better respond to our current felt emotions in a way that is useful or beneficial to us.

Accurately identifying an individual's current emotion is not trivial. One could simply ask an individual, or a group of individuals, what emotions they are feeling after exposure to a stimulus. However, individuals respond to a stimulus in a way that depends on their lived experiences, memories, current mood, recent levels or rest, and other emotional and physiological wants and desires (which themselves change from moment to moment) (Barrett et al., 2007). Consequently, the way an individual emotionally responds to a situation is specific to them and may vary considerably from the way another individual may respond to the same situation.

Furthermore, individuals may also not be able to accurately evaluate or describe their own emotions. The particular choice of words that an individual uses to describe their current emotions are largely dependent on their background and can change radically, even between individuals with very similar backgrounds (e.g., between siblings). One could use a standard scale to allow individuals to report their emotions (for example, by providing a discrete set of pre-chosen keywords as used in the Geneva Emotional Music Scales system (Vuoskoski & Eerola, 2011)), but even this approach is not without flaws as each individual may interpret the chosen words or scale differently.

Finally, individuals may not be honest when reporting some of their emotional responses. This is particularly the case with emotions at the more extreme ends of the scale or in response to stimuli that may have some associated socially dictated negative or embarrassing connotations for an individual.

These problems are increasingly being tackled by making use of our growing understanding of the neural basis of emotions. Specifically, neural technology is being developed and used to produce systems that are able to measure and respond to an individual's current emotional (affective) state.

Neural technology allows for the development of a set of metrics that are able to provide objective measures of an individual's current felt emotions. Such metrics have numerous advantages. For example, they can allow objective measurement of an individual's felt response to a stimulus or situation. This has numerous applications, including psychology or neuroscience experiments to overcome the problems with traditional emotion assessment methods outlined above.

Neural technology-based affect metrics may also be used within a neuroergonomic context to allow a system to respond to neural measures of an individual's current felt affective state. For example, an application could identify when a user is becoming frustrated with some aspect of the application and adjust accordingly.

Recent research on the development of neural technology-based metrics for measuring emotions makes use of measures of an individual's physiological state. Some initial applications have been developed that make use of these metrics for applications designed to aid people in their interactions with the world.

Of course, the use of neural technology to measure physiological correlates of emotion introduces its own specific considerations. For example, physiological correlates of emotion differ between individuals as a result of inter-person differences in physiology, signal noise, and differences in measured signal strength. These factors mean identifying a common signal component that can be used as a reliable indicator of emotion presents a considerable challenge.

In this chapter, we first introduce how neural technology, and in particular neural technology-based on the EEG, can be used to provide a metric of an individual's affective state before introducing some existing neural technology-based methods that have been developed to monitor affective states. We then go on to describe two case studies that illustrate how these metrics may be used to monitor affect and the applications that may be built around these metrics. Finally, we discuss some possible future directions in the development of this technology and the associated challenges that still need to be overcome.

2 Neural Technology-Based Measures of Emotions

There are several neuroimaging technologies available to measure activity in the brain and these vary in resolution (in space, time, and frequency), as well as in their cost and portability. Many of these technologies can be used to provide an objective measure of felt emotion (Canli, Desmond, Zhao, Glover, & Gabrieli, 1998; Ishii et al., 2014; Skouras, Gray, Critchley, & Koelsch, 2013). However, of the available neuroimaging technologies, the electroencephalogram (EEG) is generally considered to be the most suitable for providing real-time, portable, and accurate measures of an individual's current affective state due to its relative low cost, portability, high temporal and spectral resolutions, and its ability to be used directly without invasive surgery (Niedermeyer & Silva, 2005).

EEG is a measurement of summed electrophysiological activity from large numbers of cortical neurons. It is recorded from the surface of the scalp by either passive or active electrodes and has a time resolution in the order of hundreds or thousands of samples per second (Niedermeyer & Silva, 2005).

When using the EEG to measure an individual's affective state, there are a few key features that have been reliably identified and shown to allow accurate real-time measurement and tracking of an individual's current affective state. The most widely used of these metrics are

1. The prefrontal asymmetry,
2. Midline theta band power, and
3. Functional connectivity.

Prefrontal EEG asymmetry is a measure of the relative difference in EEG activity in the left- and right-hand prefrontal cortices (Coan & Allen, 2004). The ratio of EEG activity in this region is known to vary within specific frequency bands—including, but not limited to, the alpha (8–13 Hz), beta (13–25 Hz), and theta (4-8 Hz) frequency bands—as a function of approach–withdrawal behavior in response to numerous different types of stimuli, including images (Winkler, Jäger, and Vojkan Mihajlović, 2010), videos (Jones & Fox, 1992), and music (Schmidt & Trainor, 2001). This is expressed in the "hemispheric valence hypothesis", which states emotions related to approach to a stimulus (i.e., positive responses to a stimulus) are processed more in the left prefrontal cortex while withdrawal (negative) responses to a stimulus are processed more in the right prefrontal cortex (Canli et al., 1998; Silberman & Weingartner, 1986). Additionally, the parietotemporal cortex in the right hemisphere has been shown to be involved in autonomic behavior while arousal is also involved in suppression of alpha power in the right posterior cortex and increases in theta EEG power in the left prefrontal lobe (Rogenmoser, Zollinger, Elmer, & Jäncke, 2016).

A less frequently explored, yet robust, metric of felt affect is the midline theta, the change in power of the EEG within the theta frequency band as measured along the central sulcus (the midline) (Aftanas, Varlamov, Pavlov, Makhnev, & Reva, 2001). This metric has been reported to be positively correlated with the valence of an affect (Sammler, Grigutsch, Fritz, & Koelsch, 2007). Specifically, the more pleasant a stimulus or situation is felt to be by the individual, the greater their midline theta activity.

Finally, functional connectivity (a measure of the statistical relationships between different regions of the brain (Sporns, 2007)) has also been reported to be used as a metric of the felt emotional response to a stimulus (Daly et al., 2014). Functional connectivity refers to the statistical relationships between neural activity signals recorded in different locations within, or on the surface of, the brain (Raichle, 2011). Larger statistical relationships between different signals suggest a meaningful relationship between the measured regions of the brain.

Functional connectivity may be differentiated from anatomical connectivity, which measures anatomical connections between different brain regions (Sporns,

2007). For example, in the case of functional connectivity a statistical relationship may exist as a result of a common anatomical connection—both measured brain regions share with a third, unmeasured, brain region (Sporns, 2007).

Functional connectivity has been shown to change a result of changes in felt affect in a growing number of studies, including responses to music (Nicolaou et al., 2017) and after watching film clips chosen to explicitly specific emotional response (Lee & Hsieh, 2014). Furthermore, functional connectivity patterns have also been observed to vary significantly as a result of major depression (Fingelkurts et al., 2007) and other disorders of mood (Davidson, Abercrombie, Nitschke, & Putnam, 1999) within clinical populations.

Human emotion is increasingly understood to be an embodied experience (Effron, Niedenthal, Gil, & Droit-Volet, 2006). Felt emotional response to a stimulus or situation is not just a product of the activity in the brain but also a product of the autonomic response system (Kreibig, 2010). Thus, accurate physiological metrics of affective states may be identified by incorporating other physiological measures recorded from the body. These may include measures such as the heart rate, galvanic skin response, respiration rate, and blood oxygenation levels (Kreibig, 2010).

The relationships between physiological responses to emotion and specific categorization systems for emotion are varied and complex. For example, the heart rate (as measured by the ECG) not only is well-known to correlate with felt arousal but also relates to felt valence responses to some types of stimuli, such as music (Daly, Malik, et al., 2015; Daly, Williams, et al., 2015).

A system which combines some or all of these physiological measures along with neural measures of EEG indices of affect is defined as a "hybrid measure" and is increasingly used in, among other applications, hybrid brain–computer interface systems for monitoring and measuring affective states in individuals by recording and classifying combinations of neural and physiological activity (Fazli et al., 2012; Müller-Putz et al., 2011; Nicolas-Alonso & Gomez-Gil, 2012; Pfurtscheller et al., 2010).

It may be argued that measures of the autonomic systems alone may be sufficient to accurately measure affective responses to stimuli and that, in fact, there is no need for measures of the neural response to stimuli. However, this has been shown not to be the case (Daly, Malik, et al., 2015; Daly, Williams, et al. 2015). Indeed hybrid measures incorporating both neural and physiological measures of affective states have been shown to be able to more accurately measure an individual's current affective state than either set of measures alone (Daly, Malik, et al., 2015; Daly, Williams, et al. 2015).

We conclude that there is a growing range of neural technology-based measures of emotions. These may be used in a wide variety of applications, as we will illustrate in our case study below.

3 Case Study 1: Affective Brain–Computer Music Interfacing

3.1 Introduction

In this section, we introduce and describe a case study of a brain–computer music system (BCMI) developed to monitor and respond to an individual's current affective state. This system was originally described in full in (Daly et al., 2016a) and is presented here in brief.

Music is a unique and powerful method for influencing an individual's current affective state (Sacks, 2006). Even very short clips of music can produce varied and powerful emotional responses in the listener. For example, something as simple as a few piano keys played over a few seconds can produce fear, excitement, contentment, or melancholy depending on the exact nature of the music, the current mood of the listener, and other extraneous factors.

Because of this, music is increasingly being explored as a tool in therapy (Bradt, Magee, Dileo, Wheeler, & McGilloway, 2010; Erkkilä et al., 2011; Livingstone & Thompson, 2009). Indeed, music therapy is increasingly used as a therapeutic technique to help facilitate emotional communication in patients with a range of emotional disorders. The BCMI described in this case study is an attempt to provide a form of automated music therapy.

3.2 Outline

An overview of the BCMI developed and used in this case study is illustrated in Fig. 2.

The following physiological signals are recorded from the users at a sample rate of 1,000 Hz: EEG, electrocorticogram (ECG), galvanic skin response, and respiration. These are then used to attempt to detect the user's current affective state, as described in Sect. 3.3 below. A case-based reasoning system (Webber, Aha, Muñoz-Ávila, & Breslow, 2000) is then employed to identify the ideal set of music to play to the participant in order to move them from their current affective state to a new target affective state. Finally, a music generator (Williams et al., 2015, 2017) is used to generate the music identified by the case-based reasoning system and play it to the BCMI user.

A typical use case envisioned for this BCMI is modulating the affective state of an individual who is feeling mildly depressed, i.e., in a low valence, low arousal state (Russell, 1980). In this case, the affective state detection system would identify the user's current affective state as "low valence", "low arousal". The case-based reasoning system would then identify the set of musical features that would be most effective at improving that user's affective state, before the music generator is used to generate and play this music to the user.

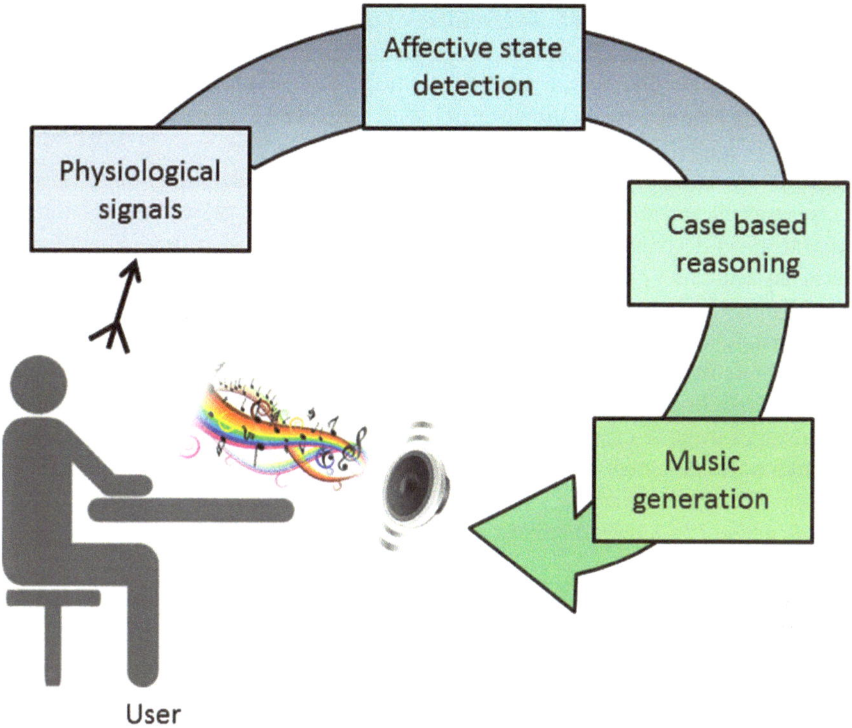

Fig. 2 Schematic of the BCMI system developed in case study 1

Of note, this system does not make any a priori assumptions about what musical transitions would be most effective for each user. This is motivated by evidence that each individual responds to music differently (Eerola, 2012). For example, a piece of music that makes one person happy may bore or frustrate another. Instead, the optimal set of music pieces to produce a target affect in an individual is determined in a training session with each individual prior to the use of the BCMI.

3.3 *Affective State Detection*

The affective state detection method used in this BCMI makes use of EEG band powers in a range of frequency and spatial regions. Specifically, the EEG is subdivided into channel sets placed over 10 key regions of the scalp, including the prefrontal, central, and occipital regions. These candidate features were chosen based on their reported successful use in classifying affective states in prior work. For example, a system proposed in (Stikic, Johnson, Tan, & Berka, 2014) uses these spatial regions to train a feature selector and classifier. This use of a wide range of different spatial regions to detect an individual's affective state is supported by work from several authors that suggests a broad involvement of different spatial regions in affective states (Daly et al., 2014; Heller, 1993; Rogenmoser et al., 2016). The EEG within each region is also filtered into 10 different frequency bands, encompassing all the key frequency bands widely used in EEG analysis and their sub-bands (Schomer & Lopes de Silva, 2011).

In total 10 different spatial regions and 10 different frequency bands are used in the study, resulting in 100 candidate EEG features. These were complemented by additional features describing each of the other physiological signals used. Specifically, the mean heart rate (peak-to-peak intervals), mean blood oxygenation rate (peak-to-peak intervals), dominant frequency of the respiration rate, and mean amplitude of the change in galvanic skin response were also used as descriptive features. This resulted in a total of 104 candidate features.

A stepwise linear regression feature selection method was then used to identify the subset of these features that was most effective at classifying the user's current affective state. Affective states were quantized into nine different values. Each of these quantized affective states is represented by a tuple composed of valence and arousal, each of which can take one of three values: high, medium, and low.

A support vector machine (SVM) classifier was trained and used to attempt a classification of the affective states. Specifically, we attempt to identify a participant's reported felt affective state via an SVM classifier trained on that participant's EEG and physiological features.

Full details of the development of this metric are described in Daly et al. (2016a, 2016b) and its implementation in the BCMI is described in more detail in Daly et al. (2016b).

3.4 Evaluation

The BCMI was evaluated with a cohort of 22 healthy participants over a series of experimental sessions (Daly et al., 2016b). The first four sessions were used to train the case-based reasoning system and the affective state detection method while the final session was used to evaluate the system.

Training and evaluation sessions took place over several, nonconsecutive days. The training sessions were used to map relationships between music features and individual affective responses in order to train the case-based reasoning system. They were also used to train the feature selection and classification steps of the affective state recognition system.

Additionally, the BCMI was also evaluated in a follow-up case study with a single individual with Huntington's disease to demonstrate its ability to also work for users with medical conditions who may stand to benefit from BCMI use (Daly et al., 2017).

Huntington's disease affects 7–12 in 100,000 people. It is a progressive central nervous system disorder that, over 10–12 years, reduces movement control and impairs cognition and behavior. It also impairs an individual's emotion regulation and, it is suggested, music therapy techniques (such as BCMI), may be beneficial (Bates, Tabrizi, & Jones, 2014; Pridmore, 1990).

Evaluation—with both the healthy user cohort and the individuals with Huntington's disease—was performed by asking the system to attempt to make a series of specific modulations to the user's affective states. Four objectives were selected, based upon potential music therapy use case scenarios. These objectives were

1. Make the user happy (move the user from a low valence to a high valence state).
2. Calm the user (reduce the user's arousal).
3. Reduce stress (increase valence while simultaneously reducing arousal).
4. Excite the user (increase the users felt arousal).

In each trial, a target change in affective state was selected at random and participants were not informed about which affective state transition was being targeted. The music was then generated according to the rules of the case-based reasoning system determined in the training sessions of the experiment.

The success of the BCMI at each of these objectives was measured by comparing the resulting trajectories of changes in user-reported affective states with trajectories that would be achieved by a random system, i.e., a system that just played randomly selected music to the users.

Users reported their current felt affect throughout the experiment using the real-time affect reporting tool FEELTRACE, which allows users to report, in real-time, their current felt affective states as a coordinate trajectory on the two-dimensional valence–arousal circumplex as they listen to the music (Cowie et al., 2000). We hypothesized that if the BCMI was not working, users' affective state trajectories, as reported via FEELTRACE, would be random. On the other hand, if the BCMI was working correctly, changes in valence and arousal that correspond to the target objectives would be observed. Full details of the experiment design are described in Daly et al. (2016b).

3.5 *Results*

The affective state detection method developed in the BCMI was able to identify affective states with statistically significant classification accuracies for the majority of participants. This is best illustrated in the work described in Daly, Malik, et al. (2015), Daly, Williams, et al. (2015), which evaluates the affective state detection method in isolation with a separate population of healthy participants.

The results are listed in Table 1 for valence classification and in Table 2 for arousal classification.

It may be observed that for the majority of the affective state detection attempts, the affective state detection method was able to identify the correct affective states with statistically significant accuracies ($p < 0.05$). This demonstrates how neural engineering may be employed to accurately identify an individual's current emotions in the context of a BCMI system.

Table 1 Classification accuracies resulting from our affective state detection method for differentiating high and low valence conditions. Statistical significance (p) is estimated from the binomial distribution. "Acc." denotes accuracy and "STD" denotes standard deviation

Participant	Acc. (mean)	Acc. (STD)	P
1	0.753	0.066	<0.001
2	0.597	0.059	0.032
3	0.619	0.084	0.019
4	0.609	0.087	0.055
5	0.628	0.069	0.014
6	0.633	0.079	0.034
7	0.629	0.081	0.034
8	0.599	0.057	0.018
9	0.671	0.063	0.009
10	0.537	0.122	0.160
11	0.629	0.048	0.006
12	0.649	0.054	0.002
13	0.647	0.067	0.006
14	0.705	0.072	<0.001
15	0.549	0.068	0.067
16	0.604	0.069	0.021
17	0.568	0.059	0.053
Avg.	0.624	0.071	–

Table 2 Classification accuracies resulting from our affective state detection method for differentiating high and low arousal conditions. Statistical significance (P) is estimated from the binomial distribution. "Acc." denotes classification accuracy and "STD" denotes standard deviation

Participant	Acc. (mean)	Acc. (STD)	P
1	0.724	0.048	<0.001
2	0.726	0.074	<0.001
3	0.694	0.065	<0.001
4	0.779	0.061	<0.001
5	0.746	0.061	<0.001
6	0.561	0.060	0.061
7	0.591	0.099	0.081
8	0.701	0.050	<0.001
9	0.795	0.056	<0.001
10	0.598	0.099	0.097
11	0.700	0.051	<0.001
12	0.659	0.053	<0.001
13	0.677	0.059	0.002
14	0.733	0.057	<0.001
15	0.694	0.052	<0.001
16	0.729	0.055	<0.001
17	0.688	0.144	0.003
Avg.	0.694	0.067	–

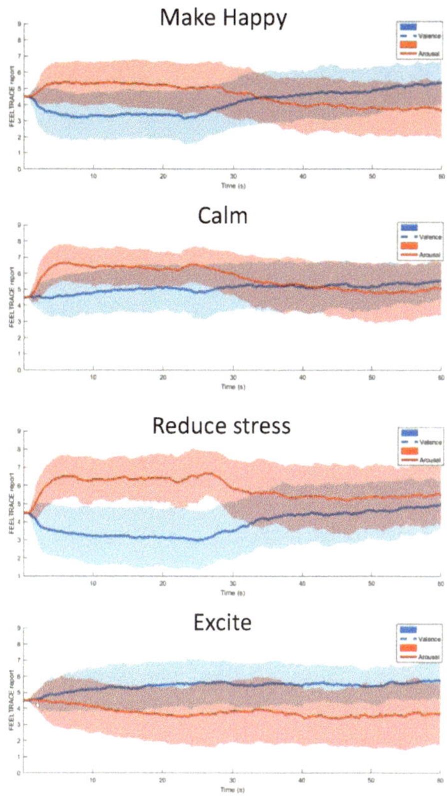

Fig. 3 Examples of affective state trajectories achieved by the BCMI system. FEELTRACE report indicates the output of the FEELTRACE interface used to allow participants to report their current affective states

The BCMI system was able to significantly modulate the user's affective states under all scenarios in the majority of cases. Examples of the affective trajectories achieved by the system are illustrated in Fig. 3. Valence and arousal are plotted as separate trajectories to allow easier interpretation of the results.

It may be observed that, in each case, the BCMI was able to move the users from their starting affective state (the first 20s of the trial) to the target affective state (the last 20s of the trial). The BCMI begins to attempt to change each user's affective state from 20s into each trial and this can be clearly seen in the plots of the mean affective state trajectories across all participants.

Specifically, there is a clear, visible change in reported affective state as the BCMI is used. For example, the third objective is to reduce the participant's level of stress ("de-stress" in the figure), which corresponds to increasing the felt valence and reducing the felt arousal (i.e., making the participant happier and calmer). This result may clearly be seen in the figure; the mean valence increases, and the mean arousal decreases. These changes in affective state are statistically significantly different from random affective state changes ($p < 0.05$) as assessed by comparing to distributions of random trajectories generated under the null hypothesis.

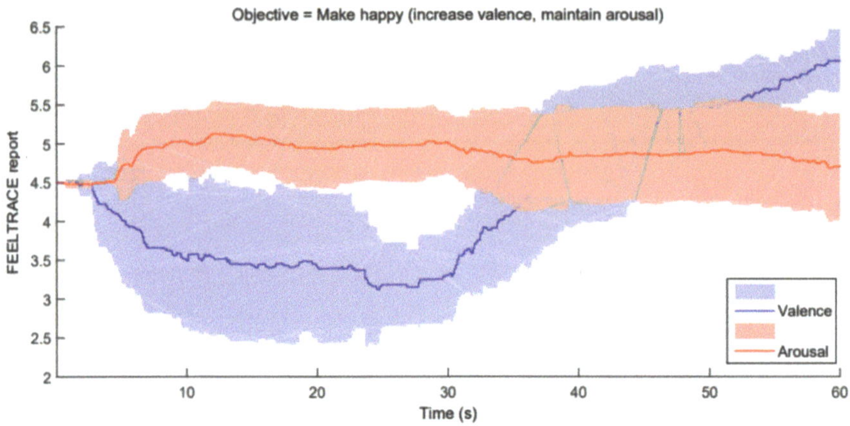

Fig. 4 Affective state trajectories achieved by the BCMI system when it was employed with an individual with Huntington's disease to attempt to increase the individual's valence. FEELTRACE report indicates the output of the FEELTRACE interface used to allow participants to report their current affective states

When the BCMI is employed with an individual with Huntington's disease in our follow-up case study, significant changes in valence are achieved with results comparable to those delivered with healthy participants. However, no significant changes in arousal are achieved. These results are illustrated in Fig. 4, which illustrates the change in FEELTRACE trajectories reported by our participant for each of the BCMI evaluation goals.

A full description of the results of the evaluation of the BCMI with the individual with Huntington's disease is available here (Daly et al., 2017).

3.6 Discussion

This case study provides an example of one of the first efforts to develop an automated BCMI system for aiding music therapy. The system uses a real-time hybrid metric of affective states to monitor an individual's emotions and respond to them. The metric uses a combination of EEG and other physiological processes and is able to accurately identify and respond to a user's current affective state.

There are some considerations to be made when evaluating the results of this system. First, there is considerable variability in the reported affective states. This is likely to be a result of a large amount of known inter-person variability in individual affective responses to stimuli and large inter-person differences in neural and physiological signal properties, upon which the affective state detection method that underpins the BCMI is based.

Additionally, it may be argued that the use of the FEELTRACE interface itself is interfering with individual felt affective states. FEELTRACE is controlled by a joystick or mouse and the act of concentrating on the control of this interface may affect an individual's affective state. However, as FEELTRACE is used throughout all parts of the experiment its effect on felt affective states may be treated as a continuous background noise signal and is unlikely to adversely affect the classification performance of the system.

This case study provides an illustration of how neural and physiological metrics may be developed and employed to allow computer systems to monitor and respond to a user's emotions.

The results highlight how a particular neural technology may be used to detect and monitor an individual's emotions over time. The resulting system has the potential for use in automated music therapy for individuals with emotion regulation problems. This is highlighted by the case study demonstrating the system's potential with an individual with Huntington's disease.

4 Case Study 2: Driving

Neural technology for measuring emotions has many other potential applications. In this section, we present a second such possible application; the use of neural technology to measure the emotional states of drivers.

Driving is an area of day-to-day life that can be negatively influenced by emotional state (stress, road rage, and so forth). Some recent studies have used biophysiological measurement to attempt to help quantify drivers' emotional states and to consider whether there are any particular independent variables which might help drivers to be more mindful of their emotional state. The study outlined below considered acoustic cues according to two different types of engines, and aimed to evaluate the link between acoustic environment, driver mood, and biophysiological response, to the new electric taxi.

Four London taxi drivers participated in an experiment using discrete brain and body sensors to measure biophysiological cues as they drove an electric and a diesel taxi.

Acoustic cues extracted from a stereo recording suggested that the electric vehicle cab was not necessarily quieter, but had an increased dynamic range and lower spectral centroid, both correlates of fatigue in music listening exercises. Drivers described their own experience in the electric vehicle as happier and less stressful than the diesel vehicle, and heart rate and heart rate variability metrics corroborated these descriptions. Brain responses suggested a degree of emotional interaction, and an unexpected finding: correlates of higher concentration when driving the electric vehicle.

4.1 Background

Quentin Willson of Top Gear/Fifth Gear described the experience of electric vehicles as "silent speed. A totally new concept for today's drivers". There are a number of proposed acoustic correlates for mental states, and many people in Britain describe their experience of the environment as excessively noisy.

Previous driving metrics have suggested that heart rate and skin resistance are most correlated with driver stress (Healey & Picard, 2005).

The ratio and duration of alpha and beta waves in brain activity have been shown to be correlated to stress (Puglisi-Allegra & Oliverio, 1990) and used as an analogue to stress in evaluation of driver mental state using virtual reality driving simulations (Benoit et al., 2009; Schier, 2000). However, recent improvements in portability, combined sampling, and durability of analog to digital conversion allow for real-world testing (Ollander, Godin, Charbonnier, & Campagne, 2016; Rigas, Goletsis, & Fotiadis, 2012). This experiment combines acoustic measurement (stereo recording) with biophysiological and self-report metrics conducted on professional London taxi drivers in two conditions: electric and diesel vehicles, across a series of randomized case series trials.

4.2 Method

Experiment information sheets were explained, and copies given to participants before undertaking the tests. Consent forms in accordance with the University of York ethics approval process were signed and collected from all participants.

Four professional taxi drivers undertook 12 trials around the Regent's Park area in the TX4 (diesel) and TX (electric) taxi cabs. Each trial took approximately 20 min (Fig. 5).

4.3 Data Collected

Stereo recordings of audio files were taken from the cab at a sampling rate of 44.1 kHz, with synchronous 720p video recording.

Galvanic skin response (electrodermal activity) was recorded from the left ear lobe of the driver (chosen for safety reasons to allow drivers to look to the right to check their blind spot). Although galvanic skin response has been previously shown to be useful in evaluating stress, it was difficult to place the sensors required in situ in the accepted position for this experiment. The sensors are traditionally placed over the finger and wrist, which would be impractical for drivers in the real-world data collection scenario. The GSR sensor was, therefore, placed behind the ear lobe.

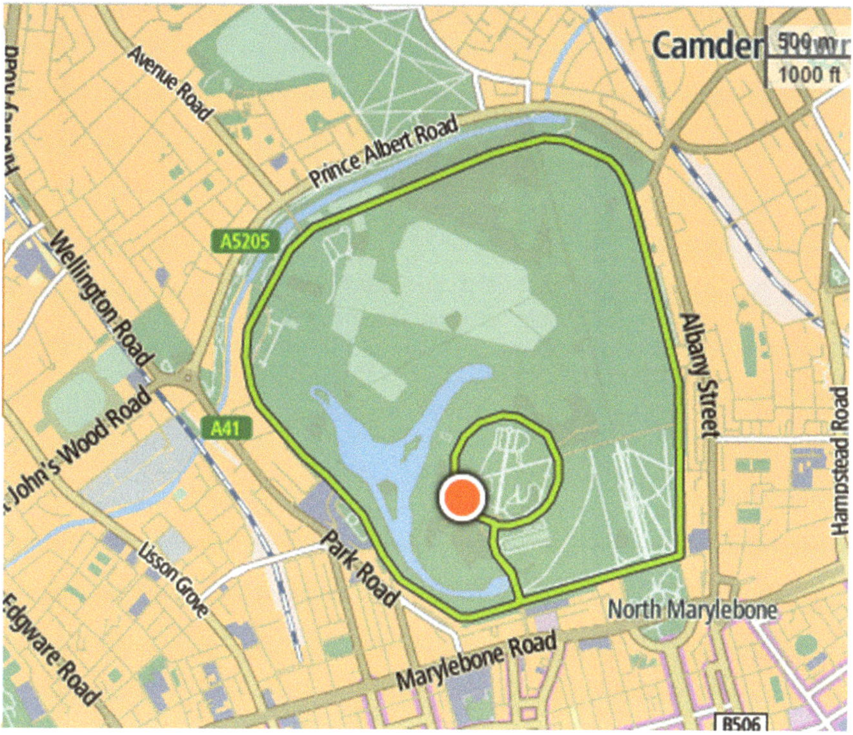

Fig. 5 Map of the route taken by the drivers around Regent's Park, London (approx. 3.5 miles)

The heart rate of each driver was also measured via an optical wrist meter. The wrist meter measured levels of oxyhemoglobin; heart rate was then inferred from this metric by peak counting.

Electroencephalogram (EEG) was recorded from each driver via dry electrodes with eight channels placed at positions Fp1, Fp2, C3, C5, P7, P8, O1, and O2 in the 10/20 configuration. The EEG was sampled at 250 Hz and measured at an impedance of 10 kΩ.

Self-reports were taken from each driver after each trial across five descriptors drawn from a survey of the psychological evaluation of driving conditions, using a 9-point Likert scale, with 1 being "least" and 10 being "most". The evaluation responses were stress, anger, distraction, fear, and happiness. Additionally, a 3-channel accelerometer was used to record driver movement, with data recorded at a sampling rate of 250 Hz. The accelerometer was mounted on a small 3D printed plastic frame just above the nape of the neck.

4.4 Analysis Techniques

Readings for each vehicle type were summed and passed through an acoustic feature analysis pipeline (custom software, MATLAB). The values of the five features were calculated as follows: peaks, RMS, dynamic range, autocorrelation, crest factor, and spectral centroid.

The EEG was low pass filtered at 50 Hz and then Butterworth filtered in the alpha and beta ranges. EEG was then inverse convolved with the accelerometer data to smooth the data with significant head movement. This approach was inspired by work in (Luu & Dinh, 2018) and interested readers may also wish to consult (Daly, Billinger, Scherer, & Mueller-Putz, 2013) for an alternative approach to solving the same problem.

Additionally, heart rate variability was also calculated on a per-participant and per-trial basis.

Across each self-report and descriptor per vehicle, the mean, standard deviation, and variance of each of the above features were taken. These were compared between both types of vehicles.

4.5 Results

Self-report measures are shown in Table 3 and indicate that participants described their experience of driving the electric vehicle as

1. Less stressful than driving the diesel vehicle.
2. Making them neither more or less angry than driving the diesel vehicle.
3. Less distracting than driving the diesel vehicle.
4. Making them neither more or less afraid than driving the diesel vehicle.
5. Making them happier than driving the diesel vehicle.

Table 3 Self-reported responses to the electric and diesel vehicle conditions made by drivers using a 9-point Likert scale (with 1 being "least" and 10 being "most") across five descriptors. Collected after each trial. "Var" denotes the variance of the reports

Diesel	Stress	Anger	Distraction	Fear	Happiness
Mean	3.50	1.33	3.83	1.17	6.83
STD	0.89	0.52	1.97	0.00	1.41
Var.	0.80	0.27	3.90	0.00	2.00
Electric	Stress	Anger	Distraction	Fear	Happiness
Mean	2.17	1.33	2.67	1.50	8.17
STD	1.47	0.52	1.51	1.84	1.47
Var.	2.17	0.27	2.27	0.70	2.17

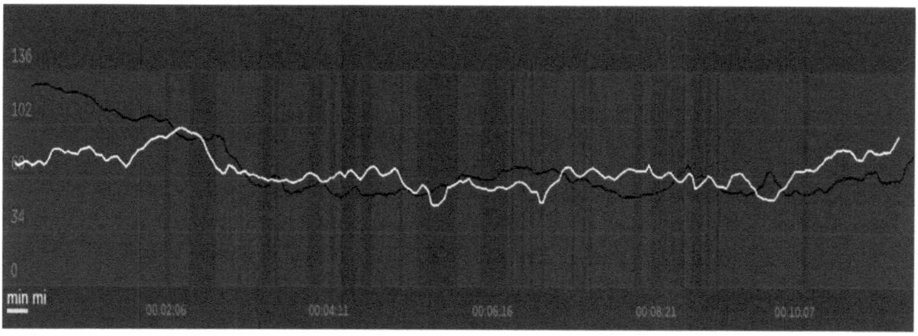

Fig. 6 Heart rate across diesel (white line) and electric (black line) care driving trials for driver A

Heart rate: Drivers generally exhibited a slightly lower mean heart rate in the electric vehicle driving trials, as shown in Fig. 6.

Note, the overall lower heart rate when comparing the electric trial to the diesel trial for the same driver. This pattern can be seen in plots for 3 out of 4 of the driver participants (drivers A, B, and C). However, driver D showed a markedly similar mean heart rate when comparing diesel and electric trials, as shown in Fig. 7.

While data regarding participant hearing was not collected, it was noted that the participant in question (driver D above) did wear a hearing aid. This suggests that while the sound world may have minimal influence on the mean heart rate, there was nevertheless a degree of heart rate variability increase in the diesel trials in comparison to the electric vehicle trials in this case, as was also seen in the other participant trials. All drivers exhibited lower heart rate variability in the electric vehicle driving trials.

The drivers in the diesel trials frequently exhibited frontal asymmetry, which has been correlated with both negative and positive affect (Schaffer, Davidson, & Saron, 1983). Asymmetry in itself is not necessarily a useful indicator of affect in this case (Chu, Tranel, & Damasio, 1994) and is to some extent a default, though

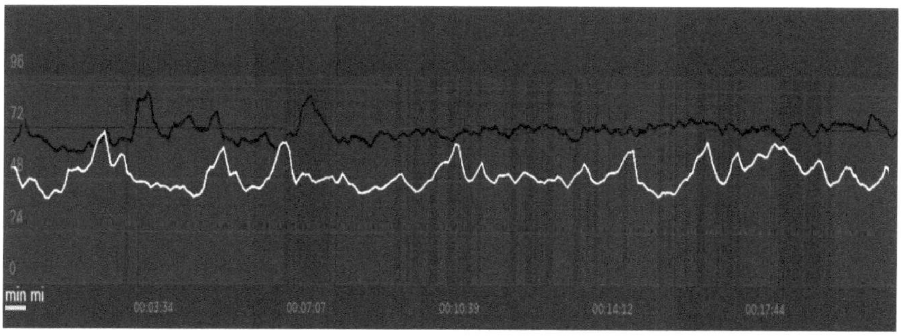

Fig. 7 Heart rate across diesel (white) and electric (black) vehicle driving trials for driver D. Note only a slight reduction was observed in the range of the heart rate and increased variability in diesel driving trials, with large peaks and troughs occurring throughout the trial

frontal and parietal asymmetry has been shown more recently in EEG studies to be well correlated with the valence of the emotion (Palmiero & Piccardi, 2017). Perhaps more unusually, the electric trials exhibit less asymmetry and higher levels of beta frequency band power in comparison to alpha frequency band power. Alpha frequencies tend to be correlated with restful or relaxed states of mind, and beta frequencies with concentration. This finding is thus quite unexpected for the electric trials—drivers reported more happiness, less stress, in the electric vehicle—so it may be that the electric vehicle was less distracting and in fact allowed drivers to concentrate more, resulting in less overall fatigue at the end of a driving session.

The GSR showed little difference between either vehicle type across the trials but these values were also below the statistical significance threshold. A distribution is shown in Fig. 8.

Six acoustic features (peak values, root mean square amplitude, dynamic range, autocorrelation, crest factor, and spectral centroid) were extracted using the signal processing toolbox in MATLAB and are shown in Table 4.

The electric vehicle samples indicate lower peak values in comparison to the diesel vehicle recordings, with higher spectral centroid in some samples in the diesel vehicle, and slightly lower mean amplitude (RMS) in the electric vehicle. Surprisingly, some electric vehicle trials indicated a larger dynamic range, though the variance across the dynamic range in all trials was minimal, consistently less than 5 dB full scale. In layman's terms, these descriptors would typically correlate to a slightly sharper (higher spectral centroid), louder (higher RMS and lower dynamic range) perceived sound in the diesel vehicle. The lower dynamic range might be directly correlated to the idle engine sound of the diesel vehicle as these analyses were extracted from completed trials.

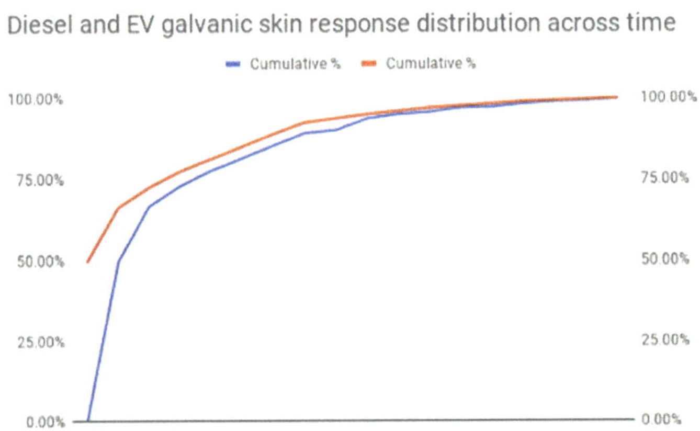

Fig. 8 Distribution of galvanic skin response values across timestamp values for recordings (electric vehicle in blue, diesel vehicle in red)

Table 4 A comparison of acoustic features extracted from stereo recordings made inside the diesel and electric vehicles during each trial. Although the order of the trials was randomized during the collection, for ease of comparison here they are presented as follows: odd numbers are recordings from the diesel cab (shaded grey), even numbers from the electric cab (no shading)

Peaks (max, min)	RMS	Range (dBFS)	Autocorrelation	Crest factor (dB)	Spectral centroid
Max 0.944 Min -0.943	0.140	89.811	2.802	16.560	4112.533
Max 0.723 Min -0.708	0.108	87.493	2.807	16.519	3290.366
Max 0.856 Min -0.877	0.164	88.962	10.255	14.364	6789.525
Max 0.914 Min -0.929	0.172	89.528	7.383	14.485	6704.825
Max 0.997 Min – 1.000	0.176	90.286	4.192	15.057	6772.741
Max 0.635 Min -0.614	0.113	86.367	15.710	14.974	7980.007
Max 0.909 Min -0.866	0.169	89.482	15.710	14.565	6750.248
Max 0.909 Min -0.866	0.169	89.482	15.710	14.565	6750.241
Max 0.999 Min -1.000	0.276	90.309	0.145	11.186	11194.000
Max 0.916 Min -0.893	0.177	89.551	1.432	14.268	5684.410
Max 0.856 Min -0.877	0.164	88.962	10.255	14.364	6789.525
Max 0.914 Min -0.929	0.172	89.528	7.383	14.485	6704.825

5 Conclusions

Drivers self-reported little difference in anger or fear between either vehicle type, with no variation in response to anger and minimal variation in response to fear across the two vehicle conditions. Given these are professional drivers with many decades of experience on London roads, it is perhaps unsurprising that the drivers felt little sense of anger or fear undergoing these trials. The electric vehicle was reported to be less stressful and less distracting, as well as a happier driving experience by the drivers than the diesel vehicle—with the largest improvement being in self-reported happiness. The only diesel vehicle ratings with significant inter-participant variation were for the distraction descriptor. There was a greater degree of inter-participant variation among the electric vehicle ratings than the diesel ratings for stress, and to a lesser extent, happiness.

The EEG showed correlates in various regions, particularly asymmetry, suggesting different emotional responses in each trial, but more significantly a greater level of beta frequency band power in the electric vehicle trials than in the diesel vehicle trials. Alpha and beta frequencies have well documented and agreed upon correlates in relaxation (alpha), and active concentration (beta). We hypothesized, based on the self-report measures, that the electric vehicle, which drivers described as more fun and less stressful, would be accompanied by higher levels of alpha than beta—but the opposite appears to be the case. This might be because drivers were, in fact, able to concentrate more without the noise of the diesel engine, which becomes most noticeable when the vehicle is at a standstill (e.g., waiting at lights).

For the majority of trials, there was a marked difference in mean heart rate, though this is not in itself a prominent correlate of mental state. Heart rate variability (HRV), however, is a well understood correlate for stressed mental states (Kim, Cheon, Bai, Lee, & Koo, 2018), and, even in the case of the driver who exhibited little difference in mean heart rate between the two conditions, there was a marked increase in HRV in the diesel vehicle trials. The driver who exhibited no significant difference in mean rate between the two trials did wear a hearing aid device—which suggests there might be a correlation between the sound world and the resulting influence on heart rate—but this would be impossible to establish without a larger number of participants.

While the biophysiological responses are not necessarily what would be expected given the self-report results, there seems to be a strong case to suggest that the drivers in the electric vehicle trials were less mentally and physically fatigued by the drive.

6 Conclusions and Future Research Directions

Computer systems that are able to measure their users' current emotional (affective) states have tremendous potential to provide improvements in therapeutic applications for a wide range of clinical populations, as well as their potential to improve interactions with computer systems and other technologies for general users. Neural technology that is able to provide such a metric, therefore, has great potential.

This chapter introduced and discussed some of the most well-known and frequently investigated neural technology-based metrics for measuring emotion. These metrics most frequently not only make use of the electroencephalogram (EEG) but can also make use of other neuroimaging technologies, such as functional magnetic resonance imaging or functional near-infrared spectroscopy. These metrics provide a quantifiable objective measure of an individual's current felt emotion.

Neural technology-based measures of an individual's current emotions have been shown to be able to identify an individual's current level of valence and arousal. Furthermore, when they are coupled with physiological measures, they may provide relatively more accurate measures of a wide range of different emotional states from an individual in a variety of different contexts.

Neural technology-based measures of affect have a large number of potential applications. Emotion is an intrinsic part of the human experience and, increasingly, our experience of the world is mediated through information communications technology. Thus, accurate metrics that allow our technology to understand how we are feeling are increasingly important in building systems that adapt and react optimally to our needs.

However, there are numerous challenges to overcome before neural technology can be most effectively used in a wide range of applications. Specifically, neural technology for measuring emotions faces challenges related to its accuracy, setup time and convenience, and its reliability.

The accuracy of any neural technology-based metrics of emotion is largely mediated by the quality of the signals used, the variability of those signals, the quality of the signal processing and machine learning techniques used to interpret those signals, and the unreliability of the ground-truth measure upon which those methods are calibrated. For example, EEG suffers from artifact contamination from a variety of sources and, thus, constitutes a noisy, nonlinear, and nonstationary measure of brain activity (Schomer & Lopes de Silva, 2011). Furthermore, the machine learning methods trained to interpret this signal are, typically, trained via a supervised learning approach and, the ground-truth labels used in this approach can be unreliable as individuals may not be able, or willing, to accurately quantify their emotional responses in every situation.

Setup time and convenience of neural technology is also a key impediment to its widespread adoption. EEG is one of the most suitable forms of neural technology for measuring emotion due to its relative low cost, relative portability, and high time resolution. However, it still takes between 5 and 20 min to set up an EEG recording, a process that typically involves mounting electrodes within a cap and forming a conductive bridge with the scalp via conductive gel, a process that can be uncomfortable and inconvenient. Modern EEG systems seek to overcome these issues via the use of "dry electrode" technology, which does away with gel, and wireless systems, that remove the need for attaching multiple wires to the user. However, further advances in setup speed and convenience are still needed.

Furthermore, other physiological measures are not well suited to all situations. An example of this is seen in our driver case study, in which GSR could not be used in a conventional way (recording from the hands) as this would be unsafe in a driving context.

Finally, the reliability of neural technology is hampered by factors including the inherent non-stationarity of brain activity, as well as artifact contamination of the signal. Signal processing techniques seek to overcome these via automated cleaning of the EEG and more accurate and robust classification methods. However, considerable advances are still needed in this area before neural technology can be sufficiently reliable.

Future research in a wide range of fields will benefit from overcoming these problems and from identifying and building neural technology that is able to accurately, reliably, and conveniently identify an individual's current emotional state.

References

Aftanas, L. I., Varlamov, A. A., Pavlov, S. V., Makhnev, V. P., & Reva, N. V. (2001). Affective picture processing: Event-related synchronization within individually defined human theta band is modulated by valence dimension. *Neuroscience Letters, 303*(2), 115–118. https://doi.org/10.1016/S0304-3940(01)01703-7.

Barrett, L. F., Mesquita, B., Ochsner, K. N., & Gross, J. J. (2007). The experience of emotion. *Annual Review of Psychology, 58,* 373–403. https://doi.org/10.1146/annurev.psych.58.110405.085709.

Bates, G., Tabrizi, S. J., & Jones, L. (2014). *Huntington's disease*. Oxford: Oxford University Press. Retrieved from https://books.google.co.uk/books?id=i-4kAwAAQBAJ&printsec=frontcover&dq=huntington%27s+disease&hl=en&sa=X&ved=0ahUKEwih9rHI287RAhUBVxoKHY7TANwQ6AEIKDAC#v=onepage&q=huntington'sdisease&f=false.

Benoit, A., Bonnaud, L., Caplier, A., Ngo, P., Lawson, L., Trevisan, D. G., … Chanel, G. (2009). Multimodal focus attention and stress detection and feedback in an augmented driver simulator. *Personal and Ubiquitous Computing*, *13*(1), 33–41. https://doi.org/10.1007/s00779-007-0173-0.

Blood, A. J., & Zatorre, R. J. (2001). Intensely pleasurable responses to music correlate with activity in brain regions implicated in reward and emotion. *Proceedings of the National Academy of Sciences of the United States of America, 98*(20), 11818–11823. https://doi.org/10.1073/pnas.191355898.

Bradt, J., Magee, W. L., Dileo, C., Wheeler, B. L., & McGilloway, E. (2010). Music therapy for acquired brain injury. *The Cochrane Database of Systematic Reviews*, (7), CD006787. https://doi.org/10.1002/14651858.CD006787.pub2.

Canli, T., Desmond, J. E., Zhao, Z., Glover, G., & Gabrieli, J. D. (1998). Hemispheric asymmetry for emotional stimuli detected with fMRI. *Neuroreport*, *9*(14), 3233–3239. Retrieved from http://www.ncbi.nlm.nih.gov/pubmed/9831457.

Chu, C.-C., Tranel, D., & Damasio, H. (1994). How reliable are occipital asymmetry measurements? *Neuropsychologia, 32*(12), 1503–1513. https://doi.org/10.1016/0028-3932(94)90122-8.

Coan, J. A., & Allen, J. J. (2004). Frontal EEG asymmetry as a moderator and mediator of emotion. *Biological Psychology, 67*(1–2), 7–50. https://doi.org/10.1016/j.biopsycho.2004.03.002.

Cowie, R., Douglas-Cowie, E., Savvidou, S., McMahon, E., Sawey, M., & Schröder, M. (2000). "FEELTRACE": An instrument for recording perceived emotion in real time. In *Proceedings of the ISCA Workshop on Speech and Emotion* (pp. 19–24). Retrieved from http://citeseerx.ist.psu.edu/viewdoc/summary?doi=10.1.1.58.7528.

Daly, I., Billinger, M., Scherer, R., & Mueller-Putz, G. (2013). On the automated removal of artifacts related to head movement from the EEG. *IEEE Transactions on Neural Systems and Rehabilitation Engineering, 21*(3), 427–434.

Daly, I., Ho, A., Marcon, J., Hwang, F., Williams, D., Kirke, A., … Nasuto, S. (2017). Affective brain computer music interfacing: A case study of use by an individual with Huntington's disease. In *Proceedings of the Graz Brain-Computer Interfacing Conference 2017.*

Daly, I., Malik, A., Hwang, F., Roesch, E., Weaver, J., Kirke, A., … Nasuto, S. J. (2014). Neural correlates of emotional responses to music: An EEG study. *Neuroscience Letters*, *573*, 52–57. https://doi.org/10.1016/j.neulet.2014.05.003.

Daly, I., Malik, A., Weaver, J., Hwang, F., Nasuto, S. J., Williams, D., … Miranda, E. (2015). Identifying music-induced emotions from EEG for use in brain-computer music interfacing. In *6th Affective Computing and Intelligent Interaction.*

Daly, I., Williams, D., Hallowell, J., Hwang, F., Kirke, A., Malik, A., … Nasuto, S. J. (2015). Music-induced emotions can be predicted from a combination of brain activity and acoustic features. *Brain and Cognition*, *101*, 1–11. https://doi.org/10.1016/j.bandc.2015.08.003.

Daly, I., Williams, D., Kirke, A., Weaver, J., Malik, A., Hwang, F., … Nasuto, S. J. (2016a). Affective brain–computer music interfacing. *Journal of Neural Engineering*, *13*(4), 046022. https://doi.org/10.1088/1741-2560/13/4/046022.

Daly, I., Williams, D., Kirke, A., Weaver, J., Malik, A., Hwang, F., ... Nasuto, S. J. (2016b). An affective brain-computer music interface. In *Proceedings of the 6th International Brain-Computer Interface Meeting, organized by the BCI Society*. https://doi.org/10.3217/978-3-85125-467-9-227.

Darwin, C. (1998). *The expression of the emotions in man and animals*. Oxford University Press.

Davidson, R. J., Abercrombie, H., Nitschke, J. B., & Putnam, K. (1999). Regional brain function, emotion and disorders of emotion. *Current Opinion in Neurobiology, 9*(2), 228–234. https://doi.org/10.1016/S0959-4388(99)80032-4.

Eerola, T. (2012). Modeling listeners' emotional response to music. *Topics in Cognitive Science, 4*(4), 607–624. https://doi.org/10.1111/j.1756-8765.2012.01188.x.

Effron, D. A., Niedenthal, P. M., Gil, S., & Droit-Volet, S. (2006). Embodied temporal perception of emotion. *Emotion, 6*(1), 1–9. https://doi.org/10.1037/1528-3542.6.1.1.

Erkkilä, J., Punkanen, M., Fachner, J., Ala-Ruona, E., Pöntiö, I., Tervaniemi, M., ... Gold, C. (2011). Individual music therapy for depression: Randomised controlled trial. *The British Journal of Psychiatry : The Journal of Mental Science, 199*(2), 132–139. https://doi.org/10.1192/bjp.bp.110.085431.

Fazli, S., Mehnert, J., Steinbrink, J., Curio, G., Villringer, A., Müller, K.-R., et al. (2012). Enhanced performance by a hybrid NIRS-EEG brain computer interface. *NeuroImage, 59*(1), 519–529. https://doi.org/10.1016/j.neuroimage.2011.07.084.

Fingelkurts, A. A., Fingelkurts, A. A., Rytsälä, H., Suominen, K., Isometsä, E., & Kähkönen, S. (2007). Impaired functional connectivity at EEG alpha and theta frequency bands in major depression. *Human Brain Mapping, 28*(3), 247–261. https://doi.org/10.1002/hbm.20275.

Fontaine, J. R. J., Scherer, K. R., Roesch, E. B., & Ellsworth, P. C. (2007). The world of emotions is not two-dimensional. *Psychological Science, 18*(12), 1050–1057. https://doi.org/10.1111/j.1467-9280.2007.02024.x.

Healey, J. A., & Picard, R. W. (2005). Detecting stress during real-world driving tasks using physiological sensors. *IEEE Transactions on Intelligent Transportation Systems, 6*(2), 156–166. https://doi.org/10.1109/TITS.2005.848368.

Heller, W. (1993). Neuropsychological mechanisms of individual differences in emotion, personality, and arousal. *Neuropsychology, 7*(4), 476–489. https://doi.org/10.1037/0894-4105.7.4.476.

Hou, J., Song, B., Chen, A. C. N., Sun, C., Zhou, J., Zhu, H., et al. (2017). Review on Neural correlates of emotion regulation and music: Implications for emotion dysregulation. *Frontiers in Psychology, 8*, 501. https://doi.org/10.3389/fpsyg.2017.00501.

Hunter, P. G., & Schellenberg, E. G. (2010). Music and emotion. In *Music perception, springer handbook of auditory research* (pp. 129–164).

Ishii, R., Canuet, L., Ishihara, T., Aoki, Y., Ikeda, S., Hata, M., ... Takeda, M. (2014). Frontal midline theta rhythm and gamma power changes during focused attention on mental calculation: An MEG beamformer analysis. *Frontiers in Human Neuroscience, 8*, 406. https://doi.org/10.3389/fnhum.2014.00406.

Jones, N. A., & Fox, N. A. (1992). Electroencephalogram asymmetry during emotionally evocative films and its relation to positive and negative affectivity. *Brain and Cognition, 20*(2), 280–299. Retrieved from http://www.ncbi.nlm.nih.gov/pubmed/1449758.

Kassam, K. S., Markey, A. R., Cherkassky, V. L., Loewenstein, G., & Just, M. A. (2013). Identifying emotions on the basis of neural activation. *PLoS ONE, 8*(6), e66032. https://doi.org/10.1371/journal.pone.0066032.

Kim, H.-G., Cheon, E.-J., Bai, D.-S., Lee, Y. H., & Koo, B.-H. (2018). Stress and heart rate variability: A meta-analysis and review of the literature. *Psychiatry Investigation, 15*(3), 235–245. https://doi.org/10.30773/pi.2017.08.17.

Koelsch, S. (2010). Towards a neural basis of music-evoked emotions. *Trends in Cognitive Sciences, 14*(3), 131–137. https://doi.org/10.1016/j.tics.2010.01.002.

Koelsch, S. (2014). Brain correlates of music-evoked emotions. *Nature Reviews Neuroscience, 15*(3), 170–180. https://doi.org/10.1038/nrn3666.

Kreibig, S. D. (2010). Autonomic nervous system activity in emotion: A review. *Biological Psychology, 84*(3), 394–421. https://doi.org/10.1016/j.biopsycho.2010.03.010.

Lee, Y.-Y., & Hsieh, S. (2014). Classifying different emotional states by means of EEG-based functional connectivity patterns. *PLoS ONE, 9*(4), e95415. https://doi.org/10.1371/journal.pone.0095415.

Livingstone, R. S., & Thompson, W. F. (2009). The emergence of music from the theory of mind. *Musicae Scientiae, 13*(2 Suppl), 83–115. https://doi.org/10.1177/1029864909013002061.

Luu, L., & Dinh, A. (2018). Artifact noise removal techniques on seismocardiogram using two tri-axial accelerometers. *Sensors, 18*(4), 1067. https://doi.org/10.3390/s18041067.

Mehrabian, A. (1996). Pleasure-arousal-dominance: A general framework for describing and measuring individual differences in Temperament. *Current Psychology, 14*(4), 261–292. https://doi.org/10.1007/BF02686918.

Mier, D., Lis, S., Neuthe, K., Sauer, C., Esslinger, C., Gallhofer, B., & Kirsch, P. (2010). The involvement of emotion recognition in affective theory of mind. *Psychophysiology, 47*(6), no-no. https://doi.org/10.1111/j.1469-8986.2010.01031.x.

Müller-Putz, G. R., Breitwieser, C., Cincotti, F., Leeb, R., Schreuder, M., Leotta, F., et al. (2011). Tools for brain-computer interaction: A general concept for a hybrid BCI. *Front Neuroinform, 5*, 30. https://doi.org/10.3389/fninf.2011.00030.

Nicolaou, N., Malik, A., Daly, I., Weaver, J., Hwang, F., Kirke, A., … Nasuto, S. J. (2017). Directed motor-auditory EEG connectivity is modulated by music tempo. *Frontiers in Human Neuroscience, 11*, 502. https://doi.org/10.3389/fnhum.2017.00502.

Nicolas-Alonso, L. F., & Gomez-Gil, J. (2012). Brain computer interfaces, a review. *Sensors (Basel, Switzerland), 12*(2), 1211–1279. https://doi.org/10.3390/s120201211.

Niedermeyer, E., & Da Silva, F. H. L. (2005). *Electroencephalography: Basic principles, clinical applications, and related fields*. Lippincott Williams & Wilkins.

Ollander, S., Godin, C., Charbonnier, S., & Campagne, A. (2016). Selection of the most relevant physiological features for classifying emotion. In *3rd International Conference on Physiological Computing Systems (PhyCS 2016)*. Retrieved from https://hal.archives-ouvertes.fr/hal-01378328.

Ortony, A., Clore, G. L., & Collins, A. (1988). *The cognitive structure of emotions*. Cambridge: Cambridge University Press. https://doi.org/10.1017/CBO9780511571299.

Palmiero, M., & Piccardi, L. (2017). Frontal EEG asymmetry of mood: A mini-review. *Frontiers in Behavioral Neuroscience, 11*, 224. https://doi.org/10.3389/fnbeh.2017.00224.

Pfurtscheller, G., Allison, B. Z., Brunner, C., Bauernfeind, G., Solis-Escalante, T., Scherer, R., … Birbaumer, N. (2010). The hybrid BCI. *Frontiers in Neuroprosthetics, 4*(30).

Pridmore, S. A. (1990). The prevalence of Huntington's disease in Tasmania. *The Medical Journal of Australia, 153*(3), 133–134. Retrieved from http://www.ncbi.nlm.nih.gov/pubmed/2142982.

Puglisi-Allegra, S., & Oliverio, A. (1990). *Psychobiology of stress*. Netherlands: Springer.

Raichle, M. E. (2011). The restless brain. *Brain Connectivity, 1*(1), 3–12. https://doi.org/10.1089/brain.2011.0019.

Reeve, J. (2014). *Understanding motivation and emotion* (Vol. 3). Wiley. Retrieved from https://books.google.com/books?hl=en&lr=&id=zfOyBwAAQBAJ&pgis=1.

Rigas, G., Goletsis, Y., & Fotiadis, D. I. (2012). Real-time driver's stress event detection. *IEEE Transactions on Intelligent Transportation Systems, 13*(1), 221–234. https://doi.org/10.1109/TITS.2011.2168215.

Rogenmoser, L., Zollinger, N., Elmer, S., & Jäncke, L. (2016). Independent component processes underlying emotions during natural music listening. *Social Cognitive and Affective Neuroscience, 11*(9), 1428–1439. https://doi.org/10.1093/scan/nsw048.

Russell, J. A. (1980). A circumplex model of affect. *Journal of Personality and Social Psychology, 39*(6), 1161–1178.

Sacks, O. (2006). The power of music. *Brain: A Journal of Neurology, 129*(Pt 10), 2528–2532. https://doi.org/10.1093/brain/awl234.

Sammler, D., Grigutsch, M., Fritz, T., & Koelsch, S. (2007). Music and emotion: Electrophysiological correlates of the processing of pleasant and unpleasant music. *Psychophysiology, 44*(2), 293–304. https://doi.org/10.1111/j.1469-8986.2007.00497.x.

Schaffer, C. E., Davidson, R. J., & Saron, C. (1983). Frontal and parietal electroencephalogram asymmetry in depressed and nondepressed subjects. *Biological Psychiatry, 18*(7), 753–762. Retrieved from http://www.ncbi.nlm.nih.gov/pubmed/6615936.

Schier, M. A. (2000). Changes in EEG alpha power during simulated driving: A demonstration. *International Journal of Psychophysiology : Official Journal of the International Organization of Psychophysiology, 37*(2), 155–162. Retrieved from http://www.ncbi.nlm.nih.gov/pubmed/10832002.

Schimmack, U., & Grob, A. (2000). Dimensional models of core affect: A quantitative comparison by means of structural equation modeling. *European Journal of Personality, 14*(4), 21.

Schmidt, L. A., & Trainor, L. J. (2001). Frontal brain electrical activity (EEG) distinguishes valence and intensity of musical emotions. *Cognition and Emotion, 15*(4), 487–500. https://doi.org/10.1080/02699930126048.

Schomer, L., & Lopes de Silva, F. (Eds.). (2011). *Niedermeyer's electroencephalography: Basic principles, clinical applications, and related fields* (6th ed.). Lippincott Williams & Wilkins.

Silberman, E. K., & Weingartner, H. (1986). Hemispheric lateralization of functions related to emotion. *Brain and Cognition, 5*(3), 322–353. https://doi.org/10.1016/0278-2626(86)90035-7.

Skouras, S., Gray, M., Critchley, H., & Koelsch, S. (2013). FMRI scanner noise interaction with affective neural processes. *PLoS ONE, 8*(11), e80564. https://doi.org/10.1371/journal.pone.0080564.

Sporns, O. (2007). Brain connectivity. *Scholarpedia, 2*(10), 4695. Retrieved from http://www.scholarpedia.org/article/Brain_connectivity.

Stikic, M., Johnson, R. R., Tan, V., & Berka, C. (2014). EEG-based classification of positive and negative affective states. *Brain-Computer Interfaces, 1*(2), 99–112. https://doi.org/10.1080/2326263X.2014.912883.

Vuoskoski, J. K., & Eerola, T. (2011). Measuring music-induced emotion: A comparison of emotion models, personality biases, and intensity of experiences. *Musicae Scientiae, 15*(2), 159–173. https://doi.org/10.1177/1029864911403367.

Webber, R., Aha, D. W., Muñoz-Ávila, H., & Breslow, L. A. (2000). Advances in case-based reasoning. *Advances in Case-Based Reasoning, 1898,* 322–334. https://doi.org/10.1007/3-540-44527-7.

Williams, D., Kirke, A., Miranda, E., Daly, I., Hwang, F., Weaver, J., et al. (2017). Affective calibration of musical feature sets in an emotionally intelligent music composition system. *ACM Transactions on Applied Perception, 14*(3), 1–13. https://doi.org/10.1145/3059005.

Williams, D., Nasuto, S., Kirke, A., Miranda, E., Daly, I., Hallowell, J., ... Hwang, F. (2015). Investigating perceived emotional correlates of rhythmic density in algorithmic music composition. *ACM Transactions on Applied Perception, 12*(3), 1–21. https://doi.org/10.1145/2749466.

Winkler, I., Jäger, M., & Vojkan Mihajlović, T. T. (2010). Frontal EEG asymmetry based classification of emotional valence using common spatial patterns. *World Academy of Science, Engineering and Technology, 69.* Retrieved from http://citeseerx.ist.psu.edu/viewdoc/summary?doi=10.1.1.190.6511.

Zentner, M., Grandjean, D., & Scherer, K. R. (2008). Emotions evoked by the sound of music: Characterization, classification, and measurement. *Emotion, 8*(4), 494–521. https://doi.org/10.1037/1528-3542.8.4.494.

The Neural Basis of Cognitive Efficiency in Motor Skill Performance from Early Learning to Automatic Stages

Maarten A. Immink, Willem B. Verwey and David L. Wright

Abstract Skill acquisition represents a progression from high to low reliance on the conscious control of the action. The ability to produce action without drawing upon limited attentional resources has traditionally been the defining characteristic of skill automaticity. As such, learning represents a progression from low to high efficiency in the cognitive processes needed to plan, execute, and update skilled movement. In this chapter, we summarize neuroimaging findings that illustrate the evolution of such efficiency in terms of the neural adaptations that underlie skill learning automatization. As a backdrop to these findings, we first review the cognitive characteristics of skill automaticity as well as a contemporary theoretical framework for how we perform action based on sequencing movement elements. This provides a vantage point from which neural basis of skill automaticity can be considered in terms of associative and sensorimotor learning processes that provide for more efficient action in terms of cognitive requirements. We then contrast this with a summary of the contextual interference effect, which represents a cautionary account for the negative learning consequences associated with training protocols that appear to expedite skill automaticity.

1 Introduction

As we acquire a motor skill, two general characteristics typically emerge (Fitts 1964; Fitts and Posner 1967; Gentile 1972, 1998). First, performance improves. Second, skill performance becomes automated (Fitts 1964). Of these two characteristics, it is the latter that has particularly fascinated society. This is not surprising given that the time constraints under which we must perform many of our everyday behaviors

M. A. Immink (✉)
University of South Australia, Adelaide, Australia
e-mail: maarten.immink@unisa.edu.au

W. B. Verwey
University of Twente, Enschede, The Netherlands

W. B. Verwey · D. L. Wright
Texas A&M University, College Station, USA

C. S. Nam (ed.), *Neuroergonomics*, Cognitive Science and Technology,
https://doi.org/10.1007/978-3-030-34784-0_12

demand that we automate action (e.g., driving a car on a busy street). Moreover, it is the transition from the effortful to the effortless that is the quintessential subjective experience of transitioning from learning to automaticity. The experience of reduced effort coincides with increased mechanical (Bernstein 1967) and metabolic (Almåsbakk et al. 2001) efficiencies. However, the concept of automaticity has been more aligned with gaining efficiency with respect to diminishing the cognitive demands required to perform skilled action. Automaticity implies the absence of conscious awareness and, therefore, the absence of attention, the fundamental process for cognition in the sense that cognition and associated terms such as cogitation and recognition related to knowing, cognoscere in Latin (Chaney 2013).

The neural underpinnings for the transition from costly, cognizant skill learning to efficient relatively attention-free behavior are the central theme of this chapter. To describe neural adaptations associated with establishing skill automaticity, we review findings from studies that have utilized fMRI or PET neuroimaging approaches to assess brain activity associated with early learning and automatic stages of skill learning. While automaticity might reflect learning-associated changes in neurocognitive processes that contribute to skilled action, we then present the contextual interference effect (Shea and Morgan 1979) as a case where neural patterns of automaticity can be established quite early in learning to provide temporary performance gains as opposed to long-term skill learning. First, though, we provide a brief historical context of automaticity followed by an overview of the cognitive characteristics of skill automaticity and a theoretical framework for the progression of skill performance strategies employed across the stages of skill learning.

1.1 Developing Views on Skill Automaticity

One of the earliest considerations of autonomous processes in human function is attributed to Aristotle (Korsgaard 2008; Bernacer and Murillo 2014). According to Aristotle, repetition of a task gave rise to enhanced task performance owing to a developed disposition that rendered action less subservient to conscious control and thus, more spontaneous (Bernacer and Murillo 2014). Thus, rather than Descarte's view of a fixed dichotomy between controlled and automatic behavior (Jonides et al. 1985), that automaticity relied on repetition was suggestive of progression across a continuum of controlled and automatic performance (Whitaker 1983), which is in line with contemporary theories of motor skill learning (e.g., Fitts and Posner 1967).

In the late nineteenth century, two students working in the Harvard Psychological Laboratory under William James, Gertrude Stein, and Leon Solomons evaluated the automaticity of writing (Solomons and Stein 1896; Stein 1898). That well-practiced writing was performed "outside of awareness" (Solomons and Stein 1896) was interpreted as being necessary to overcome resource capacity limits, particularly with respect to attention (James 1890). As with many cognitive theories (e.g., Miller 1956; Kahneman 1973; Posner 1979), the theme of limited capacity became the central raison d'être for automaticity. For example, LaBerge and Samuels (1974) stipulated that skilled performance exceeds limited attention resources and so at least a subset of cognitive and motor processes must be automated to allow for successful performance. Limited capacity refers to information processing restrictions due to the finite availability of cognitive resources being insufficient to meet the current demand from ongoing processes. Skilled behavior can potentially overwhelm these limits given the high information load associated with processes responsible for sensation and perception as well as skilled action preparation, execution, monitoring, and updating. Reduced reliance on limited resources meant that automated skills could be performed in parallel with other tasks with little or no interference (Jastrow 1891; Bahrick et al. 1954; Brown and Carr 1989; Cohen et al. 1992). Diminished resource demands have been attributed to less reliance on sensory feedback processing (Keele 1968) as automaticity allows movement to be performed under feedforward control as opposed to feedback control (Wolpert et al. 1998; Miall et al. 2001; Puttemans et al. 2005). Feedforward control, where production of skilled action does not rely on concurrent sensory input to guide the action, is possible when movement production is based on an internal model (Wolpert et al. 1998; Kawato 1999), which has been described as involving a motor program (Schmidt 1975, 2003) or motor chunk (Verwey 1999).

1.2 Extended Practice Requirements for Skill Automaticity

There is no singular agreed-upon definition of motor skill automaticity, which introduces difficulty in determining the point at which a skill is automated. Several approaches in this respect have been presented in the literature. One approach focuses on the level of practice associated with the skill. Specifically, automaticity is thought to arise from extended practice. However, extended practice itself is not defined in terms of actual practice volume. Moreover, extended practice requirements would be expected to depend on the skill as well as the individual learner. The popular 10,000 h of deliberate practice quanta for skill expertise has been commingled with skill automaticity even though automaticity is but only one dimension of expertise (Singer 2002). It is likely that the level of extended practice necessary for automaticity is well below 10,000 h as skill automaticity has been demonstrated with far less practice (e.g., Karni et al. 1995). Quantification issues aside, the provision of extended practice does not necessarily guarantee that skill automaticity will be achieved (Lang and Bastian 2002). For this reason, some propose that overtraining must be implemented to ensure that extended practice leads to automaticity (Puttemans et al. 2005; Moors and De Houwer 2006). Overtraining involves the provision of additional practice beyond the point at which skill performance reaches asymptote.

1.3 Cognitive Characteristics of Skill Automaticity

More formal, cognition-oriented, criteria of skill automaticity have been drawn from Schneider and Shiffrin's (1977) description of an automatic mode of information processing (for example, see Ashby and Crossley 2012). One criterion is that an automated skill can be performed simultaneously with another task without any loss of performance in either the skill or the paired task (Passingham 1996; Wu et al. 2004; Lehericy et al. 2005; Poldrack et al. 2005; Wu and Hallett 2005; Wu et al. 2008; for an example in sport, see Leavitt 1979). An extension of this criterion involves assessing motor skill attention demands using dual-task methodology (Posner and Keele 1969; McLeod 1980; Abernethy 1988; Wright and Kemp 1992; Verwey et al. 2010). Here, the primary task, the presumed automated skill, is performed with a secondary probe reaction time (RT) task. Reduced attention demand from the primary task is inferred when there is little increase in probe RT under dual-task conditions in comparison to performing probe RT alone.

Skill automaticity might also be characterized by the degree of performance inflexibility that is observed following a period of training. Performance inflexibility refers to the extent to which performance of the skill deteriorates when the sensory or task environment is changed from the conditions that were present during learning. For example, Proteau et al. (1992) demonstrated that manual aiming performance was impaired by introducing visual feedback after a period of practice without visual feedback availability. Automated skills are thought to be inflexible to the extent that they will not be modifiable when new conditions are introduced during additional practice (Schneider and Chein 2003). In Experiment 2 of Helie et al. (2010), extended practice involving about 11,000 trials was provided for a visual stimulus categorization task that involved key-press responses. When new stimulus–key response mappings were introduced following extended practice, performers demonstrated significant performance losses in terms of increased error and RT. These losses persisted even after the provision of 600 additional practice trials with the new stimulus–response mappings.

1.4 Cognitive Framework for Skill Automatization

While performance criteria might be useful to test for skill automaticity, they provide little to the understanding of the underlying neurocognitive changes that contribute to observed characteristics of automaticity. Thus, there has been some effort devoted to developing cognitive frameworks of skill automatization based on research addressing how we sequence basic movement elements into skillful action. Specifically, findings from studies involving discrete key press sequences have led to the notion that motor skill automaticity is associated with the development of sequence knowledge at different processing levels (Hikosaka et al. 1999; Abrahamse et al. 2013; Verwey et al. 2015). According to the cognitive framework for sequential motor behavior (C-SMB; Verwey et al. 2015, see Fig. 1), a motor skill can be represented perceptually, symbolically, and motorically and these representations are processed, respectively, by perceptual, central, and motor processors. Moreover, the manner in which the skill is represented and consequently processed changes as the skill is acquired. The idea is that sequential motor skills are initially based on spatial and perhaps even verbal sequence forms of central–symbolic representations, which are relatively slow and cognitively demanding to apply. Extensive practice allows elements to be chunked into motor sequence representations. These allow for a skill to be performed with much less cognitive demand because these motor chunks are executed by a motor processor that loads cognitive processing resources only for selecting and initiating these motor routines. So, as sequential motor skills evolve with practice, the role of central cognitive processing resources changes from selecting and initiating individual movements to selecting and initiating integrated movement patterns. This processing change would be responsible for the rapid movement execution along with reduced flexibility and limited interference with other tasks.

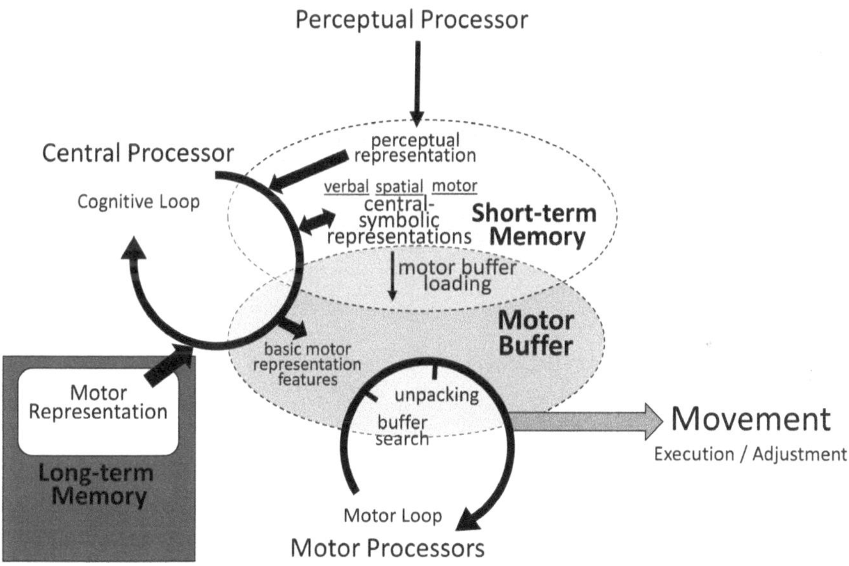

Fig. 1 Key components of the cognitive framework of sequential motor behavior (C-SMB; Verwey et al. 2015). This model describes the collaboration of perceptual, central, and motor processors in using (spatial, verbal) central–symbolic, and motor representations to execute motor sequences

1.5 Historical and Contemporary Views on Neural Changes Associated with Skill Automaticity

The presence of permanent adaptations in cognitive processes that underlie enhanced performance are expected to be accompanied by structural and functional changes in the neurobiological substrates of motor skill automaticity (Matsuzaka et al. 2007; Ashby et al. 2010). Work by the Neurophysiologist and Nobel Laureate, Sir Sherrington (1974), offered some of the earliest examinations of the neural correlates of movement automaticity (Ashby et al. 2007). Sherrington's perspective was that skill automaticity represented "spontaneous" behavior underpinned by reflexive action, which arose from extensive practice (Burke 2007). While skill automaticity is no longer viewed as being reflexive, Sherrington's idea that neural ensembles could be formed from experience (i.e., neuroplasticity) to enable movement production at any level of the central nervous system (e.g., Sherrington 1910) gave rise to the concept of hierarchical movement control (Pew 1966; Bernstein 1967; Koechlin et al. 2003; Fuster 2004). This, in turn, gave rise to the claim that novel motor skills require control at the resource-demanding and least efficient executive cortical levels (e.g., prefrontal cortex), but as they are practiced, their control is transferred to more efficient cortical regions such as the premotor and primary motor cortices (e.g., Ashby et al. 2010), or subcortical regions such as the basal ganglia (e.g., Wu et al. 2004). Thus, the hallmark feature of skill automaticity, efficiency, is derived from practice-dependent neuroplastic changes. These changes do not necessarily reflect diminished neural involvement in skilled action but rather the development of neural connections that facilitate performance speed and accuracy and in parallel, reduce the cognitive demands of performance.

2 Neural Changes from Early Skill Learning to Automatization

The autonomous stage of motor skill performance (Fitts 1964) is preceded by learning stages that involve cognitive demanding and effortful associative and motor adaptation processes (Fitts 1964; Fitts and Posner 1967; Gentile 1972, 1998; Hikosaka et al. 2002). Early skill learning relies on activity in the primary motor cortex (M1), the basal ganglia, and the cerebellum (Hikosaka et al. 2002; Krakauer and Mazzoni 2011; Penhune and Doyon 2002; Shadmehr and Krakauer 2008; Shmuelof and Krakauer 2011; Penhune and Steele 2012; Hardwick et al. 2013—see Fig. 2). In addition, several frontal regions including the dorsolateral prefrontal (DLPFC), premotor (PMC), and supplementary motor area (including the supplementary motor area proper, SMA, and the pre-supplementary motor area, preSMA) have been shown to contribute to early skill learning (Grafton et al. 1992; Jenkins et al. 1994; Jueptner et al. 1997b; Toni et al. 1998; Ghilardi et al. 2000; Grafton et al. 2002; Verwey et al. 2018). Frontal lobe cortical activity underlying skill learning follows an

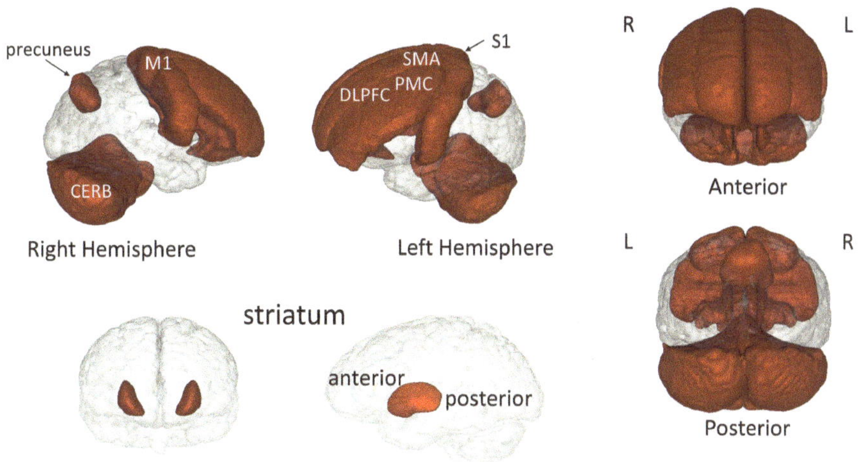

Fig. 2 Key regions associated with early motor skill practice. M1 = primary motor cortex, CERB = cerebellum, DLPFC = dorsolateral prefrontal cortex, SMA = supplementary motor area (proper and preSMA), PMC = premotor cortex, S1 = somatosensory cortex

anterior to posterior direction according to an associative-motor hierarchy for movement preparation and control (Koechlin et al. 2003; Fuster 2004). As an example, within the associative-motor hierarchy, associative processes might relate to the selection of desired responses according to task goals, stimulus conditions, or preceding responses (Toner and Moran 2015) whereas motor processes are involved in producing the action itself, which might include adapting muscle force specification to meet the skill goal (Taylor and Ivry 2012). Early in learning, resource-demanding executive prefrontal cortices are relied on to inform lower motor levels (i.e., premotor and primary motor cortices; Fuster 2004). However, as learning progresses, cortical activity in the frontal lobe diminishes in an anterior to posterior direction (Wu et al. 2004; Lehericy et al. 2005; Poldrack et al. 2005; Puttemans et al. 2005; Wu et al. 2008), which corresponds to a transition from executive to motor levels of control and consequently, a reduction in the central resource demands. This process does not appear to be irreversible though as automated skills can be consciously attended to by once again invoking activity in the prefrontal cortex (Jueptner et al. 1997b; Kubler et al. 2006). Heightened cortical activity associated with early skill learning has also been reported in parietal (specifically the somatosensory region, S1), temporal, and occipital regions along with the anterior and posterior cortices and the precuneus (Jenkins et al. 1994; Toni et al. 1998; Schendan et al. 2003; Robertson 2007; Shadmehr and Krakauer 2008; Laird et al. 2011; Lohse et al. 2014).

2.1 Fast and Slow Skill Learning Stages

Progression of skill learning from early to later levels of experience is thought to follow fast and slow stages (Karni et al. 1998; Dayan and Cohen 2011). Both fast and slow learning stages are experience-dependent since the performance improvements only occur with actual task practice. An intermediate stage highlights offline forms of motor learning and includes motor memory consolidation processes (Robertson et al. 2004; Walker and Stickgold 2004; Immink 2016). The fast learning stage is characterized by large performance improvements during initial skill practice. Fast learning is thought to involve attention-demanding, associative processes (Hikosaka et al. 1999, 2002; Lohse et al. 2014; Verwey et al. 2015, 2018). As skilled motor tasks are comprised of a series of independent movement elements, associative processes contribute to sequence learning where individual movement "building blocks" (Verwey and Dronkers 2018) are associated to form movement representations based on "chunked" sequence elements (Grafton et al. 1995; Clegg et al. 1998; Verwey 1999; Abrahamse et al. 2010; Verwey and Wright 2014; Wright et al. 2016; Immink et al. 2017). Early in learning, particularly with discrete key-pressing sequences (Verwey 1999; Abrahamse et al. 2013), performance of movement sequences is slow and demanding because initiation of each response element relies on stimulus information processing (Abrahamse et al. 2013; Verwey et al. 2015, 2018; see Fig. 3, panel b, reaction mode). To render sequential performance less stimulus-dependent, sequence learning can then progress to increased involvement of premotor processes (Lohse et al. 2014) which utilize verbal or symbolic representations (Gentile 1972, 1998; Verwey et al. 2015, 2018) such as when a phone number is verbally recited to enter the requisite sequence on the phone keypad (Fendrich and Arengo 2004; see Fig. 3, panel b, central–symbolic mode). Forming associations between successive movement elements render faster and less resource-demanding sequential behavior as a sequence representation or "motor chunk" provides for more efficient motor preparatory processes (Abrahamse et al. 2010; Verwey et al. 2015, 2018; see Fig. 3, panel c, chunk mode). "Chunked" sequence representations are then further refined in the slow stage of learning, which emphasizes effector-specific (Hikosaka et al. 2002) mechanisms that contribute to motor representation formation (Lohse et al. 2014; Verwey et al. 2015, 2018), and overlapping activity of effectors executing successive movements (like the coarticulation found with skilled typists; Shaffer 1975; Jordan 1995; Engel et al. 1997). Motor representation driven performance is fast because it does not rely on attention-demanding sensory feedback or premotor/associative processing. The tradeoff to speeded performance is the lack of flexibility inherent in feedforward modes of performance.

Neurophysiological evidence for the concept of fast and slow learning stages is based on activity within cortico-striatal-cerebellar systems (Doyon and Ungerleider 2002; Hikosaka et al. 2002; Doyon et al. 2003; Dayan and Cohen 2011). Specifically, associative learning processes, relied upon within the fast stage (Hikosaka et al. 2002; Lohse et al. 2014; Verwey et al. 2018), have been correlated with increased activity in the DLPFC (Hikosaka et al. 2002), dorsal and ventral PMC (dPMC, vPMC,

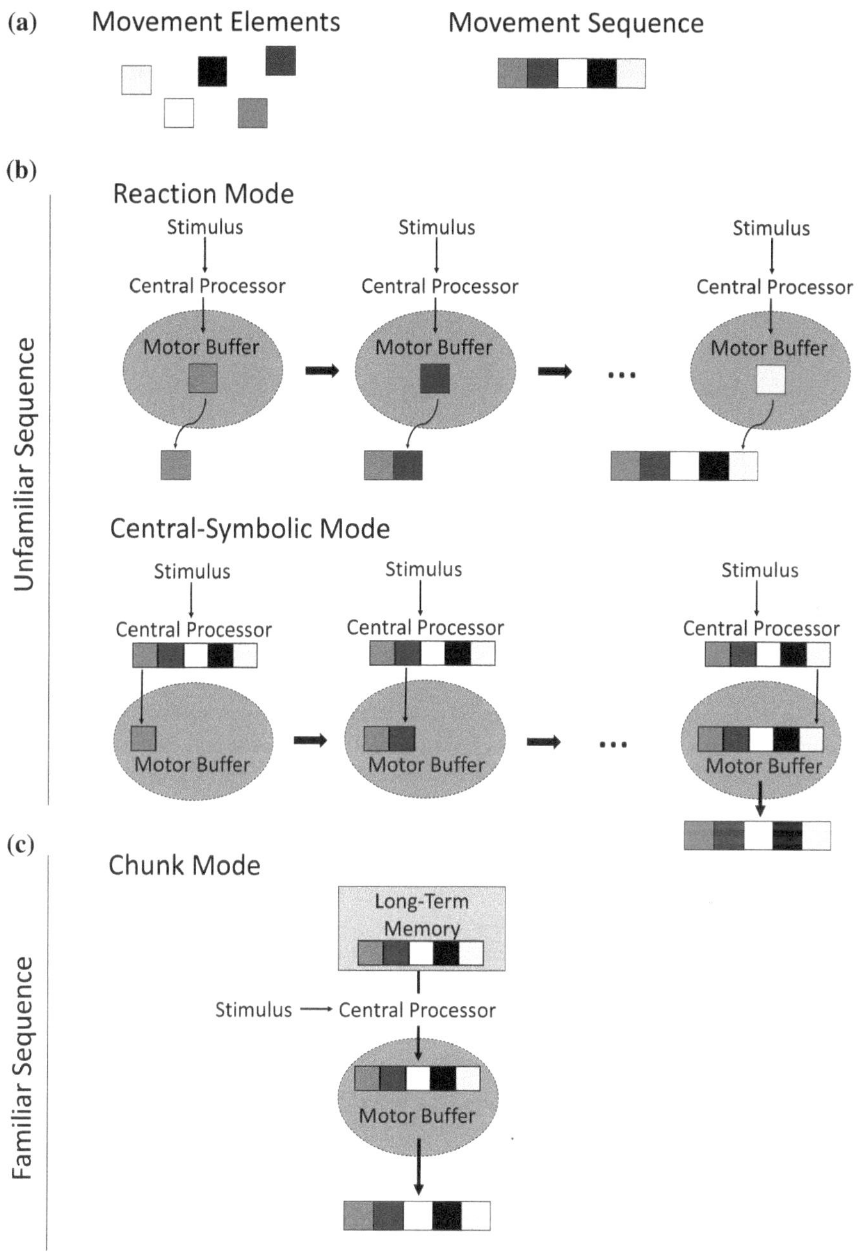
(a)
Movement Elements
Movement Sequence
(b)
Unfamiliar Sequence
Reaction Mode
Stimulus
Central Processor
Motor Buffer
Central-Symbolic Mode
Stimulus
Central Processor
Motor Buffer
(c)
Familiar Sequence
Chunk Mode
Long-Term Memory
Stimulus
Central Processor
Motor Buffer

◂**Fig. 3** The cognitive framework for sequential motor behavior (C-SMB; Verwey et al. 2015) describes how movement elements can be organized into a target motor sequence (panel **a**). Initially, the performance of an unfamiliar sequence (panel **b**) occurs via a reaction mode, where the target is reproduced based on presentation of a fixed series of movement element-specific stimuli. With practice, central–symbolic representations of the unfamiliar target sequence are developed allowing for motor buffer loading of the target sequence based on iterative translation of the symbolic representation. Both reaction and central–symbolic modes are relatively slow and place high demands on cognitive resources. Further practice allows for familiar sequences to be represented in long-term memory as an internal model of the target sequence (panel **c**). This enables the sequence to be performed via a chunk mode. The chunk mode is relatively fast and incurs little cognitive resource costs due to reduced central cognitive processing requirements

respectively; Hardwick et al. 2013), the SMA, S1, precuneus, striatum (caudate nucleus, anterior and posterior putamen), and cerebellum (Lehericy et al. 2005; Wu et al. 2008; Lohse et al. 2014). In contrast, the slow learning stage, which emphasizes sensorimotor adaptation (Hikosaka et al. 2002; Lohse et al. 2014; Verwey et al. 2018) appears to involve increased activity within M1, dPMC, superior temporal lobe, and the posterior putamen (Hikosaka et al. 2002; Lehericy et al. 2005; Coynel et al. 2010; Hardwick et al. 2013; Lohse et al. 2014). Changes associated with practice reveal decreases in SMA, vPMC, precuneus, and cerebellum activity coinciding with increases in activity in M1, posterior putamen, and globus pallidus (Grafton et al. 1995; Hazeltine et al. 1997; Lehericy et al. 2005; Wu et al. 2008; Ashby et al. 2010; Lohse et al. 2014). Overall, findings from neuroimaging studies are consistent with a transition from associative, fast learning, to motor, slow learning, processes as practice progresses. Moreover, neuroimaging studies provide evidence for decreased reliance on higher level control of movement that is concomitant with an anterior to posterior shift in frontal lobe activity (Wu et al. 2008) along with increased cortico-striatal connectivity (Toni et al. 1998).

2.2 *Differentiating Learning-Related Neural Changes Based on Skill Function and Performance Strategies*

It should be noted that there is some disagreement that exists in the literature with regard to changes in brain region activity associated with motor learning. For example, some have shown that with practice, M1 activity increases (Grafton et al. 1995; Hazeltine et al. 1997), while others have reported no change (Karni et al. 1995) or M1 activity decrease (Shadmehr and Holcomb 1997). Lohse et al. (2014) found that left M1 activity increases initially but then decreases with extended practice. Similarly, some have reported learning associated increases in cerebellar activity (Jenkins et al. 1994; Doyon et al. 1996), which contrasts with others' report of decreased cerebellum activation (Friston et al. 1992; Grafton et al. 1994; Seitz et al. 1994; Jueptner et al. 1997a; Doyon et al. 2002; Lohse et al. 2014). These mixed findings have been explained in terms of the variety of motor tasks that have been investigated (Hardwick et al. 2013). For example, some studies have involved unimanual sequence learning tasks (e.g., Jenkins et al. 1994; Karni et al. 1995) while others have involved bimanual movements requiring inter-limb coordination (e.g., Puttemans et al. 2005). Some tasks involve high motor control demands due to having to learn new spatial and temporal movement dynamics requiring adaptations to muscle recruitment and joint coordination (e.g., Pearce et al. 2000; Bezzola et al. 2011; Di Paola et al. 2013). Other tasks have involved playing musical instruments with differing temporal rhythms (e.g., Buccino et al. 2004; Herdener et al. 2010) while some tasks have low motor control demands associated with pressing sequences of keys in response to visual stimuli (e.g., Doyon et al. 1996; Verwey 1999; Seidler et al. 2002; Abrahamse et al. 2013).

Variations in reported neural activity might also be explained by the execution strategy that is employed to perform the motor skill. Specifically, using their multiple representation point of view (i.e., perceptual, central, and motor representations), Verwey et al. (2015) argued that sequence performance can involve different execution modes. These researchers explicitly distinguish between a reaction mode, a central–symbolic mode, and a chunking mode (Fig. 3, panels b and c). In line with many real-world tasks, individuals performing sequential keying tasks in the laboratory usually begin by reacting to a fixed series of key-specific stimuli in the so-called reaction mode. Quite rapidly, associations develop at various processing levels that prime the ensuing response with reduced reliance on key-specific stimuli. With practice, spatial or verbal representations develop at the central–symbolic processing level that requires cognitive processes to extract motor information. Finally, the development of tight associations between motor representations leads to the development of motor chunks that in the chunking mode can be used to rapidly execute movement sequences with little need for central cognitive processing. Importantly, after substantial practice, these execution modes can be strategically applied and may even be concurrently active. Consequently, skilled performers can switch between different processing strategies with little or no measurable performance changes.

To resolve the disagreement in neural changes associated with skill acquisition, some have proposed differentiating motor skills based on their function. Some skills place an emphasis on learning sequences based on highly practiced movement elements (Robertson 2007). An example of these types of skills involve sequences of key presses, which is common in lab-based paradigms for sequence learning such as the serial reaction time (SRT; Nissen and Bullemer 1987; Abrahamse and Noordzij 2011) and the discrete sequence production (DSP; Verwey 1999; Abrahamse et al. 2013) tasks as well as skills involving typing. Sequencing key presses contrasts with skills that emphasize development of complex movement patterns requiring sensorimotor adaptation or coordination of multiple limbs. This contrast is important as motor skills might rely on different brain networks. Specifically, during later stages of learning, discrete sequence tasks have been argued to utilize a cortico-striatal network while sensorimotor tasks rely on a cortico-cerebellar network (Doyon and Ungerleider 2002; Doyon et al. 2003). A meta-analysis conducted by Hardwick et al. (2013) which contrasted learning sequential (SRT variants) and sensorimotor skills found that both skill types share reliance on the dPMC, contralateral M1, and the primary sensorimotor zone (O'Reilly et al. 2010) of the right cerebellum (lobule VI). Acquisition of keying sequence variants was associated with activation of the preSMA and SMA along with superior parietal lobule, right lateral cerebellum and left thalamus. This pattern of activity reflects the involvement of reaction mode processes in early learning of keying sequences where selection of sequence elements is driven by visual stimuli (Abrahamse et al. 2010; Verwey et al. 2015, 2018). Instead, sensorimotor task learning elicited activation of only the SMA along with S1 and the right vermis and left and right lateral cerebellum. This demonstrates the increased complexity of motor commands that must be developed in acquisition of sensorimotor tasks (Hardwick et al. 2013). Similar to Hardwick et al. (2013), Laird et al. (2011) distinguished different motor skills when describing neural networks associated with motor skill performance. Skills that predominantly involve fixed simple movement sequences (e.g., typing) involve premotor and supplementary motor cortices. However, Laird et al. (2011) describe two additional neural networks for more complex movement types based on the locus of upper-limb control. Skills that predominantly involve sensorimotor adaptation to hand or finger movements (e.g., grasping) utilize a network consisting of M1, S1, and the cerebellum, due to high degree of error correction needed in fine control of so-called haptic skills. Prehension skills are more complex than haptic skills as they include arm movement control when objects must be reached prior to grasping them. Accordingly, these more complex skills supplement the M1-S1-cerebellar network with medial superior parietal activity.

2.3 Brain Activity Underlying Automated Skill Performance

Neuroimaging studies describing brain changes associated with motor skill learning illustrate neural adaptations that preceded skill automatization. Early experience in performing novel skills relies heavily on the activity in the anterior frontal lobe regions, which are associated with conscious attentional processes, error correction mechanisms, and the subjective experience of effortful performance. These regions are also associated with explicit forms of motor learning that are more resource demanding than implicit forms that do not require conscious awareness (Grafton et al. 1995; Doyon et al. 1996). However, most of this work has concentrated on the early stages of learning (see Lohse et al. 2014) in contrast to the automatic stage. Thus, much of what is currently understood about functional brain changes underlying skill automaticity is based on studies addressing learning, not automated performance. Only a few neuroimaging studies have directly assessed the neural correlates of skill automaticity based on dual-task interference assessment (Pashler 1994; Ashby and Crossley 2012).

In all, six neuroimaging studies have applied dual-task interference assessment to determine skill automaticity (Wu et al. 2004; Lehericy et al. 2005; Poldrack et al. 2005; Puttemans et al. 2005; Wu and Hallett 2005; Wu et al. 2008). Five of these have evaluated automaticity of sequenced finger movements (Wu et al. 2004; Lehericy

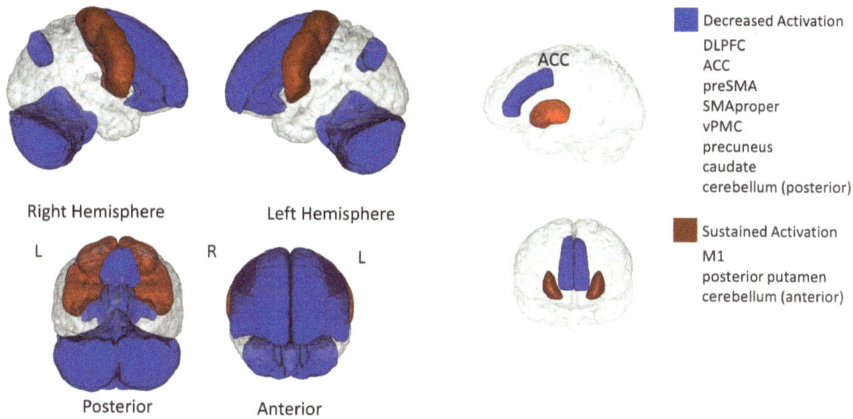

Fig. 4 Decreased and sustained functional region activation when early and automatic stages of skill learning are contrasted. DLPFC = dorsolateral prefrontal cortex, ACC = anterior cingulate cortex, SMA = supplementary motor area (proper and preSMA), vPMC = ventral premotor cortex, M1 = primary motor cortex

et al. 2005; Poldrack et al. 2005; Wu and Hallett 2005; Wu et al. 2008) while Puttemans et al. (2005) evaluated automaticity of a complex bimanual coordination task involving wrist flexion and extension. A comparison of changes in neural activity from early to automatic stages, revealed a pattern similar to that observed in early to late learning comparisons (Lohse et al. 2014). Namely, automaticity, as assessed by the dual-task interference criteria, was associated with decreased activity in the DLPFC and premotor cortices (preSMA, vPMC), cerebellum, and precuneus (see Fig. 4). Similar regional activity in late learning (Lohse et al. 2014) and skill automaticity (Wu et al. 2004; Lehericy et al. 2005; Poldrack et al. 2005; Puttemans et al. 2005; Wu and Hallett 2005; Wu et al. 2008) might give the impression that there is relatively little change in neural activity between late learning and automatic stages. However, regional activation in later learning and automaticity might be due to entirely different functional roles (Little et al. 2004). As such, similar patterns of regional activation might conceal ongoing neural changes that take place between late learning and automaticity.

Neuroimaging studies of automatic motor skill performance reported reduction in anterior cingulate cortex (ACC) activity from early to automatic skill performance stages. The ACC functions along with the DLPFC to control attention (Kondo et al. 2004) and is involved in error monitoring and correction during performance (Carter et al. 1998). To allow the acquisition of new movement patterns, the ACC is also thought to be involved in suppression of previously established movement patterns (Swinnen 2002).

In contrast to previous reports of the importance of M1 activity for performing well-learned motor skills (Grafton et al. 1995; Hazeltine et al. 1997), Puttemans et al. (2005) reported decreased M1 activity at the autonomous stage. This finding was consistent with decreased M1 activity in the late skill learning stage reported in a meta-analysis of neural changes across skill learning stages reported by Lohse et al. (2014). As discussed previously, differences in neural activity between studies might relate to different skill performance strategies (Verwey et al. 2015) or might relate to task-specific motor control demands (Laird et al. 2011; Hardwick et al. 2013). Differences in performance strategies and motor control demands might engage in functionally different cortico-striatal neural circuits (Doyon et al. 2002, 2003) resulting in disparate changes in regional activity. For example, Lehericy et al. (2005) reported reduced activity in the cortico-striatal circuit but only in relation to premotor cortices and the dorsal putamen. In contrast, activity in the posterior putamen increases (Lehericy et al. 2005). The dorsal and posterior putamen have been differentiated based on their involvement in associative and sensorimotor forms of learning (Jueptner et al. 1997a; Lehericy et al. 2005; Florio et al. 2018). Connectivity between associative and premotor cortices (vPMC) and the dorsal striatum is thought to underlie associative learning while connectivity between premotor (dPMC), motor and sensory cortices, and the posterior putamen regulates sensorimotor learning as well as production of well-learned skills (Lehericy et al. 2005; Yin and Knowlton 2006; Wu et al. 2015; see Fig. 5, panels a and b). There is some debate on how associative and sensorimotor learning processes occur. In contrast to a serial shift from associative to sensorimotor processes (Hikosaka et al. 1999; Floyer-Lea and Matthews 2005; Poldrack et al. 2005; Yin et al. 2009; Lohse et al. 2014; Fig. 5, panel c), others have argued that associative and sensorimotor processes run in parallel during motor skill learning (Thorn et al. 2010; Thorn and Graybiel 2014; Kupferschmidt et al. 2017; Fig. 5, panel d).

3 Expediting Automaticity at the Cost of Learning: A Cautionary Tale from the Contextual Interference Effect

A natural extension of the discussion so far is that automaticity results from exposure to considerable amounts of training and coincides with the development of expertise. As noted earlier, this is typically manifested as significant improvements in performance while also exhibiting an absence of secondary task interference and the development of a motor representation, motor chunks, that exhibit some rigidity in the manner in which they are used (see Sakai et al. 2004). One might assume then that training environments that quickly move individuals to display characteristics of automatic skilled performance would be desirable. Indeed, the considerable experimental effort has been exerted to identify features of the training environment that might be modified in order to move the performer to a more automated behavior. One feature, in particular, that has attracted attention over the past 40 years, studied under the rubric of the contextual interference (CI) effect, focuses on how training schedules for a set of related motor skills influences learning outcomes (Shea and Morgan 1979; Magill and Hall 1990; Brady 1998, 2004).

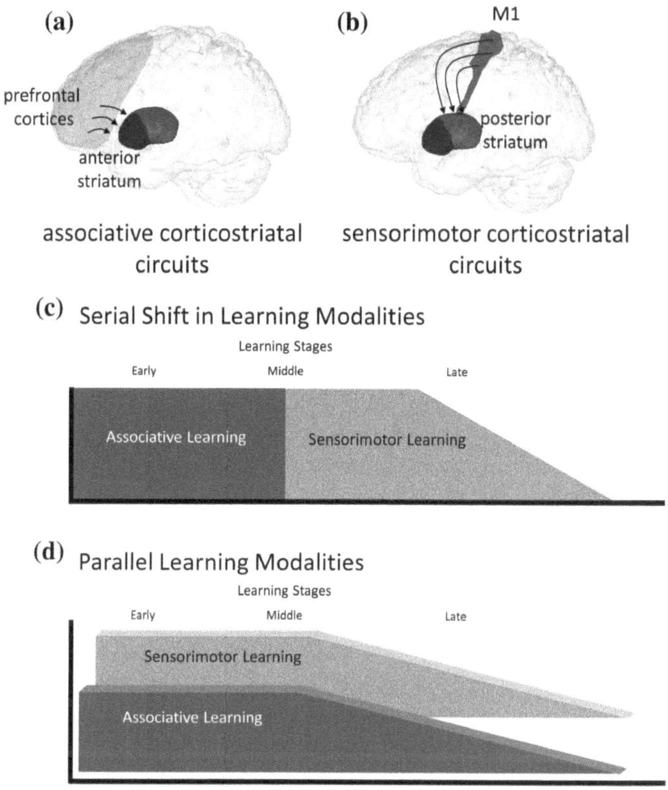

Fig. 5 The dorsal and posterior putamen are functionally differentiated according to associative and sensorimotor forms of learning. Associative learning (panel **a**) is based on circuits between the prefrontal cortices and the dorsal striatum while sensorimotor learning (panel **b**) is based on circuits between the motor cortices and the posterior putamen. Motor skill learning has been described as involving a serial shift from associative to sensorimotor processes (panel **c**) or alternatively, parallel implementation of associative and sensorimotor processes (panel **d**)

During experiments addressing the impact of CI for learning, greater CI is often engineered by arranging training such that the learner executes a set of skills in an interleaved format. Interleaved training, as it is called, creates relatively high interference throughout training because of the rapid changes in task demands across training trials. Conversely, repetitive training creates less interference because it entails executing the same skill repeatedly prior to the introduction of other skills that will be practiced later in training (see Fig. 6, panel a). Repetitive training provides for very rapid gains in performance (see Fig. 6, panel b). This initial practice benefit of repetitive training which is very robust has been observed in a wide variety of laboratory tasks (Shea and Morgan 1979; Wright et al. 2005; Pauwels et al. 2014) as well as applied situations (Goode and Magill 1986; Smith and Davies 1995; Ollis et al. 2005). In addition, the advantage of repetitive training for facilitating quick performance gains has been demonstrated for different subject populations (Del Rey

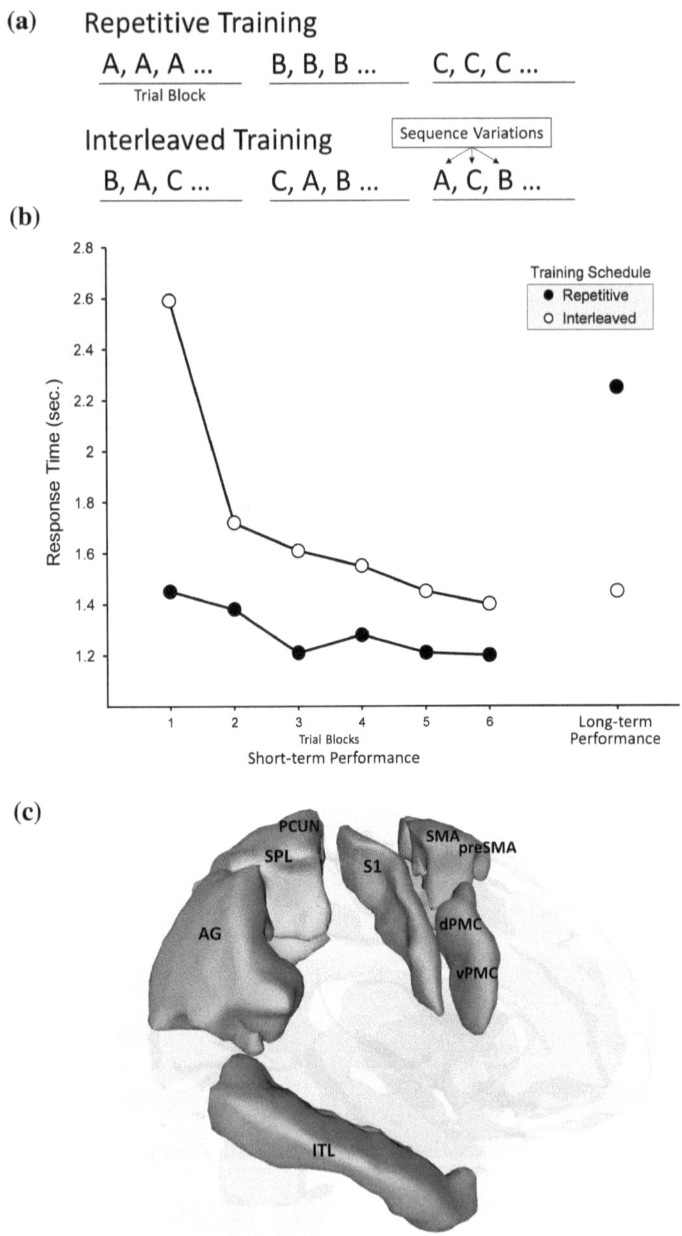
(a)
Repetitive Training
A, A, A ...
Trial Block
B, B, B ...
C, C, C ...
Interleaved Training
Sequence Variations
B, A, C ...
C, A, B ...
A, C, B ...
(b)
Training Schedule
Repetitive
Interleaved
Response Time (sec.)
2.8
2.6
2.4
2.2
2
1.8
1.6
1.4
1.2
1
2
3
4
5
6
Trial Blocks
Short-term Performance
Long-term Performance
(c)
PCUN
SPL
AG
S1
SMA
preSMA
dPMC
vPMC
ITL

◀**Fig. 6** Under the contextual interference (CI) effect, practice scheduling of motor sequence variations under repetitive and interleaved schedules (panel **a**) results in distinct short-term and long-term performance profiles (panel **b**) as well as differences in neural region recruitment (panel **c**). As illustrated in panel **b**, repetitive training affords short-term performance gains but then results in long-term performance costs, particularly when long-term performance is tested under interleaved schedules (data is based on the seminal study by Shea and Morgan 1979). Panel **c** illustrates neural regions that have been associated with the CI effect. There is reduced recruitment of these regions under repetitive training in comparison to interleaved training. However, when long-term performance is tested following repetitive training, these neural regions exhibit increased activity when compared to those previously exposed to interleaved training. ITL = inferior temporal lobe, AG = angular gyrus, SPL = superior parietal lobe, PCUN = precuneus, S1 = somatosensory cortex, SMA = supplementary motor area proper, preSMA = pre-supplementary motor area, dPMC = dorsal premotor cortex, vPMC = ventral premotor cortex

1982; Porretta and O'Brien 1991), as well as in the clinical domain (Adams and Page 2000; Knock et al. 2000; Wambaugh et al. 2014).

3.1 Early Performance Automatization Under Repetitive Training

Importantly, for the purpose of the present discussion, individuals that are trained in a repetitive rather than interleaved training environment not only reveal superior skilled performance much earlier in training but also display characteristics of the automated performance. For example, Li and Wright (2000) used a dual-task paradigm to pair a secondary choice reaction time (CRT) task that involved tone discrimination with the acquisition of a set of motor sequence skills acquired in either an interleaved or repetitive training format. During both training formats, the secondary CRT task was periodically presented shortly after the participant was informed about the particular motor sequence to be performed on the next trial presumably while the performer engaged in task preparation. Li and Wright hypothesized that the observed efficacy of repetitive training for skill acquisition would coincide with reduced attention demands reflected in lower dual-task cost relative to the alternative, interleaved training. The results supported this hypothesis indicating roughly 20% lower attention demand in conjunction with significantly greater performance improvement during repetitive as opposed to interleaved training. This finding was consistent with earlier reports for the need of much shorter inter-trial intervals during repetitive as opposed to interleaved training (see, Immink and Wright 1998). Taken together, the implication of these data is that repetitive schedules can be used as an efficient format for training because performance improves at a relatively fast rate and at low cognitive cost.

It was proposed earlier that part of the reduction in attentional load during the learning of perceptual-motor skills was the development of units of action, referred to as motor chunks (Verwey 1999; Verwey et al. 2015). This has been reported to involve the compilation of individual movement elements being compiled into motor sequence representation that allows a skill to be executed with little cognitive demand (Klapp 1995; Sakai et al. 2004; Verwey et al. 2015). This is a central feature of the C-SMB (Verwey et al. 2015) described earlier in which the motor chunks developed from extended practice are executed by a motor rather than central (cognitive) processor. Thus, as sequential motor skills evolve with practice, the role of central cognitive processing resources changes from selecting and initiating individual movements to selecting and initiating integrated movement patterns. This processing change would be responsible, at least in part, for the evolution of reduced attention demands reflected in limited interference with other tasks (Verwey et al. 2015). It is not surprising then that it has been reported that individuals exposed to repetitive training exhibit characteristics of motor chunk development during the earliest phase of training. Wright et al. (2004) had individuals practice both simple and complex motor sequences in either a repetitive or interleaved training format across 4 days of practice. As expected, reaction time (RT) was initially longer when initiating the more complex sequences (Henry and Rogers 1960; Klapp 1995; Verwey 1999). Participants assigned to interleaved training took a considerable portion of the 4 days of training to eventually initiate the more complex sequences with RT performance similar to that observed for the simpler sequence, indicative of motor chunk development. In contrast, for the individuals that experienced repetitive training, the RT did not differ as a function of sequence complexity from a very early stage in training, suggesting that integration of the individual movement elements in the more complex sequence had occurred affording faster initiation of the skill. Thus, a repetitive training format appears to be an effective training protocol since it is associated with characteristics previously ascribed to automated skill production while also associated with the rapid attainment of successful performance.

3.2 Performance Automatization Versus Learning

Given this brief overview of the merits of repetitive training, one would be forgiven for assuming that this training approach should result in successful skill acquisition and as such should demand widespread acceptance and use. However, when addressing learning, a critical goal is the establishment of a "relatively permanent changes in the capability for skilled behavior", which is commonly assessed through the use of long-term performance evaluation (Schmidt and Lee 2005, p. 302). This focus on long-term retention was initially addressed as the performance-learning distinction (Salmoni et al. 1984) but has been formally revisited many times since (Schmidt and Bjork 1992; Kantak and Winstein 2012). This distinction highlights the difference between transient influences on current behavior and those that reflect "relatively permanent" learning effects. Immediate performance during the early phase of skill acquisition is very likely influenced by any number of transient features such as the availability of rich sources of feedback, motivation, and/or attention (Kantak and Winstein 2012). To infer learning, it is argued that performance is assessed after sufficient time delays to allow temporary influences to dissipate. This is often accomplished in the motor neuroscience literature by the inclusion of retention tests after varying temporal delays (e.g., 24 h, 7 days, or even months) after training has concluded.

The performance-learning distinction turns out to be critical when evaluating the CI effect. As detailed earlier, repetitive training results in very rapid, and in some cases large gains in performance beyond that observed from engaging in interleaved training. This benefit appears in part to be fostered by the development of motor chunks (Wright et al. 2004), which in turn, leads to reduced cognitive effort reflected in lower dual-task costs (Li and Wright 2000). However, when delayed tests are administered, individuals exposed to repetitive training exhibit significant forgetting from the end of practice accompanied by the return of significant conscious effort. This is not the case for learners that encounter interleaved training. Sizeable demands are placed on attention throughout a significant portion of interleaved training (Li and Wright 2000) which results in the development of resilient motor chunks that continue to be used up to 72 h after training supporting performance levels that are consistent with those observed at the end of training or in some cases slightly enhanced (Kim et al. 2018). Herein lies the cautionary note. While repetitive training affords rapid improvements in performance in addition to revealing characteristics of automaticity (i.e., reduced conscious contribution and rapid adoption of motor representations), one would surely be less convinced of the effectiveness of this training format for facilitating learning in terms of long-term performance (see Fig. 6, panel b).

3.3 Neural Correlates of the CI Effect

An examination of the neural adaptations that occur across both interleaved and repetitive forms of practice are informative and may provide some guidance in thinking about automaticity in the context of skill acquisition and long-term retention. Initial neurophysiologic evidence revealed that interleaved training participants maintained a constant level of activation within some motor planning regions while increasing the activity in other regions that have been reported to be central to improvement in motor sequence performance across the practice (Cross et al. 2007). Such changes have been described in detail in earlier sections (see Floyer-Lea and Matthews 2005; Doyon et al. 2009; Dayan and Cohen 2011; Penhune and Steele 2012; Hardwick et al. 2013). Specifically, dPMC, vPMC, preSMA, and the SMA premotor areas exhibited a blood-oxygenated level-dependent (BOLD) signal that was at least 50% greater later in practice than earlier, implying significant action preparation continued late into interleaved training (Cross et al. 2007; see Fig. 6, panel c). The lateral premotor areas (dPMC, vPMC) have been recognized for the development of associations between arbitrary sensory stimuli and actions or the acquisition of novel stimulus–response rules which would certainly be an early challenge when learning a new set of motor sequences. The medial premotor areas, including preSMA and SMA, have been associated with selecting, retrieving, and organizing subsets or whole movement sequences at different points of skill acquisition (Gerloff et al. 1997; Verwey et al. 2002; Nachev et al. 2008). Heightened recruitment of the medial premotor areas is consistent with the claim that interleaved as opposed to repetitive training encourages the development and use of motor chunks (Immink and Wright 2001; Wright et al. 2004; Abrahamse et al. 2013; Verwey et al. 2015). Moreover, interleaved training results in a bilateral increase in activation in angular gyrus, superior parietal lobe, inferior temporal lobe, and precuneus with extended practice (Wymbs and Grafton 2009; see Fig. 6, panel c). These neural areas have been associated with the construction of novel visual–spatial mappings that would presumably be important for the acquisition of motor sequences (Hikosaka et al. 1999, 2002).

In the case of repetitive training, Wymbs and Grafton (2009) reported more limited activation as fewer regions were recruited for a shorter proportion of the total training duration. The behavioral findings from Immink and Wright (1998) suggest that this limited activation reflects disengagement from response preparation early in repetitive training. Consequently, there is inadequate development of motor response information storage and retrieval processes needed to support long-term performance. Neurophysiological and behavioral findings provide an account for the reduced attentional costs observed early in repetitive training by Li and Wright (2000). While reduced attentional costs could be interpreted as reflecting a form of skill automaticitySkill automaticity, it is clear that this form of automaticity has come at a significant cost to learning. The premature automaticity afforded by repetitive practice will eventually require greater investment of resources needed to perform in the long term. For example, during delayed retention test, Lin et al. (2011) reported more extensive recruitment of prefrontal, premotor, and parietal regions by individuals trained in the repetitive format than those trained under the interleaved format. Thus, in contrast to repetitive practice, interleaved practice appears to have exhibited greater automaticity during delayed test performance and this advantage may be associated with the diverse set of other neural adaptations that are instigated by training under high contextual interference. For example, Lin et al. (2011) revealed heightened M1 excitability long after interleaved training was complete (i.e., for up to 72 h) which has been argued to be a relatively permanent neuroplastic change that enhances memory retrieval possibly via an impact of post-practice consolidation (see Breton and Robertson 2014; Tunovic et al. 2014). A persistent change in M1 excitability resulting from interleaved training appears to occur concomitantly with changes in resting state connectivity between the premotor cortex and other critical neural regions involved in consolidation of motor memories such as dorsolateral prefrontal cortex, inferior parietal lobule, hippocampus, putamen, and cerebellum.

4 Summary

The aim of this chapter was to describe the neural changes that afford the progression of skill performance from an early learning stage associated with slow, inefficient performance to the automatic stage, which is characterized by speeded performance and the absence of central resource demands. Based on the neuroimaging findings reviewed here, one consistent change in neural activation associated with achieving skill automaticity occurs in an anterior to posterior direction within the frontal lobe. Specifically, while early performance relies on activity in anterior frontal regions such as the DLPFC along with premotor cortices (vPMC, preSMA, and SMA) and M1, skill automaticity only elicits sustained activation in most posterior region of the frontal lobe, the M1 (Wu et al. 2004; Lehericy et al. 2005; Poldrack et al. 2005; Puttemans et al. 2005; Wu and Hallett 2005; Wu et al. 2008). The anterior–posterior frontal lobe axis has been described as representing an associative-motor hierarchy for movement preparation and control (Koechlin et al. 2003; Fuster 2004). Thus, the anterior to posterior reduction in neural activity associated with automaticity can be interpreted as reflecting reduced dependence on associative processes. Within the motor levels of the hierarchy, reduced PMC and SMA activity also indicates changes in how automated skills are prepared.

The neural changes underlying skill automaticity can be further understood through the C-SMB cognitive framework (Verwey et al. 2015). Early learning activation of anterior frontal regions suggests initial reliance-associative processes that utilize central–symbolic representations to prepare the response. Reliance on the central processor to prepare the response is slow and imparts high demands on limited central resources. Finally, exposure to extensive practice is central to most accounts of the attainment of automaticity. In the present chapter, we highlighted the importance of designing practice that is geared toward recruiting the network of neural regions associated with early skill learning to ensure that the acquired sequential motor behavior can then be executed relatively attention free for an extended period of time.

References

Abernethy, B. (1988). Dual-task methodology and motor skills research: Some methodological constraints. *Journal of Human Movement Studies, 14,* 101–132.

Abrahamse, E. L., Jimenez, L., Verwey, W. B., & Clegg, B. A. (2010). Representing serial action and perception. *Psychonomic Bulletin & Review, 17*(5), 603–623. https://doi.org/10.3758/PBR.17.5.603.

Abrahamse, E. L., & Noordzij, M. L. (2011). Designing training programs for perceptual-motor skills: Practical implications from the serial reaction time task. *European Review of Applied Psychology-Revue Europeenne De Psychologie Appliquee, 61*(2), 65–76. https://doi.org/10.1016/j.erap.2010.12.001.

Abrahamse, E. L., Ruitenberg, M. F. L., de Kleine, E., & Verwey, W. B. (2013). Control of automated behavior: Insights from the discrete sequence production task. *Frontiers in Human Neuroscience, 7.*

Adams, S. G., & Page, A. D. (2000). Effects of selected practice and feedback variables on speech motor learning. *Journal of Medical Speech-Language Pathology, 8*(4), 215–220.

Almåsbakk, B., Whiting, H. T. A., & Helgerud, J. (2001). The efficient learner. *Biological Cybernetics, 84*(2), 75–83.

Ashby, F. G., & Crossley, M. J. (2012). Automaticity and multiple memory systems. *Wiley Interdisciplinary Reviews-Cognitive Science, 3*(3), 363–376. https://doi.org/10.1002/wcs.1172.

Ashby, F. G., Ennis, J. M., & Spiering, B. J. (2007). A neurobiological theory of automaticity in perceptual categorization. *Psychological Review, 114*(3), 632–656. https://doi.org/10.1037/0033-295x.114.3.632.

Ashby, F. G., Turner, B. O., & Horvitz, J. C. (2010). Cortical and basal ganglia contributions to habit learning and automaticity. *Trends in Cognitive Sciences, 14*(5), 208–215. https://doi.org/10.1016/j.tics.2010.02.001.

Bahrick, H. P., Noble, M., & Fitts, P. M. (1954). Extra-task performance as a measure of learning a primary task. *Journal of Experimental Psychology, 48,* 298–302.

Bernacer, J., & Murillo, J. I. (2014). The Aristotelian conception of habit and its contribution to human neuroscience. *Frontiers in Human Neuroscience, 8.*

Bernstein, N. (1967). *The Coordination and Regulation of Movements*. London: Pergamon Press.

Bezzola, L., Merillat, S., Gaser, C., & Jancke, L. (2011). Training-induced neural plasticity in golf novices. *Journal of Neuroscience, 31*(35), 12444–12448. https://doi.org/10.1523/Jneurosci.1996-11.2011.

Brady, F. (1998). A theoretical and empirical review of the contextual interference effect and the learning of motor skills. *Quest, 50*(3), 266–293.

Brady, F. (2004). Contextual interference: A meta-analytic study. *Perceptual and Motor Skills, 99*(1), 116–126. https://doi.org/10.2466/pms.99.1.116-126.

Breton, J., & Robertson, E. M. (2014). Flipping the switch: Mechanisms that regulate memory consolidation. *Trends in Cognitive Sciences, 18*(12), 629–634. https://doi.org/10.1016/j.tics.2014.08.005.

Brown, T. L., & Carr, T. H. (1989). Automaticity in skill acquisition—mechanisms for reducing interference in concurrent performance. *Journal of Experimental Psychology-Human Perception and Performance, 15*(4), 686–700.

Buccino, G., Vogt, S., Ritzl, A., Fink, G. R., Zilles, K., Freund, H. J., et al. (2004). Neural circuits underlying imitation learning of hand actions: An event-related fMRI study. *Neuron, 42*(2), 323–334. https://doi.org/10.1016/S0896-6273(04)00181-3.

Burke, R. E. (2007). Sir Charles Sherrington's the integrative action of the nervous system: A centenary appreciation. *Brain, 130,* 887–894. https://doi.org/10.1093/brain/awm022.

Carter, C. S., Braver, T. S., Barch, D. M., Botvinick, M. M., Noll, D., & Cohen, J. D. (1998). Anterior cingulate cortex, error detection, and the online monitoring of performance. *Science, 280*(5364), 747–749.

Chaney, D. W. (2013). An overview of the first use of the terms cognition and behavior. *Behavioral Sciences, 3*(1), 143–153.

Clegg, B. A., Digirolamo, G. J., & Keele, S. W. (1998). Sequence learning. *Trends in Cognitive Sciences, 2*(8), 275–281.

Cohen, J. D., Servanschreiber, D., & Mcclelland, J. L. (1992). A parallel distributed-processing approach to automaticity. *American Journal of Psychology, 105*(2), 239–269. https://doi.org/10.2307/1423029.

Coynel, D., Marrelec, G., Perlbarg, V., Pelegrini-Issac, M., Van de Moortele, P. F., Ugurbil, K., … Lehericy, S. (2010). Dynamics of motor-related functional integration during motor sequence learning. *Neuroimage, 49*(1), 759–766. https://doi.org/10.1016/j.neuroimage.2009.08.048.

Cross, E. S., Schmitt, P. J., & Grafton, S. T. (2007). Neural substrates of contextual interference during motor learning support a model of active preparation. *Journal of Cognitive Neuroscience, 19*(11), 1854–1871.

Dayan, E., & Cohen, L. G. (2011). Neuroplasticity subserving motor skill learning. *Neuron, 72*(3), 443–454.

Del Rey, P. (1982). Effects of contextual interference on the memory of older females differing in levels of physical activity. *Perceptual and Motor Skills, 55*(1), 171–180.

Di Paola, M., Caltagirone, C., & Petrosini, L. (2013). Prolonged rock climbing activity induces structural changes in cerebellum and parietal lobe. *Human Brain Mapping, 34*(10), 2707–2714. https://doi.org/10.1002/hbm.22095.

Doyon, J., Bellec, P., Amsel, R., Penhune, V., Monchi, O., Carrier, J., ... Benali, H. (2009). Contributions of the basal ganglia and functionally related brain structures to motor learning. *Behavioral Brain Research, 199*(1), 61–75.

Doyon, J., Owen, A. M., Petrides, M., Sziklas, V., & Evans, A. C. (1996). Functional anatomy of visuomotor skill learning in human subjects examined with positron emission tomography. *European Journal of Neuroscience, 8*(4), 637–648. https://doi.org/10.1111/j.1460-9568.1996.tb01249.x.

Doyon, J., Penhune, V., & Ungerleider, L. G. (2003). Distinct contribution of the cortico-striatal and cortico-cerebellar systems to motor skill learning. *Neuropsychologia, 41*(3), 252–262. https://doi.org/10.1016/s0028-3932(02)00158-6.

Doyon, J., Song, A. W., Karni, A., Lalonde, F., Adams, M. M., & Ungerleider, L. G. (2002). Experience-dependent changes in cerebellar contributions to motor sequence learning. *Proceedings of the National Academy of Sciences of the United States of America, 99*(2), 1017–1022. https://doi.org/10.1073/pnas.022615199.

Doyon, J., & Ungerleider, L. G. (2002). Functional anatomy of motor skill learning. In R. E. Squire & D. L. Schacter (Eds.), *Neuropsychology of memory*. New York: Guilford.

Engel, K. C., Flanders, M., & Soechting, J. F. (1997). Anticipatory and sequential motor control in piano playing. *Experimental Brain Research, 113*(2), 189–199.

Fendrich, D. W., & Arengo, R. (2004). The influence of string length and repetition on chunking of digit strings. *Psychological Research, 68*(4), 216–223. https://doi.org/10.1007/s00426-003-0142-9.

Fitts, P. M. (1964). Perceptual-motor skills learning. In A. W. Melton (Ed.), *Categories of human learning* (pp. 243–285). New York: Academic Press.

Fitts, P. M., & Posner, M. I. (1967). *Human performance*. Belmont, CA: Brooks/Cole.

Florio, T. M., Scarnati, E., Rosa, I., Di Censo, D., Ranieri, B., Cimini, A., ... Alecci, M. (2018). The basal ganglia: More than just a switching device. *Cns Neuroscience & Therapeutics, 24*(8), 677–684. https://doi.org/10.1111/cns.12987.

Floyer-Lea, A., & Matthews, P. M. (2005). Distinguishable brain activation networks for short- and long-term motor skill learning. *Journal of Neurophysiology, 94*(1), 512–518.

Friston, K. J., Frith, C. D., Passingham, R. E., Liddle, P. F., & Frackowiak, R. S. (1992). Motor practice and neurophysiological adaptation in the cerebellum: A positron tomography study. *Proceedings of the Royal Society B: Biological Sciences, 248*(1323), 223–228. https://doi.org/10.1098/rspb.1992.0065.

Fuster, J. M. (2004). Upper processing stages of the perception-action cycle. *Trends in Cognitive Sciences, 8*(4), 143–145. https://doi.org/10.1016/j.tics.2004.02.004.

Gentile, A. M. (1972). A working model of skill acquisition with application to teaching. *Quest, 17*, 3–23.

Gentile, A. M. (1998). Implicit and explicit processes during acquisition of functional skills. *Scandinavian Journal of Occupational Therapy, 5*, 7–16.

Gerloff, C., Corwell, B., Chen, R., Hallett, M., & Cohen, L. G. (1997). Stimulation over the human supplementary motor area interferes with the organization of future elements in complex motor sequences. *Brain, 120*, 1587–1602.

Ghilardi, M., Ghez, C., Dhawan, V., Moeller, J., Mentis, M., Nakamura, T., ... Eidelberg, D. (2000). Patterns of regional brain activation associated with different forms of motor learning. *Brain Research, 871*(1), 127-145.

Goode, S., & Magill, R. A. (1986). Contextual interference effects in learning three badminton serves. *Research Quarterly for Exercise and Sport, 57*(4), 308–314.

Grafton, S. T., Hazeltine, E., & Ivry, R. (1995). Functional mapping of sequence learning in normal humans. *Journal of Cognitive Neuroscience, 7*(4), 497–510. https://doi.org/10.1162/jocn.1995.7.4.497.

Grafton, S. T., Hazeltine, E., & Ivry, R. B. (2002). Motor sequence learning with the nondominant left hand. A PET functional imaging study. *Experimental Brain Research, 146*(3), 369–378. https://doi.org/10.1007/s00221-002-1181-y.
Grafton, S. T., Mazziotta, J. C., Presty, S., Friston, K. J., Frackowiak, R. S., & Phelps, M. E. (1992). Functional anatomy of human procedural learning determined with regional cerebral blood flow and PET. *Journal of Neuroscience, 12*(7), 2542–2548.
Grafton, S. T., Woods, R. P., & Tyszka, M. (1994). Functional imaging of procedural motor learning: Relating cerebral blood flow with individual subject performance. *Human Brain Mapping, 1*(3), 221–234. https://doi.org/10.1002/hbm.460010307.
Hardwick, R. M., Rottschy, C., Miall, R. C., & Eickhoff, S. B. (2013). A quantitative meta-analysis and review of motor learning in the human brain. *Neuroimage, 67,* 283–297. https://doi.org/10.1016/j.neuroimage.2012.11.020.
Hazeltine, E., Grafton, S. T., & Ivry, R. (1997). Attention and stimulus characteristics determine the locus of motor-sequence encoding—A PET study. *Brain, 120,* 123–140. https://doi.org/10.1093/brain/120.1.123.
Helie, S., Waldschmidt, J. G., & Ashby, F. G. (2010). Automaticity in rule-based and information-integration categorization. *Attention, Perception, & Psychophysics, 72*(4), 1013–1031. https://doi.org/10.3758/App.72.4.1013.
Henry, F. M., & Rogers, D. E. (1960). Increased response latency for complicated movements and a "memory drum" theory of neuromotor reaction. *Research Quarterly. American Association for Health, Physical Education and Recreation, 31*(3), 448–458.
Herdener, M., Esposito, F., di Salle, F., Boller, C., Hilti, C. C., Habermeyer, B., … Cattapan-Ludewig, K. (2010). Musical training induces functional plasticity in human hippocampus. *Journal of Neuroscience, 30*(4), 1377–1384. https://doi.org/10.1523/jneurosci.4513-09.2010.
Hikosaka, O., Nakahara, H., Rand, M. K., Sakai, K., Lu, X., Nakamura, K., … Doya, K. (1999). Parallel neural networks for learning sequential procedures. *Trends Neurosci, 22*(10), 464–471.
Hikosaka, O., Nakamura, K., Sakai, K., & Nakahara, H. (2002). Central mechanisms of motor skill learning. *Current Opinion in Neurobiology, 12*(2), 217–222.
Immink, M. A. (2016). Post-training meditation promotes motor memory consolidation. *Frontiers in Psychology, 7*. https://doi.org/10.3389/fpsyg.2016.01698.
Immink, M. A., Colzato, L. S., Stolte, M., & Hommel, B. (2017). Sequence learning enhancement following single-session meditation is dependent on meta control mode and experienced effort. *Journal of Cognitive Enhancement, 1*(2), 127–140.
Immink, M. A., & Wright, D. L. (1998). Contextual interference: A response planning account. *The Quarterly Journal of Experimental Psychology Section A, 51*(4), 735–754.
Immink, M. A., & Wright, D. L. (2001). Motor programming during practice conditions high and low in contextual interference. *Journal of Experimental Psychology: Human Perception and Performance, 27*(2), 423–437.
James, W. (1890). *Principles of psychology* (Vol. 1). New York: Holt.
Jastrow, O. (1891). The interference of mental processes. *American Journal of Psychology, 1891*(4), 219–223.
Jenkins, I. H., Brooks, D. J., Nixon, P. D., Frackowiak, R. S., & Passingham, R. E. (1994). Motor sequence learning: A study with positron emission tomography. *Journal of Neuroscience, 14*(6), 3775–3790.
Jonides, J., Naveh-Benjamin, M., & Palmer, J. (1985). Assessing automaticity. *Acta Psychologica, 60*(2–3), 157–171. https://doi.org/10.1016/0001-6918(85)90053-8.
Jordan, M. I. (1995). The organization of action sequences—Evidence from a relearning task. *Journal of Motor Behavior, 27*(2), 179–192.
Jueptner, M., Frith, C. D., Brooks, D. J., Frackowiak, R. S., & Passingham, R. E. (1997a). Anatomy of motor learning. II. Subcortical structures and learning by trial and error. *Journal of Neurophysiology, 77*(3), 1325–1337. https://doi.org/10.1152/jn.1997.77.3.1325.

Jueptner, M., Stephan, K. M., Frith, C. D., Brooks, D. J., Frackowiak, R. S., & Passingham, R. E. (1997b). Anatomy of motor learning. I. Frontal cortex and attention to action. *Journal of Neurophysiology, 77*(3), 1313–1324. https://doi.org/10.1152/jn.1997.77.3.1313.

Kahneman, D. (1973). *Attention and effort*. Englewood Cliffs, N.J.: Prentice-Hall.

Kantak, S. S., & Winstein, C. J. (2012). Learning-performance distinction and memory processes for motor skills: A focused review and perspective. *Behavioural Brain Research, 228*(1), 219–231. https://doi.org/10.1016/j.bbr.2011.11.028.

Karni, A., Meyer, G., Jezzard, P., Adams, M. M., Turner, R., & Ungerleider, L. G. (1995). Functional MRI evidence for adult motor cortex plasticity during motor skill learning. *Nature, 377*(6545), 155–158. https://doi.org/10.1038/377155a0.

Karni, A., Meyer, G., Rey-Hipolito, C., Jezzard, P., Adams, M. M., Turner, R., et al. (1998). The acquisition of skilled motor performance: Fast and slow experience-driven changes in primary motor cortex. *Proceedings of the National Academy of Sciences of the United States of America, 95*(3), 861–868.

Kawato, M. (1999). Internal models for motor control and trajectory planning. *Current Opinion in Neurobiology, 9*(6), 718–727.

Keele, S. W. (1968). Movement control in skilled motor performance. *Psychological Bulletin, 70*(6), 387–403.

Kim, T., Chen, J., Verwey, W. B., & Wright, D. L. (2018). Improving novel motor learning through prior high contextual interference training. *Acta Psychol (Amst), 182,* 55–64. https://doi.org/10.1016/j.actpsy.2017.11.005.

Klapp, S. T. (1995). Motor response programming during simple choice reaction time: The role of practice. *Journal of Experimental Psychology: Human Perception and Performance, 21*(5), 1015.

Knock, T. R., Ballard, K. J., Robin, D. A., & Schmidt, R. A. (2000). Influence of order of stimulus presentation on speech motor learning: A principled approach to treatment for apraxia of speech. *Aphasiology, 14*(5–6), 653–668.

Koechlin, E., Ody, C., & Kouneiher, F. (2003). The architecture of cognitive control in the human prefrontal cortex. *Science, 302*(5648), 1181–1185. https://doi.org/10.1126/science.1088545.

Kondo, H., Osaka, N., & Osaka, M. (2004). Cooperation of the anterior cingulate cortex and dorsolateral prefrontal cortex for attention shifting. *NeuroReport, 23*(2), 670–679.

Korsgaard, C. M. (2008). Aristotle's function argument. *The constitution of agency* (pp. 129–150). Oxford: Oxford University Press.

Krakauer, J. W., & Mazzoni, P. (2011). Human sensorimotor learning: Adaptation, skill, and beyond. *Current Opinion in Neurobiology, 21*(4), 636–644. https://doi.org/10.1016/j.conb.2011.06.012.

Kubler, A., Dixon, V., & Garavan, H. (2006). Automaticity and reestablishment of executive control-an fMRI study. *Journal of Cognitive Neuroscience, 18*(8), 1331–1342. https://doi.org/10.1162/jocn.2006.18.8.1331.

Kupferschmidt, D. A., Juczewski, K., Cui, G. H., Johnson, K. A., & Lovinger, D. M. (2017). Parallel, but dissociable, processing in discrete corticostriatal Inputs encodes skill learning. *Neuron, 96*(2), 476–489.

LaBerge, D., & Samuels, S. J. (1974). Toward a theory of automatic information processing in reading. *Cognitive Psychology, 6,* 293–323.

Laird, A. R., Fox, P. M., Eickhoff, S. B., Turner, J. A., Ray, K. L., McKay, D. R., ... Fox, P. T. (2011). Behavioral interpretations of intrinsic connectivity networks. *Journal of Cognitive Neuroscience, 23*(12), 4022–4037. https://doi.org/10.1162/jocn_a_00077.

Lang, C. E., & Bastian, A. J. (2002). Cerebellar damage impairs automaticity of a recently practiced movement. *Journal of Neurophysiology, 87*(3), 1336–1347. https://doi.org/10.1152/jn.00368.2001.

Leavitt, J. L. (1979). Cognitive demands of skating and stickhandling in ice hockey. *Canadian Journal Applied Sport Sciences, 4*(1), 46–55.

Lehericy, S., Benali, H., Van de Moortele, P. F., Pelegrini-Issac, M., Waechter, T., Ugurbil, K., et al. (2005). Distinct basal ganglia territories are engaged in early and advanced motor sequence learning. *Proceedings of the National Academy of Sciences of the United States of America, 102*(35), 12566–12571.

Li, Y., & Wright, D. L. (2000). An assessment of the attention demands during random-and blocked-practice schedules. *The Quarterly Journal of Experimental Psychology Section A, 53*(2), 591–606.

Lin, C.-H. J., Knowlton, B. J., Chiang, M.-C., Iacoboni, M., Udompholkul, P., & Wu, A. D. (2011). Brain–behavior correlates of optimizing learning through interleaved practice. *Neuroimage, 56*(3), 1758–1772.

Little, D. M., Klein, R., Shobat, D. M., McClure, E. D., & Thulborn, K. R. (2004). Changing patterns of brain activation during category learning revealed by functional MRI. *Cognitive Brain Research, 22*(1), 84–93. https://doi.org/10.1016/j.cogbrainres.2004.07.011.

Lohse, K. R., Wadden, K., Boyd, L. A., & Hodges, N. J. (2014). Motor skill acquisition across short and long time scales: A meta-analysis of neuroimaging data. *Neuropsychologia, 59*, 130–141. https://doi.org/10.1016/j.neuropsychologia.2014.05.001.

Magill, R. A., & Hall, K. G. (1990). A review of the contextual interference effect in motor skill acquisition. *Human Movement Science, 9*(3–5), 241–289.

Matsuzaka, Y., Picard, N., & Strick, P. L. (2007). Skill representation in the primary motor cortex after long-term practice. *Journal of Neurophysiology, 97*(2), 1819–1832. https://doi.org/10.1152/jn.00784.2006.

McLeod, P. D. (1980). What can probeRT tell us about the attentional demands ofmovements? In G. E. Stelmach & J. Requin (Eds.), *Tutorials in motor behavior* (pp. 579–589). Amsterdam: North-Holland.

Miall, R. C., Reckess, G. Z., & Imamizu, H. (2001). The cerebellum coordinates eye and hand tracking movements. *Nature Neuroscience, 4*(6), 638–644. https://doi.org/10.1038/88465.

Miller, G. A. (1956). The magical number seven, plus or minus two: Some limits on our capacity for processing information. *Psychological Review, 63*(2), 81–97.

Moors, A., & De Houwer, J. (2006). Automaticity: A theoretical and conceptual analysis. *Psychological Bulletin, 132*(2), 297–326. https://doi.org/10.1037/0033-2909.132.2.297.

Nachev, P., Kennard, C., & Husain, M. (2008). Functional role of the supplementary and pre-supplementary motor areas. *Nature Reviews Neuroscience, 9*(11), 856–869.

Nissen, M. J., & Bullemer, P. (1987). Attentional requirements of learning—Evidence from performance-measures. *Cognitive Psychology, 19*(1), 1–32. https://doi.org/10.1016/0010-0285(87)90002-8.

O'Reilly, J. X., Beckmann, C. F., Tomassini, V., Ramnani, N., & Johansen-Berg, H. (2010). Distinct and overlapping functional zones in the cerebellum defined by resting state functional connectivity. *Cerebral Cortex, 20*(4), 953–965. https://doi.org/10.1093/cercor/bhp157.

Ollis, S., Button, C., & Fairweather, M. (2005). The influence of professional expertise and task complexity upon the potency of the contextual interference effect. *Acta Psychologica (Amst), 118*(3), 229–244. https://doi.org/10.1016/j.actpsy.2004.08.003.

Pashler, H. (1994). Dual-task interference in simple tasks—Data and theory. *Psychological Bulletin, 116*(2), 220–244. https://doi.org/10.1037/0033-2909.116.2.220.

Passingham, R. E. (1996). Attention to action. *Philosophical Transactions of the Royal Society of London B: Biological Sciences, 351*(1346), 1473–1479. https://doi.org/10.1098/rstb.1996.0132.

Pauwels, L., Swinnen, S. P., & Beets, I. A. (2014). Contextual interference in complex bimanual skill learning leads to better skill persistence. *PLoS ONE, 9*(6), e100906. https://doi.org/10.1371/journal.pone.0100906.

Pearce, A. J., Thickbroom, G. W., Byrnes, M. L., & Mastaglia, F. L. (2000). Functional reorganisation of the corticomotor projection to the hand in skilled racquet players. *Experimental Brain Research, 130*(2), 238–243. https://doi.org/10.1007/s002219900236.

Penhune, V. B., & Doyon, J. (2002). Dynamic cortical and subcortical networks in learning and delayed recall of timed motor sequences. *Journal of Neuroscience, 22*(4), 1397–1406.

Penhune, V. B., & Steele, C. J. (2012). Parallel contributions of cerebellar, striatal and M1 mechanisms to motor sequence learning. *Behavioural Brain Research, 226*(2), 579–591. https://doi.org/10.1016/j.bbr.2011.09.044.

Pew, R. W. (1966). Acquisition of hierarchical control over the temporal organization of a skill. *Journal of Experimental Psychology, 71*(5), 764–771.
Poldrack, R. A., Sabb, F. W., Foerde, K., Tom, S. M., Asarnow, R. F., Bookheimer, S. Y., et al. (2005). The neural correlates of motor skill automaticity. *Journal of Neuroscience, 25*(22), 5356–5364.
Porretta, D. L., & O'Brien, K. (1991). The use of contextual interference trials by mildly mentally handicapped children. *Research Quarterly for Exercise Sport, 62*(2), 240–244.
Posner, M. I. (1979). *Chronometric explorations of mind*. Hillsdale, N.J.: Erlbaum.
Posner, M. I., & Keele, S. W. (1969). Attentional demands of movement. In *Proceedings of the 16th Congress of Applied Psychology*. Amsterdam: Swets and Zeitlinger.
Proteau, L., Marteniuk, R. G., & Levesque, L. (1992). A sensorimotor basis for motor learning—evidence indicating specificity of practice. *Quarterly Journal of Experimental Psychology Section A-Human Experimental Psychology, 44*(3), 557–575. https://doi.org/10.1080/14640749208401298.
Puttemans, V., Wenderoth, N., & Swinnen, S. P. (2005). Changes in brain activation during the acquisition of a multifrequency bimanual coordination task: From the cognitive stage to advanced levels of automaticity. *Journal of Neuroscience, 25*(17), 4270–4278. https://doi.org/10.1523/JNEUROSCI.3866-04.2005.
Robertson, E. M. (2007). The serial reaction time task: Implicit motor skill learning? *Journal of Neuroscience, 27*(38), 10073–10075. https://doi.org/10.1523/Jneurosci.2747-07.2007.
Robertson, E. M., Pascual-Leone, A., & Miall, R. C. (2004). Current concepts in procedural consolidation. *Nature Reviews Neuroscience, 5*(7), 576–582. https://doi.org/10.1038/nrn1426.
Sakai, K., Hikosaka, O., & Nakamura, K. (2004). Emergence of rhythm during motor learning. *Trends in Cognitive Sciences, 8*(12), 547–553. https://doi.org/10.1016/j.tics.2004.10.005.
Salmoni, A. W., Schmidt, R. A., & Walter, C. B. (1984). Knowledge of results and motor learning—A review and critical reappraisal. *Psychological Bulletin, 95*(3), 355–386. https://doi.org/10.1037/0033-2909.95.3.355.
Schendan, H. E., Searl, M. M., Melrose, R. J., & Stern, C. (2003). An FMRI study of the role of the medial temporal lobe in implicit and explicit sequence learning. *Neuron, 37*(6), 1013–1025.
Schmidt, R. A. (1975). A schema theory of discrete motor skill learning. *Psychological Review, 82*, 225–260.
Schmidt, R. A. (2003). Motor schema theory after 27 years: Reflections and implications for a new theory. *Research Quarterly for Exercise and Sport, 74*(4), 366–375.
Schmidt, R. A., & Bjork, R. A. (1992). New conceptualizations of practice: Common principles in three paradigms suggest new concepts for training. *Psychological Science, 3*(4), 207–218.
Schmidt, R. A., & Lee, T. D. (2005). *Motor control and learning: A behavioural emphasis* (4 ed.). Champaign: Human Kinetics.
Schneider, W., & Chein, J. M. (2003). Controlled & automatic processing: behavior, theory, and biological mechanisms. *Cognitive Science, 27*(3), 525–559. https://doi.org/10.1016/S0364-0213(03)00011-9.
Schneider, W., & Shiffrin, R. M. (1977). Controlled and automatic human information-processing I: Detection, search, and attention. *Psychological Review, 84*(1), 1–66. https://doi.org/10.1037/0033-295x.84.1.1.
Seidler, R. D., Purushotham, A., Kim, S.-G., Uğurbil, K., Willingham, D., & Ashe, J. (2002). Cerebellum activation associated with performance change but not motor learning. *Science, 296*(5575), 2043–2046.
Seitz, R. J., Canavan, A. G., Yaguez, L., Herzog, H., Tellmann, L., Knorr, U., ... Homberg, V. (1994). Successive roles of the cerebellum and premotor cortices in trajectorial learning. *Neuroreport, 5*(18), 2541–2544.
Shadmehr, R., & Holcomb, H. H. (1997). Neural correlates of motor memory consolidation. *Science, 277*(5327), 821–825.
Shadmehr, R., & Krakauer, J. W. (2008). A computational neuroanatomy for motor control. *Experimental Brain Research, 185*(3), 359–381. https://doi.org/10.1007/s00221-008-1280-5.

Shaffer, L. H. (1975). Control processes in typing. *Quarterly Journal of Experimental Psychology, 27*(Aug), 419–432.
Shea, J. B., & Morgan, R. L. (1979). Contextual interference effects on the acquisition, retention, and transfer of a motor skill. *Journal of Experimental Psychology: Human Learning and Memory, 5*(2), 179–187.
Sherrington, C. S. (1910). Flexion-reflex of the limb, crossed extension-reflex and reflex stepping and standing. *Journal of Physiology, 40,* 28–121.
Sherrington, C. S. (1974). *The integrative action of the nervous system.* New Haven, C.T.: Yale University Press.
Shmuelof, L., & Krakauer, J. W. (2011). Are we ready for a natural history of motor learning? *Neuron, 72*(3), 469–476. https://doi.org/10.1016/j.neuron.2011.10.017.
Singer, R. N. (2002). Preperformance state, routines, and automaticity: What does it take to realize expertise in self-paced events? *Journal of Sport & Exercise Psychology, 24*(4), 359–375. https://doi.org/10.1123/jsep.24.4.359.
Smith, P. J., & Davies, M. (1995). Applying contextual interference to the Pawlata roll. *Journal of Sports Sciences, 13*(6), 455–462.
Solomons, L., & Stein, G. (1896). Normal motor automatism. *Psychological Review, 3,* 492–512.
Stein, G. (1898). Cultivated motor automatism: A; a study of character in its relation to attention. *Psychological Review, 5*(3), 295–306.
Swinnen, S. P. (2002). Intermanual coordination: From behavioural principles to neural-network interactions. *Nature Reviews Neuroscience, 3*(5), 350–361. https://doi.org/10.1038/nrn807.
Taylor, J. A., & Ivry, R. B. (2012). The role of strategies in motor learning. *Year in Cognitive Neuroscience, 1251,* 1–12. https://doi.org/10.1111/j.1749-6632.2011.06430.x.
Thorn, C. A., Atallah, H., Howe, M., & Graybiel, A. M. (2010). Differential dynamics of activity changes in dorsolateral and dorsomedial striatal loops during learning. *Neuron, 66*(5), 781–795. https://doi.org/10.1016/j.neuron.2010.04.036.
Thorn, C. A., & Graybiel, A. M. (2014). Differential entrainment and learning-related dynamics of spike and local field potential activity in the sensorimotor and associative striatum. *Journal of Neuroscience, 34*(8), 2845–2859. https://doi.org/10.1523/Jneurosci.1782-13.2014.
Toner, J., & Moran, A. (2015). Enhancing performance proficiency at the expert level: Considering the role of' somaesthetic awareness'. *Psychology of Sport and Exercise, 16,* 110–117. https://doi.org/10.1016/j.psychsport.2014.07.006.
Toni, I., Krams, M., Turner, R., & Passingham, R. E. (1998). The time course of changes during motor sequence learning: A whole-brain fMRI study. *Neuroimage, 8*(1), 50–61. https://doi.org/10.1006/nimg.1998.0349.
Tunovic, S., Press, D. Z., & Robertson, E. M. (2014). A physiological signal that prevents motor skill improvements during consolidation. *Journal of Neuroscience, 34*(15), 5302–5310. https://doi.org/10.1523/Jneurosci.3497-13.2014.
Verwey, W. B. (1999). Evidence for a multistage model of practice in a sequential movement task. *Journal of Experimental Psychology-Human Perception and Performance, 25*(6), 1693–1708. https://doi.org/10.1037/0096-1523.25.6.1693.
Verwey, W. B., Abrahamse, E. L., & de Kleine, E. (2010). Cognitive processing in new and practiced discrete keying sequences. *Frontiers in Psychology, 1.* https://doi.org/10.3389/fpsyg.20https://doi.org/10.00032.
Verwey, W. B., & Dronkers, W. J. (2018). Skill in discrete keying sequences is execution rate specific. *Psychological Research-Psychologische Forschung, [Epub ahead of print].* https://doi.org/10.1007/s00426-017-0967-2.
Verwey, W. B., Jouen, A. L., Dominey, P. F., & Ventre-Dominey, J. (2018). Explaining the neural activity distribution associated with discrete movement sequences: Evidence for parallel functional systems. *Cognitive, Affective, & Behavioral Neuroscience.* https://doi.org/10.3758/s13415-018-00651-6.
Verwey, W. B., Lammens, R., & van Honk, J. (2002). On the role of the SMA in the discrete sequence production task: A TMS study. *Neuropsychologia, 40*(8), 1268–1276.

Verwey, W. B., Shea, C. H., & Wright, D. L. (2015). A cognitive framework for explaining serial processing and sequence execution strategies. *Psychonomic Bulletin & Review, 22*(1), 54–77. https://doi.org/10.3758/s13423-014-0773-4.

Verwey, W. B., & Wright, D. L. (2014). Learning a keying sequence you never executed: Evidence for independent associative and motor chunk learning. *Acta Psychologica (Amst), 151,* 24–31. https://doi.org/10.1016/j.actpsy.2014.05.017.

Walker, M. P., & Stickgold, R. (2004). Sleep-dependent learning and memory consolidation. *Neuron, 44*(1), 121–133. https://doi.org/10.1016/j.neuron.2004.08.031.

Wambaugh, J. L., Nessler, C., Wright, S., & Mauszycki, S. C. (2014). Sound production treatment: effects of blocked and random practice. *American Journal of Speech-Language Pathology, 23*(2), 225–245.

Whitaker, H. A. (1983). Towards a brain model of automatization: A short essay. In R. A. Magill (Ed.), *Memory and control of action* (pp. 199–214). Amsterdam: North-Holland.

Wolpert, D. M., Miall, R. C., & Kawato, M. (1998). Internal models in the cerebellum. *Trends in Cognitive Sciences, 2*(9), 338–347.

Wright, D., Verwey, W., Buchanen, J., Chen, J., Rhee, J., & Immink, M. (2016). Consolidating behavioral and neurophysiologic findings to explain the influence of contextual interference during motor sequence learning. *Psychonomic Bulletin & Review, 23*(1), 1–21. https://doi.org/10.3758/s13423-015-0887-3.

Wright, D. L., Black, C. B., Immink, M. A., Brueckner, S., & Magnuson, C. (2004). Long-term motor programming improvements occur via concatenation of movement sequences during random but not during blocked practice. *Journal of Motor Behavior, 36*(1), 39–50.

Wright, D. L., & Kemp, T. (1992). The dual-task methodology and assessing the attentional demands of ambulation with walking devices. *Physical Therapy, 72,* 306–312.

Wright, D. L., Magnuson, C. E., & Black, C. B. (2005). Programming and reprogramming sequence timing following high and low contextual interference practice. *Research Quarterly for Exercise and Sport, 76*(3), 258–266. https://doi.org/10.1080/02701367.2005.10599297.

Wu, T., Chan, P., & Hallett, M. (2008). Modifications of the interactions in the motor networks when a movement becomes automatic. *Journal of Physiology-London, 586*(17), 4295–4304.

Wu, T., & Hallett, M. (2005). The influence of normal human ageing on automatic movements. *Journal of Physiology-London, 562*(2), 605–615.

Wu, T., Kansaku, K., & Hallett, M. (2004). How self-initiated memorized movements become automatic: A functional MRI study. *Journal of Neurophysiology, 91*(4), 1690–1698. https://doi.org/10.1152/jn.01052.2003.

Wu, T., Liu, J., Zhang, H., Hallett, M., Zheng, Z., & Chan, P. (2015). Attention to Automatic Movements in Parkinson's Disease: Modified Automatic Mode in the Striatum. *Cerebral Cortex, 25*(10), 3330–3342. https://doi.org/10.1093/cercor/bhu135.

Wymbs, N. F., & Grafton, S. T. (2009). Neural substrates of practice structure that support future off-line learning. *Journal of Neurophysiology, 102*(4), 2462–2476. https://doi.org/10.1152/jn.00315.2009.

Yin, H. H., & Knowlton, B. J. (2006). The role of the basal ganglia in habit formation. *Nature Reviews Neuroscience, 7*(6), 464–476. https://doi.org/10.1038/nrn1919.

Yin, H. H., Mulcare, S. P., Hilario, M. R. F., Clouse, E., Holloway, T., Davis, M. I., ... Costa, R. M. (2009). Dynamic reorganization of striatal circuits during the acquisition and consolidation of a skill. *Nature Neuroscience, 12*(3), 333–341. https://doi.org/10.1038/nn.2261.

Approaches for Inserting Neurodynamics into the Training of Healthcare Teams

Ronald Stevens, Trysha Galloway and Ann Willemsen-Dunlap

Abstract Team neurodynamics is the study of the changing rhythms and organizations of teams from the perspective of neurophysiology. As a discipline, team neurodynamics is located at the intersection of collaborative learning, psychometrics, complexity theory, and neurobiology with the resulting principles and applications both drawing from and contributing to these specialties. This article describes the tools for studying team neurodynamics and illustrates the potential and the challenges these methods and models have for better understanding healthcare team training and performance. The fundamental metric is neurodynamic organization, which is the tendency of teams and its members to enter into prolonged metastable relationships when they experience and resolve uncertainty. The patterns of these relationships are resolved by symbolic modeling of electroencephalographic (EEG) power levels of the team members, and the information in these patterns are calculated using information theory tools. The topics discussed in this chapter anticipate the time when dynamic biometric data can contribute to our understanding of how to rapidly determine a team's functional status, and how to use this information to optimize outcomes and training. The rapid, dynamic, and task neutral measures make the lessons learned in healthcare applicable to other complex group and team environments, and provide a foundation for incorporating these models into machines to support the training and performance of teams.

Keywords Team neurodynamics · EEG · Debriefing · Entropy · Information theory

R. Stevens
UCLA School of Medicine, Brain Research Institute, Los Angeles, CA, USA

R. Stevens (✉) · T. Galloway
The Learning Chameleon, Inc., Culver City, CA 90230, USA

A. Willemsen-Dunlap
JUMP Simulation and Education Center, Peoria, IL, USA

C. S. Nam (ed.), *Neuroergonomics*, Cognitive Science and Technology,
https://doi.org/10.1007/978-3-030-34784-0_13

Much of our understandings about others has been learned from observation. Observational analyses have been foundational in developing our understanding of individual and team behaviors by detailing what a person is doing, how they are doing it, and what they are saying (Cooke, Gorman, & Kiekel, 2008; Entin & Serfaty, 1999; Salas, Stagl, & Burke, 2004). But training observations are time consuming and expensive, and the bandwidth of speech is limited. The information that is selected for attention, kept in working memory, and channeled out for reporting is only a partial readout of the causal information being processed by the brain. Only a small fraction of consciousness thought emerges as speech after having undergone a complex process of construction and filtration (Tononi, Boly, Massimini, & Koch, 2016). Furthermore, teams under pressure might be silent (Gardezi et al., 2009).

A shift is underway toward more dynamical measures of team performance that avoid some of the limitations of observation and speech. Multimodal biometric dynamic data streams with seconds' resolutions hold the promise for shaping existing theories of teamwork and the creation of new forms of simulations, performance measures, and practices (Abersold, 2018; Guastello & Peressini, 2018; Stevens, Galloway, Gorman et al., 2018; Zenati et al., 2019).

But how do we build on what we already know about teamwork to incorporate dynamics that are 10^3– 10^4 times more rapid and detailed, like those that are possible from heart rate or electroencephalographic (EEG) measures (Salas et al., 2015)? What metrics would be suitable for comparing observations of team behavior with the millisecond dynamics of neurologic and physiologic measures? What new questions can be approached with these tools, and how will answers improve the economics of healthcare delivery and patient safety? What are the implications for improving patient safety when communication, cognitive, and neurodynamic analysis become real time? We are far from answering these questions, but we show how these questions can be approached by couching them in terms of the flows of neural information, as information and neural rhythms pervade across all scales of team organization.

To set the context for the analyses described later, the chapter begins by describing the two main learning events during team training in health care, the scenario and debriefing segments of simulations.

Healthcare organizations rely on simulations of complex processes to provide the training needed to evolve and maintain high levels of professional competence. High-fidelity simulations provide opportunities for skill acquisition and maintenance, team training, as well as high-stakes testing, and are widely accepted today as an essential educational modality for healthcare professionals (Schmidt, Goldhaber-Fiebert, Ho, & McDonald, 2013; Staropoli et al., 2018). Simulation provides a mechanism for standardized clinical education across all learners, allowing exposure to rare, critical events that clinicians might never encounter in their career in a live patient. Simulation education is a bridge between classroom learning and real-life clinical experience. Finally, simulation provides a safe environment where learners can come together as interprofessional teams to practice critical teamwork skills that are often overlooked in clinical teaching, and then openly discuss their actions.

Debriefing plays a key role in simulation training because it bolsters the transfer of experience into learning through a process of reflection (Bond et al., 2006; Fanning

& Gaba, 2007; Husebo, Dieckmann, Rystedt, Soreide, & Friberg, 2013; Petranek, 1994). An important feature of healthcare debriefings is the group identification of significant spaces of teamwork and clinical events that serve as exemplars to guide learning and adapt current behaviors to future challenges. Debriefing models like those of Petranek (1994), or Lederman (1992), sequence elements according to events, emotions, empathy, explanation and applicability, or reflection and analysis, personalization of the experience, and generalization and application, respectively.

While providing guidance for sequencing topics (Cheng et al., 2016; Clapper, 2016; Sawyer & Deering, 2013) current debriefing structures provide few recommendations about what debriefings might look like in the future from dynamical perspectives that are more closely linked with the fundamental processes of learning. In other words, current models have practical value but provide few theoretical suggestions or hard data for accelerating reflective learning or streamlining the journey to competence, not only of the participant but also of the facilitators. There is a sense in the research community that more dynamic models, objective metrics, and theories of simulation education including debriefing are needed (Bowe, Johnson, & Pusca, 2017; Fanning & Gaba, 2007), but empiric evidence about what these models would look like is sparse.

1 Team Neurodynamics

1.1 Team Neurodynamics: An Information Organization Framework for Modeling Team Performance

Information is a common exchange currency of living systems used to support homeostasis and survival. Information can be acquired or created, processed, shared, stored, and destroyed. Forms of information have a past, a present or a future (James, Ellison & Crutchfield, 2011), while other information may have causal necessity for the system to maintain its own existence over time (Kolchinsky & Wolpert, 2018).

Many of the studies and measures of dynamical science use information theory which was originally conceived as a way of describing the fundamental limits on signal processing and communication (Shannon, 1948, 1951). A unit of information is the bit, which is the amount of information that allows the choice between two equally probable alternatives. Bits can be extracted from many sources by different means, but a common feature is that they measure the amount of uncertainty that can be removed in a variable or process. One measure is Shannon entropy which is the average surprise of outcomes sampled from a probability distribution or density. A distribution with low entropy means that the data has organization and is relatively predictable, while a system with higher entropy indicates there is less organization and the system is less predictable.

Teams are far from equilibrium, metastable, and thermodynamically open systems, and the nested organizational structures found across space and time are thought to encode information hierarchies that reduce uncertainty, facilitate extraction of energy and foster adaptive behaviors (Flack, 2017). In this way, the information at macroscales can be thought of as the summed outputs or aggregated components of the information from components below. Information, however, can flow both ways on this information hierarchy. Information from regularities (i.e., patterns) in the environment during prolonged periods of uncertainty, decision-making and learning are also fed back down to lower levels influencing lower level activities to promote adaptation.

Teams create, store and share different forms of information using speech (Cooke et al., 2008), action understandings like gestures (Schippers, Roebroeck, Renken, Nanetti, & Keysers, 2010), posture (Shockley, Santana, & Fowler, 2003), facial expressions (Anders, Heinzle, Weiskopf, Ethofer, & Hayes, 2011), and other nonverbal communications (Menoret et al. 2014).

This information is shaped by contextual cues arising from the environment (Ancona & Chong, 1999) and the interdependent acts of team members (O'Neil, Chung, & Brown, 1997; Marks, Mathieu, & Zaccaro, 2001; von Davier & Halpin, 2013) into observable behaviors reflecting individual and team cognition, shared mental models (Cooke, Salas, Cannon-Bowers, & Stout, 2000) situation awareness (Mosier & Chidester, 1991), and performance. The shaping process is being formalized into frameworks for team coordination (Gorman, 2014) that acknowledge the dynamic top-down and bottom-up interactions occurring "on the fly" that transform internalized knowledge into externalized team knowledge, i.e. team macrocognition (Fiore, Smith-Jentsch, Salas, Warner, & Letsky, 2010), and is being extended to more global social, cognitive and event-related perspectives through the lens of distributed cognition (Giere & Moffat, 2003).

Given their fundamental role in cognition and communication, neurophysiologic measures may provide additional quantitative measures that link team members and their team processes (Stevens & Galloway, 2017; Stevens, Galloway, & Willemsen-Dunlap, 2018). It is not surprising that neurophysiologic activities are the underpinnings of the information flows of social coordination dynamics seen in teams (Stephens, Silbert, & Hasson, 2010; Sanger, Muller, & Lindenberger, 2012; Tognoli & Kelso, 2015), yet it is only recently that these evolving dynamics in real-world teamwork settings have begun to be continuously, and quantitatively modeled using tasks lasting an hour or more.

At the scale of milliseconds there are predictable neurophysiological activities when a word is spoken, (Draschkow, Heikel, Võ, Fiebach, & Sassenhagen, 2018) when a gesture is made (Tognoli & Kelso, 2015; Schippers et al., 2010; Menoret et al., 2014), when attention is captured, when memories are recalled (Vergauwe & Cowan, 2014) and when the environment is scanned. But as these neurodynamic markers become aggregated overtime during teamwork, these distinctions become lost. For training purposes, these particular very rapid responses must be represented and modeled in ways that relate to what we consider as observable aspects of teamwork, yet they should also help to explain microlevels of teamwork for theory building. These measures need to accurately reflect the organizational properties of teams, for example distinguishing expert–novice differences while also describing the mechanisms at lower hierarchies that provide an explanation for these observations.

One neurodynamic marker model we have developed while studying military and healthcare teams is neurodynamic organization, which is the tendency of team members to enter prolonged (i.e., minutes) metastable neurodynamic relationships as they encounter and resolve uncertainty (Stevens & Galloway, 2014, 2015, 2017). The source signals for team neurodynamic modeling are EEG-derived brainwaves, which are recordings of the electrical activity of the brain at different regions along the scalp. The rhythmic patterns in the electrical oscillations from different brain regions contain signals representing complex facets of brain activity (Delorme, Palmer, Onton, Oostelveld, & Makeig, 2012). EEG was chosen for these studies as it provided real-time and high-resolution temporal recording of the brain's electrical activity at different scalp regions and different frequencies in an unobtrusive fashion. There is also a significant body of research that links these rhythms to cognitive functions helping to build a theoretical bridge between the micro and macro teamwork scales (Buzaki, 2006).

1.2 Modeling Team Neurodynamics

In prior studies, we developed neurodynamic models of submarine and healthcare teams with the goal of better understanding, and eventually guiding, interprofessional team-based learning across the novice–expert continuum (Stevens, Galloway, Halpin, & Willemsen-Dunlap, 2016; Stevens, Galloway, Lamb, Steed, & Lamb, 2017). The metrics derived from these measures conditioned detailed performance models that provided historical neurodynamic traces of how teams perceived and responded to evolving task events at micro and longer term observational/behavioral scales.

For these studies, we sought representation(s) that used the physical units of EEG amplitudes (in μ-volts) of the team members as inputs and then exported higher level representations of team organization (Stevens & Galloway, 2014, 2015, 2017). The resulting metric, neurodynamic organization (*NO*), is the tendency of team members to enter into prolonged (i.e., minutes) metastable neurodynamic relationships as they encounter and resolve uncertainty as a unit. Increased neurodynamic organization occurs when a team's operating rhythm no longer can support the complexity of the task (Stevens & Galloway, 2017; Stevens, Galloway, Halpin, & Willemsen-Dunlap, 2016) and the team needs to expend energy to reorganize into structures that better minimize the "surprise" of the environment (Friston, 2010). Consistent with this hypothesis, the frequency and magnitude of neurodynamic organizations are often greater in novice teams compared with experienced teams (Stevens et al., 2017).

Measures of *NO* are based on the internal patterns and organizations in EEG-derived data streams, and the analyses are grounded in information theory (Garner, Thompson, Vedral, & Gu, 2017). Symbolic transformations of discrete data can be used to detect and quantitate the fluctuating dynamics of these organizations (Daw,

Finney, & Tracy, 2003; Stevens & Galloway, 2014, 2015, 2017). Information theory provides methods for determining when and how information is created, how it is stored, shared, and destroyed (Shannon, 1948; James et al., 2011).

The modeling begins by postulating that each of the three team members has three possible energy levels, below average, average and above average, which represent the EEG signal power (in microvolts) in a frequency bin recorded from scalp sensors; these levels can change every second. These three states could easily be fifths or tenths, or other discrete bins, and teams containing a variety of team members, with the associated scale-up costs in model computation.

The one-second interval is a theoretically plausible number for modeling teamwork as periods of functional brain connectivity associated with speech or playing guitar in duets (Sanger et al., 2012; Stephens et al., 2010), and nonverbal recognition (Hari, 2006) occur in the 250–500 ms time range, or a little over a half a second for a two-person action-response round trip; in reality individuals in teams probably speed this up by predicting ahead, although at a cost of increased uncertainty (Hsu, Bhatt, Adolphs, Tranel, & Camerer, 2005).

The first modeling step separates the EEG power of each team member each second into high, medium or low ranges for the performance (Fig. 1a). For ease of visualization, these categories were assigned the symbols 3, 1 and −1, resulting in a three-element array each second which was visualized as a three-histogram symbol

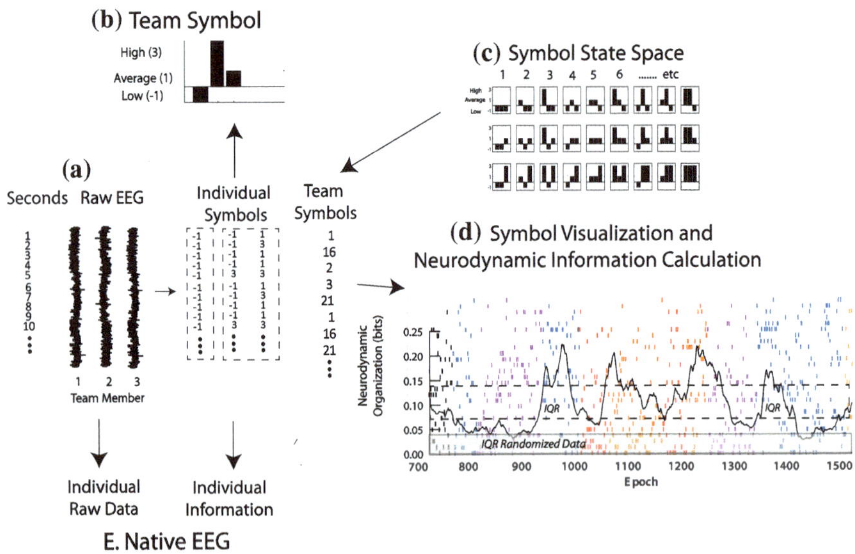

Fig. 1 Steps for generating and modeling neurodynamic data streams. **a**, **b** the pre-processed EEG signals from each person are discretized each second into low, average, and high power levels and assembled into a *NS*. **c** The symbol matching the three-person power array is determined from the *NSS* lookup table and assembled into a *NDS*, where, **d** the symbols are visually mapped and a moving average of *NI* is calculated each second. **e** To determine each team member's *NI* the symbols of the raw EEG and normalized values (i.e., −1, 1, & 3) were used for modeling

termed a neurodynamic symbol (*NS*) (see Box 1 for a Glossary of Terms). The *NS* in Fig. 1b indicates that at this second, team member 1 had below average EEG power levels, team member 2 had above average and team member 3 had average EEG power levels. The possible combinations of three persons and three EEG power levels create a 27 symbol neurodynamic state space (*NSS*) (Stevens & Galloway, 2014).

Box 1 Glossary

- *Neurodynamic symbols* (*NS*) are symbolic representations of the momentary power levels of a neurodynamic marker for each team member; at its simplest a *NS* would be the simultaneous power levels of two persons from a single EEG frequency bin at a single sensor channel.
- *Neurodynamic state space* (*NSS*) is a collection of *NS* that together describe the possible states of a team performance.
- *Neurodynamic data streams* (*NDS*) are the second-by-second sequences of *NS* that temporally encompass a team performance.
- *Neurodynamic organizations* (*NO*) are the patterns found in *NDS*, and represent the tendency of team members to organize into persistent neurodynamic relationships with each other. Quantitative measures of these relationships are neurodynamic entropy (*NE*) and *Neurodynamic information* (*NI*).
  - *Neurodynamic entropy* (*NE*) is the degree of organization that can be quantitated by measuring the Shannon entropy over moving segments (60–100 s) of the symbol data.

Low *NE* represents greater team neurodynamic organization. To clarify this relationship, entropy values are subtracted from the maximum entropy for the number of symbols being modeled, i.e., 3.17 bits for 9 symbols or 4.775 bits for 27 symbols. The result is termed *Neurodynamic Information* and abbreviated *NI*; this is the measure used throughout this chapter.

While developing the *NSS*, a topology was generated such that the symbols in the upper left corner (1–5, 10–14) represented moments when most team members had low to average levels of EEG power, while the symbols on the bottom right (numbers 9, 18 and 24–27) represented times when most team members had above average power levels (Fig. 1c). Each *NS* in the symbolic state space, therefore, situates the EEG power levels of each team member in the context of the levels of the other team members and the context of the task.

The neurodynamic profiles of each team member were created in a similar way using the 3, 1, and −1 values as symbols to create a three-symbol state space that was used for calculating individual team member *NI* (Stevens, Galloway, & Willemsen-Dunlap, 2018).

In all teams, the temporal distribution of *NS* was nonuniform, with a limited subset of symbols being expressed for a minute or more that was exchanged for another symbol subset when the task demands changed. These *NS* concentrations produced local variations in the expressions of the neurodynamic data streams (*NDS*); differences that were quantitated by measuring the entropy over a 60 s-moving window that was updated each second (Fig. 1d).

Entropy is the average surprise of outcomes sampled from a probability distribution or density. A *NS* density with low entropy means that, on average, the system is relatively predictable while a system with higher entropy would be less predictable. In this way, a dynamical and quantitative profile can be reported with a 1 s granularity for real-time modeling, or aggregated over a performance for comparisons across teams and/or team members (Stevens & Galloway, 2017). Because increased organization is shown as a decrease in entropy, in this chapter we subtract the entropy levels from the maximum entropy for the symbol system, whether it is a three or 27 symbol modeling state space.

The context-neutral property of information is useful for developing generic information organization-based models of teamwork, as analytic frameworks developed for one task can then be used for understanding other tasks. The challenge is making sense of such models (Stevens, Galloway, Gorman, Willemsen-Dunlap, Halpin, & Grimm, 2018). A series of studies spanning high school teams to military and healthcare teams have indicated that neurodynamic organizations are likely a fundamental property of teamwork (Stevens & Galloway, 2015, 2017; Stevens, Galloway, & Willemsen-Dunlap, 2017, 2018). This modeling transforms the EEG power in microvolts, which is useful for theory building, into bits of organization, which is useful for linking to performance measures like TeamSTEPPS® (Baker, Amodeo, Krokos, Slonim, & Herra, 2010; Stevens, Galloway, & Willemsen-Dunlap, 2016) or speech.

1.3 Identifying the Contributions of Team Members to the Team Organization

One of the team neurodynamics modeling goals is to be able to determine when a team, or one of its members, is experiencing difficulty. This requires deconstructing the team neurodynamics into that of each team member. An example is shown in Fig. 2 for a simulation performance that involved the preoperative ventilation of a patient by the anesthesiologist (AN), assisted by a circulating nurse (CN), and a scrub nurse (SN). Subsequent complications included an adverse response to a relative overdose of aerosolized lidocaine that caused patient seizures and cardiac dysrhythmias. The total scenario time was 1524 s (Stevens, Galloway, Gorman et al., 2018).

The events during this three-person training segment included an early unsuccessful intubation (INTB) attempt (INTB-1), Lidocaine-Induced seizures requiring a call for a Crash Cart, and a successful INTB (INTB-2) (Fig. 2a). The rows in Fig. 2b indicate when the anesthesiologist (AN), circulating nurse (CN), or nobody (NULL) was speaking; the scrub nurse (SN) did not speak during the scenario.

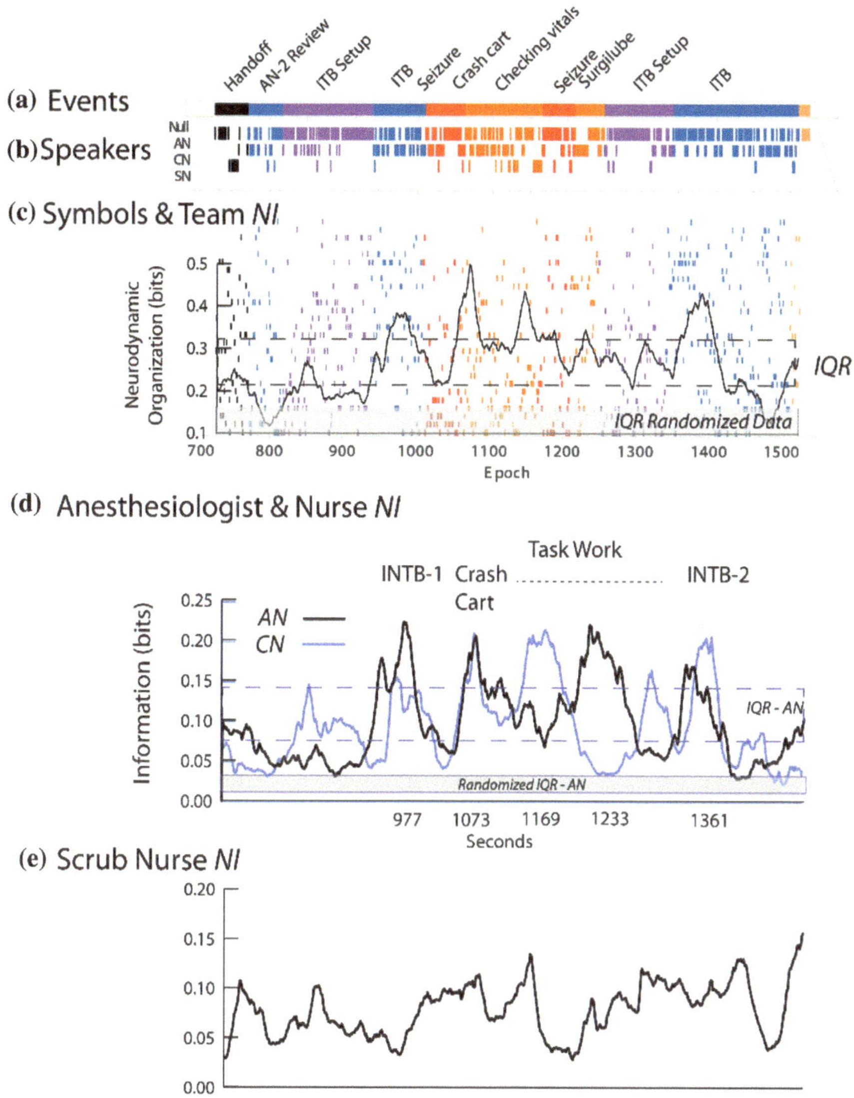

Fig. 2 Team and individual neurodynamics of an experienced team in the context of the events of a healthcare simulation. **a** The task event segments. **b** Times when each person was speaking. **c** The *NS* (*Y*-axis) in the neurodynamic data stream of the team were sequentially plotted each second on the *X*-axis. A quantitative profile of the variability in the symbol distribution over a 60 s time segment was obtained as *NI* and is a profile of the bits of information. The dotted lines indicate the inter quartile range (IQR), and the gray line indicates the IQR for randomized data which served as an internal control. The extensive patterning in the performance is shown by the comparison between the IQR of the randomized and non-randomized data. **d** The *NI* traces of the AN (dark) and CN (light) during the simulation with selected events labeled. **e** The *NI* trace of the SN. The values shown are averages across all sensors and frequencies of a 9-sensor EEG headset. Scalp electrodes were placed according to the International 10–20 System at: F3, F4, C3, C4, Fz, Cz, P3, P4 and P0z

Team neurodynamics (Fig. 2c) were separated into the neurodynamics of each team member along with the information shared across the team using the three and 27 neurodynamic symbol states as shown in Fig. 1, and described previously (Stevens, Galloway, & Willemsen-Dunlap, 2018). Figure 2d plots the *NI* of the AN and CN who worked together, and the *NI* of the SN who stood to the side (Fig. 2e). For three of the four primary Scenario events (i.e., INTB-1 & 2, and when the crash cart was called for), the *NI* of the AN and CN who were working closely together, showed similar *NI* dynamics while the neurodynamics of the SN who waited to be called to assist in surgery were unrelated.

The team and individual neurodynamics are shown in Fig. 3 for a medical student team performing similar ventilation procedures also on a simulated mannequin. The medical students did not have defined roles as did the experienced team in Fig. 3.

The medical student team neurodynamic profile lacked sharp *NI* fluctuations during the three intubation attempts (Intub-1, Intub-2, Intub-3). While isolated *NI* peaks were predominantly missing in the neurodynamic profile of the medical student team, the overall *NI* levels were greater, this was a significant trend that was seen when comparing the performance aggregated *NI* levels of five experienced and three medical student teams (0.37 bits $\pm$ 0.08 bits versus 0.89 $\pm$ 0.27 bits, $Z = -1.94$, $p = 0.05$).

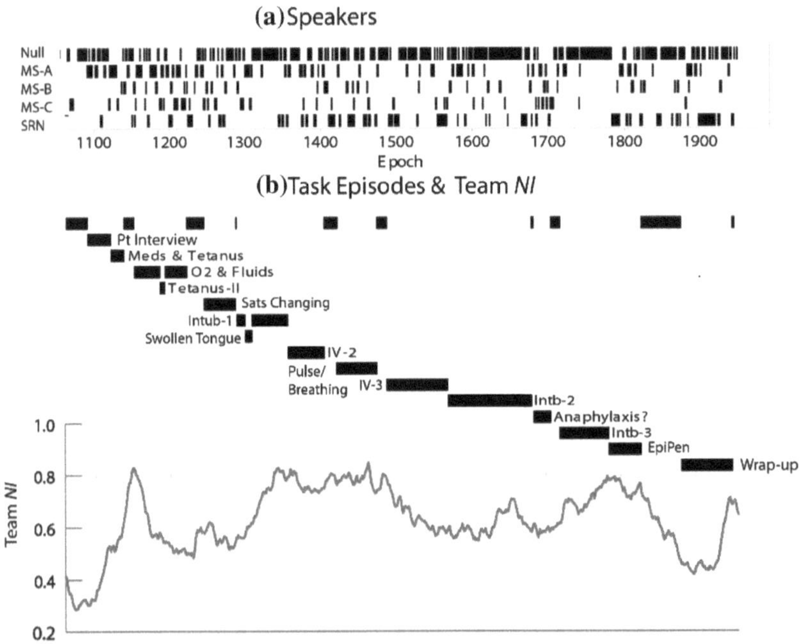

Fig. 3 Team neurodynamics of a medical student team. **a** Periods of different team member's speech are marked. **b** Task events are shown by sequential bars and labels, and the profile of the overall team neurodynamics is shown in light gray. Scalp electrodes were placed according to the International 10–20 System at: F3, F4, C3, C4, Fz, Cz, P3, P4 and P0z

2 Neurodynamics of Simulation Debriefings

The quantitative decomposition of team neurodynamic information into that of the individual which is shared by two or more individuals and the overall team, provides a unique multilevel perspective of team dynamics at different scales. The next examples move the focus from the scenario to the debriefing. Currently, simulation debriefings are labor intensive as well as expensive, and training institutions need recurring investments in training experienced personnel to be debriefers who can effectively facilitate student recall and prompt discussions for fostering the development of clinical reasoning.

Developing recall skills, both as individuals and as a team, are important for maximizing learning opportunities, yet studies of novice physicians have indicated that their reflection levels were generally in the lower half of scale average (Kihlgren, Spange, & Dieckmann, 2014). The second-by-second neurodynamics of different team-related data streams may provide new opportunities to purposefully develop recall skills.

The next studies modeled the second-by-second neurodynamics of the team and each team member as shown in Fig. 2, and plotted their dynamics in relation to the debriefing topic discussions. These studies took advantage of the ability to separate the team *NI* into that of each team member (Stevens, Galloway, & Willemsen-Dunlap, 2018).

The speech during the debriefing was transcribed and boundaries for the discussion topics were identified by the facilitator. The simulation involved the preoperative ventilation of a patient by an anesthesiologist (AN), assisted by a circulating nurse (CN), and a scrub nurse (SN), where the patient had an adverse response to a relative overdose of aerosolized lidocaine which caused convulsions and cardiac dysrhythmias. The five debriefing discussion topics resulting from this simulation were (1) Why was the patient experiencing convulsions? (2) What was the nature of the blockage the AN felt when intubating the patient? (3) Sharing stories of the delivery of anesthesia in the operating room; (4) Team situation awareness; and (5) Fidelity of the simulation/room setting (Fig. 4b, Event Descriptions).

The speech of the AN, CN, SN and the facilitator (INST) and observer (OBS) is shown in Fig. 4a, where the person speaking at any time appears as a mark in that speaker's row.

The discussions during the debriefing showed fluctuations that suggested that the periods of increased *NI* represented meaningful events for the team members. For instance, during the 1700–1900s discussion segment SN's *NI* was elevated as she related the tension she felt during the simulation regarding whether or not to break scrub during the patient's seizures.

There was also increased *NI* for the three team members during discussion segments 2360–2450s where the SN remarked how well coordinated the AN & CN were. The AN & CN confirmed they had worked together on multiple occasions in the operating room.

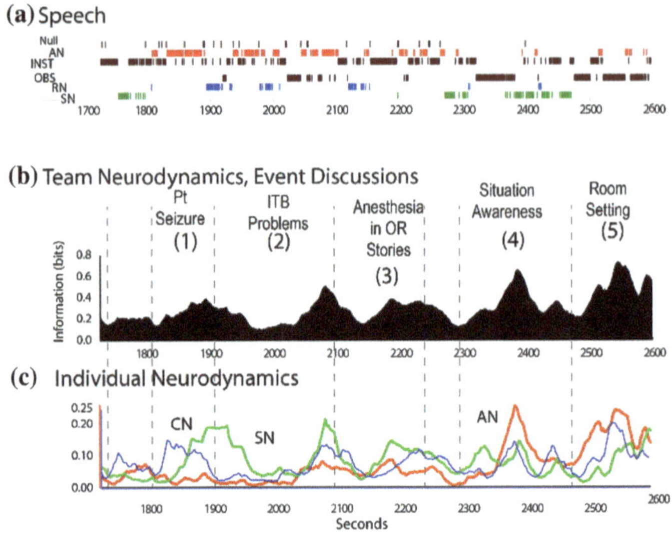

Fig. 4 Team and team member's neurodynamic organizations during debriefing. **a** A plot of the second-by-second speech of each member of the team. Shown in Fig. 2. **b** The five main discussion topics are delimited by dotted lines. The profile of the team *NI* is shown by a surface plot. **c** The individual *NI* of the anesthesiologist (AN), circulating nurse (CN), and scrub nurse (SN) are labeled

The quantitative measurement of individual, shared and team information will support ready comparisons across team members, teams and training as well as debriefing protocols. The quantitative nature of the different information dynamics will also provide new opportunities for more personalized and targeted debriefings. First, multimodal tracings of information changes in the scenario, resembling those in Figs. 2 and 3, could be provided to the facilitator prior to the debriefing, thusly providing focal points and events of importance to the team or different team members. This resource carries promise for targeting discussion and exploration toward points where team members and the team were focused or experiencing uncertainty; such resources could similarly support un-facilitated debriefings by the team members. Finally, the neurodynamic traces in conjunction with detailed audio and video analyses could shorten the time required for a facilitator to develop the ability to ask questions that prompt deeper levels of reflection in ways most likely to catalyze useful behavioral change on the part of learners (Husebo et al., 2013; Kihlgren et al., 2014).

2.1 How Well Do Team Neurodynamics from Simulations Reflect Those from Live Patient Encounters?

Much of the team training in healthcare occurs during simulations using medical mannequins. Inherent in this process is the belief, generally founded on macroscale measures such as observations and workplace-based assessments, that simulations provide the degree of psychological fidelity needed to accomplish this goal. It is expected that initial dynamical understandings of teams will be derived from simulation studies, much like those we have described. However, it is currently unknown at the neural/physiologic/cognitive level how well simulation training elicits the types of dynamic thinking that is actually used by operating room teams during live patient surgery, i.e., the ecological validity of simulation environments is unknown for dynamic neural and physiologic measures of team performance. Ecological validity is the extent to which the conclusions of a research study can be generalized to the settings and situations in which the phenomenon would naturally occur. In healthcare settings ecologic validity approximates the actual look, feel and procedure of a real medical encounter/procedure (Vaskinn, Sergi, & Green, 2009).

A final example directly explored the question of how well the simulation-derived team dynamics reflected those that occurred while the surgical team removed the pressure on the peroneal nerve of a live patient (Stevens, Galloway, & Willemsen-Dunlap, 2019).

At 48 min the live patient surgery was longer than the eighteen <20 min simulation scenarios we have studied. The team members were more mobile with the two neurosurgeons (NS1 & NS2) entering and leaving the operating room several times prior to the surgery, and because of a breakdown in communication the SN was absent until the gowning and gloving procedure began (Fig. 5). As expected, the team neurodynamic profile was more characteristic of the experienced teams previously studied during simulations, with isolated *NI* peaks.

We have recently published an expanded view of the team members' neurodynamic profile for the surgery (from 1700–2890 seconds) that highlights the dynamics from the time the nerve was identified until just before closing (Stevens, Galloway, & Willemsen-Dunlap, 2019).

One finding was that when the simulation and live patient neurodynamic measures were compared at progressively smaller time scales, differences in the levels of neurodynamic organization emerged across scalp locations and EEG frequencies, the most significant being the suppression of gamma rhythms in the frontal regions during the live patient ventilation. Gamma suppression is in part due to deactivation of the default mode network (Farooqui & Manly, 2018; Raichle et al., 2001).

Gamma oscillations occur alongside, and in proportion to perceptual processes/salience (Sedley & Cunningham, 2013) and are thought to: (1) be pivotal in the search for information, or the refreshing of information within the brain (Vergauwe & Cowan, 2014) and (2) the communication of this information across regions of the brain (Bonnefond & Jensen, 2015; Satoo & Mizuhara, 2018).

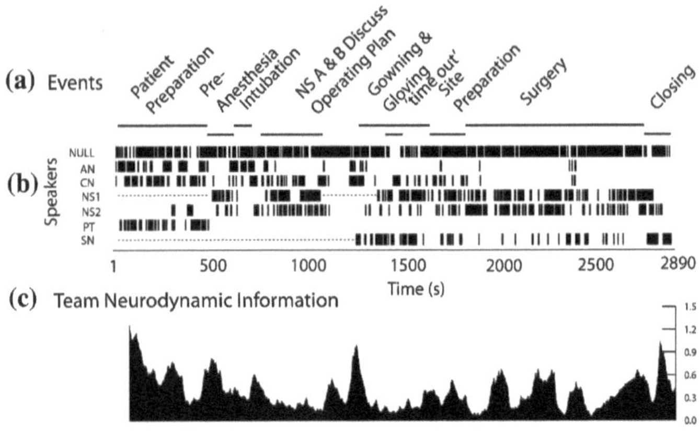

Fig. 5 The speech events and team neurodynamic profile for a live patient surgery. **a** The different events during the surgery. **b** The marks indicate when the different team members were speaking; the dashed lines indicate when that person was out of the operating room. **c** The *NI* profile was obtained by averaging across ten EEG sensors (Fpz, Fp1, Fp2, F7, F8, P7, P8, O1, Oz, and O2) and the 1–40 Hz frequency bands

As frontal brain regions are involved in the detection of unfavorable outcomes, error correction and resolution of uncertainty, and gamma rhythm suppression is associated with better outcomes on complex tasks (Ossandon et al., 2011; Murray & Rudebeck, 2017), we hypothesize that while ventilating the live patient the anesthesiologist was more attentive, therefore, performing finer error detection/correction than when ventilating a mannequin. If so, this would suggest current state of the art simulations might not yet have the fidelity to engage akin balances of cognitive elements such as those during live patient procedures like intubation, therefore, biometric models built from only simulation data should be viewed with this in mind.

Simulation education is a bridge between classroom learning and real-life clinical experience and it likely facilitates learning in a different way from the other learning modalities; therefore, it is not surprising that differences in the microlevel neurodynamics were seen between simulated patient and live patient performances. The nature of the differences is perhaps more important as it suggests that suppression of gamma rhythms may be amenable for study during complex real-world situations, and not confined to studies using intracerebral electrodes (Lachaux et al., 2008).

The topics discussed anticipate the time when dynamic biometric data can contribute to our understanding of how to rapidly determine a team's functional status, and how to use this information to optimize training and outcomes. There is no question that simulation education can improve the skills of healthcare workers, but there are limited theoretical insights available to make proactive predictions of the learning protocols for any individual. The rapid, dynamic, and task neutral measures that can be viewed through different temporal and spatial lenses should provide clarity about what these insights should look like. They will also provide a foundation for incorporating these models into machines to support the training and the performance of teams (Stevens & Galloway, 2019).

Such comparable quantitative measures of individual, shared, and team information will support ready comparisons across team members, teams, and training as well as debriefing protocols. The quantitative nature of the different a bridge between classroom learnin dynamics will also provide new opportunities for more personalized and targeted debriefings around moments where team members and the team were focused or experiencing uncertainty. Such multimodal experiences would also shorten the period of time required for a facilitator to develop the abilities to ask questions that prompt deeper levels of reflection in ways most likely to catalyze useful behavioral change on the part of learners (Kihlgren et al., 2014; Huesbo et al., 2013). Eventually such resources could support independent debriefings by the team members who could use the neurodynamic traces in conjunction with detailed audio and video analyses to change behavior which in turn, leads to an improvement in patient safety.

3 Key Points

- Neurodynamic organization is a scalable neurodynamic representation that links to organizational metrics of teamwork but can be taken apart for theoretical understandings of observational meanings.
- It is now possible to quantitatively compare the organizational dynamics of different teams and team members across teams and time.
- Neurodynamic organizations can become closely coordinated among interacting teammates during complex tasks and with a team's/instructor's speech on more directed tasks like debriefings.
- The neurodynamics of simulated and live patient surgery are similar at more macroscales of teamwork, but differences become apparent as the hierarchy moves to lower levels.

Appendix 1: Processing EEG Data Streams

The data acquisitions began shortly after the EEG sensors were adjusted for good contact (<10 Ω). The EEG data streams were aligned for the three team members using electronic markers inserted into the EEG data streams as well as the events observed in videos. Signals from outside the brain can be a confounder when interpreting models built from EEG signals, especially signals obtained in complex environments. Commonly found artifacts are generated from speech, eyeblinks, heartbeats, breathing rhythms and other electromyography sources. As neurodynamic organizations regularly occur during silence, speech is an unlikely source for most organizations (Stevens & Galloway, 2014). EEG processing included separate high and low bandpass filters, the rejection of bad channels and regular rhythms associated with eyeblinks and heartbeats were identified and removed during data preprocessing (Delorme & Makeig, 2004; Delorme et al., 2012) by the interactive Matlab® toolboxes EEGLAB and FieldTrip (oostenveld, Fries, Maris, & Schoffelen, 2011).

References

Abersold, M., (2018). Simulation-based learning: No longer a novelty in undergraduate education. *OJIN: The Online Journal of Issues in Nursing 23*(2). https://doi.org/10.3912/ojin.vol23no02ppt39.

Ancona, D., & Chong, C. L. (1999). Cycles and synchrony: The temporal role of context in team behavior. *Research on Managing Groups and Teams, 2,* 33–48.

Anders, S., Heinzle, J., Weiskopf, N., Ethofer, T., & Haynes, J. (2011). Flow of affective information between communicating brains. *Neuroimage, 54,* 439–446.

Baker, D. P., Amodeo, A. M., Krokos, K. J., Slonim, A., & Herrera, H. (2010). Assessing teamwork attitudes in healthcare: Development of the TeamSTEPPS® teamwork attitudes questionnaire. *Quality and Safety Health Care, 19,* e49.

Bond, W. F., Deitrick, L. M., Eberhardt, M., et al. (2006). Cognitive versus technical debriefing after simulation training. *Academy of Emergency Medicine, 13,* 276–283.

Bonnefond, M., & Jensen, O. (2015). Gamma activity coupled to alpha phase as a mechanism for top-down controlled gating. *PLoS ONE, 10,* e012866.

Bowe, S., Johnson, K., & Puscas, L. (2017). Facilitation and debriefing in simulation education. *Otolaryngologic Clinics of North America, 50*(5), 989–1001.

Buzaki, G. (2006). *Rhythms of the brain*. New York, NY: Oxford University Press.

Cheng, A., Morse, K. J., Rudolph, J., Arab, A., Runnacles, J., & Eppich, W. (2016). Learner-centered debriefing for health care simulation education: Lessons for faculty development. *Simulation in Healthcare, 11,* 32–40.

Clapper, T. C. (2016). Proposing a new debrief checklist for team STEPPS to improve documentation and clinical debriefing. *Simulation and Gaming, 47,* 710–719.

Cooke, N. J., Gorman, J. C., & Kiekel, P. (2008). Communication as team-level cognitive processing. In M. P. Letsky, N. W. Warner, S. M. Fiore & C. A. P. Smith (Eds.), *Macrocognition in teams*. Ashgate: Burlington, VT, USA.

Cooke, N. J., Salas, E., Cannon-Bowers, J., & Stout, R. (2000). Measuring team knowledge. *Human Factors, 42,* 151–173.

Daw, C. S., Finney, C. E. A., & Tracy, E. R. (2003). A review of symbolic analysis of experimental data. *Review of Scientific Instruments, 74,* 915. https://doi.org/10.1063/1.1531823.

Delorme, A., Palmer, J., Onton, J., Oostenveld, R., & Makeig, S. (2012). Independent EEG sources are dipolar. *PLoS ONE, 7*(2), e30135.

Draschkow, D., Heikel, E., Võ, M., Fiebach, C. J., & Sassenhagen, J., (2018). No evidence from MVPA for different processes underlying the N300 and N400 incongruity effects in object-scene processing. *Neuropsychologia, 120,* 9–17. ISSN 0028-3932. https://doi.org/10.1016/j.neuropsychologia.

Entin, E. E., & Serfaty, D. (1999). Adaptive team coordination. *Human Factors: The Journal of the Human Factors and Ergonomics Society, 41*(2), 312–325.

Fanning, R., & Gaba, D. (2007). The role of debriefing in simulation-based learning. *Simulation in Healthcare, 2,* 115–125.

Farooqui, A. A., & Manly, T. (2018). Hierarchical cognition caused task related deactivations but not just in default mode regions. *eNeuro, 5*(6), ENEURO.0008-18.2018.
Fiore, S. M., Smith-Jentsch, K. A., Salas, E., Warner, N., & Letsky, M. (2010). Towards an understanding of macro cognition in teams: Developing and defining complex collaborative process and product. *Theoretical Issues in Ergonomics Science, 11*(4), 250–271.
Flack, J. (2017). Life's information hierarchy. In S. I. Walker, P. C. W. Davies & G. F. R. Ellis (Eds.), *From matter to life: Information and causality*. Cambridge: Cambridge University Press.
Friston, K. (2010). The free-energy principle: A unified brain theory? *Nature Reviews Neuroscience, 11,* 127–138. https://doi.org/10.1007/s11229-007-9237-y.
Gardezi, F., Lingard, L., Espin, S. L., Whyte, S., Orser, B., & Baker, G. R. (2009). Silence, power and communication in the operating room. *Journal of Advanced Nursing, 65,* 1390–1399.
Garner, A. J. P., Thompson, J., Vedral, V., & Gu, M. (2017, April 16). Thermodynamics of complexity and pattern manipulation. arXiv:1510:00010v3[quant.ph].
Giere, R. N., & Moffatt, B. (2003). Distributed cognition: Where the cognitive and social merge. *Social Studies of Science, 33,* 301–310.
Gorman, J. C. (2014). Team coordination and dynamics: Two central issues. *Current Directions in Physiological Science, 23,* 355–360.
Grimm, D., Gorman, J. C., Stevens, R. H., Galloway, T., Willemsen-Dunlap, A. M., & Halpin, D. J. (2017). Demonstration of a method for real-time detection of anomalies in team communication. In *Proceedings of the Human Factors and Ergonomics Society 59th Annual Meeting* (pp. 282–286). Santa Monica, CA: Human Factors and Ergonomics Society.
Guastello, S. J., & Peressini, A. F. (2018). Development of a synchronization coefficient for biosocial interactions in groups and teams. *Nonlinear Dynamics, Psychology, Life Sciences, 48,* 3–33.
Hari, R. (2006). Action perception connection and the cortical mu-rhythm. *Progress in Brain Research, 159,* 253–260.
Hsu, M., Bhatt, M., Adolphs, R., Tranel, D., & Camerer, C. F. (2005). Neural systems responding to degrees of uncertainty in human decision making. *Science, 310,* 1680–1683.
Husebo, S. E., Dieckmann, P., Rystedt, H., Soreide, E., & Friberg, F. (2013). The relationship between facilitators' questions and the level of reflection in postsimulation debriefing. *Simulation in Healthcare, 8,* 135–141.
James, R. G., Ellison, C. J., & Crutchfield, J. P. (2011). *Anatomy of a bit: Information in a time series observation* (Santa Fe Institute Working Paper 11–05). arXiv:1105.2988v1[cs.IT].
Kihlgren, P., Spanger, L., & Dieckmann, P. (2014). Investigating novice doctors' reflections in debriefings after simulation scenarios. *Journal of Medical Teacher, 37,* 437–443.
Kolchinsky, A., & Wolpert, D. H. (2018). Semantic information, autonomous agency and non-equilibrium statistical physics. *Interface Focus, 8,* 20180041.
Lachaux, J. P., Jung, J., Dreher, J. C., Bertrand, O., Minotti, L., Hoffman, D., et al. (2008). Silence is golden: Transient neural deactivation in the prefrontal cortex during attentive reading. *Cerebral Cortex, 18,* 443–450.
Lederman, L. (1992). Debriefing: toward a systemic assessment of theory and practice. *Simulation and Gaming, 23,* 145–160.
Marks, M., Mathieu, J. E., & Zaccaro, S. J. (2001). A temporally-based framework and taxonomy of teamwork. *The Academy of Management Review, 26,* 356–376.
Menoret, M., Varnet, L., Fargier, R., Cheylus, A., Curie, A., des Portes, V., et al. (2014). Neural correlates of non-verbal social interactions: A dual-EEG study. *Neurophyschologia, 55,* 85–91.
Mosier, K. H., & Chidester, T. R. (1991). Situation assessment and situation awareness in a team setting. In Y. Queinnec & F. Daniellou (Eds.), *Designing for everyone* (pp. 798–800). London: Taylor & Francis.
Mullen, T. (2012). NITRC: *CleanLine: Tool/Resource Info*. Retrieved from http://www.nitrc.org/projects/cleanline.
Murray, E. A., & Rudebeck, P. H. (2017). Specializations for reward-guided decision-making in the primate ventral prefrontal cortex. *Nature Reviews Neuroscience, 19,* 404–417.

Oostenveld, R., Fries, P., Maris, E., & Schoffelen, M. (2011). FieldTrip: Open source software for advanced analysis of MEG, EEG, and invasive electrophysiological data. *Computational Intelligence and Neuroscience*, Article ID 156869, 9 pp. https://doi.org/10.1155/2011/156869.

Ossandon, T., Jerbi, K., Vidal, J. R., Bayle, D. J., Henaff, M.-A., Jung, J., et al. (2011). Transient suppression of broadband gamma power in the default mode network is correlated with task complexity and subject performance. *Journal of Neuroscience, 31,* 14521–14530.

O'Neil, H. F., Jr., Chung, G. K., & Brown, R. (1997). Use of networked simulations as a context to measure team competencies. In H. F. O'Neil Jr. (Ed.), *Workforce readiness: Competencies and assessment* (pp. 411–452). Mahwah, NJ: Erlbaum.

Petranek, C. (1994). Maturation in experiential learning: Principles of simulation and gaming. *Simulation in Gaming, 25,* 513–522.

Raichle, M. E., MacLeod, A. M., Snyder, A. Z., Powers, W. J., Gusnard, D. A., & Shulman, G. L. (2001). A default mode of brain function. *Proceedings of National Academy of Science of USA, 98,* 676–682.

Salas, E., Stagl, K.C., & Burke, C. S. (2004). *25 Years of team effectiveness in organizations: Research themes and emerging needs.* In C. L. Cooper & I. T. Robertson (Eds.), *International review of industrial and organizational psychology* (Vol. 19, pp. 56–101). New York: Wiley.

Salas, E., Stevens, R., Gorman, J., Cooke, N. J., Guastello, S., & von Davier, A. (2015, September). What will quantitative measures of teamwork look like in 10 years? In *Proceedings of the Human Factors and Ergonomics Society Annual Meeting, 59* (1), 235–239 (Sage Publications).

Sanger, J., Muller, V., & Lindenberger, U. (2012, November). Intra-and interbrain synchronization and network properties when playing guitar in duets. *Frontiers in Human Neuroscience*, Article 312. https://doi.org/10.3389/fnhum.201200312.

Sato, N., & Mizuhara, H. (2018). Successful encoding during natural reading is associated with fixation related potential and large-scale network activation. *eNeuro.* https://doi.org/10.1523/eneuro.0122-18.2018.

Sawyer, T., & Deering, S. (2013). Adaptation of the US Army's after-action review for simulation debriefing in healthcare. *Simulation in Healthcare, 8,* 388–397.

Schippers, M., Roebroeck, A., Renken, R., Nanetti, L., & Keysers, C. (2010). Mapping the information flows from one brain to another during gestural communication. *Proceedings of National Academy of Science of USA, 107,* 9388–9393.

Schmidt, E., Goldhaber-Fiebert, S. M., Ho, L. A., & McDonald, D. (2013). Simulation exercises as a patient safety strategy: A systematic review. *Annals of Internal Medicine, 158,* 426–432.

Sedley, W., & Cunningham, M. O. (2013). Do cortical gamma oscillations promote or suppress perception? An under-asked question with an over-assumed answer. *Frontiers in Human Neuroscience. 7,* Article 595.

Shockley, K., Santana, M.-V., & Fowler, C. A. (2003). Mutual interpersonal postural constraints are involved in cooperative conversation. *Journal of Experimental Psychology: Human Perception and Performance, 29,* 326–332.

Shannon, C. E. (1948). A mathematical theory of communication. *The Bell System Technical Journal, 27,* 379–423.

Shannon, C. E. (1951). Prediction and entropy of printed English. *The Bell System Technical Journal, 30,* 50–64.

Staropoli, P. C., Gregori, N. Z., Junk, A. K., Galor, A., Goldhardt, R., Goldhagen, B. E., et al. (2018). Surgical simulation training reduces intraoperative cataract surgery complications among residents. *Simulation in Healthcare, 13,* 11–15. https://doi.org/10.1097/SIH.0000000000000255.

Stephens, G., Silbert, L., & Hasson, U. (2010). Speaker-listener neural coupling underlies successful communication. *Proceedings of the National academy of Sciences of the United States of America.* Retrieved from www.pnas.org/cgi/doi/10.1073/pnas.1008662107.

Stevens, R. H., & Galloway, T. (2014). Toward a quantitative description of the neurodynamic organizations of teams. *Social Neuroscience, 9*(2), 160–173.

Stevens, R. H., & Galloway, T. (2015). Modeling the neurodynamic organizations and interactions of teams. *Social Neuroscience, 11,* 123–139. https://doi.org/10.1080/17470919.2015.1056883.

Stevens, R. H., & Galloway, T. (2017, May). Are neurodynamic organizations a fundamental property of teamwork? *Frontiers in Psychology*. https://doi.org/10.3389/fpsyg.2017.00644.
Stevens, R., & Galloway, T. L. (2019). Teaching machines to recognize neurodynamic correlates of team and team member uncertainty. *Journal of Cognitive Engineering and Decision Making, 13*, 310–327. https://doi.org/10.1177/1555343419874569
Stevens, R. H., Galloway, T., Halpin, D., & Willemsen-Dunlap, A. (2016). Healthcare teams neurodyamically reorganize when resolving uncertainty. *Entropy, 18*, 427. https://doi.org/10.3390/e18120427.
Stevens R. H., Galloway T., Lamb J., Steed R., & Lamb C. (2017). Linking team neurodynamic organizations with observational ratings of team performance. In A. von Davier, M. Zhu & P. Kyllonen (Eds.), *Innovative assessment of collaboration*. New York, NY: Springer.
Stevens, R.H., Galloway, T., & Willemsen-Dunlap, A., (2016). Intermediate neurodynamic representations: A pathway towards quantitative measurements of teamwork? *Human Factors & Ergonomics National Meeting, 60*(1), 1996–2000.
Stevens, R. H., Galloway, T. & Willemsen-Dunlap, A. (2017). Low level predictors of team dynamics: A neurodynamic approach. In E. Salas, B. Vessey & L. B. Landon (Eds.), *Research in managing groups and teams*. Research series: Team dynamics over time, Vol. 18, pp. 71–92. Emerald Insight.
Stevens, R.H., Galloway, T., & Willemsen-Dunlap. (2018). Quantitative modeling of individual, shared and team neurodynamic information. *Human Factors, 60*, 1022–1034.
Stevens, R.H., Galloway, T., Gorman, J., Willemsen-Dunlap, A., Halpin, D., & Grimm, D. (2018, October). *Making sense of team information*. Paper presented at 2018 Human Factors & Ergonomics National Meeting, Philadelphia, PA.
Stevens, R., Galloway, T. L., & Willemsen-Dunlap, A. (2019). Advancing our understandings of healthcare team dynamics from the simulation room to the operating room: A neurodynamic perspective. *Frontiers in Psychology, 10*, 1660. https://doi.org/10.3389/fpsyg.2019.01660
Tognoli, E., & Kelso, J. A. (2015). The coordination dynamics of social neuromarkers. *Frontiers of Human Neuroscience, 9*, 563. https://doi.org/10.3389/fnhum.2015.00563.
Tononi, G., Boly, M., Massimini, M., & Koch, C. (2016). Integrated information theory: From consciousness to its physical substrate. *Nature Reviews in Neuroscience, 17*, 450–461.
Vaskinn, A., Sergi, M. J., & Green, M. F. (2009). The challenges of ecological validity in the measurement of social perception in schizophrenia. *The Journal of Nervous and Mental Disease, 197*, 700–702.
Vergauwe, E., & Cowan, N. (2014, March). A common short-term memory retrieval rate may describe many cognitive procedures. *Frontiers of Human Neuroscience*, Article 126. https://doi.org/10.3389/fnhum.2014.00126.
von Davier, A. A., & Halpin, P. F. (2013). Collaborative problem solving and the assessment of cognitive skills: Psychometric considerations. *ETS Research Report Series, 2013*(2), 1–36. https://doi.org/10.1002/j.2333-8504.2013.tb02348.x.
Zenati, M. A., Leissner, K. B., Zorca, S., Kennedy-Metz, L., Yule, S., & Dias, R. (2019). First reported use of team cognitive workload for root cause analysis in cardiac surgery. *Seminar Thoracic Surgery*.

The Neuroergonomics of Music Proficiency and Performance

Thomas J. Smith and Joshua McNiven

Without music, life would be a mistake.
Nietzsche.

Abstract This chapter addresses the nature of the interaction between musical behavior and underlying brain mechanisms. We address a series of relevant questions: (1) does neuroergonomics offer a legitimate scientific framework for understanding variability in musical behavior? (2) are there other scientific perspectives that complement or contradict the neuroergonomic perspective, in terms of understanding musical behavior? and (3) does musical behavior conform to the predictions of neuroergonomics and as such, represent an appropriate topic for this text? Our analysis supports the following conclusions: (1) regarding Questions 1 and 3, evidence for cognitive specialization, based on brain laterality, in the expression of musical behavior and talent, is aligned with the focus of neuroergonomics; but (2) nevertheless, based on studies of other modes of behavior, we also argue that there are a series of environmental design factors that also have the potential to influence variability in musical behavior, a phenomenon termed context specificity mediated by brain plasticity. Neuroergonomics, arguably, has not broadly embraced the idea that context specificity influences observed variability in behavior, including musical behavioral expression. Nevertheless, music also represents the only mode of artistic expression for which concrete empirical evidence exists for the neuroergonomic assumption that proficiency in musical talent must be understood in terms of cognitive specialization.

T. J. Smith (✉)
University of Minnesota, Minneapolis, USA
e-mail: Smith293@umn.edu

J. McNiven
Royal Welsh College of Music and Drama, Cardiff, UK

C. S. Nam (ed.), *Neuroergonomics*, Cognitive Science and Technology,
https://doi.org/10.1007/978-3-030-34784-0_14

1 Introduction

This chapter addresses the nature of the interaction between musical behavior and underlying brain mechanisms, a field of ergonomic/human factors (E/HF) study termed *neuroergonomics* (Parasuraman 2003). This focus raises three relevant questions: (1) does neuroergonomics offer a legitimate scientific framework for understanding variability in musical behavior? (2) are there other scientific perspectives that complement or contradict the neuroergonomic perspective, in terms of understanding musical behavior? and (3) does musical behavior conform to the predictions of neuroergonomics and as such, represent an appropriate topic for this text? Question 1 is addressed in this introductory section. Questions 2 and 3 are addressed in the third and fourth sections, respectively, of this chapter.

Merriam-Webster offers seven different definitions for the term "music" (Merriam-Webster 2019), some of which will be referenced in Sect. 3. A perspective favored by some observers assumes that music represents a type of language (Patel 2008; Slevc and Okada 2015), although others have expressed skepticism about this idea (Mithen 2005; Pinker 1997). The music as a language idea is relevant to both Questions 2 and 3 above.

Question 1: *Does neuroergonomics offer a legitimate scientific framework for understanding variability in musical behavior?* Various viewpoints, plus available evidence, offer a mixed perspective on this question. The idea that the brain plays a central, and critical, role in mediating behavior seems self-evident. Parasuraman (IBID), recognized by the E/HF community as a central figure in originating the field of neuroergonomics, characterizes the field as follows.

> Neuroergonomics focuses on investigations of the neural bases of mental functions and physical performance in relation to technology, work, leisure, transportation, health care and other settings in the real world. The two major goals of neuroergonomics are to use knowledge of brain function and human performance to design technologies and work environments for safer and more efficient operation, and to advance understanding of brain function underlying real-world human performance.

Despite the fact that this characterization links the neural basis of performance to design factors in the performance environment (work, transportation, etc.), its formulation leaves little doubt that neuroergonomics is conceived as a brain-centered understanding—neural determinism if you will—of how behavioral performance is controlled.

In contrast to these brain-centered views of how behavior is controlled, the alternate viewpoint is that design factors in the performance environment prominently influence observed variability in behavior. The terms applied to this perspective are *brain plasticity* and *context specificity* in behavioral expression, based on a control systems interpretation of reciprocal performance–design interaction. Smith is the first author of an entire text providing empirical evidence (dating back over a century), supported by conceptual analysis, for the idea of context specificity in behavior, including cognition (Smith et al. 2015). Fischer and colleagues (2016) argue that neural plasticity has guided adaptive changes observed in evolution itself. Plasticity thus extends neural control capabilities to encompass a closed-loop interaction—and reciprocal influence—with environmental design factors subject to such control.

As applied to musical behavior, the idea of context specificity extends the neuroergonomic perspective to a more comprehensive frame of reference, addressed in Sect. 3, that design factors prominently influence all modes of observed variability in musical behavioral expression. Nevertheless, as addressed in Sect. 4, music also represents the only mode of artistic expression for which concrete empirical evidence exists for the neuroergonomic assumption that proficiency in musical talent must be understood in terms of cognitive specialization.

2 Historical Perspective

An obvious obstacle to our understanding of the nature of musical behavioral expression in history and prehistory is that sensory perception of music is mediated primarily by control of auditory feedback. As depicted in Fig. 1, effector control of the behavioral environment relies upon three modes of visual feedback, namely reactive, instrumental and operational. Reactive feedback is generated by movements of the effector itself (during passive listening, these effectors comprise postural and head movements, and movements of the tiny muscles controlling the ossicles of the middle ear). Instrumental feedback is generated by movements of the tool controlled by the effector (the musical instrument during musical behavioral expression). Operational feedback is generated by the action of the effector (in the figure, used to actuate a tool) on the environment.

Passive listening to music, and vocal generation of music, both involve control of reactive feedback from (1) effectors mediating postural and head movements, and movements of the tiny muscles controlling the ossicles of the middle ear, in the case of listening and (2) additional vocal cord, diaphragm and chest and abdominal muscles with vocal generation of music. In neither case, before recording technology, are we left with any concrete evidence as to how music of the past actually sounded (some groups, the Academy of Ancient Music being one of the most prominent [origins in the early eighteenth century], perform on period instruments in an attempt to replicate music composed hundreds of years ago).

Instrumental feedback also must be controlled during use of some sort of instrument to generate music. Thus, only with instrumental music might we obtain evidence, through fossilized musical instruments, of how music was generated in the distant past. The oldest musical instruments so far discovered are flutes (in appearance), made from bird bone and from mammoth ivory, found in a cave in southern Germany, and dated to 42–43 thousand years ago (Higham et al. 2012; Wilford 2012).

Given its discovery in northern Europe, we may infer from this find that music must have originated earlier in Africa. No direct evidence for this idea has yet been documented. However, symbolic behavior in the form of abstract art, discovered in Blombos Cave in South Africa, has been dated to 73,000 years (Moore 2019), and the oldest hominid tools, dated to 3.3 million years ago, have been discovered on

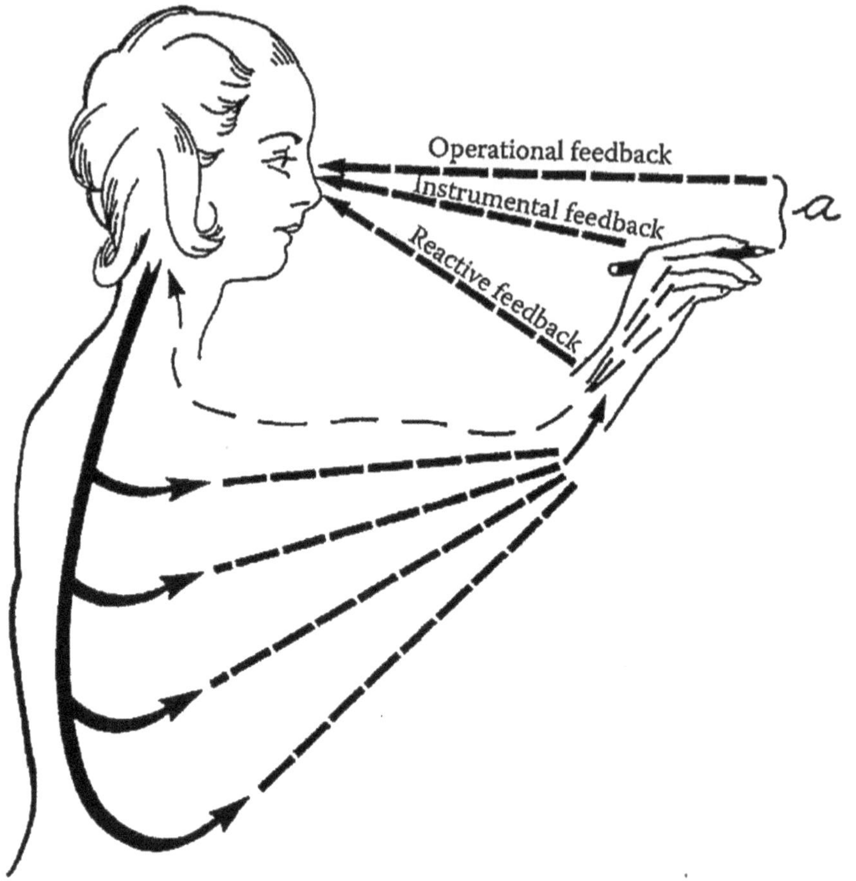

Fig. 1 Sensory feedback control demands inherent to tool use (Smith et al. 2015, Chap. 1, p. 23). Use of a tool entails control of reactive feedback from effector movement and control of instrumental feedback from movement of the tool. Tool use also may modify the external environment in a manner that persists after tool use has ceased, in which operational feedback from this action of the tool also must be controlled (Courtesy K. U. Smith)

the shores of Lake Turkana in Kenya (Harmand et al. 2015). An African origin for musical behavior thus seems plausible—that a revamped museum termed the Africa Museum in Belgium includes a display of 8,000 musical instruments from Africa likewise is suggestive of this idea (The Economist 2018a).

Finally, arguments supporting the centrality of musical expression to the human condition have been advanced. Bispham (2006) summarizes the following points favoring this conclusion.

- Every known human society has what trained musicologists would describe as "music"; hence, we can expect that music accompanied Homo sapiens out of Africa.

- We spend a lot of time doing it [e.g., Mekranoti Indians].
- 6-month-old infants are capable listeners and display proto-musical abilities from birth.
- There is evidence for the existence of specialized brain structures for music. The existence of Amusics who retain emotive and referential linguistic abilities is suggestive.
- Music has been around for a long time, as noted above (Higham et al. 2012; Wilford 2012). We should expect, however, that the use of non-fossiling instruments (e.g., bamboo, wood, etc.) and the voice preceded considerably any archaeological evidence.

The fourth bullet supports a neuroergonomic perspective on music, a viewpoint further addressed in Sect. 4.

Support for the idea of an intimate relationship between music and human behavior also is offered by Goodall (2015), who devotes an entire book to exploring the theme that music has shaped civilization.

3 Context Specificity in Musical Behavior

This section offers an affirmative answer to the second question posed at the outset of the Introduction. Question 2: *Are there other scientific perspectives that complement or contradict the neuroergonomic perspective, in terms of understanding musical behavior?* Observations and evidence point to the conclusion that context specificity represents an appreciable source of variability in musical behavior and performance. As noted previously, this term refers to the phenomenon of plasticity in brain function, such that there is a closed-loop relationship between musical behavioral expression and design factors in the performance environment in which music behavior occurs.

The list in Table 1 summarizes a series of such factors, each of which is discussed in paragraphs below.

Table 1 Context specificity in musical behavior. Design factors in the musical performance environment with a likely influence on variability in musical expression

Design factors likely to influence variability in musical behavior	
Social structure/social tracking	Musculoskeletal demands
Culture	Natural environmental sounds
Entertainment	Musical structure
Genre	Religion
Instruments and instrumentation	Rhythm
Human context	Training
Sensory feedback control	

Social Structure/Social Tracking. It is difficult to overestimate the significance of social behavior for the evolution, expansion, organization, and technological progress of the human species and human civilization. At the most fundamental level, we may cite the Darwinian theory of evolution: both reproduction and survival (reciprocal predator–prey tracking; group coordination) rely upon social interaction and cooperation. From a behavioral cybernetic (closed-loop control) perspective (Smith 1972), social interaction is understood as a social tracking process (Smith 1974; Smith and Smith 1987; Smith et al. 1994, 2015 (Chap. 8)).

The term *social tracking* is used to describe the feedback-controlled process by which an individual follows or tracks a social target. Social interaction is conceived as a dynamic linking of the motor-sensory processes of two or more people in a social tracking relationship. In cybernetic terms, the activities of one person in a social context effect behavioral–physiological changes in others, whose activities in turn feedback influence the ongoing behavior and physiology of the first. These effects arise as a consequence of control by each participant of sensory feedback generated by the other social tracking partners.

Figure 2 illustrates this behavioral cybernetic concept of social interaction using both interpersonal and group social tracking as examples. As indicated by the arrows and muscle symbols in the figure, the movements of one individual generate stimuli that are controlled as sensory feedback by the movements of the social partner(s), whose actions generate compliant sensory stimuli back to the first person, who in turn controls this sensory feedback by further movement, and so on. Social tracking typically requires each participant to control multiple motor, sensory, and cognitive modalities (vision, speech, writing, etc.) and transformations (displacements, delays, etc.) of sensory feedback. The social partners thus become dynamically yoked or interlocked behaviorally and physiologically as a result of mutual body movement tracking and control of each other's sensory feedback. Through such interlocks, the participants in a social group begin to operate as an integrated system, with definite system feedback parameters and feedback control characteristics, indicated in the figure. In particular, social cybernetic systems can involve many different types of social tracking modes (direct pursuit, compensatory, parallel, series linked, etc.) in a variety of different social interactive contexts (interpersonal, group, and/or institutional social systems).

Our premise with introducing social tracking as a design factor key to musical behavioral variability is that social cooperation, if not universal, is extremely common in the production and expression of music. Music composition itself may involve social interaction—McCartney and Lennon with the Beatles (Somers 2018, p. 30), or Rogers and Hammerstein, represent two examples. During orchestral or vocal performance, performers typically track a conductor and perhaps the concertmaster as well. During ensemble performance with or without a conductor, members may track one another. It is not uncommon for musical performers also to track the audience and vice versa. The emergence, development, maturation, and elaboration of music could not have occurred without social cooperation, and arguably all of the other context-specific features of musical behavior discussed in paragraphs below rest fundamentally on social tracking.

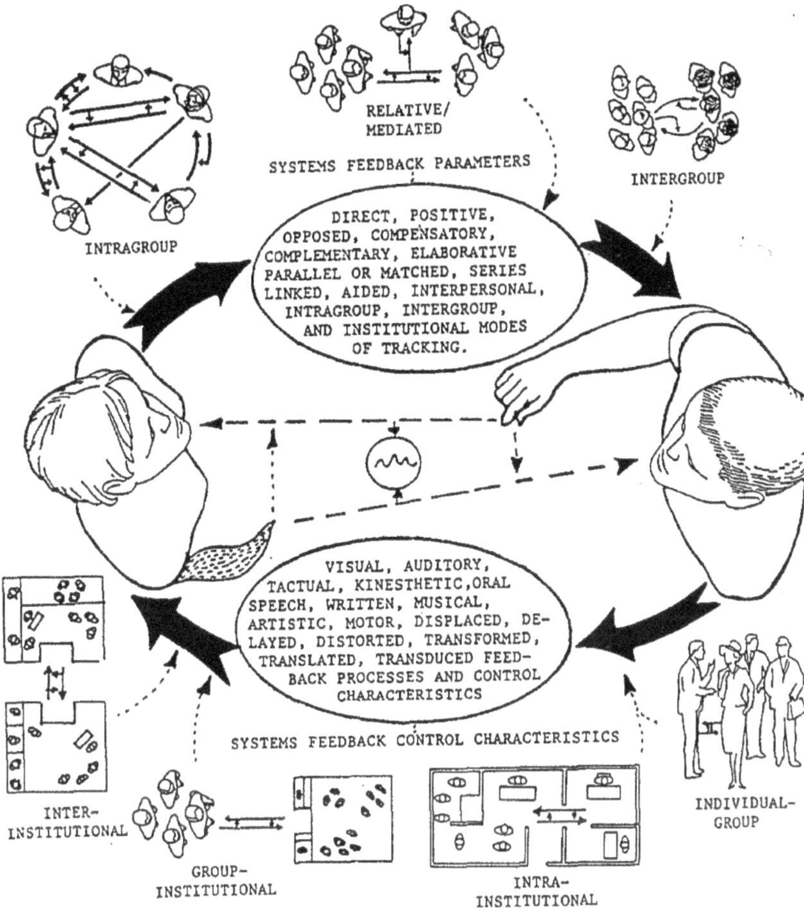

Fig. 2 Cybernetic theory of social behavior, indicating systems feedback parameters and feedback control characteristics (Smith et al. 2015, Chap. 8) governing both interpersonal and group social interaction (Courtesy K. U. Smith)

Culture. This term has geographical, generational and ethnic connotations for musical behavioral expression. Geographically, it is self-evident that different musical styles and modes of expression have emerged in different geographical areas. Thus, Indian music differs from Greek music, that in turn differs from Lebanese music—and so forth. Even within a given geographical area (India for example), musical differences in different areas may persist.

The review by Bilger (2019) of the musical range of the experimental choral group, Roomful of Teeth, dramatizes the scope of these differences. This group specializes in performing music from cultures worldwide. Bilger cites music from 10 different cultures that the group performs, from Bulgarian belting to Swiss yodeling (one of three yodeling cultures).

Generational evolution of musical differences also represents a common theme. In the U.S. for example, the twentieth century has seen an evolution from European influences early in the century (baroque, renaissance, romantic styles), to U.S. hymn traditions (early American; Shaker), to emergence of a seemingly endless progression of blues, jazz, rock and roll and rap styles, and so forth. There also are ethnic connotations to this progression—with black influences on both blues and jazz (a carryover from the era of slavery) well documented.

Our interpretation of these patterns is that they represent cultural influences, in relation to differential effects on musical behavior at geographical, generational, and ethnic levels. A clear parallel is language. Papua New Guinea for example, with an area a bit larger than California, has an estimated 832 living languages, making it the most linguistically diverse place on earth. Although natives across the island have biologic and ethnic similarities, this remarkable pattern is attributed to different tribal cultures living in geographically—and therefore culturally—isolated valleys across the island.

Entertainment. Much entertainment relies on music, and music in turn has been strongly influenced by the demands of entertainment. During the sixth and fifth centuries B.C., for example, a solitary actor in Greek theater typically was accompanied by a Greek chorus singing or chanting. A more recent example is that 78% of the tunable FM radio stations in the Twin Cities of Minnesota area are music stations.

The neurophysiological key to the entertainment value of music—a neuroergonomic connection—is its emotional appeal. Daly and Williams in this text (2019) provide an excellent review of this topic.

Genre The history of music has been marked by the unceasing emergence, and evolution, of new genres of music. With classical music (of which we both are familiar), we can cite the baroque, renaissance, romantic, and modern eras as examples. Each of these eras also feature different compositional genres, such as orchestra, concerto, sonata, serenade, and so forth. As an example, we refer to a series of free 24/7 television music stations broadcast by Comcast, the local carrier for the Twin Cities area. There are 45 such stations, each broadcasting a different music genre.

Instruments and Instrumentation. For centuries, the type of music composed has depended in part on the type of instruments available, and a reciprocal influence also has been the case. The first instrument of course was the human voice, and we may speculate that the first type of music was the human chant. Nevertheless, discovery of fossilized flute-like instruments dated to 40,000 years ago (above) suggests that the influence of musical instrumental designs on musical expression has ancient origins.

Human Context. This term refers to the influence on music compositional style by certain individuals or groups. One well-known example is the leitmotif, a short musical phrase used to evoke a particular person, and popularized through its use in Wagner's operas. Another example is the ballad. An entire movie by the Coen Brothers is entitled The Ballad of Buster Scruggs; the Ballad of Jesse James has been performed by both Johnny Cash and the Bruce Springsteen Band.

Musculoskeletal Demands. Arguably, most modes of musical behavioral expression involve execution of visuomotor tasks of different types. For example, in his review of the musical proficiency of an experimental choral group (specializing in performing music from a broad range of different cultures), Bilger (2019) points to the diverse types of skilled facial, tongue, laryngeal, chest, and diaphragmatic muscle control required for high-level musical expression. One remarkable illustration is maintaining a steady tone through both exhalation and inhalation.

Repetitive movements of different effectors are central to the performance of some of these tasks, such as piano or violin playing. As a consequence, the musculoskeletal demands of such tasks pose an appreciable risk of musculoskeletal injuries to the performer (Boyette 2003; Chesky et al. 2002; Horvath 2014).

Horvath's study of negative practice habits notes that the formation of these habits is a larger issue than we think. In her research, she gives surprising statistics about the pain a large number of pianists feel while playing, with over 30% reporting pain in their right side. She pinpoints the cause to simple fundamentals that many musicians ignore when playing at professional levels. In particular, reducing tension, and pain while playing is linked to correct posture and regular breaks. Horvath also talks about the great amount of musculoskeletal repetition required of musicians. As both Boyette and Chesky note, continued repetitive strain can lead to serious repetitive strain injury.

Horvath goes over correct posture for pianists, such as keeping shoulders back and tension free, keeping the head level, and, if sitting, making sure the knees are below the hips to avoid awkward bending of the body. Given that many children who are introduced to music begin with piano, her focus on piano is a major positive in terms of reaching a wider audience of young musicians and their parents. These simple guidelines also work for other instrumentalists who may experience pain in their practice. For one of us (McNiven), a low brass player, following these simple instructions helps improve breathing and keeps shoulder and back fatigue from setting in too quickly in a longer practice sessions.

This summary of how musical context places demands on musculoskeletal control accentuates a link with both context specificity and neuroergonomics of musical expression.

Natural Environmental Sounds. It is tempting—yet reasonable—to suggest that sounds in the natural environment represented the original inspiration for human musical behavior. Bird song represents an obvious example (after all, the term "song birds" is applied to a number of species). Rood and colleagues (2017) suggest that bird song, as well as calls of other animals, can (at least in some cases) plausibly be equated with music. Ottorino Respighi referenced his composition, "The Birds," to bird song. Other natural environmental sounds may provide inspiration. Thus, Beethoven used the sounds of a thunderstorm as a source for his 6th (Pastoral) symphony.

Musical Structure. A major design influence for many centuries on musical expression has been so-called "rules" of musical structure (Benward and Saker 2015; Burholder et al. 1960). These arguably date back some 2500 years to Pythagoras, who conceived and elaborated the mathematical rules of musical harmony. Maor (2018) devotes an entire book to exploring the relationships between music and mathematics. He quotes Thelonious Monk: "*All musicians are subconsciously mathematicians.*" Harmonic musical expression more or less held sway until the twentieth century, although dissonance in music has been documented as early as the late sixteenth century (The Economist 2018b, p. 40). A major musical departure from harmony was the 12-tone scale, pioneered by composers such as Schoenberg. These days, a debatable assertion regarding musical structure is, "anything goes."

Religion. Religion, as a human systems design, has influenced music in two major related ways, through sponsorship of musical composers, and through inspiration for religious music. Bach is an example of the former—he spent all of his career in the employ of religious institutions. As for its inspirational influence, religious belief has played a central role in musical expression. During Judaic religious services, a musician (the Cantor) participates in leading the service; tradition has it that since earliest times, the Torah was chanted not spoken during recitation. Throughout history (and likely prehistory) music has assumed near-universal importance in the religious sphere (Ellingson 1987).

Rhythm. Rhythm is considered here as musical design feature distinct from structure (above) because of its remarkable effect not only musical behavioral expression, but on kinesiological and affective behavior. In 1159, John of Salisbury offered an early observation of the latter by noting, "effete emotings [of polyphonic singers] can more easily occasion titillation between the legs than a sense of devotion in the brain" (The Economist 2018b, p. 39). Mithen (2005) observes that music compels us to move. Music is the intimate soulmate of most dance. Stravinsky's visceral rhythms in his ballet "The Rite of Spring" imagine primitive rituals celebrating the advent of spring, after which a young girl is chosen as a sacrificial victim and dances herself to death—who are we to say that this is pure fiction. One only has to attend a concert by a modern rock star, with its thousands of screaming fans, to experience this effect directly. No other art form, or indeed no other mode of auditory feedback, evokes a comparable effect. Why should this be? The next section offers some insight, but no definitive answer.

Training. In popular imagination, musical genius is marked by the almost magical ability to produce fully realized music without apparent a priori effort. In the movie Amadeus, the Prince-Archbishop of Salzburg asks Mozart where the opera he has commissioned is. Mozart's reply: "In my noodle." Recent musicology casts doubt on this account. Much of Mozart's work was sketched out, or even improvised on a keyboard; he is thought to have done little composition without one (The Economist 2018c, p. 98). Gladwell (2008, pp. 53–56), in his book Outliers, notes that prior to their exceptionally smooth and accomplished performance on the Ed Sullivan Show (that launched their remarkable success), the Beatles had played an estimated twelve hundred concerts in Hamburg, Germany, five or more hours per night. Gladwell's rule of thumb is that for anyone to achieve recognition as an elite performer (in music or any other profession), at least ten thousand repetitions or hours of training will be required. In other words, popular imagination about musical genius likely is just that—training is key to success as a musical performer or composer.

Sensory Feedback Control. Our premise in introducing this factor is that neural plasticity relies not only on the provision of sensory feedback from the performance environment (input), but central nervous system control of that feedback (closed-loop control). There is an instructive observation from music performance supporting this idea, related to control of auditory feedback. Using the first laboratory computer system implemented for closed-loop control of different modes of sensory feedback to human subjects, K. U. Smith explored the effect of delayed auditory feedback (Smith 1962, Chap. 3) on musical performance. A small instrumental ensemble was asked to perform under conditions of computerized auditory feedback delay on the order of 200 ms. Within a minute, the group stopped and refused to continue, claiming that further performance under delay would inflict lasting damage on their ability to perform (K. U. Smith, personal communication) (disclosure: K. U. Smith was the father of the first author of this chapter). Their concerns likely were well founded, given that stuttering (often persisting for a period of time) is a well-documented aftereffect of exposure to delayed auditory feedback of speech (Smith 1962).

The larger implication of this finding relates to how we are to understand the interaction of the brain and musical performance. Basic neurophysiological principles suggest that closed-loop neural control of proprioceptive feedback should not be possible during performance of piano or violin compositions requiring extremely rapid finger and/or hand-arm effector movements. A prevailing alternative interpretation of rapid movement control is the motor program hypothesis, that assumes that a motor schema in the brain, in a one-way manner, controls rapid motor movements (Schmidt and Lee 1999). The findings described above imply that auditory feedback represents a critical source of sensory information for control of musical performance, regardless of the rapidity of effector movements required. The broader implication is that the phenomenon of context specificity in musical behavior and performance is inextricably linked to the nature of the auditory behavioral and performance environment.

In summary, this section addresses the proposition that variability in the expression of musical behavior is influenced by design factors in the environment in which such behavior occurs. A series of such factors are introduced in preceding paragraphs. The underlying functional basis for this idea is termed context specificity mediated by neural plasticity, an adaptive mechanism by means of which the brain is able to update its functional capabilities to adjust to and control behavioral interaction with the environmental exigencies confronting the behaving individual. Since environmental design conditions are continuously in flux, neural plasticity represents a highly dynamic process.

The idea of neural plasticity is not new—an early advocate was William James well over a century ago (Berlucchi and Buchtel 2009). As noted previously, Smith et al. (2015) survey a series of types of behaviors for which context specificity in behavior has been documented. Of particular relevance for this chapter are task behavior during actual work situations, and learning behavior. Regarding the former, Jones (1966, 1969) concludes that in the case of actual work performance, the variance attributable to the specific design of the task situation is even greater than that found for experimental test situations, amounting usually to at least 75% of the total variance, frequently to as much as 90%. Regarding the latter, Chap. 3 in Smith et al. (IBID) reviews cognitive performance and differential learning research dating back over a century pointing to a similar conclusion that cognitive specialization during learning also is highly context specific. These findings are directly applicable to musical behavior, in that the work of musical composition or performance is a task-based process, and learning during training has been linked to the achievement of musical proficiency.

Nevertheless, based on contemporary accounts, the paradigm of neuroergonomics has yet to recognize the centrality of context specificity and neural plasticity for purposes of understanding the nature and sources of variability in behavior. Studies of musical behavior to date display a comparable lack of understanding. Based on these conclusions, it might be argued that the perspective introduced in this section is both necessary and sufficient to redress these shortcomings. Nevertheless, musical behavior is of particular interest because there is compelling evidence pointing to a high degree of cognitive specialization—brain centeredness if you will—among individuals who exhibit exceptional musical proficiency and talent. In other words, musical behavior displays the properties of both context specificity and centralized neural control in terms of its expression. Evidence for the latter is in line with the predictions of neuroergonomics, and is reviewed in the next section.

4 Neuroergonomics of Musical Behavior—Cognitive Specialization in Music Proficiency and Talent

This section offers a rather nuanced answer to the third question posed at the outset of the introduction. Question 3: *Does musical behavior conform to the predictions of neuroergonomics and as such, represent an appropriate topic for this text?* In particular, available evidence points to a role of brain specialization in musical behavior, but evidence bearing on this question is not completely definitive.

Our consideration of the role of brain specialization in musical behavior rests on the phenomenon of brain lateralization. It has long been known that linguistic functions—i.e., those mediating language—are highly lateralized, localized primarily (but not exclusively) to the left hemisphere (Robinson 1997, Lecture 26). As noted previously, some observers favor that notion that musical behavioral expression can be understood as a type of language. As such, is musical behavior also lateralized?

A series of analyses on this question concludes that the findings are not entirely consistent. A study over forty years ago on cerebral dominance in musicians and nonmusicians by Bever and Chiarello (1974) found that musically experienced listeners recognize simple melodies better in the right ear than the left, while the reverse is true for naive listeners.

Prompted by their insight that the nature of hemispheric specialization and locus of brain activity during rhythm processing remains poorly understood, Limb et al. (2006) studied passive rhythm perception to investigate the hypotheses that rhythm processing results in left-hemispheric lateralization of brain activity and is affected by musical training. In contrast to the Bever and Chiarello conclusions, their findings suggest that musical training leads to localized activation of left-sided brain areas, also associated with language comprehension, during passive rhythm perception.

However, a study of a patient with a particular brain lesion points to right-brain localization of music perception. Sparr (2002) reports a clinical study of a trained musician patient who was unable to identify the melody of any of a wide variety of well-known musical pieces presented by recording or live by piano. He also did not recognize relatively simple tunes played on a single instrument, or vocal music which contained the additional clues of verbal lyrics. This condition, termed amelodia (alternatively, amusia (Bispham 2006)), was linked to right temporal lobe injury.

David (1994) offers a useful perspective on the dual hemispheric nature of music perception. He begins by pointing out that different so-called mental functions also display dual hemispheric laterality (Table 2). Likewise, the same is true for different parameters of music perception (Table 3). Note that this analysis of left-brain laterality for rhythm perception agrees with the later results of Limb et al. (2006).

When it comes to understanding the neural basis of exceptional musical talent, studies by K. U. Smith on facial laterality offer interesting insight. As part of a general program of movement research, and inspired by the studies by Duchenne on the human face over 150 years ago (Duchenne 1862), Smith initiated analysis of facial movements in the 1950s. This work led to recognition that all persons have distinctive patterns of facial laterality, just as they have distinctive patterns of handedness. The term "facedness" was adopted to describe relatively greater muscular activity of one side of the face during speech (Smith 1984; Smith et al. 1991a). Thus, a person is designated as being right faced if muscles on the right side of the face are more active in speaking.

Table 2 Hemispheric laterality for selected mental functions

Right hemisphere	Left hemisphere
Color	Logic
Images	Numbers
Insight	Language
Dimension	Analysis
Imagination	Linearity
Daydreaming	Reasoning
Left-hand control	Right-hand control

Table 3 Hemispheric laterality for different parameters of music perception

Right hemisphere	Left hemisphere
Melody	Form
Timbre	Tempo
Emotion	Rhythm
Loudness	Reading
Intervals	Writing
Creativity	Sequence
Intonation	Analysis

In addition, these observations also indicated that dynamic movement indicators of facial laterality typically are accompanied by characteristic static features (Smith 1984; Smith et al. 1991b). In particular, relative to the subordinate side of the face, the dominant side of the face:

(1) is larger (wider and longer) between the brow and the jaw;
(2) is fatter and more developed muscularly;
(3) is more active in articulating lip-produced consonants (m, p, b, etc.);
(4) is smoother and less furrowed by wrinkles;
(5) has a higher elevated brow;
(6) shows the mouth drawn further back during smiling; and
(7) is displayed (turned) more toward the listener during talking.

This observational protocol was used to examine the distribution of facedness among various groups of people. Extensive surveys of sample populations representing six different nationalities (U.S., Canada, Spain, Mexico, Norway, Germany) established that 65–88% of individuals are judged to be right faced, whereas 12–30% are judged to be left faced.

However, talented musicians—opera singers, symphony instrumentalists, jazz artists, composers, conductors—are almost without exception judged to be left faced. Among the great classical composers, Beethoven, Brahms, Handel, Liszt, Rachmaninoff, Schubert, Tchaikovsky, and Wagner all are judged to have been left faced.

To enable a more objective assessment of facial laterality, William Smith at Dartmouth University implemented a computerized method for analysis of video images of the face (disclosure: William Smith was the uncle of the first author of this chapter). The method measured differences in the lateral dimensions of the two sides of the face, based on the hypothesis that bilateral facial asymmetry arises from differential muscular development of the right and left sides of the face (just as the dominant hand-arm typically is more muscular than the subordinate hand-arm). Photographs of 16 individuals were evaluated using this technique. Subjects were selected primarily on the basis of the availability of useable frontal photographs; they include ordinary individuals, some well-known public figures, and eight established opera singers.

Measures were made by digitizing photographs of individuals and displaying the images on a high-resolution monitor. The midpoint between the pupils of the eyes is identified on the monitor display, and image manipulation techniques are used to guide halving of the face in the vertical dimension, creating two half-face images (termed hemi-faces) for each individual.

Using these half-face images, two versions of each face in addition to the normal image were composed on the monitor: a "left" composite image consisting of the two left sides of the face, and a "right" composite image consisting of the two right sides of the face. The width of each of the three facial images is measured across the tip of the nose to the two sides of the face. Measurements are taken from images displayed on the center of the monitor and thus do not represent actual facial dimensions.

Figures 3 and 4 illustrate representative results from this facial laterality video analysis technique. In each figure, the normal face (N) is in the center, the composite

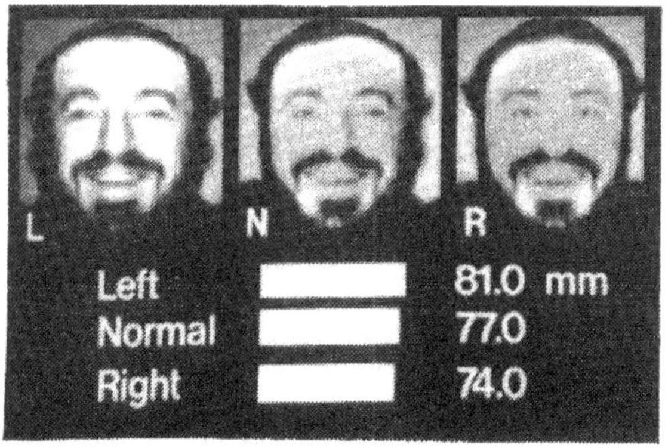

Fig. 3 "Left" (L) composite, normal (N), and "right" (R) composite faces of Luciano Pavarotti, with bar graph indication of the lateral width of each image at the level of the nose (Courtesy W. M. Smith)

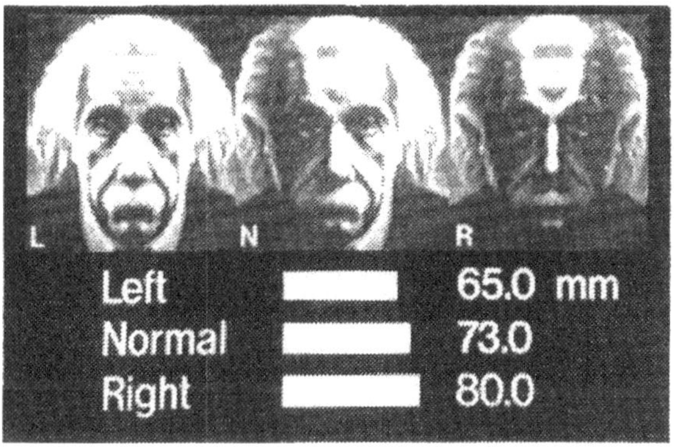

Fig. 4 "Left" (L) composite, normal (N), and "right" (R) composite faces of Albert Einstein, with bar graph indication of the lateral width of each image at the level of the nose (Courtesy W. M. Smith)

"left" hemi-face (L) to the left, and the composite "right" hemi-face (R) to the right. The bar graphs show the lateral widths of each image, in mm (measured on the monitor), across the tip of the nose.

Luciano Pavarotti, depicted in Fig. 3, one of the great operatic tenors of our age, is distinctly left faced—the left composite face is 7 mm wider than the right composite face. Albert Einstein, depicted in Fig. 4, a renowned physicist and mathematician, is distinctly right faced—the right composite face is 15 mm wider than the left. This 20.5% right laterality represents the greatest right-facedness found among the 16 individuals evaluated.

In each set of images, the fuller and more robust appearance of the composite image of the dominant side of the face is readily apparent.

The Pavarotti images suggest an intimate feedback link between neuromuscular control of the left face and musical talent, widely considered to represent one of the most sublime of cognitive skills. They also are in agreement with the summary tabulation by David (1994) of hemispheric laterality for different parameters of musical perception—as Table 3 indicates, perceptions of melody, timbre, and intonation are lateralized to the right hemisphere.

The Einstein images suggest a direct feedback link between the cognitive skills of speech fluency and mathematical ability, and neuromuscular control of the right face. They also are in agreement with the summary tabulation by David (1994) of hemispheric laterality for selected cognitive functions (Table 2)—language, numbers, and logic capabilities are lateralized to the left hemisphere. That almost all U.S. presidents, leading politicians at the state and national levels, and leading television newscasters are right faced, based on subjective judgments of facedness (Smith 1974; Smith et al. 1991b), provides additional support for hemispheric lateralization of speech (language).

More generally, the facedness research described above confirms and extends the early viewpoint of Duchenne (1862) that muscular actions of the face mirror the cognitive workings of the brain.

5 Discussion and Conclusions

The analysis offered in this chapter introduces two perspectives regarding the origins and nature of variability in musical behavior. A series of design factors are specified in Sect. 3 that are assumed to influence the expression of music behavior and performance, a process termed context specificity linked to brain plasticity. The following section, in line with the assumptions of neuroergonomics, addresses a series of observations supporting the role of cognitive specialization mediated by brain lateralization in influencing music behavior and performance, as well as the expression of musical talent.

This chapter thus offers conditional answers to the three questions posed in the Introduction (Sect. 1). Various viewpoints, plus available evidence, offer a mixed perspective on the question of whether neuroergonomics offers a legitimate scientific framework for understanding variability in musical behavior? Design factors prominently influence all modes of observed variability in musical behavioral expression—the idea of context specificity thus complements the neuroergonomic perspective. Nevertheless, music also represents the only mode of artistic expression for which concrete empirical evidence exists for the neuroergonomic assumption that proficiency in musical talent must be understood in terms of brain specialization.

What academic discipline is best equipped to enhance our scientific understanding of music behavior? Our answer to this question is the field of ergonomics/human factors (E/HF). Yet for a number of observers, the answer is the field of psychology (Bispham 2006; Mithen 2005; Patel 2008; Slevc and Okada 2015). In support of our view, we point out that:

- Psychology has been consistently devoted to a brain-centered understanding of learning, an approach termed cognitive theory. Educators who embrace cognitive theory study the learner rather than their environment (Mandler 2002). This approach has given rise to educational neuroscience, an emerging scientific field that brings together researchers in cognitive neuroscience, developmental cognitive neuroscience, educational psychology, educational technology, education theory, and other related disciplines to explore the interactions between biological processes and education (Ansari and Coch 2006). Yet the cognitive theory of learning disregards a variety of evidence that design factors in the learning environment prominently influence observed variability in learning performance, a field of study termed *educational ergonomics* (Smith and Smith 1966; Smith 2007, 2013; Smith et al. 2015, Chap. 4).

- In contrast, since its inception, the field of E/HF has emphasized an understanding of behavior and performance that emphasizes the interaction of performance and design, an integrated approach termed context specificity, articulated in the following characterization of the field:

> Ergonomics (or human factors) is the scientific discipline concerned with the understanding of interactions among humans and other elements of a system, and the profession that applies theory, principles, data, and other methods to design in order to optimize human well-being and overall system performance (International Ergonomics Association 2019).

The scientific framework of E/HF thus lends support to our assumption that context specificity offers an important, if not essential, perspective for understanding musical behavior and performance. Indeed, Moray (1994) has asserted that "It is very doubtful whether there are any context-free laws of behavior…" Aligned with this view, a 1/10/2019 search of the journal *Human Factors* documented 1181 references to the terms "context" and "context specificity" in all issues dating back to 1958, and 3801 references to the same two terms in the journal *Proceedings of the Human Factors and Ergonomics Society Annual Meeting* in all issues dating back to 1974.

Nevertheless, it is fair to say that music behavior has not represented a major focus of concern for the field of E/HF. Thus, in the titles of published reports, the term "music" is only mentioned: (1) 4 times in a total of 245 issues of the journal *Human Factors*, published between 1958 and 2019; (2) 15 times in a total of 54 issues of the journal *Proceedings of the Human Factors and Ergonomics Society Annual Meeting*, published between 1974 and 2018; (3) zero times in 109 issues of the journal *Ergonomics in Design: The Quarterly of Human Factors Applications*, published between 1993 and 2019; and (4) zero times in 10 issues of *Reviews of Human Factors and Ergonomics*, published between 2005 and 2015. Furthermore, we are not aware of any publication prior to this chapter that considers the relevance of context specificity to the analysis of music behavior and performance.

This seems a rather curious shortcoming, considering the ubiquity of music behavior and performance for human experience. Music has ancient origins, extending back tens of thousands of years. It spans cultures, continents, ethnicities, and civilizations around the world. It evokes an extraordinary range of kinesiological and emotional responses, ranging from the most rabid raves to the most tender expressions of affection. In posing the question, "Why music?" The Economist (2008) cites three hypotheses as to the origins of music: (1) it serves as a sexual attractant (Shakespeare claimed that music is one of the foods of love); (2) it serves to promote group cohesion (see discussion of social tracking in Sect. 3); and (3) it emerged as a consequence of abilities that emerged for other purposes. The first two hypotheses have evolutionary implications. In fact, Bispham (2006) concludes his report by noting that "...music is an essential biological and cultural component of who we are. No evolutionary story can be complete without it!" No other mode of artistic expression comes close to matching the scope of this influence and of these effects.

Collectively, considering the argument in this chapter that music behavior has both context-specific and neuroergonomic connotations, coupled with the foregoing observations, we conclude that the time is long overdue for the field of E/HF to embark on an intensive and comprehensive investment in research and analysis on music behavior and performance.

Our consideration of music in this chapter is by no means definitive. Indeed, it can be argued that our understanding of music itself is by no means definitive (The Economist 2008). A series of unanswered questions that occur to us include: (1) Can an empirical case be made for the role of context specificity in influencing variability in music behavior and performance? (2) If so, what environmental design factor has the greatest influence on such variability? (3) Are there specific genetic and neural developmental substrates that account for the expression of musical behavior, and particularly the capability for extraordinary musical talent; (4) Given evidence for right hemispheric laterality for a number of parameters of music perception (associated with left-facedness), why is perception of rhythm (a key musical element) associated with left hemispheric laterality? (5) What are the underlying biological mechanisms that account for the dramatic kinesiological and emotional effects of music? (6) Does music represent a type of language?—And if so, is there a universal musical grammar, in parallel with Chomsky's claim for a universal linguistic grammar? And (7) Will we ever arrive at a definitive account for the evolutionary emergence of music?

We close this chapter with a consideration of an arguably important attribute of music that is not addressed in the three reviews of music cited in the references (Bispham 2006; David 1994; The Economist 2008), namely the idea of musical quality. Quality is one of those terms for which we have definite perceptual understanding, but we cannot really define what it means (Smith 1999). The second author of this chapter, an accomplished musician who has dedicated his career both to musical performance and to academic inquiry, offers the following perspective.

> How does one judge the quality of music? Before that question can be answered we must understand music in its current form today. Until Beethoven's time, music was written mainly for religious services, entertainment, or for patrons who could give composers sustainability in life. However, Beethoven is one of the most popular names associated with the idea of absolute music, though the term wasn't coined until 1846 in Wagner's description of Beethoven's 9th symphony. This idea opened a new way of writing music. Simply put, music was written purely to be music; "Music for Music's Sake." This is when we get composers who begin to write true masterworks and virtuosic pieces which we would refer to as quality music today.
>
> With the daunting amount of musical styles that exist in the world, this question seems near impossible to answer in a way that appeases all genres. As a classically trained musician, I've found that defining the quality of music cannot be done with a single characteristic. As such, quality music, regardless of genre, has many defining traits; two of the most important I

find myself focusing on being consistency and expression. When exposed to virtuosos in the field of music, we look at them as people who can pick up an instrument and play whatever and however they please. Although, we only see the tip of the iceberg. The impressive performances we're graced with by musicians are preceded by countless hours of practice and hard work.

One major discovery I've had during my Master's work has been the importance of consistency. As impressive as it is to listen to great musical feats, careers are formed by those who can perform at a level equal to those feats always. In the competitive performance world, bad performances can mean the loss of a job in an instant with 30 people at the ready to take your position. Each day I pick up my instrument I have certain sets of warm-ups I alternate between to prepare me for the day of playing. These warm-ups range from holding long, extended notes to scale work meant to increase finger speed and flexibility between high and low range. When it comes to playing music for a performance you must approach the music with clear-cut ideas. For one recital, I began work on the pieces three months previously. Before playing the pieces, I listened to several recordings to answer a couple of questions: What do I want the music to sound like? and How do I make myself sound like that all the time? I've spent countless hours working on a 30-minute program to achieve these goals with work still yet to come.

Amateurs practice until they get it right; professionals practice until they can't get it wrong."—Unknown. This quote, which has many iterations, describes what consistency in music looks like. However, consistency in and of itself does not lead to quality music. Expression is what helps makes music more than dots and lines on a page. Expression in music leads to standing ovations after concerts. Whether playing program music (telling a specific story) or absolute music, the goal of a musician is to evoke a response from the listener. Music is filled with markings by composers and performers that help create these responses such as dynamics, articulations, expressive words and so on. However, the musician can include anything from physical movement to facial expressions to help in their ambitions. I attended a concert where a very fine violinist performed Dmitri Shostakovich's Violin Concerto, a piece not to be played by the faint of heart. After his final note the auditorium immediately shot up to applaud him. As an encore, he performed a movement from one of Bach's sonatas. While much easier and simpler than Shostakovich's concerto, he performed with the same energy and intention, which made the piece as enjoyable and impressive as the 40-minute concerto he had just completed. It just as easily could have ruined his performance had he not put in the same level of effort.

This only begins to scratch the surface of what makes quality music, but I see it as the same answer to the question of what makes a quality musician. In a world where musicians must continually adapt, I see consistency and expression as necessary components which help create a strong foundation of quality in music.

References

Ansari, D., & Coch, D. (2006). Bridges over troubled waters: Education and cognitive neuroscience. *Trends in Cognitive Sciences, 10*(4), 146–151.

Benward, B., & Saker, M. (2015). *Music in theory and practice* (9th ed., Vol. I). New York: McGraw-Hill.

Berlucchi, G., & Buchtel, H. A. (2009). Neuronal plasticity: Historical roots and evolution of meaning. *Experimental Brain Research, 192*(3), 307–319.

Bever, T. G., & Chiarello, R. J. (1974). Cerebral dominance in musicians and nonmusicians. *Science, 185*(4150), 537–539.

Bilger, B. (2019). Extreme range. The vocal experiments of Roomful of Teeth. *New Yorker* (February 11), *XCIV*(48), 44–53.
Bispham, J. (2006). "Music" means nothing if we don't know what it means. *Journal of Human Evolution, 50*(2), 587–593.
Boyette, J. (2003). Instrument ergonomics: treatment and prevention of injury in the musician. In *Proceedings of the Human Factors and Ergonomics Society 47th Annual Meeting* (pp. 1582–1585). Santa Monica, CA: Human Factors and Ergonomics Society.
Burholder, J. P., Grout, D. J., & Palisca, C. V. (Eds.). (1960). *A history of western music* (9th ed.). New York: Norton.
Chesky, K., Devroop, K., & Ford, J., III. (2002). Medical problems of brass instrumentalists: prevalence rates for trumpet, trombone, French horn, and low brass. *Medical Problems of Performing Artists, 17,* 93–98.
Daly, I., & Williams, D. (2019). Case studies of applications that use neural technology to measure emotions. In C. Nam (Ed.), *Neuroergonomics: Principles and practice*. New York: Springer.
David, J. H. Jr. (1994). *The two sides of music*. Austin, TX: Austin Community College. Retrieved January 09, 2019 from http://jackhdavid.thehouseofdavid.com/papers/brain.html.
Duchenne, G. B. A. (1862). *M ecanisme de la physionomie humaine: ou analyse electro-physiologique de/'expression des passions applicàbble a la pratique des arts plastiques*. Paris: Libraire de Jules Renouard.
The Economist. (2008). Why music? (Biologists are addressing one of humanity's strangest attributes, its all-singing, all-dancing culture). *The Economist* (December 20), 41–44.
The Economist. (2018a). The burden of history. *The Economist* (December 8), 79–80.
The Economist. (2018b). Dona nobis pacem. *The Economist* (December 22), 39–40.
The Economist. (2018c). Sweet and serious songs. *The Economist* (December 22), 95–98.
Ellingson, T. (1987). Music: Music and religion. *Encyclopedia of Religion*. Farmington Hills, MI: Thomson Gale. Retrieved January 07, 2019 from Encyclopedia.com: www.encyclopedia.com/environment/encyclopedias-almanacs-transcripts-and-maps/music-music-and-religion.
Fischer, E. K., Ghalombor, C. K., & Hoke, K. L. (2016). Can a network approach resolve how adaptive vs nonadaptive plasticity impacts evolutionary trajectories. *Integrative and Comparative Biology*, *56*(5), 877–888.
Gladwell, M. (2008). *Outliers. The story of success*. New York: Little, Brown and Company.
Goodall, H. (2015). *The story of music from Babylon to the Beatles: How music has shaped civilization*. Plano, TX: Pegasus Classics.
Harmand, et al. (2015). 3.3-million-year-old stone tools from Lomekwi 3, West Turkana, Kenya. *Nature, 521,* 310–315.
Higham, T., Basell, L., Jacobi, R., Wood, R., Ramsey, C. B., & Conard, N. J. (2012). Testing models for the beginnings of the Aurignacian and the advent of figurative art and music: the radiocarbon chronology of Geißenklösterle. *Journal of Human Evolution, 62*(6), 664–676.
Horvath, J. (2014). Playing healthy staying healthy. No pain, all gain: Strategies for healthy and happy musicians. *American Music Teacher* (October/November), 26–30.
International Ergonomics Association. (2019). Retrieved January 11, 2019 from www.iea.cc/whats/index.html.
Jones, M. B. (1966). Individual differences. In E. A. Bilodeau (Ed.), *Acquisition of skill* (pp. 109–146). New York: Academic Press.
Jones, M. B. (1969). Differential processes in acquisition. In E. A. Bilodeau & I. M. Bilodeau (Eds.), *Principles of skill acquisition* (pp. 141–170). New York: Academic Press.
Limb, et al. (2006). Left hemispheric lateralization of brain activity during passive rhythm perception in musicians. *The Anatomical Record Part A, 288A,* 382–389.
Mandler, G. (2002). Origins of the cognitive (r)evolution. *Journal of the History of the Behavioral Sciences, 38,* 339–353.
Maor, E. (2018). *Music by the numbers: From pythagoras to schoenberg*. Princeton, NJ: Princeton University Press.

Merriam-Webster. (2019). Retrieved January 06, 2019 from www.merriam-webster.com/dictionary/music.

Mithen, S. (2005). *The singing Neanderthals: The origins of music, language, mind and body*. London: Weidenfeld and Nicolson.

Moore, K. (2019). First abstract art. *Natural History, 127*(1), 7.

Moray, N. (1994). "De maximis non curat lex"; or How context reduces science to art in the practice of human factors. In *Proceedings of the Human Factors and Ergonomics Society 38th Annual Meeting* (pp. 526–530). Santa Monica, CA: Human Factors and Ergonomics Society.

Parasuraman, R. (2003). Neuroergonomics: Research and practice. *Theoretical Issues in Ergonomics Science, 4*, 5–20.

Patel, A. D. (2008). *Music, language, and the brain*. Oxford, UK: Oxford University Press.

Pinker, S. (1997). *How the mind works*. London: Allen Lane.

Robinson, D. N. (1997). *The great ideas of psychology*. Chantilly, VA: The Teaching Company Great Courses.

Rood, J., et al. (2017). Song around the animal kingdom. *The Scientist, 31*(3), 34.

Schmidt, R. A., & Lee, T. D. (1999). *Motor control and learning. A behavioral emphasis* (3rd ed.). Champaign, IL: Human Kinetics.

Slevc, L. R., & Okada, B. M. (2015). Processing structure in language and music: A case for shared reliance on cognitive control. *Psychonomic Bulletin & Review, 22*, 637–652.

Smith, K. U. (1962). *Delayed sensory feedback and behavior*. Philadelphia: Saunders.

Smith, K. U. (1972). Cybernetic psychology. In R. N. Singer (Ed.), *The psychomotor domain* (pp. 285–348). New York: Lea and Febiger.

Smith, K. U. (1974). *Industrial social cybernetics*. Madison, WI: University of Wisconsin Behavioral Cybernetics Laboratory.

Smith, K. U. (1984). "Facedness" and its relation to musical talent. *Journal of the Acoustical Society of America, 75*(6), 1907–1908.

Smith, K. U., & Smith, M. F. (1966). *Cybernetic principles of learning and educational design*. New York: Holt, Rinehart and Winston.

Smith, K. U., Smith, T. J., & Smith, W. M. (1991a). Computerized research on the human face (part I of 2 part article). *Electric Quarterly, 13*(1), 2–3.

Smith, K. U., Smith, T. J., & Smith, W. M. (1991b). Computerized research on the human face (part II of 2 part article). *Electric Quarterly, 13*(2), 2–3.

Smith, T. J. (1999). Synergism of ergonomics, safety, and quality—A behavioral cybernetic analysis. *International Journal of Occupational Safety and Ergonomics, 5*(2), 247–278.

Smith, T. J. (2007). The ergonomics of learning: Educational design and learning performance. *Ergonomics, 50*(10), 1530–1546.

Smith, T. J. (2013). Designing learning environments to promote student learning–ergonomics in all but name. *Work: A Journal of Prevention, Assessment & Rehabilitation, 44*, S39–S60.

Smith, T. J., Henning, R. H., & Smith, K. U. (1994). Sources of performance variability. In G. Salvendy & W. Karwowski (Eds.), *Design of work and development of personnel in advanced manufacturing* (Chapter 11, pp. 273–330). New York: Wiley.

Smith, T. J., & Smith, K. U. (1987). Feedback-control mechanisms of human behavior. In G. Salvendy (Ed.), *Handbook of human factors* (Chapter 2.9, pp. 251–293). New York: Wiley.

Smith, T. J., Wade, M. G., Henning, R., & Fisher, T. (2015). *Variability in human performance*. Boca Raton, FL: CRC Press.

Sparr, S. A. (2002). Receptive amelodia in a trained musician. *Neurology, 59*(10), 1659–1660.

Somers, J. (2018). Binary stars. *The New Yorker, XCIV*(40), 28–35 (December 10).

Wilford, J. N. (2012). Flute's revised age dates the sound of music earlier. *New York Times*, D4 (May 29).

Brain-Computer Interfaces and Neuroergonomics

Hybrid EEG–fTCD Brain–Computer Interfaces

Aya Khalaf, Ervin Sejdic and Murat Akcakaya

Abstract This chapter introduces two novel hybrid brain–computer interfaces (BCIs) based on electroencephalography (EEG) and functional transcranial Doppler ultrasound (fTCD). Noninvasive BCIs based on EEG have become popular due to cost-effectiveness, high temporal resolution, and portability of EEG recording devices. However, the performance of the EEG-based BCIs prevents them from a consistent use by the target population, which usually includes individuals with limited speech and physical abilities. Various neuroimaging modalities that measure different brain activities have been used with EEG to improve the speed and accuracy of the noninvasive BCIs. The two hybrid BCIs introduced in this chapter are developed as alternatives to the existing systems, and they were shown to balance between speed and accuracy and outperform the existing hybrid BCIs. Both systems measure electrical brain activity as well as cerebral blood flow velocity using EEG and fTCD, respectively. For these two systems, two different visual presentation paradigms are used to induce simultaneous changes in EEG and fTCD. The visual stimuli in the first system include two objects flickering with different frequencies instructing word generation (WG) and mental rotation (MR) tasks, and the presentation in the second system includes two arrows instructing left and right motor imagery (MI) cognitive tasks. Experimental results show that the flickering MR/WG presentation outperforms the MI presentation by 10% accuracy for task versus task problem. However, the MI presentation outperforms the MR/WG one in terms of transmission rates.

1 Introduction

Brain–computer interface (BCI) systems are commonly developed to aid individuals with limited speech and physical abilities, and they enable means to bypass neuromuscular activity and to mentally control external devices that can compensate for speech and/or motor impairments (Nicolas-Alonso & Gomez-Gil, 2012; Wolpaw

A. Khalaf (✉) · E. Sejdic · M. Akcakaya
University of Pittsburgh, Pittsburgh, USA
e-mail: afk17@pitt.edu

C. S. Nam (ed.), *Neuroergonomics*, Cognitive Science and Technology,
https://doi.org/10.1007/978-3-030-34784-0_15

et al., 2002). Usage of such systems as novel rehabilitation and intervention techniques to restore the lost functionalities have also been demonstrated (Blank, French, Pehlivan, & O'Malley, 2014; Fager et al., 2011). Noninvasive BCIs are usually used to avoid risks of surgical procedures needed for invasive BCIs (Nicolas-Alonso & Gomez-Gil, 2012). Electroencephalography (EEG) which measures brain electrical activity is widely used to design noninvasive BCIs due to its high temporal resolution, portability, and low cost (Farwell & Donchin, 1988; Wolpaw et al., 2002). However, EEG is highly nonstationary, and it has low signal-to-noise ratio (SNR) negatively affecting its performance (Akcakaya et al., 2014). In order to improve both the speed and accuracy performance, hybrid BCI systems have been developed (Amiri, Fazel-Rezai, & Asadpour, 2013). One way to develop such hybrid BCI systems is to exploit multimodal signals, specifically supporting EEG with another modality that is measured simultaneously as the EEG (Muller-Putz et al., 2015; Pfurtscheller et al., 2010).

Even though there are various modalities that are used to support EEG for hybrid BCI design such as functional near-infrared spectroscopy (fNIRS) (Coyle, Ward, Markham, & McDarby, 2004), functional magnetic resonance imaging (fMRI) (Weiskopf et al., 2004), and magnetoencephalography (MEG) (Mellinger et al., 2007), fNIRS is the most commonly used modality for hybrid BCI design. This is mainly because fMRI and MEG are not portable, and hybrid BCIs built based on these modalities can only operate efficiently in controlled laboratory environments (Allison, Wolpaw, & Wolpaw, 2007). For instance, EEG and fNIRS modalities were used to design asynchronous hybrid BCI in which the BCI user does not need any cue to initiate communication with the machine (Buccino, Keles, & Omurtag, 2016). In this study, EEG and fNIRS data were recorded in response to motor execution tasks. Motor imagery (MI) tasks were employed for development of an EEG–fNIRS BCI in which occurrence of MI was detected using fNIRS modality, while MI type was identified using EEG modality (Koo et al., 2015). Moreover, many researchers exploited both motor execution and MI tasks to design hybrid EEG–fNIRS BCIs, and it was found that the data from both modalities due to different tasks complement each other (Blokland et al., 2014; Fazli et al., 2012). In another EEG–fNIRS study, four commands were decoded using mental arithmetic and motor execution tasks (Khan, Hong, & Hong, 2014).

Despite its popularity, fNIRS has relatively slow target response and the infrared transmission is difficult through hair (Min, Marzelli, & Yoo, 2010). In addition, the number of fNIRS sensors to be used is variable and depends on the application (Naseer & Hong, 2015). To overcome these limitations, we introduced functional transcranial Doppler ultrasound (fTCD) as an alternative to fNIRS. In our previous study, we have shown that compared to fNIRS, fTCD has a faster response time and less setup complexity since it uses only two ultrasound sensors to measure the cerebral blood velocity; and a BCI that is based on fTCD only can achieve accuracies of approximately 80 and 60% for binary and 3-class classification of different cognitive tasks within observation period of around 5 s (Khalaf, Sybeldon, Sejdic, & Akcakaya, 2018). Therefore, we argue that fTCD response time is comparable to EEG and it is a viable candidate to support EEG in a real-time BCI design.

In this chapter, we introduce two EEG and fTCD-based hybrid BCI designs. These two designs rely on detecting simultaneously induced changes in EEG and fTCD through two different visual presentation paradigms. In the first system, a visual paradigm will induce changes in cerebral blood flow (measured by fTCD) through the presentations of two stimuli that instruct word generation (WG) and mental rotation (MR) tasks, respectively. Simultaneously, the abovementioned two stimuli are designed to flicker with different frequencies and they induce steady-state visual evoked potential in EEG (Khalaf, Sejdic, & Akcakaya, 2018b). In the second system, the visual presentation paradigm includes two stimuli directing the users to perform left or right motor imagery (MI), and MI induces changes simultaneously in EEG and fTCD (Khalaf, Sejdic, & Akcakaya, 2018a). Both visual presentation paradigms include a fixation cross to represent the baseline, see Sect. 2 for the details of the visual paradigms. Healthy participants were recruited to test both systems experimentally. For both systems, features derived from the power spectrum for both EEG and fTCD signals were calculated to identify the significant frequency bands in EEG and fTCD signals. Mutual information and linear support vector machines (SVM) were employed for feature selection and classification.

2 Visual Presentation Paradigms

We introduce two visual presentation paradigms to simultaneously induce changes in EEG and fTCD. These two visual presentation paradigms are then used to develop two different hybrid EEG and fTCD-based BCIs.

2.1 Flickering MR/WG-Based Paradigm

The visual presentation paradigm includes two tasks and a fixation cross that represents the baseline. The presented tasks have to be differentiated by both EEG and fTCD modalities in order to obtain a successful hybrid BCI system. It was shown that word generation (WG) induces higher blood flow velocity in left middle cerebral arteries (MCAs), while the mental rotation (MR) induces bilateral activation enabling different responses fTCD (Myrden et al., 2011). However, such tasks cannot be distinguished using the EEG. Therefore, they cannot be directly employed in a hybrid EEG–fTCD system design. To build an efficient hybrid BCI in which cognitive tasks can be differentiated using both brain activity sensing modalities, we propose to combine the WG and MR tasks with steady-state evoked potential (SSVEP) paradigm such that WG and MR tasks include a flickering checkerboard texture as shown in Fig. 1a (Khalaf, Sybeldon, Sejdic, & Akcakaya, 2017). SSVEP-based BCIs have been extensively investigated for communication and control purposes (Bin et al., 2009; Lesenfants et al., 2014; Wang et al., 2017). In such SSVEP systems, visual stimuli with different flickering frequencies are used to elicit temporally matching

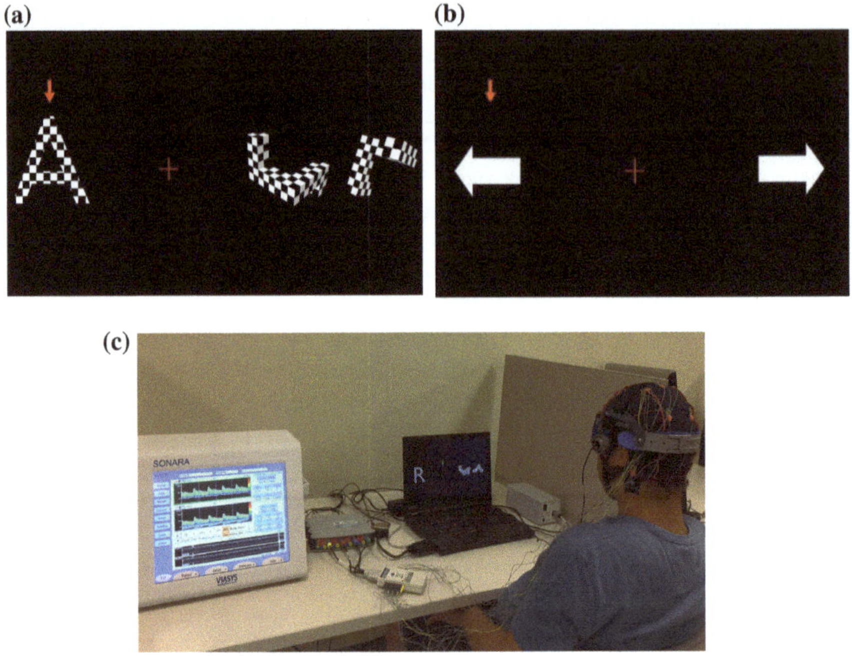

Fig. 1 Stimulus presentation for flickering MR/WG (**a**) and motor imagery (**b**) EEG–fTCD BCIs as well as the hybrid system setup captured during a data collection session (**c**). © IOP Publishing. Reproduced with permission from (Khalaf et al., 2018b)

electrical oscillations in the visual cortex (Wang et al., 2015). To issue a command to control an external device using SSVEP BCI, the user has to focus his/her attention on one of the visual stimuli shown on the screen. Recently, several studies focused on designing techniques to enhance the recognition of stimuli corresponding to the elicited SSVEPs. For instance, a joint frequency-phase modulation method was introduced to improve the differentiation between SSVEPs due to different stimuli (Chen et al., 2015). Another study suggested a novel spatial filtering approach known as task-related component analysis (TRCA) to enhance SSVEP detection (Nakanishi et al., 2018). On the other hand, other researchers focused on optimizing the stimulus time to enhance the performance of SSVEP BCI speller using fixed and dynamic optimization approaches (Yin et al., 2015a).

More specifically, in the developed presentation paradigm, the icons/visual stimuli that instruct the users to perform WG and MR tasks were textured with a flickering checkerboard pattern as shown in Fig. 1a to induce SSVEPs in EEG. For SSVEPs to be elicited, the flickering frequency of the stimuli has to be in the range from 7 to 60 Hz (Vanegas, Blangero, & Kelly, 2013). In addition, it was found that flickering frequencies higher than 20 Hz elicit SSVEPs with low amplitudes (Vanegas et al., 2013). Based on this information, WG and MR stimuli flickered with frequencies of 7 and 17 Hz. In addition, the system included a third class which is a fixation cross that represents the baseline. For flickering WG, a randomly chosen letter flickers on the screen in order to instruct the user to silently generate words starting with that letter. Flickering MR task is represented on the screen by a pair of flickering 3D similar shapes rotated with different angles, and the user is asked to mentally rotate the shapes to decide if they are identical or mirrored. These shapes were inspired from a database of 3D shapes constructed from cubes (Peters & Battista, 2008). The tasks were designed using Blender computer graphics software. During each trial, a vertical arrow points randomly to one of the three tasks for duration of 10 s and the user has to focus on performing the mental task specified by that arrow.

In summary, using the proposed presentation scheme, flickering checkerboard-textured tasks will induce SSVEPs in the EEG corresponding to the flickering frequency of each task leading to different EEG responses, while word generation and mental rotation will induce a different cerebral blood flow in the two hemispheres of the brain; therefore, they will generate distinct fTCD responses. The baseline EEG and fTCD will be recorded when the participants are performing no mental activity (while looking at the red cross located at the center of the screen).

2.2 *Motor Imagery-Based Paradigm*

Among the existing diverse EEG-based BCI categories, BCIs based on motor imagery have been intensively used in rehabilitation applications that seek assisting disabled individuals as well as restoring individual's physical and cognitive functions lost due to neural disorders (van Dokkum, Ward, & Laffont, 2015). Motor imagery (MI) is the process of movement imagination without any actual muscle activation. It was found that the MI process activates the same brain regions activated during the actual physical movement (Schlögl, Lee, Bischof, & Pfurtscheller, 2005). Therefore, during the rehabilitation process, patients with motor impairments practice MI process to activate the injured brain motor regions (Ang et al., 2010). Several studies were performed on both healthy participants and participants with physical and/or speech impairments to examine the feasibility of motor imagery for BCI applications (Ang & Guan, 2017; LaFleur et al., 2013; Schlögl et al., 2005). Toward motor recovery after stroke, several motor imagery BCIs with robotic feedback were developed (Ang et al., 2010; Blank et al., 2014). Such systems decode the motor imagery signals into robot-assisted movements, and it was shown that such systems yielded motor improvements.

It was found that the cerebral blood velocity in left and right middle cerebral arteries (MCAs) changes depending on whether the moving arm is the left or the right one (Matteis et al., 2006). Such findings suggest that fTCD might be promising for MI-based BCIs. Inspired by these findings as well as the results we achieved previously with fTCD as a candidate for real-time BCIs (Khalaf et al., 2018), we developed MI-based hybrid BCI that uses both EEG and fTCD modalities (Khalaf et al., 2018a). Such a system acquires the electrical activity of the brain using the EEG and the vascular response of the brain using the fTCD.

In the MI-based presentation scheme, a basic motor imagery task is visualized while acquiring EEG and fTCD simultaneously. As shown in Fig. 1b, the screen shows a horizontal arrow pointing to the right representing right-arm MI and another horizontal arrow pointing to the left representing left-arm MI as well as a fixation cross that represents the baseline. Each trial lasts for 10 s. During each trial, a vertical small arrow, shown in Fig. 1b, points randomly to one of the three tasks for duration of 10 s, and the user has to take rest if the vertical arrow points to the fixation cross or to imagine moving either left or right arm depending on which MI task is specified by the vertical arrow.

3 Methods

3.1 Data Acquisition

Sixteen electrodes were used to collect the EEG data. The electrodes were positioned over frontal, central, and parietal lobes at positions Fp1, Fp2, F3, F4, Fz, Fc1, Fc2, Cz, P1, P2, C1, C2, Cp3, Cp4, P5, and P6 according to the 10–20 system. Left mastoid was used as reference. Although SSVEPs give the strongest response over occipital area, we anticipated getting responses similar to those obtained from the occipital area using the electrodes mentioned above. Especially, we have electrodes on locations P5 and P6 which are close to the occipital area given that EEG is known to have low spatial resolution. A g.tec EEG system with a g. USBamp amplifier was used in this study. The amplifier included 16 24-bit simultaneously sampled channels with an internal digital signal filtering and processing unit and sampling rate up to 38.4 kHz. The data were digitized with a sampling rate of 256 samples/s and filtered by the amplifier's eighth-order bandpass filter with corner frequencies of 2 and 62 Hz in addition to fourth-order notch filter with corner frequencies of 58 and 62 Hz. Processed data were transferred from the amplifiers to a laptop via USB 2.0.

A SONARA TCD system that utilizes two 2 MHz transducers was used to record fTCD signals. These transducers were placed on the left and right sides of the transtemporal window located above the zygomatic arch (Alexandrov et al., 2007). Since middle cerebral arteries (MCAs) provide approximately 80% of the brain with blood (Stroobant & Vingerhoets, 2000), the fTCD depth was set to 50 mm which is the depth of the mid-point of the MCAs (Monsein et al., 1995). See Fig. 1c for the experimental setup.

3.2 Participants

Twenty-one healthy individuals participated in this study including 11 individuals (3 females and 8 males) with ages ranging from 25 to 32 years who participated in testing flickering MR/WG paradigm and 10 subjects (4 males and 6 females) with ages ranging from 23 to 32 who participated in testing MI paradigm. None of the participants had a history of migraines, concussions, strokes, heart murmurs, or other brain-related injuries. The experiment lasted for approximately 1 h and 15 min including the time required for the setup. All research procedures were approved by local institutional review board (IRB) under the University of Pittsburgh IRB number of PRO16080475. Participants signed a written informed consent before starting the experiment. During the experiment, subjects were seated in a comfortable chair approximately 1 m away from the screen. Each participant attended one session. Each session included 150 trials. Each trial lasted 10 s, and the trials are randomly but equally distributed across three conditions.

3.3 Feature Extraction and Fusion

Since each session contained 150 trials with 10 s duration per each, EEG and fTCD data corresponding to each observation were segmented. For each observation, the power spectrum was obtained for the corresponding 16 EEG segments as well as the corresponding 2 fTCD segments using Welch's power spectral density estimate (Welch, 1967). Instead of using all the power spectrum values as features, the average of power spectrum values within a sliding window of specific width (in Hz) was considered as one feature. For the average power to be calculated for the next window, the original window was shifted by a value equal to its width so that there is no overlap between consecutive windows and so on. For EEG signals, the average power was calculated using a window of 2 Hz width. Since the fTCD signal has much higher bandwidth ($\approx$2.5 kHz) compared to the EEG signals ($\approx$40 Hz), and considering the need to reduce the number of features, a window of 50 Hz width was used to reduce the fTCD power spectrum. The EEG/fTCD feature vector was formed by concatenating the features from all EEG/fTCD segments. The feature vectors of both EEG and fTCD signals were concatenated to form one single feature vector that represents each observation.

3.4 Feature Selection

Due to the high dimensionality of the EEG–fTCD feature vector (420 features), we applied a filter method for feature selection (Saeys, Inza, & Larranaga, 2007). The main advantage of the filter methods compared to wrapper and embedded methods is the low computational complexity. Filtering is based on the mutual information

(Peng, Long, & Ding, 2005). Specifically, in our approach, MI measures contribution of each feature toward taking a correct decision by assigning each feature a score based on its contribution. Higher MI score implies higher contribution of that feature toward correct classification. To calculate the mutual information score, each feature is quantized adaptively such that the number of data samples is almost the same in each quantization bin so that quantization levels are equiprobable (Pohjalainen, Räsänen, & Kadioglu, 2015). Mutual information score between the discretized feature value x and the class label y is given by (1).

$$\mathrm{MI} = \sum_{x \in X} \sum_{y \in Y} p(x, y) \log\left(\frac{p(x, y)}{p(x)p(y)}\right). \tag{1}$$

In (1), x is the discretized feature value and y is the class label. In our approach, to determine the number of features to be used for each binary selection problem, the cumulative distribution function (CDF) of the mutual information scores was computed. We calculated CDF thresholds corresponding to probabilities ranging from 0.5 to 0.95 with 0.05 step in addition CDF thresholds corresponding to probabilities 0.98 and 0.99. For each CDF threshold, the features obtaining scores greater than or equal to that threshold were selected. The performance measures including accuracy, sensitivity, and specificity corresponding to each CDF threshold are computed.

3.5 Classification and Performance Measures

Support vector machine (SVM) was used to perform the classification task (Hsu & Lin, 2002). Basic SVM provides a discriminative approach that uses labeled data (supervised learning) to find the optimal hyperplane that gives the largest distance to the nearest training example. Since the BCI is intended to be used for real-time applications, the basic linear SVM is considered as the best choice especially if the BCI is designed to exploit an online adaptive update of the classifier parameters.

Six binary selection problems were formulated and classified using SVM. For the MI-based system, classification problems included right MI versus baseline, left MI versus baseline, and right MI versus left MI; and for the flickering MR/WG-based system the classification problems are MR versus baseline, WG versus baseline, and MR versus WG. For each participant, a subject-specific classifier was trained and tested using leave-one-out cross-validation. Performance measures including accuracy and information transfer rate given by (2) were computed to evaluate the hybrid system (Obermaier, Neuper, Guger, & Pfurtscheller, 2001).

$$B = \log_2(N) + P\log_2(P) + (1 - P)\log_2\left(\frac{1 - P}{N - 1}\right), \tag{2}$$

where N is the number of classes, P is the classification accuracy, and B is the data transmission rate per trial.

3.6 Evaluation of the Effectiveness of the Hybrid System

The nonparametric Wilcoxon signed-rank test was used to assess the significance of the EEG–fTCD combination by statistically comparing the resultant performance measures with those obtained using EEG data only (Blair & Higgins, 1980). Specifically, EEG–fTCD accuracy vector containing accuracies for all the participants as well as the corresponding EEG only accuracy vector for the same participants represented the two populations to be tested using Wilcoxon signed-rank test. The test returns the p-value for the null hypothesis that assumes that the difference between the two populations follows a zero-median distribution.

3.7 Temporal Analysis to Identify Trial Length

An incremental window of 1 s initial width was used to calculate performance measures for each participant. The window width is increased by 1 s increment up to 10 s which is the trial length. Performance measures are evaluated at each increment. The objective of this analysis is to identify the optimum trial length that could be used to reduce the trial length and to improve the speed of the system. For each participant, the performance measures were computed as a function of time using 12 different CDF thresholds corresponding to probabilities ranging from 0.5 to 0.95 with 0.05 step as well as 0.98 and 0.99. These CDF thresholds were computed based on the CDF of mutual information scores reflecting the relevance of EEG and fTCD features within each binary selection problem, see also Sect. 3.4. Therefore, for every participant, 12 different profiles of performance measures across time were obtained. Considering subject-specific CDF thresholds, for each participant, all the accuracies at all CDF thresholds were considered and the maximum accuracy and the corresponding performance measures as well as the corresponding CDF threshold were used to represent that participant. Therefore, in subject-specific analysis, each subject might have a different CDF threshold that corresponds to his/her maximum performance accuracy. Moreover, average of the maximum accuracy and the corresponding sensitivity, and specificity across all participants were obtained.

4 Results and Discussions

Tables 1 and 2 show the maximum accuracy achieved by each participant using the EEG–fTCD combination for MI-based and flickering MR/WG-based paradigms, respectively. In order to demonstrate the advantages of the hybrid system, classification accuracies are also calculated using EEG only and fTCD only with the same time interval (trial length) at which the EEG–fTCD combination gives the maximum

Table 1 Maximum accuracy (Acc) achieved for each subject using hybrid system, EEG only, and fTCD only for MI paradigm

Sub_ID	Right-arm MI versus baseline			Left-arm MI versus baseline			Right-arm MI versus left-arm MI		
	Acc_EEG (%)	Acc_fTCD (%)	Acc_Hybrid (%)	Acc_EEG (%)	Acc_fTCD (%)	Acc_Hybrid (%)	Acc_EEG (%)	Acc_fTCD (%)	Acc_Hybrid (%)
1	91.67	59.38	**94.79**	87.63	65.98	**97.94**	82.86	51.43	**93.33**
2	91.67	61.46	**92.71**	91.75	65.98	**93.81**	75.24	38.10	75.24
3	81.25	51.04	**86.46**	81.44	51.55	**93.81**	71.43	69.52	**81.90**
4	81.25	55.21	**87.50**	78.35	61.86	**81.44**	81.90	53.33	**85.71**
5	87.50	61.46	**90.63**	87.63	53.61	87.63	78.10	50.48	**81.90**
6	85.42	56.25	**86.46**	74.23	53.61	**86.60**	85.71	42.86	83.81
7	80.21	47.92	**86.46**	88.66	54.64	**90.72**	71.43	43.81	**79.05**
8	81.25	69.79	**91.67**	85.57	50.52	**92.78**	96.19	50.00	**97.14**
9	71.88	50.00	**76.04**	74.23	48.45	**79.38**	70.48	44.76	**71.43**
10	86.46	69.79	**90.63**	91.75	47.42	90.72	62.86	66.67	**74.29**
Mean	83.85	58.23	**88.33**	84.12	55.36	**89.48**	77.62	51.10	**82.38**

Table 2 Maximum accuracy (Acc) achieved for each subject using hybrid system, EEG only, and fTCD only for flickering MR/WG paradigm

Sub_ID	MR versus baseline			WG versus baseline			MR versus WG		
	Acc_EEG (%)	Acc_fTCD (%)	Acc_Hybrid (%)	Acc_EEG (%)	Acc_fTCD (%)	Acc_Hybrid (%)	Acc_EEG (%)	Acc_fTCD (%)	Acc_Hybrid (%)
1	91.67	50.00	89.58	82.47	55.67	**89.69**	95.24	51.43	**99.05**
2	84.38	57.29	**89.58**	63.92	43.30	**67.01**	82.86	52.38	**85.71**
3	90.63	56.25	**93.75**	73.20	53.61	**80.41**	96.19	73.33	**97.14**
4	83.33	52.08	**84.38**	68.04	42.27	**76.29**	84.76	54.29	**89.52**
5	82.29	46.88	**88.54**	68.04	57.73	**86.60**	89.52	63.81	86.67
6	90.63	60.42	**94.79**	84.54	53.61	**89.69**	90.48	68.57	**93.33**
7	93.75	39.58	**94.79**	94.85	55.67	**95.88**	94.29	57.14	**96.19**
8	88.54	47.92	**89.58**	63.92	50.00	**68.04**	88.57	43.81	**93.33**
9	83.33	50.00	**86.46**	78.35	54.64	78.35	87.62	47.62	**88.57**
10	58.33	67.71	**80.21**	68.04	62.89	**75.26**	91.43	60.95	91.43
11	88.54	38.54	88.54	80.41	52.58	**82.47**	91.43	52.38	**95.24**
Mean	85.04	51.52	**89.11**	75.07	52.91	**80.88**	90.22	56.88	**92.38**

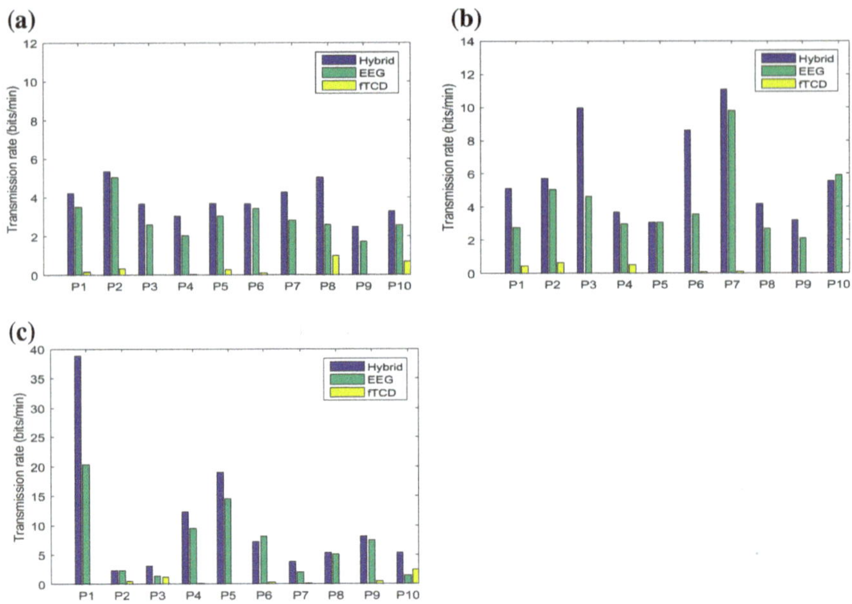

Fig. 2 Transmission rates for each participant (*p*) calculated using both EEG and fTCD, EEG only, and fTCD only for right-arm MI versus baseline (**a**) left-arm MI versus baseline problem (**b**) and right MI versus left MI (**c**)

accuracy as given in Tables 1 and 2. Transmission rates corresponding to the accuracies and times listed in Tables 1 and 2 are also calculated for each binary problem using EEG data, fTCD data, and EEG–fTCD combination as shown in Figs. 2 and 3. Figures 2 and 3 present information transfer rates for MI-based and flickering MR/WG-based systems, respectively. The *p*-values representing the significance of the EEG–fTCD hybrid system are calculated by statistically comparing the EEG–fTCD accuracy/bit rate vector with the EEG only accuracy/bit rate vector for all the binary selection problems. Moreover, Table 3 compares the paradigms of the EEG and fTCD-based hybrid BCI with the state-of-the-art EEG and fNIRS-based hybrid BCIs.

As given in Table 1, for right-arm MI versus baseline problem, it was found that the EEG–fTCD combination achieved 88.33% average accuracy within 7.7 s compared to 83.85% obtained by EEG only and 58.23% by fTCD only. The average accuracy difference between the hybrid combination and the EEG only was 4.48%. As shown in Fig. 2a, EEG–fTCD combination achieved higher bit rates compared to EEG only for all of the participants. On average, average transmission rate of 3.87 bits/min was obtained. Statistically, the EEG–fTCD combination achieved higher accuracy compared to EEG only for all the participants with a *p*-value of 0.002. As for left MI versus baseline, an average accuracy difference of 5.36% between the combination and EEG only was achieved with a *p*-value of 0.0078. For 8 out of

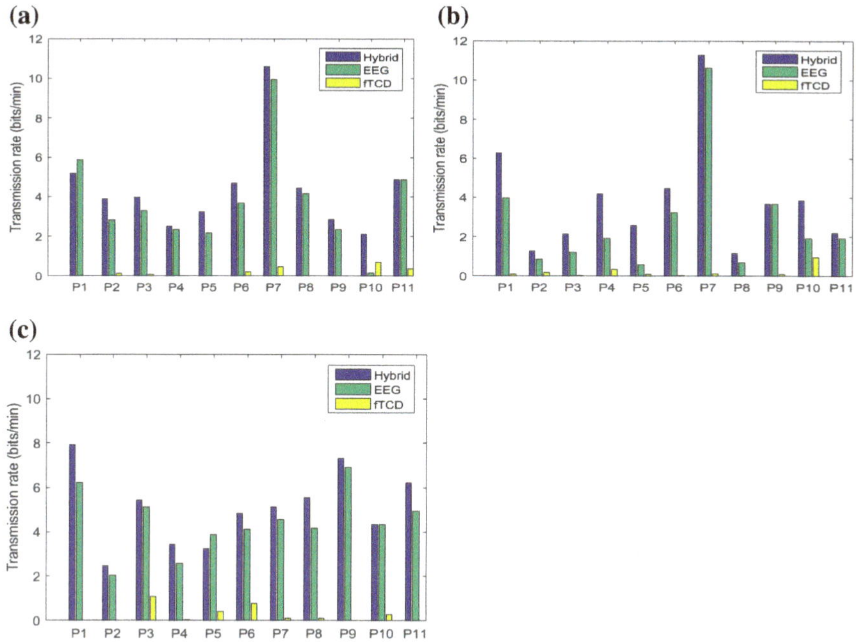

Fig. 3 Transmission rates for each participant (p) calculated using both EEG and fTCD, EEG only, and fTCD only for MR versus baseline (**a**) WG versus baseline problem (**b**) and MR versus WG (**c**)

10 participants, EEG–fTCD scored higher accuracy compared to EEG only. EEG–fTCD scored 89.48% accuracy within 6.1 s, while EEG only scored 84.12% accuracy. As shown in Fig. 2b, we obtained higher bit rates using EEG–fTCD combination compared to bit rates generated using EEG only and fTCD only. In addition, we achieved 6.02 bits/min average bit rate. Within 3.4 s, 82.38% average accuracy was achieved for right MI versus left MI problem using the EEG–fTCD combination and 77.62% using EEG only leading to a significant average accuracy difference of 4.76% with 0.0195 p-value. The EEG–fTCD combination scored higher accuracy for 8 out of 10 participants. On average, from Fig. 2c, 10.57 bits/min were achieved.

EEG–fTCD combination achieved 89.11% average accuracy in approximately 7.73 s for MR versus baseline problem as given in Table 2. The combination outperformed EEG only by an average accuracy difference of 4.06%. The EEG–fTCD combination scored higher accuracies than EEG only for 9 out of 11 participants. Performance measures obtained using fTCD only were nonsignificant. However, fTCD boosted the overall performance when it was combined with the EEG. In particular, the combination was proved to be significant compared to EEG only with a p-value of 0.0156. In terms of transmission rates, as shown in Fig. 3, for most of the participants, the combination achieved higher bit rates compared to EEG only. Average bit rate of 4.39 bits/min was achieved. Considering WG versus baseline problem, as given

Table 3 Comparison between the proposed hybrid system and the state-of-the-art hybrid BCIs. © IOP Publishing. Reproduced with permission from Khalaf et al. (2018b)

Method	BCI modalities	Accuracy (%)	Trial length (s)	
			Task	Baseline/rest
Fazli et al. (2012)	EEG + fNIRS	83.20	15	6/0
Blokland et al. (2014)	EEG + fNIRS	79.00	15	0/30 ± 3
Khan et al. (2014)	EEG + fNIRS	83.60	10	0/5
Putze et al. (2014)	EEG + fNIRS	94.70	12.5 ± 2.5	0/20 ± 5
Yin et al. (2015b)	EEG + fNIRS	89.00	10	0/21 ± 1
Koo et al. (2015)	EEG + fNIRS	88.00	15	0/60
Buccino, Keles, and Omurtag (2016)	EEG + fNIRS	72.20	6	6/0
Buccino, Keles, and Omurtag (2016)	EEG + fNIRS	94.20	6	6/0
Shin et al. (2017)	EEG + fNIRS	88.20	10	0/16 ± 1
Proposed method (Left MI/baseline)	EEG + fTCD	88.33	10	NA
Proposed method (Right MI/baseline)	EEG + fTCD	89.48	10	NA
Proposed method (Right MI/left MI)	EEG + fTCD	82.38	10	NA
Proposed method (MR/baseline)	EEG + fTCD	89.11	10	NA
Proposed method (WG/baseline)	EEG + fTCD	80.88	10	NA
Proposed method (MR/WG)	EEG + fTCD	92.38	10	NA

NA Not applicable

in Table 2, average accuracy difference of 5.81% between the EEG–fTCD combination and EEG only was achieved. Specifically, the combination achieved 80.88% average accuracy in 5.64 s while EEG only obtained 75.07%. The EEG–fTCD combination scored higher accuracies for 10 out of 11 participants using subject-specific thresholds. The hybrid system was shown to provide a significant accuracy improvement compared to EEG only with a *p*-value of 0.012. The EEG–fTCD combination achieved higher bit rates than EEG only for most of the participants as shown in Fig. 3b. Subject-specific thresholds achieved average bit rate of 3.92 bits/min. For MR versus WG problem, EEG–fTCD combination obtained 92.38% average accuracy in 7.82 s, while EEG only got 90.22%. The hybrid combination achieved higher accuracies for 9 out of 11 participants. Moreover, the hybrid combination was shown to be significant with *p*-value of 0.0078. Bit rates obtained using the EEG–fTCD combination were higher for most of the participants compared to those obtained using EEG as shown in Fig. 3c. Average bit rate of 5.07 bits/min was obtained.

Considering the MI visual presentation, by inspecting the selected significant features across all participants for the right-arm MI versus left-arm MI problem, it was found that, across all electrodes, the EEG average power spectrum values at frequencies up to 2 Hz (delta frequency band) are the most common selected features across participants. Moreover, it was found that the common selected features belonging to theta (5–8 Hz) and mu (8–13 Hz) bands are coming from electrodes Fp1, Fp2, F3, and F4 while the common features belonging to beta (16–28 Hz) band are associated with electrodes C1, C2, Cp3, Cp4, P5, and P6. As for the fTCD, the common significant features were found at frequency bands 550–600, 2000–2050, and 2150–2200 Hz for the right fTCD channel and at frequency band 2050–2100 Hz for the left fTCD channel.

As for the flickering MR/WG visual presentation, through investigation of the common EEG and fTCD significant features across all participants for the classification problem that yielded the highest accuracy (WG Vs. MR), as expected, the top common EEG power spectrum features were found approximately around the 1st, 2nd, and 3rd harmonics of the 7 Hz and around 1st and 2nd harmonics of the 17 Hz. Considering the fTCD features, the most common selected significant power spectrum features were found at frequency bands of 0–50, 1200–1250, and 1350–1400 for the left fTCD channel and at frequency bands of 0–100, 300–350, and 1950–2000 Hz for the right fTCD channel.

In Table 3, we compared the proposed hybrid BCI with the state-of-the-art EEG–fNIRS hybrid BCIs (Blokland et al., 2014; Buccino et al., 2016; Fazli et al., 2012; Khan et al., 2014; Koo et al., 2015; Putze et al., 2014; Shin et al., 2017; Yin et al., 2015b). Accuracies of the above-listed six binary selection problems are listed in Table 3. Comparisons were performed in terms of trial length and accuracy. For the EEG- and fTCD-based hybrid BCI, comparing the MI-based presentation paradigm to the flickering MR/WG paradigm, we observe that the latter achieved an average of 10% increase in accuracy. However, the hybrid system based on MR/WG paradigm was slower since it required an average time of 6.36 s to achieve 92.38% accuracy, while the system based on MI paradigm required only 3.5 s on average to achieve 82.38% accuracy. In line with the differences in speed, flickering MR/WG presentation achieved maximum bit rate of 5.6 bits/min, while MI visual presentation obtained 10.57 bits/min. On the other hand, right/left-arm MI versus baseline achieved higher accuracies compared to right-arm versus left-arm MI. In contrast, it was found that MR/WG versus baseline problems achieved lower accuracies compared to MR versus WG problem. Since the location of the baseline cross is very close to the flickering MR and WG tasks as shown in Fig. 1, the flickering affected the subject attention even during focusing at the baseline cross and thus caused reduction in accuracy for MR/WG versus baseline problems.

Compared to the other BCIs listed in Table 3, in terms of trial length, the EEG and fTCD-based system has the shortest trial length of maximum 10 s. In addition, the proposed system is faster since it requires no baseline/rest period before/after each task. In terms of accuracy, the proposed hybrid BCI outperforms most of the methods in comparison. However, the systems introduced by Putze et al. (2014) and Buccino et al. (2016) achieved higher accuracy compared to ours as they obtained 94.70%

and 94.20% average accuracy, respectively. Yet, these systems are slower than our system since the one introduced by Putze et al. (2014) requires at least 12.5 s as a task period and 20 s as a resting period, while the one presented by Buccino et al. (2016) requires a baseline period of 6 s before starting each task.

Note here that despite the high accuracies obtained in a previous study with the fTCD data only (Khalaf et al., 2018), the presented hybrid system obtained lower accuracies with fTCD data only. It is well known that fTCD can differentiate imagery and analytical tasks since analytical tasks induce higher blood velocity in left MCAs, while imagery tasks induce bilateral activation. However, for the MI-based system, both tasks are imagery tasks which makes the classification problem harder to solve. On the other hand, even though flickering MR/WG-based system utilizes both imagery and analytical tasks, the tasks were flickering to elicit SSVEPs. Such flickering reduced the concentration of each subject on the mental task to be performed. Moreover, in the previous study, a 15-min baseline period was recorded before starting the tasks to stabilize the cerebral blood flow. Also, a resting period of 45 s was inserted between consecutive tasks. During the design of the two hybrid systems presented in this chapter, in order to improve the speed of the systems and also to reduce the trial length, no baseline/rest periods were added to stabilize the cerebral blood. In fact, in the EEG and fTCD-based hybrid BCI systems, the baseline was shown at random times since it was designed as a task that resembles the condition in which the BCI user does not intend to produce a command.

5 Conclusions

In this chapter, we introduced two novel hybrid BCI systems that use EEG as the primary sensing modality to measure electrical brain activity and the fTCD as the secondary sensing modality to measure cerebral blood flow velocity. One of the systems was based on motor imagery to induce changes simultaneously in EEG and fTCD, while the other system simultaneously utilized mental rotation and word generation as stimuli for fTCD and flickering checkerboards to induce SSVEP in EEG. The experimental results that are obtained through the participation of 21 individuals were promising to show that the presented hybrid systems are feasible candidates for real-time BCI applications. Compared to the existing work on hybrid BCI that combines EEG with other modalities, the two hybrid systems show important progress toward making such systems real-world worthy in terms of speed and accuracy. However, the systems still have limitations such that the temporal resolution of fTCD is lower than EEG resulting in longer trial lengths and decreasing the speed of the system. For example, such mismatch between the temporal resolutions of these modalities can be minimized by introducing advanced analysis techniques for fTCD data to improve the obtained accuracy within the minimum possible task period. In particular, Wavelet analysis can be employed for fTCD analysis since it was used in a recent study (Khalaf et al., 2018) to prove that fTCD is as a viable candidate for real-time BCIs and it achieved accuracies of approximately 80% and 60% for binary and

3-class BCIs within 3 and 5 s, respectively. Further future directions could include investigation of analysis techniques for both EEG and fTCD signals to increase accuracy and decrease trial length for the hybrid system so that the information transfer rate for the EEG–fTCD combination is increased. Moreover, the generalization of the system across subjects could be explored using transfer learning techniques to decrease calibration requirements. On the other hand, based on the feedback from the BCI users, a bigger screen will be used to run the experiment to reduce the flickering effect on the subject's attention while focusing on the baseline cross. Moreover, the 25-min session will be divided into two sessions to reduce the user fatigue due to the flickering.

References

Akcakaya, M., et al. (2014). Noninvasive brain–computer interfaces for augmentative and alternative communication. *IEEE Reviews in Biomedical Engineering, 7,* 31–49. Retrieved February 12, 2019, from http://ieeexplore.ieee.org/document/6684304/.

Alexandrov, A. V., et al. (2007). Practice standards for transcranial doppler ultrasound: Part I-test performance. *Journal of Neuroimaging, 17*(1), 11–18. https://doi.org/10.1111/j.1552-6569.2006.00088.x.

Allison, B. Z., Wolpaw, E. W., & Wolpaw, J. R. (2007). Brain-computer interface systems: progress and prospects. *Expert review of Medical Devices, 4*(4): 463–474. Retrieved August 28, 2016, from http://www.ncbi.nlm.nih.gov/pubmed/17605682.

Amiri, S., Fazel-Rezai, R., & Asadpour, V. (2013). A review of hybrid brain-computer interface systems. *Advances in Human-Computer Interaction, 2013,* 1–8. Retrieved December 10, 2017, from http://www.hindawi.com/journals/ahci/2013/187024/.

Ang, K. K., et al. (2010). Clinical study of neurorehabilitation in stroke using EEG-based motor imagery brain-computer interface with robotic feedback. In *2010 Annual International Conference of the IEEE Engineering in Medicine and Biology* (pp. 5549–5552). IEEE. Retrieved October 20, 2017, from http://www.ncbi.nlm.nih.gov/pubmed/21096475.

Ang, K. K., & Guan, C. (2017). EEG-based strategies to detect motor imagery for control and rehabilitation. *IEEE Transactions on Neural Systems and Rehabilitation Engineering, 25*(4), 392–401. Retrieved October 20, 2017, from http://ieeexplore.ieee.org/document/7802578/.

Bin, G., et al. (2009). An online multi-channel SSVEP-based brain–computer interface using a canonical correlation analysis method. *Journal of Neural Engineering, 6*(4), 046002. Retrieved December 9, 2017, from http://stacks.iop.org/1741-2552/6/i=4/a=046002?key=crossref.501c725e9b2f8545feaf7401703ed8c1.

Blair, R. C., & Higgins, J. J. (1980). A comparison of the power of Wilcoxon's rank-sum statistic to that of student's T statistic under various non-normal distributions. *Journal of Educational Statistics, 5*(4), 309. Retrieved August 8, 2016, from http://www.jstor.org/stable/1164905?origin=crossref.

Blank, A. A., French, J. A., Pehlivan, A. U., & O'Malley, M. K. (2014). Current trends in robot-assisted upper-limb stroke rehabilitation: Promoting patient engagement in therapy. *Current Physical Medicine and Rehabilitation Reports, 2*(3), 184–195. Retrieved October 20, 2017, from http://link.springer.com/10.1007/s40141-014-0056-z.

Blokland, Y., et al. (2014). Combined EEG-fNIRS decoding of motor attempt and imagery for brain switch control: An offline study in patients with tetraplegia. *IEEE Transactions on Neural Systems and Rehabilitation Engineering, 22*(2), 222–229. Retrieved October 21, 2017, from http://www.ncbi.nlm.nih.gov/pubmed/24608682.

Buccino, A. P., Keles, H. O., & Omurtag, A. (2016). Hybrid EEG-fNIRS asynchronous brain-computer interface for multiple motor tasks. In B. He (ed.), *PLOS ONE, 11*(1), e0146610. Retrieved October 21, 2017, from http://dx.plos.org/10.1371/journal.pone.0146610.

Chen, X., et al. (2015). High-Speed Spelling with a Noninvasive Brain-Computer Interface. *Proceedings of the National Academy of Sciences of the United States of America, 112*(44), E6058–E6067. Retrieved January 29, 2018, from http://www.ncbi.nlm.nih.gov/pubmed/26483479.

Coyle, S., Ward, T., Markham, C., & McDarby, G. (2004). On the suitability of near-infrared (NIR) systems for next-generation brain–computer interfaces. *Physiological Measurement, 25*(4), 815–822. Retrieved September 6, 2016, from http://stacks.iop.org/0967-3334/25/i=4/a=003?key=crossref.2b3bec58c1fb74ea68c7e4e57a6ac10d.

Fager, S., et al. (2011). Access interface strategies. *Assistive Technology : The Official Journal of RESNA, 24*(1), 25–33. Retrieved November 1, 2017, from http://www.ncbi.nlm.nih.gov/pubmed/22590797.

Farwell, L. A., & Donchin, E. (1988). Talking off the top of your head: Toward a mental prosthesis utilizing event-related brain potentials. *Electroencephalography and Clinical Neurophysiology, 70*(6), 510–523. Retrieved November 1, 2017, from http://www.ncbi.nlm.nih.gov/pubmed/2461285.

Fazli, S., et al. (2012). Enhanced performance by a hybrid NIRS–EEG brain computer interface. *NeuroImage, 59*(1), 519–529. Retrieved October 21, 2017, from http://www.ncbi.nlm.nih.gov/pubmed/21840399.

Hsu, C.-W., & Lin, C.-J. (2002). A comparison of methods for multiclass support vector machines. *IEEE Transactions on Neural Networks, 13*(2), 415–425. Retrieved August 7, 2016, from http://ieeexplore.ieee.org/lpdocs/epic03/wrapper.htm?arnumber=991427.

Khalaf, A., Sejdic, E., & Akcakaya, M. (2018a). A novel motor imagery hybrid brain computer interface using EEG and functional transcranial doppler ultrasound. *Journal of Neuroscience Methods, Under Rev.*

Khalaf, A., Sejdic, E., & Akcakaya, M. (2018b). Towards optimal visual presentation design for hybrid EEG—fTCD brain–computer interfaces. *Journal of Neural Engineering, 15*(5), 056019. Retrieved September 7, 2018, from http://stacks.iop.org/1741-2552/15/i=5/a=056019?key=crossref.48956e67c2a5bbd451b5d7f08db2ce08.

Khalaf, A., Sybeldon, M., Sejdic, E., & Akcakaya, M. (2017). An EEG and fTCD based BCI for control. In *Conference Record—Asilomar Conference on Signals, Systems and Computers*.

Khalaf, A., Sybeldon, N., Sejdic, E., & Akcakaya, M. (2018). A brain-computer interface based on functional transcranial doppler ultrasound using wavelet transform and support vector machines. *Journal of Neuroscience Methods, 293*, 174–182. Retrieved October 20, 2017, from http://www.sciencedirect.com/science/article/pii/S0165027017303515.

Khan, M. J., Hong, M. J., & Hong, K. S. (2014). Decoding of four movement directions using hybrid NIRS-EEG brain-computer interface. *Frontiers in Human Neuroscience, 8*, 244. Retrieved January 28, 2018, from http://www.ncbi.nlm.nih.gov/pubmed/24808844.

Koo, B., et al. (2015). A hybrid NIRS-EEG system for self-paced brain computer interface with online motor imagery. *Journal of Neuroscience Methods, 244*, 26–32. Retrieved October 21, 2017, from http://www.ncbi.nlm.nih.gov/pubmed/24797225.

LaFleur, K., et al. (2013). Quadcopter control in three-dimensional space using a noninvasive motor imagery-based brain–computer interface. *Journal of Neural Engineering, 10*(4), 046003. Retrieved October 20, 2017, from http://stacks.iop.org/1741-2552/10/i=4/a=046003?key=crossref.ac48eeb233dce6b8bef5ccda9dae0dd2.

Lesenfants, D., et al. (2014). An independent SSVEP-based brain–computer interface in locked-in syndrome. *Journal of Neural Engineering, 11*(3), 035002. Retrieved November 1, 2017, from http://www.ncbi.nlm.nih.gov/pubmed/24838215.

Matteis, M., et al. (2006). Cerebral blood flow velocity changes during meaningful and meaningless gestures—A functional transcranial doppler study. *European Journal of Neurology, 13*(1), 24–29. Retrieved October 20, 2017, from http://www.ncbi.nlm.nih.gov/pubmed/16420390.

Mellinger, J., et al. (2007). An MEG-based brain–computer interface (BCI). *NeuroImage, 36*(3), 581–593.
Min, B.-K., Marzelli, M. J., & Yoo, S.-S. (2010). Neuroimaging-based approaches in the brain–computer interface. *Trends in Biotechnology, 28*(11), 552–560. Retrieved October 21, 2017, from http://www.ncbi.nlm.nih.gov/pubmed/20810180.
Monsein, L. H., et al. (1995). Validation of transcranial doppler ultrasound with a stereotactic neurosurgical technique. *Journal of Neurosurgery, 82*(6), 972–975. Retrieved August 7, 2016, from http://thejns.org/doi/abs/10.3171/jns.1995.82.6.0972.
Muller-Putz, G., et al. (2015). Towards noninvasive hybrid brain-computer interfaces: Framework, practice, clinical application, and beyond. *Proceedings of the IEEE, 103*(6), 926–943. Retrieved October 21, 2017 from http://ieeexplore.ieee.org/document/7109824/.
Myrden, A. J. B., et al. (2011). A brain-computer interface based on bilateral transcranial doppler ultrasound. In G. Cymbalyuk (Ed.), *PLoS ONE, 6*(9), e24170. Retrieved October 21, 2017, from http://dx.plos.org/10.1371/journal.pone.0024170.
Nakanishi, M., et al. (2018). Enhancing detection of SSVEPs for a high-speed brain speller using task-related component analysis. *IEEE Transactions on Biomedical Engineering, 65*(1), 104–112. Retrieved January 29, 2018, from http://ieeexplore.ieee.org/document/7904641/.
Naseer, N., & Hong, K.-S. (2015). fNIRS-based brain-computer interfaces: A review. *Frontiers in Human Neuroscience, 9*, 3. Retrieved January 29, 2018, from http://www.ncbi.nlm.nih.gov/pubmed/25674060.
Nicolas-Alonso, L. F., & Gomez-Gil, J. (2012). Brain computer interfaces, a review. *Sensors (Basel, Switzerland), 12*(2), 1211–1279. Retrieved October 20, 2017, from http://www.ncbi.nlm.nih.gov/pubmed/22438708.
Obermaier, B., Neuper, C., Guger, C., & Pfurtscheller, G. (2001). Information transfer rate in a five-classes brain-computer interface. *IEEE Transactions on Neural Systems and Rehabilitation Engineering, 9*(3), 283–288. Retrieved January 24, 2018, from http://ieeexplore.ieee.org/document/948456/.
Peng, H., Long, F., & Ding, C. (2005). Feature selection based on mutual information criteria of max-dependency, max-relevance, and min-redundancy. *IEEE Transactions on Pattern Analysis and Machine Intelligence, 27*(8), 1226–1238. Retrieved October 21, 2017, from http://ieeexplore.ieee.org/document/1453511/.
Peters, M., & Battista, C. (2008). Applications of mental rotation figures of the Shepard and Metzler type and description of a mental rotation stimulus library. *Brain and Cognition, 66*(3), 260–264.
Pfurtscheller, G., et al. (2010). The hybrid BCI. *Frontiers in Neuroscience, 4*, 3. Retrieved January 24, 2018, from http://journal.frontiersin.org/article/10.3389/fnpro.2010.00003/abstract.
Pohjalainen, J., Räsänen, O., & Kadioglu, S. (2015). Feature selection methods and their combinations in high-dimensional classification of speaker likability, intelligibility and personality traits. *Computer Speech & Language, 29*(1), 145–171. Retrieved December 9, 2017, from http://linkinghub.elsevier.com/retrieve/pii/S0885230813001113.
Putze, F., et al. (2014). Hybrid fNIRS-EEG based classification of auditory and visual perception processes. *Frontiers in Neuroscience, 8*, 373. Retrieved January 28, 2018, from http://www.ncbi.nlm.nih.gov/pubmed/25477777.
Saeys, Y., Inza, I., & Larranaga, P. (2007). A review of feature selection techniques in bioinformatics. *Bioinformatics, 23*(19), 2507–2517. Retrieved October 31, 2017, from https://academic.oup.com/bioinformatics/article-lookup/doi/10.1093/bioinformatics/btm344.
Schlogl, A., Lee, F., Bischof, H., & Pfurtscheller, G. (2005). Characterization of four class motor imagery EEG data for the BCI-competition 2005. *Journal of Neural Engineering, 2*(4), L14–L22. Retrieved October 20, 2017, from http://stacks.iop.org/1741-2552/2/i=4/a=L02?key=crossref.b48710597fa273469a1d48e6b80b7177.
Shin, J., et al. (2017). Evaluation of a compact hybrid brain-computer interface system. *BioMed Research International, 2017*, 1–11. Retrieved November 1, 2017, from https://www.hindawi.com/journals/bmri/2017/6820482/.

Stroobant, N., & Vingerhoets, G. (2000). Transcranial doppler ultrasonography monitoring of cerebral hemodynamics during performance of cognitive tasks: A review. *Neuropsychology Review, 10*(4), 213–231. Retrieved October 21, 2017, from http://www.ncbi.nlm.nih.gov/pubmed/11132101.

van Dokkum, L. E. H., Ward, T., & Laffont, I. (2015). Brain computer interfaces for neurorehabilitation—Its current status as a rehabilitation strategy post-stroke. *Annals of Physical and Rehabilitation Medicine, 58*(1), 3–8. Retrieved October 21, 2017, from http://www.ncbi.nlm.nih.gov/pubmed/25614021.

Vanegas, M. I., Blangero, A., & Kelly, S. P. (2013). Exploiting individual primary visual cortex geometry to boost steady state visual evoked potentials. *Journal of Neural Engineering, 10*(3), 036003. Retrieved January 28, 2018, from http://www.ncbi.nlm.nih.gov/pubmed/23548662.

Wang, M., et al. (2015). A new hybrid BCI paradigm based on P300 and SSVEP. *Journal of Neuroscience Methods, 244,* 16–25. Retrieved October 31, 2017, from http://www.ncbi.nlm.nih.gov/pubmed/24997343.

Wang, Y.-T., et al. (2017). An online brain-computer interface based on SSVEPs measured from non-hair-bearing areas. *IEEE Transactions on Neural Systems and Rehabilitation Engineering, 25*(1), 14–21. Retrieved November 1, 2017, from http://ieeexplore.ieee.org/document/7480820/.

Weiskopf, N. et al. (2004). Principles of a brain-computer interface (BCI) based on real-time functional magnetic resonance imaging (fMRI). *IEEE Transactions on Biomedical Engineering, 51*(6), 966–970. Retrieved September 6, 2016, from http://ieeexplore.ieee.org/lpdocs/epic03/wrapper.htm?arnumber=1300789 ().

Welch, P. (1967). The use of fast fourier transform for the estimation of power spectra: A method based on time averaging over short, modified periodograms. *IEEE Transactions on Audio and Electroacoustics, 15*(2), 70–73. Retrieved February 27, 2017, from http://ieeexplore.ieee.org/document/1161901/.

Wolpaw, J. R., et al. (2002). Brain-computer interfaces for communication and control. *Clinical Neurophysiology : Official Journal of the International Federation of Clinical Neurophysiology, 113*(6), 767–791. Retrieved October 31, 2017, from http://www.ncbi.nlm.nih.gov/pubmed/12048038.

Yin, E., et al. (2015a). A dynamically optimized SSVEP brain–computer interface (BCI) speller. *IEEE Transactions on Biomedical Engineering, 62*(6), 1447–1456. Retrieved May 27, 2018, from http://www.ncbi.nlm.nih.gov/pubmed/24801483.

Yin, X., et al. (2015b). A hybrid BCI based on EEG and fNIRS signals improves the performance of decoding motor imagery of both force and speed of hand clenching. *Journal of Neural Engineering, 12*(3), 036004. Retrieved October 21, 2017, from http://stacks.iop.org/1741-2552/12/i=3/a=036004?key=crossref.fe66dac345d35e64effc9b537cbab91c.

Brain–Computer Interfaces for Spinal Cord Injury Rehabilitation

Alyssa Merante, Yu Zhang, Satyam Kumar and Chang S. Nam

Abstract Brain–computer interface (BCI) is an emerging tool that has a variety of practical applications, including rehabilitation. BCIs are systems that extract and classify features in neural data, and then produce an output when a specific feature is detected. Motor imagery-based BCIs (MI BCIs), a more specific form of BCI, detect features that indicate the user is imagining a specific motor action, such as moving their arm or leg. There have been several studies released discussing the potential for BCIs to be used in a clinical setting for applications like rehabilitation. Spinal cord injuries (SCIs) are a form of injury that damages the spinal cord and causes either partial or total paralyzation. Those with SCI typically undergo rehabilitation for many years after the injury, and BCIs have begun to be tested for their benefits when included in SCI rehabilitation sessions. There are several ways for BCI systems to be used in SCI rehabilitation, which include virtual reality, exoskeletons, and neuroprosthesis. When using these methods as an output for a BCI system, SCI patients experience numerous benefits, most notably being an increase in mobility in the paralyzed region of their body. While there are several advantages to using BCIs for SCI rehabilitation, there are also several challenges that need to be addressed. In this chapter, we will discuss the current potential of BCIs for SCI rehabilitation, as well as what areas of this field need to be improved in the future.

A. Merante
Department of Biomedical Engineering, North Carolina State University, Raleigh, NC 27695-7906, USA

Y. Zhang
Department of Psychiatry and Behavior Sciences, Stanford University, Stanford, CA 94304, USA

S. Kumar
Department of Electrical and Computer Engineering, University of Texas at Austin, Austin, TX, USA

C. S. Nam (✉)
Edward P. Fitts Department of Industrial & Systems Engineering, North Carolina State University, Raleigh, NC 27695-7906, USA
e-mail: csnam@ncsu.edu

C. S. Nam (ed.), *Neuroergonomics*, Cognitive Science and Technology,
https://doi.org/10.1007/978-3-030-34784-0_16

1 Introduction

Electroencephalogram (EEG)-based brain–computer interfaces (BCIs) are systems that extract and classify features in EEG data while generating an output when these features are detected. The output is then further used to control an external device directly without the use of motor pathways and muscular control. One of the promising uses of BCIs is in patients with spinal cord injuries (SCIs) since they lose some of their motor control.

As an advanced communication technique, BCI provides an elegant way to build a direct connection between human brain and computer (Chaudhary et al. 2016). By decoding the brain activity pattern from an EEG recorded during a specific mental task, BCI is able to translate a task-related neural response into a computer command. Due to this capability, BCI has been proven to serve as an assistive tool for severely disabled people, such as those with spinal cord injuries (SCIs), to recover their abilities for controlling of the external environment (Ang et al. 2015).

One of the most studied BCIs is based on EEG pattern of sensorimotor rhythm (SMR) associated with motor imagery task. SMR is typically characterized as a change in band power at a specific frequency band (e.g., μ or β band) appearing over the sensorimotor areas of brain when the subject is performing a motor imagery task, i.e., imagining body movement (e.g., left hand, right hand, and foot) (Pfurtscheller et al. 2006). Accordingly, a motor imagery-based BCI system can be designed by decoding the SMR pattern as target signal for neuroprosthesis and exoskeletons control as well as virtual reality operation, which could provide potentially great benefits for rehabilitation applications (Pichiorri et al. 2015). In recent years, rehabilitative MIBCI has been increasingly studied and shown a great potential in restoring motor function by inducing activity-dependent neuroplasticity, especially for neurological rehabilitation reference (Frolov et al. 2017; Pichiorri et al. 2017).

Despite these pioneer studies, no systematic study about the state-of-the-art progress for MIBCI-based SCI rehabilitation has been presented in the previous literature. Such a systematic review with comprehensive analysis and discussion is considerably important for researchers to better understand advantages of using BCI for rehabilitation and challenges to be addressed.

This chapter focuses on how the outputs of motor imagery-based BCIs can be geared toward rehabilitation purposes of those with spinal cord injuries (SCIs). It will also discuss the advantages and disadvantages of using BCIs for SCIs, as well as the current challenges surrounding this topic. This chapter can be an important tool for future researchers to better understand the needs in this field so that they can be addressed in future literature.

1.1 How Do Brain–Computer Interfaces (BCIs) Work?

Brain–computer interfaces are an emerging tool being used for a variety of purposes in the medical field. A brain–computer interface, or BCI, is a system comprised of a computer and an electroencephalograph (EEG) acquisition device that analyzes various features in brain activity from a subject and then produces an output which can be used for controlling an external device. These features are created by concentrated thought when a desired stimulus is presented to a subject, or when the system's output is desired. BCIs first emerged when Jacques Vidal of UCLA discussed the possibility of creating a system that could determine brain features and create an output (Vidal 1973). By the 2000s, studies were being conducted on monkeys and rats; however, BCIs were soon identified as a potential for helping those with communication or motor disabilities (Lebedev and Nicolelis 2017).

One form of BCI is one that is motor imagery-based, meaning that the features extracted from neural signals are obtained by a subject's concentration on the imagination of a motor movement, such as moving their arm, leg, or tongue. This is done internally, as no body parts actually move to create the brain signals required for the system. An imagination of movement causes a significant change in the amplitude of the μ and β frequency bands in a subject's EEG, which can be detected by the system. This form of BCI can be very useful in those with paralyzation, since they are unable to actually move parts of their body, but have retained the ability to imagine the movement of paralyzed body parts.

BCI systems have two general steps: calibration and decoding. In the first step, neural signals are acquired from a user. Specific cues are presented to the users to generate data of various brain states, such as relaxation, leg motor imagery, and hand motor imagery. In the next step, these signals are decoded by a computational model. In an offline system, this is done all at once with a full collection of data. In an online system, this is done in real time as neural data enters the system and then has the ability to be used for an output.

There are various steps and modules in the development and practice of a brain–computer interface. The first step is the data acquisition model, in which EEG data from a subject is obtained. The EEG should be collected from an adequate number of (usually noninvasive) electrodes, to ensure that there is sufficient resolution and information received from all areas of the brain. For the calibration phase of a BCI, a paradigm has to be created that will give cues to a subject so they can perform motor imagery in a controlled way that can be used for analysis. A paradigm for a BCI system includes a system that provides the subject with a visual or aural cue, which then initiates motor imagery. An example paradigm would be one in which the subject is presented with a 4 s visual warning, then a 4 s cue on a screen that indicates them to perform a motor imagery of one body part, and then a 4 s rest that then returns to the beginning of the paradigm. The cues are randomly presented as to what motor imagery action, or class, is going to be displayed. This process is continued for a set number of trials, and the data will be used to create a computational model for classification of the neural signals. The next module is signal processing, where the signal goes through preprocessing to remove irrelevant frequency components, such as line noise and high-frequency components. Usually, the signal is bandpass filtered between 8 and 30 Hz, which encompasses μ and β ranges bearing most information associated with motor imagery.

After this, the preprocessed signal goes through feature extraction. There are several algorithms that could be used for this, with the most common being common spatial patterns (CSP) for motor imagery BCI. This method aims at finding the optimal spatial filters to project the multichannel EEG into a new feature space, so that the variance ratio of one class versus another one at the new feature space is maximized. As a result, CSP identifies the most discriminative features in the power of the EEG signal. After the features are extracted, they are then used for classifier training and neural pattern decoding. If there are only two classes being examined, linear discriminant analysis (LDA) and support vector machine (SVM) are common methods used to classify the data.

In an online BCI system, there will be a closed human–machine loop. The system detects and characterizes the EEG signal in real time and produces an output. The possibilities for outputs include moving a computer character in a game, turning on a light, or even triggering the movement of an exoskeleton. Other output options include navigating a wheelchair and operating a robot arm, especially for rehabilitation applications.

1.2 Current Spinal Cord Injuries (SCIs) Rehabilitation

Spinal cord injuries are a debilitating disorder that negatively affects the quality of life for those who suffer from them. SCIs are caused by direct physical damage to the spinal cord and cause paralyzation below the injury. This can be caused by traumatic events such as car accidents and falls, cancer, and infections (Alam et al. 2016).

There are currently three many methods of SCI rehabilitation that are used in hospitals and rehabilitation centers. One of these is active therapy, where the patient practices sitting, walking, and completing tasks with the assistance of a medical professional. Another is functional electric stimulation or FES, where stimulating electrodes are placed on the muscles of paralyzed body parts to induce movement, such as walking (Wirz and van Hedel 2018). This method has a few drawbacks, including that it is very difficult to place the high amount of electrodes in the correct locations, and that it also causes muscle fatigue. A third current method of SCI rehabilitation is the use of powered exoskeletons (Louie et al. 2015). These exoskeletons are placed on the paralyzed body part and use electrical stimulation to move. There is emerging data that suggests that BCIs that have an output function of moving an exoskeleton can serve as a useful tool in rehabilitation (Lazarou et al. 2018; Slutzky 2018).

1.3 Advantages of Using BCI Systems for Spinal Cord Injuries (SCIs)

One advantage of using BCIs for SCIs is that there is quite a bit of available software, tutorials, toolboxes, and existing data. The BCI competition data, which can be accessed online for free, is an abundant source of various EEG recordings from subjects, all using various paradigms to complete the data acquisition (Blankertz et al. 2006). There are various toolboxes with MATLAB functions from BCI lab that give users the ability to quickly and easily create BCIs. NeuroPype is a platform that can perform online BCI analysis. Software options like OpenVibe and BCI 2000 are functions BCIs that can be used for online and offline analyses.

Many of the studies in which the subjects were able to evaluate the enjoyability of their BCI rehabilitation responded positively to the treatment (Donati et al. 2016; Thomschewski et al. 2017; Yoshida et al. 2016). Because patients are likely to enjoy the process of therapy with BCIs, it is safe to assume this is an area that should continue to be pursued. If the patients did not enjoy the therapy at all, this might not be the right path to take. However, since it is an enjoyable experience for those with SCI, this area should continue to be pursued.

A study that tested the feasibility of BCIs on SCI patients noted that the subjects enjoyed the virtual game and rehabilitation therapy that was presented to them (Salisbury et al. 2016a, b). The study also found that the subjects that had head injuries performed essentially the same as those without head injuries. More testing needs to be done in this area, but it can be concluded right now that there is a possibility that BCI usage for SCI rehabilitation can be expanded to reach those with head injuries.

A study that surveyed 156 individuals with SCIs found that, in general, SCI patients were receptive and accepting the idea of using a BCI system to help with their rehabilitation (Blabe et al. 2015). 91% of the subjects with C1–4 level SCI said that they would try BCI rehabilitation if it could help with grasping motions. Additionally, 80–95% of the subjects said they would accept using eye-tracking glasses or some form of product that could help improve the accuracy of a BCI. The subjects were also twice as likely to use an implanted, wireless EEG for everyday use. This information is helpful in understanding that SCI patients are likely to make some sacrifices for the sake of rehabilitation performance. These sacrifices include wearing the eye-tracking glasses or enduring an operation for the implantable EEG system. Since these pieces of technology are crucial in implementing BCIs for everyday use, it is extremely important that the target population is on board with using them.

Another study with a similar goal also found that the majority of their subjects had a positive attitude toward implantable EEGs for BCI rehabilitation (Lahr et al. 2015). They also asked the individuals to discuss how they gained most information regarding assistive technology devices (ATDs), like the implantable EEG, BCI, exoskeletons, etc. It turns out that most subjects relied on the Internet for information regarding ATD, which could be beneficial to the progression of BCIs and integration into everyday rehabilitation if articles and important information can be made easily available to those with SCI and paralyzation.

Traditionally, these systems are big and bulky; however, there is research going into small and portable BCI systems (McCrimmon et al. 2016). A team designed a BCI system, specifically for motor rehabilitation, that was battery powered and portable. The system cost less than $200 and had respectable accuracy percentages. This area needs to be pursued more, as big bulky systems might not be easy to use and could take require large amounts of money, training, and space in order for them to work properly. An easy-to-use, small, and cheap system could prove to be the future for BCIs for SCI rehabilitation.

2 Motor Imagery Ability in Those with SCI

One concern with using MIBCIs with SCI patients is that their motor imagery ability might be diminished. Depending on how long the patient has had paralysis, they might not remember what moving their paralyzed limbs feels like. There might also be changes in neural plasticity among those with SCI, which can impact the accuracy of a BCI system.

One study attempted to assess, via questionnaires, how easy motor imagery is for SCI patients and healthy control subjects (Thomschewski et al. 2017). They instructed all subjects to perform motor imagery and then gave them a visual imagery questionnaire. It was determined that the SCI patients had a more difficult time completing the MI than the healthy subjects. In terms of mental health, the authors found that there was no statistical difference between the healthy subjects and SCI patients. This study, however, had a limitation of having only 7 patients and 22 controls; therefore, it is important that future studies include a more even number of patients and control subjects.

Another study also utilized questionnaires to assess the impact of MI on SCI patients. This study found that those with SCI had more neuropathic pain than the control group (Scandola et al. 2017). They also determined that those with neuropathic pain had a decreased performance on MI over time, and those without the pain did better over time. This indicates that more studies need to be done in how to decrease neuropathic pain in SCI individuals, since the decrease in pain will likely result in better MI performance for rehabilitation purposes.

In a study done to assess whether high-density EEGs improved the accuracy of MI classification in SCI patients, it was determined that the accuracies from the control group and the SCI group were very similar (Höller et al. 2018). The data showed that the high-density EEG did not specifically improve the classification accuracy for those with SCI. However, it is possible that testing more HD-EEG systems on SCI patients could be an important tool in improving MI classification accuracy.

2.1 BCIs and Virtual Reality (VR)

It is possible that combining BCIs and virtual reality (VR) could be an option in helping those with SCIs recover. In a case study done on a 25-year-old male patient suffering from tetraplegia, he was given BCI-based rehabilitation in which he had to move a virtual block either forward, to the left, or to the right by using a motor imagery-based BCI (Salisbury et al. 2015). The patient had a 92% success rate and identified that it was an enjoyable experience, rating it as 80/100, with 100 being most enjoyable and 0 unenjoyable. The subject had an easy time learning the paradigm and found it was a nice change from his typical rehabilitation routine. The experimenters stated that there is an interest in exploring how BCIs in SCI rehabilitation could be geared toward pain management for patients by using biofeedback and muscle relaxation techniques.

A study that tested the ability of SCI patients with hand paralyzation found a significant accuracy in MI to open and close a virtual hand (Foldes et al. 2015). In this study, three individuals with hand paralyzation due to SCI were able to accurately control the virtual hand 63% of the time.

Another study tested how well a virtual reality and implanted microelectrode arrays to regain movement in a paralyzed limb (Ajiboye et al. 2017). The study centered around a 53-year-old man with a C-4 SCI. He was implanted with a microelectrode array over the hand portion of his motor cortex, as well as in his right arm to stimulate muscle movement. The subject completed several sessions of training, with and without a virtual three-dimensional arm. After 463 days from the date of implantation, the subject was able to successfully drink from a mug, and after 717 days the subject was able to feed himself.

2.2 BCIs and Exoskeletons/Neuroprosthesis

The use of exoskeletons for SCI rehabilitation has been used in the past, but coupling it with a MIBCI can be used to make it an even stronger tool. There are, however, some areas that need to be explored and addressed. In an experiment where the experimenters were testing the accuracy of HD-EEG classification in SCI patients, it was discovered that MI accuracy for SCI patients decreased when compared to healthy subjects when they were asked to imagine foot movement (Höller et al. 2018). This was mostly because many of the SCI subjects were unable to move their feet.

Coupling exoskeletons and BCIs might also be useful in creating positive changes in the brain due to neuroplasticity. A study found that combining BCIs and robotic exoskeletons improved the localization of pain in SCI patients, as well as the proprioceptive sensing for SCI patients in the areas below their injury (Wirz and van Hedel 2018).

There is evidence that using electrical stimulation to the spinal cord and muscles can create movement (Capogrosso et al. 2016). A study using neuroprosthetics (Kilgore et al. 2018) showed that there is potential in using controllable algorithms to influence muscle contraction to create movement. Thirteen patients with cervical-level spinal cord injuries that resulted in tetraplegia were implanted with myoelectrically controlled neuroprosthesis. These devices were then controlled using various computer algorithms to affect the current flowing through. The study concluded that there is a significant amount of potential algorithms that can be used for these devices, as well as "improved active range of motion, grip strength, and the ability to pick up and release objects were improved in all arms tested" (Kilgore et al. 2018). From this information, it can be inferred that these algorithms could be combined with a BCI in order to allow the user more control over their neuroprosthesis, allowing them more independence in their daily lives (Bockbrader et al. 2018).

Continuing off of using BCIs and electrical stimulation, a study was conducted that combined a brain–computer interface with functional electrical stimulation therapy (FEST) and computer vision (CV) (Likitlersuang et al. 2018). This study included six subjects, one of which had a cervical spinal cord injury. The results found that this method for BCI had an accuracy of approximately 90% in the healthy subjects and 87.5% in the SCI subject. The purpose of using the FEST with BCI is to improve the movement ability of those with SCI by practicing grasping movements. With the SCI subject accuracy so close to the healthy subject accuracy, it is possible that these BCI and CV methods could be used to rehabilitate those with SCI.

A study done on a 26-year-old male SCI patient concluded that there is a significant accuracy when using BCIs with an implanted FES (Colachis et al. 2018). A FES was attached to the subject's right forearm and stimulated when the BCI system detected motor imagery. There were seven various handgrasping motions that were generated using the system. The accuracy for each type of movement was between 96 and 99% for the trials completed.

3 Challenges of Using BCI Systems for Spinal Cord Injuries

While there is promise in using BCIs for SCIs, there are several challenges that need to be tackled, with the first being the need for BCIs to be conducted in a quiet, distraction-free area for the highest accuracy. In one virtual reality case study (Salisbury et al. 2015), it was difficult to find areas in a hospital setting that were adequate for this form of rehabilitation. It was also noted that it is important to balance the patient's life outside the BCI rehabilitation with the mental and physical fatigue produced from altering a virtual environment.

There are several ethical dilemmas that are associated with using BCIs for SCI rehabilitation (Burwell et al. 2017). The first being that there is a possibility that using BCIs on children with SCI injuries could have negative effects due to their developing brains. This is an area that has not really been explored, especially because most studies in this area are performed on people above 20 years old. There is a possibility that it could cause irreversible damage; however, this needs to be explored more.

There is also worry that invasive BCI systems might cause physical damage. In the future, BCI systems will probably be made to be invasive so they can be used constantly in a patient. However, there is a possibility this can cause acute brain damage. Implanting an electrocorticographic (ECoG) device also requires surgery, which could lead to further complications during the operation.

These systems are never 100% accurate and, when used in a practical setting, could frustrate the user when they do not work properly. If a patient relies on their BCI system for movement of their exoskeleton, for example, and it does not work when they need it to, it can be extremely frustrating for the patient. There is also a concern for dangerous scenarios if the system does not work properly. For instance, if a tetraplegic patient relies on a wheelchair for movement and it is controlled by a BCI system, it is safest if it works all the time. However, if the patient is trying to cross a busy street and their system stops working, it can be extremely dangerous. Therefore, if systems like this are to be implemented in daily living, the accuracy needs to be very close to 100% to prevent events like this from occurring.

Looking at just the exoskeleton, there are some challenges regarding the complexity of their design. According to a review on exoskeletons (Gorgey 2018), exoskeletons are very expensive and require training for the individuals that plan to place them on the patients and supervise their use. The patient-specific fitting of these exoskeletons takes up to 30 min, and on top of that requires an hour to put on. Exoskeletons are also slow and do not accurately reflect the speed of human movement. Also, they are prone to damaging the bones and joints of those who wear them due to the unnatural movement they produce (Tariq et al. 2018).

Many of the studies discussed in this chapter had a fairly low number of subjects that actually suffered from SCI. Since those that suffer from SCI have some form of paralysis, they experience fairly dramatic neural plasticity. Because of this, traditional BCI systems may not have adequate accuracy when compared to those that were calibrated with subjects without SCI. There is evidence that BCI systems for SCI patients could help them regain movement through changes in neural plasticity (Mateo et al. 2015); however, the changes in neural plasticity amongst different types of SCI injuries could show that this is not a possibility for all patients.

A majority of these studies has mostly male subjects, with some being exclusively male (Thomschewski et al. 2017). The lack of female subjects poses an issue in future application for female SCI patients that use BCI rehabilitation. A study found that there are significant connectivity differences between the male and female brain (Glezerman 2016; Madhura et al. 2014). Therefore, there is potential that these differences could result in higher accuracy for one sex over the other. It is crucial that more studies are done that evaluate the accuracy of current methods for both sexes equally to see if different methods need to be done for each sex.

One study analyzed how easy it was for SCI patients to perform motor imagery versus a control group of healthy individuals. All subjects were asked to perform several motor imagery tasks and then assess their difficulty with a questionnaire. They noted that it was significantly more difficult for SCI patients to imagine movement when compared to the control group (Thomschewski et al. 2017). They also noted that it was more difficult for older SCI patients to perform motor imagery compared to younger patients. This indicates that in order for a MIBCI system to work well for SCI rehabilitation, there might need to be some preliminary tasks that help strengthen their motor imagery skills.

Moreover, one of the major obstacles in the ubiquitous use of BCIs is the need for subject-specific calibration. Generally, computation models built during the calibration phase perform poorly in the case of inter-subject analysis, meaning that a BCI system calibrated on one subject has a low chance of being good for some other subject. Recently, there has been a lot of ongoing work (Jayaram et al. 2016; Zanini et al. 2018) in transfer learning and adaptive frameworks for BCIs. A future direction of work could be the analysis of subject-to-subject transfer learning for BCIs in case of SCI patients as it will help in a considerable reduction of time in BCI calibration for patients.

4 Summary

To sum up, brain–computer interfaces can be a very useful tool in the rehabilitation for those with spinal cord injuries. From virtual reality training to exoskeleton and neuroprosthesis control, novel methods of improving the daily lives of those with SCI are becoming more understood. With more research, BCIs will hopefully be able to be implemented in the rehabilitation of several patients with SCI.

There are several points that need to be improved upon in order for this to become a reality. First, more studies need to be conducted on SCI patients, with even more being done in a hospital setting. The neural plasticity of SCI patients and their ability to imagine motor actions are different from able-bodied subjects. Therefore, in order to develop a system that is geared toward those with SCI and paralyzation, more information must be obtained from actual SCI patients. Furthermore, since hospitals are typically more noisy and fast-paced, noise and other forms of disruptions can affect the signals obtained. The data can be skewed from noise in the signal itself, making signal processing more challenging and possibly leading to a poorer system performance. It can also cause disruptions for the user, which in turn can produce poor signals that are not categorized properly by the system. Therefore, it is important that future BCI research starts taking these concepts into account when testing BCI systems.

Online BCIs have shown suboptimal performances in outside-lab settings due to low signal-to-noise ratio and the presence of nonstationarity in signals. One of the important directions of future works would be to design adaptive machine learning and signal processing frameworks for BCI which can take into account the current mental workload and state of the mind (i.e., fatigue) which can eventually lead to increase in performance of online BCIs. Portability and time taken to set up a BCI is one of the limitations behind its pervasive use. Consider a case when the rehabilitation is at such a stage where subjects need to walk, a portable BCI amplifier mounted on the head can be much more effective in making the rehabilitation experience real.

Future BCI research also needs to be more geared toward testing on more equal concentrations of male and female subjects. The majority of the studies reviewed had very high male to female ratios, with only one study having a greater concentration of females than males. Male and female neural plasticity is different, which could affect the accuracy of these systems. It seems that, at the moment, they are geared more toward males with SCI. Hopefully, in the future, more research can be done on women with SCI to ensure that the system is useful to both genders.

Extensive studies in the perspective of neuroscience with a large number of clinical trials in BCI-based rehabilitation are needed to understand the functional neuroplastic mechanisms of BCI learning. These research efforts would be considerably crucial for finally realizing motor recovery.

On the other hand, we consider that designing more demanding rehabilitation therapies based on virtual or augmented reality platforms could significantly benefit to enhance the motivation and intensity of rehabilitative training. This is also worth studying with different BCI-based rehabilitation systems to achieve the optimal physiotherapy effect. Developing portable EEG-based systems with dry electrodes and visual feedback may also be important to move BCI learning outside the laboratory, thereby further extending the practical applications of BCI-based rehabilitation.

Below are our recommendations for future BCI-assisted rehabilitation for SCIs:

- ***More Testing in Hospital Settings***: Currently, using BCIs for rehabilitation is mostly studied in lab settings, and systems calibrated in lab settings perform poorly in hospital settings. This is because hospitals have a low signal-to-noise ratio that decreases the accuracy of BCI systems. More studies focusing on using these systems in hospitals would improve their effectiveness.
- ***Study the Effects of BCIs on Underdeveloped Brains***: There is not enough literature on using BCIs for SCI rehabilitation on people under 20 years old. More studies need to be conducted on underdeveloped brains to see if there are any negative impacts.
- ***Improve Exoskeletons***: Most exoskeletons used for this purpose need significant training for physical therapists to set up and use. Making them simpler yet equally effective would increase the possibility of using them with BCIs for rehabilitation.
- ***Conduct more Studies on SCI Patients***: The literature that studies the effectiveness of BCIs for SCI rehabilitation usually have a very high ratio of controls to patients. Evening this out more would improve the information gained on how BCIs impact a variety of patients, due to the differences in neural plasticity between controls and patients, as well as between patients.

With continued and more specialized research using BCIs for those with SCI, rehabilitation with these systems is a very real possibility.

References

Ajiboye, A. B., Willett, F. R., Young, D. R., Memberg, W. D., Murphy, B. A., Miller, J. P., … & Peckham, P. H. (2017). Restoration of reaching and grasping movements through brain-controlled muscle stimulation in a person with tetraplegia: a proof-of-concept demonstration. *The Lancet, 389*(10081), 1821–1830.

Alam, M., Rodrigues, W., Pham, B. N., & Thakor, N. V. (2016). Brain-machine interface facilitated neurorehabilitation via spinal stimulation after spinal cord injury: Recent progress and future perspectives. *Brain Research, 1646,* 25–33.

Ang, K. K., Chua, K. S. G., Phua, K. S., Wang, C., Chin, Z. Y., Kuah, C. W. K., et al. (2015). A randomized controlled trial of EEG-based motor imagery brain-computer interface robotic rehabilitation for stroke. *Clinical EEG and Neuroscience, 46*(4), 310–320.

Blabe, C. H., Gilja, V., Chestek, C. A., Shenoy, K. V., Anderson, K. D., & Henderson, J. M. (2015). Assessment of brain-machine interfaces from the perspective of people with paralysis. *Journal of Neural Engineering, 12*(4), 043002.

Blankertz, B., Muller, K. R., Krusienski, D. J., Schalk, G., Wolpaw, J. R., Schlogl, A., … & Birbaumer, N. (2006). The BCI competition III: Validating alternative approaches to actual BCI problems. *IEEE Transactions on Neural Systems and Rehabilitation Engineering, 14*(2), 153–159.

Bockbrader, M. A., Francisco, G., Lee, R., Olson, J., Solinsky, R., & Boninger, M. L. (2018). Brain computer interfaces in rehabilitation medicine. *PM&R, 10*(9), S233–S243.

Burwell, S., Sample, M., & Racine, E. (2017). Ethical aspects of brain computer interfaces: A scoping review. *BMC Medical Ethics, 18*(1), 60.

Capogrosso, M., Milekovic, T., Borton, D., Wagner, F., Moraud, E. M., Mignardot, J., … Courtine, G. (2016). A brain-spine interface alleviating gait deficits after spinal cord injury in primates. *Nature, 539*(7628), 284–288.

Chaudhary, U., Birbaumer, N., & Ramos-Murguialday, A. (2016). Brain–computer interfaces for communication and rehabilitation. *Nature Reviews Neurology, 12*(9), 513.

Colachis, S. C., Bockbrader, M. A., Zhang, M., Friedenberg, D. A., Annetta, N. V., Schwemmer, M. A., … Sharma, G. (2018). Dexterous control of seven functional hand movements using cortically-controlled transcutaneous muscle stimulation in a person with tetraplegia. *Frontiers in Neuroscience, 12,* 208.

Donati, A. R. C., Shokur, S., Morya, E., Campos, D. S. F., Moioli, R. C., Gitti, C. M., … Nicolelis, M. A. L. (2016). Long-term training with a brain-machine interface-based gait protocol induces partial neurological recovery in paraplegic patients. *Scientific Reports, 6,* 30383.

Foldes, S. T., Weber, D. J., & Collinger, J. L. (2015). MEG-based neurofeedback for hand rehabilitation. *Journal of Neuroengineering and Rehabilitation, 12,* 85. https://doi.org/10.1186/s12984-015-0076-7.

Frolov, A. A., Mokienko, O., Lyukmanov, R., Biryukova, E., Kotov, S., Turbina, L., et al. (2017). Post-stroke rehabilitation training with a motor-imagery-based brain-computer interface (BCI)-controlled hand exoskeleton: a randomized controlled multicenter trial. *Frontiers in neuroscience, 11,* 400.

Glezerman, M. (2016). Yes, there is a female and a male brain: Morphology versus functionality. *Proceedings of the National Academy of Sciences of the United States of America, 113*(14), E1971.

Gorgey, A. S. (2018). Robotic exoskeletons: The current pros and cons. *World Journal of Orthopedics, 9*(9), 112–119.

Höller, Y., Thomschewski, A., Uhl, A., Bathke, A. C., Nardone, R., Leis, S., … Höller, P.(2018). HD-EEG based classification of motor-imagery related activity in patients with spinal cord injury. *Frontiers in Neurology, 9,* 955.

Jayaram, V., Alamgir, M., Altun, Y., Scholkopf, B., & Grosse-Wentrup, M. (2016). Transfer learning in brain-computer interfaces. *IEEE Computational Intelligence Magazine, 11*(1), 20–31.

Kilgore, K. L., Bryden, A., Keith, M. W., Hoyen, H. A., Hart, R. L., Nemunaitis, G. A., et al. (2018). Evolution of neuroprosthetic approaches to restoration of upper extremity function in spinal cord injury. *Topics in Spinal Cord Injury Rehabilitation, 24*(3), 252–264.

Lahr, J., Schwartz, C., Heimbach, B., Aertsen, A., Rickert, J., & Ball, T. (2015). Invasive brain-machine interfaces: A survey of paralyzed patients' attitudes, knowledge and methods of information retrieval. *Journal of Neural Engineering, 12*(4), 043001.

Lazarou, I., Nikolopoulos, S., Petrantonakis, P. C., Kompatsiaris, I., & Tsolaki, M. (2018). EEG-based brain-computer interfaces for communication and rehabilitation of people with motor impairment: A novel approach of the 21st century. *Frontiers in Human Neuroscience, 12,* 14.

Lebedev, M. A., & Nicolelis, M. A. L. (2017). Brain-machine interfaces: From basic science to neuroprostheses and neurorehabilitation. *Physiological Reviews, 97*(2), 767–837.

Likitlersuang, J., Koh, R., Gong, X., Jovanovic, L., Bolivar-Tellería, I., Myers, M., ... Márquez-Chin, C. (2018). EEG-controlled functional electrical stimulation therapy with automated grasp selection: A proof-of-concept study. *Topics in Spinal Cord Injury Rehabilitation, 24*(3), 265–274.

Louie, D. R., Eng, J. J., & Lam, T. (2015). Gait speed using powered robotic exoskeletons after spinal cord injury: a systematic review and correlational study. *Journal of neuroengineering and rehabilitation, 12*(1), 82.

Madhura, I., Alex, S., Drew, P., Theodore, D. S., Mark, A. E., Kosha, R., ..., Ragini, V. (2014). Sex differences in the structural connectome of the human brain. *Proceedings of the National Academy of Sciences of the United States of America,* 111(2), 823-828. https://doi.org/10.1073/pnas.1316909110.

Mateo, S., Di Rienzo, F., Bergeron, V., Guillot, A., Collet, C., & Rode, G. (2015). Motor imagery reinforces brain compensation of reach-to-grasp movement after cervical spinal cord injury. *Frontiers in Behavioral Neuroscience, 9,* 234.

McCrimmon, C. M., Ming Wang, n., Silva Lopes, L., Wang, P. T., Karimi-Bidhendi, A., Liu, C. Y., ... Do, A. H. (2016). A small, portable, battery-powered brain-computer interface system for motor rehabilitation. Conference Proceedings: ... Annual International Conference of the IEEE Engineering in Medicine and Biology Society. *IEEE Engineering in Medicine and Biology Society. Annual Conference*, 2016, 2776–2779.

Pfurtscheller, G., Brunner, C., Schlögl, A., & Da Silva, F. L. (2006). Mu rhythm (de) synchronization and EEG single-trial classification of different motor imagery tasks. *NeuroImage, 31*(1), 153–159.

Pichiorri, F., Morone, G., Petti, M., Toppi, J., Pisotta, I., Molinari, M., et al. (2015). Brain–computer interface boosts motor imagery practice during stroke recovery. *Annals of Neurology, 77*(5), 851–865.

Pichiorri, F., Mrachacz-Kersting, N., Molinari, M., Kleih, S., Kübler, A., & Mattia, D. (2017). Brain-computer interface based motor and cognitive rehabilitation after stroke–state of the art, opportunity, and barriers: summary of the BCI Meeting 2016 in Asilomar. *Brain-Computer Interfaces, 4*(1–2), 53–59.

Salisbury, D. B., Driver, S., & Parsons, T. D. (2015). Brain-computer interface targeting non-motor functions after spinal cord injury: A case report. *Spinal Cord, 53*(S1), S25.

Salisbury, D. B., Dahdah, M., Driver, S., Parsons, T. D., & Richter, K. M. (2016). Virtual reality and brain computer interface in neurorehabilitation. *Proceedings (Baylor University. Medical Center), 29*(2), 124–127.

Salisbury, D. B., Parsons, T. D., Monden, K. R., Trost, Z., & Driver, S. J. (2016b). Brain-computer interface for individuals after spinal cord injury. *Rehabilitation Psychology, 61*(4), 435–441.

Scandola, M., Aglioti, S. M., Pozeg, P., Avesani, R., & Moro, V. (2017). Motor imagery in spinal cord injured people is modulated by somatotopic coding, perspective taking, and post-lesional chronic pain. *Journal of Neuropsychology, 11*(3), 305–326.

Slutzky, M. W. (2018). Brain-machine interfaces: powerful tools for clinical treatment and neuroscientific investigations. *The Neuroscientist*, 1073858418775355.

Tariq, M., Trivailo, P. M., & Simic, M. (2018). EEG-based BCI control schemes for lower-limb assistive-robots. *Frontiers in Human Neuroscience, 12,* 312.

Thomschewski, A., Ströhlein, A., Langthaler, P. B., Schmid, E., Potthoff, J., Höller, P., ... Höller, Y. (2017). Imagine there is no plegia. mental motor imagery difficulties in patients with traumatic spinal cord injury. *Frontiers in Neuroscience*, *11*, 689.

Vidal, J. J. (1973). Toward direct brain-computer communication. *Annual review of Biophysics and Bioengineering, 2*(1), 157–180.

Wirz, M., van Hedel, & Hubertus, J. A. (2018). Balance, gait, and falls in spinal cord injury. *Handbook of Clinical Neurology, 159,* 367–384.

Yoshida, N., Hashimoto, Y., Shikota, M., & Ota, T. (2016). Relief of neuropathic pain after spinal cord injury by brain-computer interface training. *Spinal Cord Series and Cases, 2,* 16021.

Zanini, P., Congedo, M., Jutten, C., Said, S., & Berthoumieu, Y. (2018). Transfer learning: a Riemannian geometry framework with applications to brain–computer interfaces. *IEEE Transactions on Biomedical Engineering, 65*(5), 1107–1116.

A Sensorimotor Rhythm-Based Brain–Computer Interface Controlled Functional Electrical Stimulation for Handgrasp Rehabilitation

Inchul Choi, Na Young Kim and Chang S. Nam

Abstract Each year, 795,000 stroke patients suffer a new or recurrent stroke and 235,000 severe traumatic brain injuries (TBIs) occur in the US. These patients are susceptible to a combination of significant motor, sensory, and cognitive deficits, and it becomes difficult or impossible for them to perform activities of daily living due to residual functional impairments. Recently, sensorimotor rhythm (SMR)-based brain–computer interface (BCI)-controlled functional electrical stimulation (FES) has been studied for restoration and rehabilitation of motor deficits. To provide future neuroergonomists with the limitations of current BCI-controlled FES research, this chapter presents the state-of-the-art SMR-based BCI-controlled FES technologies, such as current motor imagery (MI) training procedures and guidelines, an EEG-channel montage used to decode MI features, and brain features evoked by MI.

1 Introduction

Each year, 795,000 stroke patients suffer a new or recurrent stroke and 235,000 severe traumatic brain injuries (TBIs) occur in the US. These patients are susceptible to a combination of significant motor, sensory, and cognitive deficits, and it becomes difficult or impossible for them to perform activities of daily living due to residual functional impairments. Recently, sensorimotor rhythm (SMR)-based brain–computer interface (BCI)-controlled functional electrical stimulation (FES) has been studied for restoration and rehabilitation of motor deficits. However, there still remain many questions to be addressed. To provide neuroergonomists with the limitations of current BCI-controlled FES research, this chapter presents the state-of-the-art SMR-based BCI-controlled FES technologies, such as current motor imagery (MI) training procedures and guidelines, an EEG-channel montage used to decode MI features, and brain features evoked by MI.

I. Choi · N. Y. Kim · C. S. Nam (✉)
Edward P. Fitts Department of Industrial & Systems Engineering, North Carolina State University, Raleigh, NC 27695, USA
e-mail: csnam@ncsu.edu

C. S. Nam (ed.), *Neuroergonomics*, Cognitive Science and Technology,
https://doi.org/10.1007/978-3-030-34784-0_17

The remainder of this chapter is organized as follows. Section 2 reviews SMR-based BCI-controlled FES systems in order to identify limitations of current SMR-based BCIs for FES studies. Section 3 presents the state-of-the-art SMR-based BCIs for FES in terms of motor imagery instruction and tasks in current research and signal processing methods for SMR-based BCIs, along with limitations of current research. Finally, Sect. 4 concludes by presenting a list of recommendations for studies on BCI-controlled FES systems.

2 FES Systems for SMR-Based BCI

2.1 Systematic Review Procedure

For the systematical review, the preferred reporting items for systematic reviews and meta-analyses (PRISMA) method was applied (Moher et al. 2015). The search engines used in this review were Web of Science, PubMed, and Engineering Village, because these search engines cover both engineering and medical publications (Powers et al. 2015). The keywords used in the search engines were a combination of either "brain–computer interface" or "brain–machine interface" and either "functional electrical stimulation" or "neuromuscular electrical stimulation." After searching for keywords, 94, 51, and 42 articles were searched by each search engine, respectively, and 97 articles were left after removing duplicate articles. Afterward, the remaining articles were screened out based on the titles and abstracts related to BCI-controlled FES systems only. Prescreened 53 articles were checked for eligibility based on full-text screening with some exclusion criteria, such as (1) articles for non-journal, book chapter, or review papers; (2) experiments for nonhuman primates; (3) invasive brain imaging methods and implanted electrical stimulation systems; (4) articles for neuroscience studies that did not focus on motor function restoration; and (5) studies not related to hand and wrist control. After eligibility screening, only 18 studies remained for the review of the FES systems. Figure 1 shows the screening procedures and results.

2.2 FES Systems in Current Research

To determine the extent of application of critical features that should be applied when building a customized FES system, the FES systems used in 18 selected papers were analyzed. As a result, nine different FES systems were discovered, three of which did not support real-time computer control and were customized to meet the authors' specific research objectives. The number of electrode channels supported by these FES units is between 2 and 8 channels, except for universal external control unit, which supports up to 32 channels. Pulse rate of these devices are up to 150 Hz, but

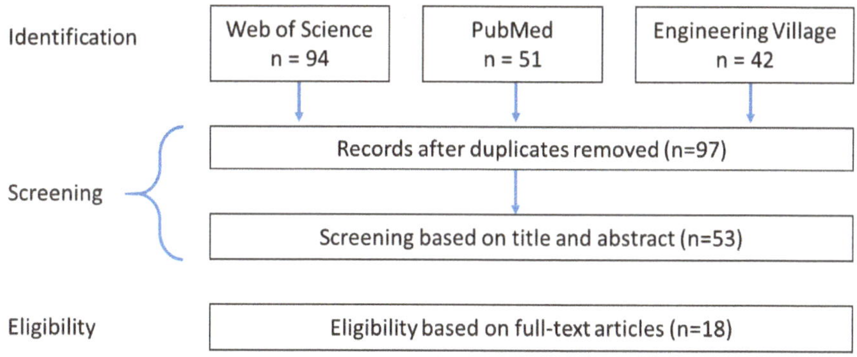

Fig. 1 PRISMA flow diagram for FES systems used in BCI studies

most can cover from 1 to 99 Hz. The range of pulse width is between 10 and 500 μs except for Compex Motion, which supports up to 16,000 μs, and the range of current amplitude is up to 150 mA. Table 1 summarizes the names and parameters of each FES system used in these studies.

2.3 *FES Electrode Placement and Parameters in Current Research*

FES electrodes should be placed on the correct position according to the target motions in order to avoid unnatural and uncomfortable movements (Lawrence 2009; Lyons et al. 2004). Therefore, precise electrode placement is important for generating natural hand-wrist motions. However, the proper electrode positions depend on target hand motions. Furthermore, appropriate electrode placement also varies from person to person due to anatomical differences in length and mass of the forearm (Gordon et al. 1989). Despite of the importance, 2 of 18 articles did not mention information about electrode placement, and the other articles also ambiguously described the name of muscle without referring the exact location as given in Table 2, which summarizes the authors' comments from the literature review. For example, Young et al. (2014) mentioned that a pair of electrodes was placed on the extensor carpi radialis longus/brevis and on the portion of the extensor digitorum to generate the extension of wrist and fingers.

In addition, there were only two studies mentioning the size and shape of the electrode pads used, and only five of them specified the number of electrodes utilized. Although Forrester and Petrofsky (2004) disputed the effects of electrode pad size and shape on the tolerance of electrical stimulation, sufficient studies support the size and shape of electrodes as important factors for motor functionality and user comfort (Lyons et al. 2004). As mentioned earlier, FES parameters are also important factors, and pulse rate, current amplitude, and pulse width are the most important to

Table 1 FES systems used in the previous studies

FES system	# of channels	Pulse rate (Hz)	Pulse width (μs)	Current amplitude (mA)	Used research studies
Universal external control unit	32	1–80	~255	0–60	Daly et al. (2009)
RehaStim	8	1–140	20–500	0–130	Elnady et al. (2015), Reynolds et al. (2015), Vuckovic et al. (2015)
Compex motion	4	1–100	75–16,000	0–125	McGie et al. 2015, Tan et al. (2011)
MEB-2200	2	–	–	–	Mukaino et al. (2014)
MOTIONSTIM 8	8	1–99	10–500	0–125	Pfurtscheller et al. (2003, 2005), Rohm et al. (2010, 2013)
NESS H200	5	18 or 36	10–500	0–150	Roset et al. (2014)
Microstim (customized)	2	1–99	~250	1–99	Kim et al. 2016)
EMPI 300PV (customized)	2	1–99	10–500	0–100	Looned et al. (2014)
LG-7500 (customized)	2	1–150	30–260	0–90	Young et al. (2014, 2015)

achieve the proper motor function. Most studies used a fixed pulse rate but varied between 16 and 83.3 Hz. The pulse width was also different in each study, and all but one study (McGie et al. 2015) adopted a fixed pulse width between 100 and 300 μs. Current amplitude was usually set to be sufficient for muscle contraction but not uncomfortable. Vuckovic et al. (2015) clearly stated the use of ramp time to reduce muscle fatigue, but not other research studies. Table 3 summarizes detailed information on the parameters utilized, with the exception of three studies that did not report information on FES parameters.

Table 2 Comments for electrode placement in the previous studies

Article	Electrode placement
Daly et al. (2009)	Placed over the indicis proprius muscle and the portion of the extensor digitorum communis muscle serving the index finger
Elnady et al. (2015)	Attached to the stroke-affected extensor digitorum
Tan et al. (2011)	One electrode was placed proximally over the forearm below the elbow, and the other was placed distally on the forearm (positioned for optimally balanced joint movement)
Kim et al. (2016)	FES was triggered and stimulated wrist extensor muscles of the affected upper extremity
Liu et al. (2014)	FES is triggered and delivered to patients' extensor carpus radialis muscles, which causes real movement of their hands or arms
Looned et al. (2014)	Hand opening was achieved by placing two electrodes on the distal and proximal ends of the extensor digitorum muscles of the forearm [35]
McGie et al. (2015)	Stimulate the following muscles: flexor carpi radialis, flexor digitorum superficialis, and flexor digitorum profundus (FCR/FDS/FDP), abductor pollicis brevis/flexor pollicis brevis (APB/FPB/OP), and extensor digitorum (for wrist/finger extension)
Mukaino et al. (2014)	Placed over the muscle belly of the extensor digitorum communis (EDC) on the paralyzed side
Pfurtscheller et al. (2003)	The finger (M. ext. digitorum communis EDC) and thumb (M. ext. pollicis longus EPL) extensors for hand opening, the finger flexors (M. flex. digitorum superficialis FDS, M. flex. digitorum profundus) hand closing, the thumb flexor (M. flex. pollicis longus FPL) for grasping, and the wrist extensors (M. ext. carpi radialis longus/brevis ECRL/ECRB) for stabilization of the hand
Pfurtscheller et al. (2005)	An opening of the hand (phases 1 and 4, Fig. 1) by extension of all fingers joints and the thumb could be achieved by stimulation of the finger extensors (M. extensor digitorum communis) and the thumb extensor muscle (M. extensor pollicis longus) with electrodes on the radial side of the proximal forearm. For the actual grasping (phase 2, Fig. 1), we simultaneously stimulated the finger flexors (M. flexor digitorum superficialis, less the M. flexor digitorum profundus) by one pair of electrodes on the ulnar side of the proximal forearm and the intrinsic hand muscles with two further electrodes on the dorsal side of the hand
Reynolds et al. (2015)	Placed on the right-hand extensor muscles
Rohm et al. (2010)	–
Rohm et al. (2013)	The finger extensors (extensor digitorum communis; electrode pair (EP) 1 in Fig. 1a), the finger flexors (flexor digitorum superficialis, flexor digitorum profundus; EP 2 in Fig. 1b), and the thumb extensor (extensor pollicis longus; EP 3 in Fig. 1a) and flexor (flexor pollicis longus; EP 4 in Fig. 1b) of the right hand were stimulated via four separate pairs of surface electrodes

(continued)

Table 2 (continued)

Article	Electrode placement
Roset et al. (2014)	Extensor muscles (extensor digitorum communis and extensor pollicis brevis)
Vuckovic et al. (2015)	The wrist extensor muscles (extensor carpi radialis longus)
Tan et al. (2011)	Extensor carpi radialis
Young et al. (2014)	Extensor carpi radialis brevis and extensor digitorum muscles
Young et al. (2015)	Placed over the indicis proprius muscle and the portion of the extensor digitorum communis muscle serving the index finger

Table 3 FES parameters in the previous studies

Article	Pulse rate (Hz)	Pulse width (μs)	Current amplitude
Daly et al. (2009)	83.3	255	Comfortable amplitude
Elnady et al. (2015)	35	150	Comfortable amplitude
Tan et al. (2011)	25	250	Comfortable amplitude
Kim et al. (2016)	60	150	20–27 mA
Liu et al. (2014)	60	150	25 mA
Looned et al. (2014)	25	200	Comfortable amplitude
McGie et al. (2015)	40	0–250	Comfortable amplitude
Mukaino et al. (2014)	20	100	15–20 mA
Pfurtscheller et al. (2003)	16	300	–
Pfurtscheller et al. (2005)	18	–	–
Reynolds et al. (2015)	30	200	10–17 mA
Rohm et al. (2010)	–	–	–
Rohm et al. (2013)	16	–	–
Roset et al. (2014)	35	300	Amplitude for the maximal contraction * 1.25
Vuckovic et al. (2015)	30	300	15 mA
Tan et al. (2011)	40	300	50–80 mA
Young et al. (2014)	–	–	–
Young et al. (2015)	–	–	–

2.4 *Limitations of Current Research*

From the systematic literature review of the FES systems for SMR-based BCI, three main issues have been identified and are described below.

First, most FES units used in the previous research were either not commercially (readily) available in the United States or did not support real-time computer control. Also, some functions, such as ramp time (Doucet et al. 2012), were not clearly specified in the articles and specification sheets provided by manufacturers. Second, most of the surveyed research studies did not precisely define the exact electrode positions but vaguely described the name of the muscles where FES electrodes were placed. Furthermore, there was no clear guidance to help identify the proper electrode location with respect to anthropometric data, such as the length and mass of forearm, which varies for each individual. Third, there were a few studies that considered the HF/E perspective by applying user-specific pulse rate and pulse width to increase user satisfaction, but other studies arbitrarily selected fixed parameters. Therefore, the range of parameters was very wide. In addition, only a few studies applied proper ramp time for current amplitude and pulse rate to achieve natural hand and wrist movements with minimum muscle pain and discomfort.

The summary of limitations and findings in current research studies is as follows:

- FES units with real-time control are not readily available in the United States. Therefore, a customized FES platform is required.
- A proposed FES platform should support a sufficient number of FES electrode channels (up to 8) and a wide range of FES parameters.
- Most of the studies did not precisely define the exact electrode location. There were no procedures to help identify the proper electrode position with respect to anthropometric data.
- There were only a few studies that considered FES parameter setting procedure for user-specific pulse rate and pulse width to improve user satisfaction by minimizing perceived muscle pain and discomfort.

3 Review on SMR-Based BCIs for FES

3.1 Systematic Review Procedure

For the systematic review, the PRISMA method was applied in the same way as in Sect. 2. After keyword searching, 35, 20, and 24 articles were found from each search engine, respectively, with 47 articles left after duplicates were removed. The remaining articles were then screened out based on the titles and abstracts related to SMR-based BCI studies.

Prescreened articles were checked for eligibility based on full-text screening with some inclusion criteria, such as (1) review articles for SMR-based BCI studies and (2) using EEG brain imaging technology. After eligibility screening, 26 studies including 18 studies from the previous review were selected for a review of SMR-based BCI. Figure 2 shows the steps and results of screening.

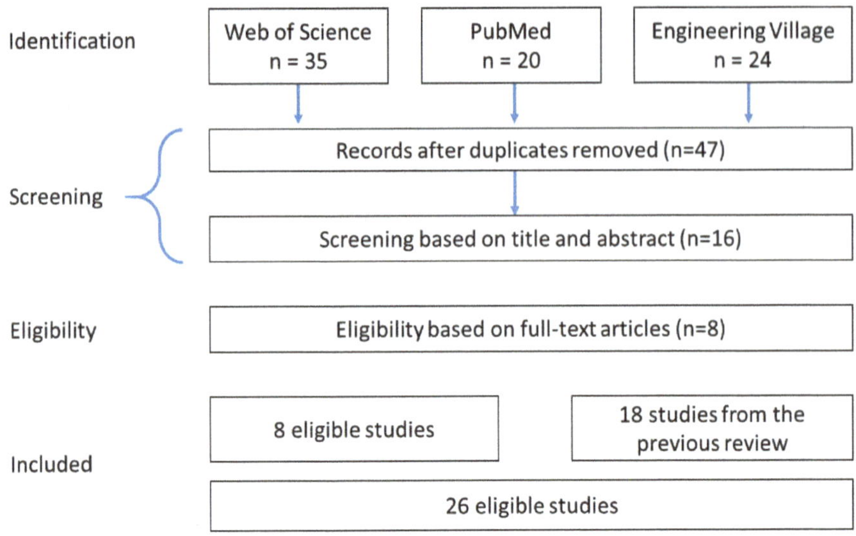

Fig. 2 PRISMA flow diagram for SMR-based BCI studies

3.2 MI Instruction and Tasks in Current Research

An MI task is a mental imagination that does not involve physical movements or physiological feedback (Ang and Guan 2015). This makes SMR-based BCI difficult because (1) experimenters or physical therapists cannot know whether the subject or patient completed MI tasks in the right way; (2) it is somewhat difficult to explain to users how to correctly perform MI tasks; and (3) it could be hard to maintain exactly the same MI tasks during the experiment. To address these issues, clear instructions are required to help both participants and experimenters during training. However, according to the literature review, only two studies (Kim et al. 2016; Looned et al. 2014) provided video clips as visual guidance to help subjects mimic MI tasks, while other researchers verbally explained MI tasks.

MI tasks also varied between studies. Some studies had asked participants to imagine wrist or finger movements (Daly et al. 2009; Tan et al. 2011), while others asked them to imagine waving or reaching motions (Reynolds et al. 2015; Young et al. 2015). Moreover, the frequency of MI tasks varied from study to study. Three studies (Elnady et al. 2015; Mukaino et al. 2014; Reynolds et al. 2015) applied repetitive MI tasks, but other studies applied one-time MIs or did not clearly mention the frequency of MI tasks. Table 4 shows authors' instructions for MI tasks mentioned in the surveyed research studies, and what the authors had asked the users to do.

Although not found in the systematic literature review, there were few studies that utilized distinct SMR features induced by different rhythmic MIs. Gu et al. (2009) examined the identifiability of four MI tasks consisting of two MI types (wrist extension and rotation) and two speeds (fast and slow), and the authors reported that

Table 4 Instruction for MI tasks in the previous studies

Article	Instruction
Daly et al. (2009)	"…attempted finger movement and relax conditions or imagined finger movement and relax conditions."
Elnady et al. (2015)	"…asked to perform different repetitive imagery tasks according to the stimulus or visual cues…"
Tan et al. (2011)	"…subjects used hand grasping and wrist flexion/extension movements to elicit motor imagery."
Kim et al. (2016)	"…were advised to keep focusing on their affected arm/hand action observational tasks." (with video)
Liu et al. (2014)	"…a bold arrow … instructing patients to imagine left or right."
Looned et al. (2014)	"…asked to imagine the corresponding motion for each displayed action…"
McGie et al. (2015)	"…asked to perform motor imagery pertaining to a grasping movement, …"
Mukaino et al. (2014)	"…asked to attempt finger opening for 3 s at maximal voluntary effort." (at 1 Hz frequency)
Pfurtscheller et al. (2003)	"…the subject imagines a foot movement, …"
Pfurtscheller et al. (2005)	"Imaginations of left versus right hand movements were carried out… single-foot motor imageries versus relaxing or hand movement imagination …"
Reynolds et al. (2015)	"…instructed to imagine waving with their right hand with a frequency of about 0.5 Hz as soon as they saw the execution cue."
Rohm et al. (2010)	–
Rohm et al. (2013)	–
Roset et al. (2014)	"…instructed the person to either open or close his hand." (with visual cues)
Vuckovic et al. (2015)	"…instructed to perform kinesthetic motor imagery of activities performed mentally with their left or right hand" (according to the cue)
Tan et al. (2011)	"…instructed to perform motor imagery of his or her wrist on the affected side."
Young et al. (2014)	"…taught to use attempted movements of each hand to control the BCI device."
Young et al. (2015)	"…instructed to use attempted movements of the right or left hand to drive the cursor…"

the speed variable had greater classification ability than the MI types. Similarly, Choi et al. (2016) applied slow-continuous MI and fast-transient MI to classify a 2-class MI task in a single hand such as grasping or opening, and the results were promising in terms of classification accuracy.

In terms of imagination tasks, some researchers applied two different conditions, rest and MI, to detecting the presence of SMR, and the user could control a device to initiate or stop the operation by classifying brain signals. In contrast, other researchers utilized two MIs, one hand MI and the other hand (or foot) MI, to classify the directional SMR. In this case, the most common cues that represented the imagination tasks found in the literature review were arrows according to target directions (Liu et al. 2014). However, in these two cases, it was impossible or unnatural to implement hand opening or closing operation through the FES system by classifying SMRs. As given in Table 5, there was only one study that investigated different MI tasks within the same body part, such as imagining opening and closing in a single hand (Roset et al. 2014).

In addition, there was no research that utilized the SMR feature as a criterion to stop FES with the voluntary intention of the user. This is because FES-driven passive-movement-evoked SMRs are similar to voluntary MI-evoked SMRs, and are therefore difficult to classify under electrical stimulation artifacts (Müller et al. 2003). Due to this phenomenon, most SMR-based BCI systems for controlling FES have applied synchronous methods, also known as cue-based, or added other stimulus modalities, such as visual or tactile stimulation, as a stop trigger (Pfurtscheller et al. 2010).

3.3 *Signal Processing for SMR-Based BCIs*

The signal processing procedure for SMR-based BCI system shown in Fig. 3 consists of five consecutive steps such as signal acquisition, preprocessing, feature extraction and selection, classification, and application control. In this section, the state of the art in each signal processing stage was summarized from the literature review.

3.3.1 Signal Acquisition

To measure brain signals related to SMRs, proper placement of the EEG electrodes is essential (Nam et al. 2014, 2015). Most of the reviewed research studies measured the motor cortex area, but the specific location and number of electrodes used varied between studies. More electrodes are required for higher spatial resolution (Ferree et al. 2001) but require longer computation time and higher complexity for signal processing (Wang et al. 2012). Furthermore, from an HF/E perspective, the use of many electrode channels could reduce user satisfaction by increasing the setup time and the conductive gel applied to the scalp (Zickler et al. 2011). Figure 4 shows the layout of EEG electrodes used in 18 studies that clearly identified electrode positions in the literature review. The numbers in the circles show the number of research studies used in that location, and the yellow circles indicate the positions used in more than two studies. As can be seen in Fig. 4b, most of the yellow circles are symmetrically distributed near the motor cortices, C3 and C4.

Table 5 MI classes, analyzed band, and feature extraction methods

Article	MI classes	Feature extraction	Classification methods
Daly et al. (2009)	Attempted movement versus attempted relaxation	3 Hz bins from 0 to 30 Hz	Time-averaged power, R-square
Elnady et al. (2015)	MI versus rest	CSP for each electrode	LDA, 10×10 CV
Tan et al. (2011)	MI versus Rest	Bandlimited multiple Fourier linear combiner with 1 Hz bin	Compare μ-rhythm
Kim et al. (2016)	MI or action observation versus rest	12–15 Hz (SMR), 16–20 Hz (mid-beta)	(SMR + middle beta)/theta, Threshold
Liu et al. (2014)	Left MI versus right MI	FastICA, tensor-based nearest feature line distance	SVM
Looned et al. (2014)	MI versus rest or jaw clench versus rest	7–13 Hz (mu) and 13–30 Hz (beta)	LDA
McGie et al. (2015)	MI versus rest	10–12 and 13–15 Hz	No clear
Pfurtscheller et al. (2003)	Foot MI versus right hand MI	–	fLDA
Reynolds et al. (2015)	Left-hand MI versus right-hand MI	Select significant band by t-percentile bootstrap	LDA
Rohm et al. (2010)	Feet MI versus hands MI	–	–
Rohm et al. (2013)	Foot MI versus right-hand MI	Select the most reactive band power features	LDA
Roset et al. (2014)	Hand open MI versus hand close MI	PSD from FFT with 1 Hz bin	Normalized PSD z-scores and actor-critic reinforcement learning algorithm
Vuckovic et al. (2015)	Left-hand MI versus right-hand MI	Delta, theta, mu/alpha, beta1 (12–16), beta (16–24), PSD	Threshold
Tan et al. (2011)	Left-hand MI versus right-hand MI	CSP	fLDA, adaptive (last 6 trials)
Young et al. (2014)	Attempted movements (right versus left) versus rest	Determine channels with the largest r-squared values, ERD	–
Young et al. (2015)	Left-hand MI versus right-hand MI versus rest	8–14 Hz (mu) and 18–26 Hz (beta)	–

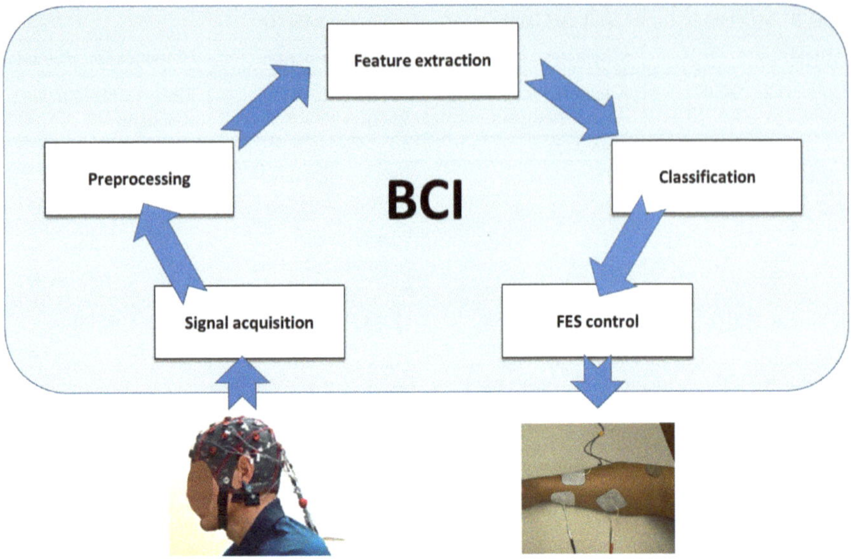

Fig. 3 Flow diagram for signal processing

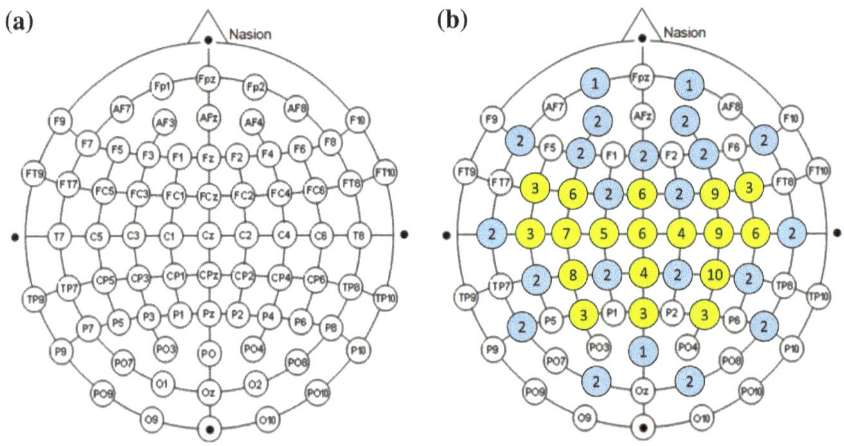

Fig. 4 EEG electrode montage. **a** Modified 10–10 system, **b** the counting of electrodes position: Circles colored as yellow indicate the positions were utilized in more than two research studies

3.3.2 Signal Preprocessing

Once the brain signal is recorded, it is important to filter out noise, artifacts, and irrelevant brain signals to extract only SMR-related features. The EEG signal must, therefore, be inspected for unexpected contamination due to electrode drift (line noise), or motion artifacts. Thereafter, the contaminated trials should be removed through signal preprocessing procedures for further processing. One of the common techniques is to use statistical methods with rejection thresholds (Delorme et al. 2001; Delorme and Makeig 2004; Nolan et al. 2010).

In addition, various spatial filtering methods to enhance the brain signal by increasing the signal-to-noise ratio are widely used, such as bipolar derivation, common average reference, and surface Laplacian estimation techniques (Choi et al. 2015; Hamedi et al. 2016; Pfurtscheller and Neuper 2006). Among them, the surface Laplacian estimation has been applied to many SMR-based BCI studies that can enhance spatial EEG trace by filtering the common spatial noise among nearest-neighbor or next-nearest electrodes (Bashashati et al. 2007; Wolpaw et al. 2002). The equation of surface Laplacian estimation is shown in Eq. 3.1, where $V_{(i)}$ is *i*th electrode signal, $d_{i,j}$ is the distance between electrodes *i* and *j*, S(*i*) is the set of adjacent electrode near *i*th electrode such as anterior, posterior, left- and right-lateral, and $w_{i,j}$ is the weight parameters which is set in inverse proportion to the distance from the *i*th electrode (Schalk and Mellinger 2010).

$$V_{\text{Laplacian}(i)} = V_{(i)} - \sum_{j \in S(i)} w_{i,j} V_{(j)} \quad \text{where } w_{i,j} = \frac{\frac{1}{d_{i,j}}}{\sum_{j \in S(i)} \frac{1}{d_{i,j}}}. \tag{3.1}$$

Independent component analysis (ICA) is also a useful technique for removing artifacts such as eye-blinking and muscle activities. The aim of ICA is to identify the best estimate of the separation matrix with the assumption the target signal and artifacts are mutually independent (Hyvärinen and Oja 2000). ICA has the advantage of not relying on a priori knowledge or hypothesis, but it also has a disadvantage that it requires a lot of calculation time to analyze high-dimensional (channel) data.

To address this issue, principal component analysis (PCA) could be utilized, which is also called decomposition of the covariance matrix. PCA is a statistical procedure that transforms correlated variables into linearly uncorrelated variables and can reduce brain signal with high dimensions (channels) into lower dimensions while maintaining important features (Kayser and Tenke 2003; Lee and Choi 2003). Once the dimension of the EEG data is reduced through the PCA process, the reduced dimensional signal is decomposed into maximally independent components through the ICA.

3.3.3 Feature Extraction and Selection

One of the major feature extraction methods for the SMR-based BCI research is to utilize event-related desynchronization and synchronization (ERD/ERS), which analyzes the alpha and beta frequency bands (Pfurtscheller and Neuper 2006; Young et al. 2015). To compute ERD/ERS, the recorded EEG signal is divided into SMR and reference periods based on predefined timing, such as the point at which MI tasks started, and then each period is transformed from time-domain data to frequency-domain data by fast Fourier transform (FFT) (Roset et al. 2014). Afterward, the relative band power between a reference period and an MI period of each frequency band is calculated by Eq. 3.2, where M is the FFT transformed absolute power of the MI period, and R is that of the reference period.

$$\text{ERD/ERS}(\%) = \frac{(M - R)}{R} * 100. \tag{3.2}$$

Although the range of each frequency band has a large impact on ERD/ERS features, the alpha and beta bands used in previous studies varied from study to study. The research studies using two bands in the literature review used a subset of 7–15 Hz for the alpha band while a subset of 13–30 Hz for the beta band. In contrast, Daly et al. (2009) utilized 10 bins divided by 3 Hz intervals of EEG data between 0 and 30 Hz rather than dividing the two bands. Reynolds et al. (2015) and Rohm et al. (2013) employed very wide frequency bands, such as 5–60 and 0.5–100 Hz, respectively, and then selected significant frequency bands by statistical methods. Pfurtscheller and Lopes (1999) suggested a guideline to determine the proper range of ERD/ERS bands for each subject such as "comparison of short-time power spectra" and "determination of frequency bands dependent on the peak frequency." The former method should plot short-time power spectra value to find the user-specific frequency band, while the latter just calculates the ERD for each frequency sub-band to avoid inter-individual differences.

The other well-known feature extraction method used in the previous studies is common spatial pattern (CSP). The CSP method transforms multichannel EEG data into a subspace using a variance matrix that can maximize discrimination between two classes (C1 and C2). Therefore, the use of CSP as a spatial filter can increase the subsequent classification accuracy (Nicolas-Alonso and Gomez-Gil 2012). The normalized covariance matrix in CSP can be calculated as shown in Eq. 3.3, where trace ($X_{\text{Cx}}X_{\text{Cx}}^{\text{T}}$) is the sum of diagonal elements of $X_{\text{Cx}}X_{\text{Cx}}^{\text{T}}$.

$$C_{C1} = \frac{X_{C1}X_{C1}^T}{\text{trace}\left(X_{C1}X_{C1}^T\right)},\ C_{C2} = \frac{X_{C2}X_{C2}^T}{\text{trace}\left(X_{C2}X_{C2}^T\right)},\ C_c = C_{C1} + C_{C2}. \tag{3.3}$$

Afterward, the composite spatial covariance, C, and CSP projection matrix, W, can be calculated as shown in Eq. 3.4, where U_{c} is a matrix of normalized eigenvectors, and λ is eigenvalues. The projection matrix W then transforms multichannel EEG signals into CSP features that can help to discriminate vague spatial information evoked by MI (Elnady et al. 2015; Tan et al. 2011).

$$C_c = C_{C1} + C_{C2} = U_c \lambda U_C^T \quad W = U^{\text{T}} \sqrt{\lambda} U_C^T. \tag{3.4}$$

Although the CSP method has shown promising results to analyze SMR features, performance of CSP was greatly influenced by which frequency bands were selected similar to the ERD/ERS method (Ang et al. 2008; Novi et al. 2007). One exhaustive approach is to find the optimal frequency band for each subject after examining all possible frequency bands, but this method may require a lot of time and effort to identify user-specific frequency bands. To address this issue, Novi et al. (2007) proposed sub-band CSP (SBCSP) that first decomposes the EEG signal into multiple sub-frequency bands, extracts CSP features from each sub-band, and then makes a decision based on the fused classification scores obtained using the sub-band features. The advantages of SBCSP are that it does not require exhaustive band selection procedure, as well as allows to use multiple frequency bands at the same time (Blanchard and Blankertz 2004; Novi et al. 2007).

3.3.4 Classification Algorithms

One of the easiest and simplest approaches to decoding classes from extracted features is the threshold detection method (Kim et al. 2016; Pfurtscheller et al. 2005; Vuckovic et al. 2015). The threshold detection algorithm detects signals that exceed a predefined value, such as EEG potentials, FFT values, and ERD/ERS values. One of the most widely used algorithms used in MI is the linear discriminant analysis (LDA) algorithms including Fisher's LDA, stepwise LDA, and shrinkage LDA (Choi et al. 2016; Elnady et al. 2015; Looned et al. 2014; Müller et al. 2003; Tan et al. 2011). LDA sets a hyperplane that maximizes the separability between classes and minimizes the variance within classes, assuming that the training data follows multivariate normal distributions with equal covariance matrices (Lotte et al. 2007). For example, Fisher's LDA can classify two classes with Eq. 3.5, where N indicates the number of training samples, (x_i) belongs to each class (C_1 and C_2), and m_1 and m_2 are class means. The slope of the hyperplane (w) is used to identify the local maxima of S, and $r(w)$ should be maximized. Therefore, the solution is as below:

$$\begin{gathered}
\underbrace{m_1 = \frac{1}{N_1}\sum_{x_i \in C_1} x_i, \quad m_2 = \frac{1}{N_2}\sum_{x \in C_2} x_i}_{\textbf{Class means}} \\
\underbrace{\tilde{m}_1 = \frac{1}{N_1}\sum_{x_1 \in C_1} w^T x_i = w^T m_1, \quad \tilde{m}_2 = \frac{1}{N_2}\sum_{x_i \in C_2} w^T x_i = w^T m_2}_{\textbf{Projected class means}} \\
\underbrace{\tilde{m}_2 - \tilde{m}_1 = w^T (m_2 - m_1)}_{\textbf{Difference between projected class means}} \\
\underbrace{J(w) = \frac{(\tilde{m}_2 - \tilde{m}_1)^2}{\tilde{s}_1^2 + \tilde{s}_2^2} = \frac{w^T S_B w}{w^T S_w w}, \text{ where } \begin{array}{l} S_w = S_1 + S_2 \\ S_B = (m_2 - m_1)(m_2 - m_1)^T \end{array}}_{\textbf{Ratio of difference of projected means over total scatter}} \\
\underbrace{w^* = S_w^{-1}(m_2 - m_1)}_{\textbf{The optimal projection matrix}} \;.
\end{gathered} \tag{3.5}$$

Support vector machine (SVM) is one of the most popular classifiers used in MI. SVM utilizes the maximum margin linear classifier, which is the width of the boundary strip between classes, to minimize the error rate. Therefore, SVM is robust to over-filtering and less sensitive to high dimensionality compared to LDA. If a training dataset with two classes is defined as in Eq. 3.6.1, then the data points on each marginal boundary X^+ and X^- satisfy Eq. 3.6.2. The margin width, M, can be calculated with Eq. 3.6.3. Equation 3.6.4 can maximize M.

$$\begin{aligned} &\text{For } y_i = +1, \quad \mathbf{w}^T\mathbf{x}_i + b \geq 1 \\ &\text{For } y_i = -1, \quad \mathbf{w}^T\mathbf{x}_i + b \leq -1 \end{aligned} \tag{3.6.1}$$

$$\begin{aligned} &\mathbf{w}^T\mathbf{x}^+ + b = 1 \\ &\mathbf{w}^T\mathbf{x}^- + b = -1 \end{aligned} \tag{3.6.2}$$

$$\begin{aligned} M &= (\mathbf{x}^+ - \mathbf{x}^-)\cdot\mathbf{n} \\ &= (\mathbf{x}^+ - \mathbf{x}^-)\cdot\frac{\mathbf{w}}{\|\mathbf{w}\|} = \frac{2}{\|\mathbf{w}\|} \end{aligned} \tag{3.6.3}$$

$$\text{maximize } \frac{2}{\|\mathbf{w}\|} \equiv \text{minimize } \frac{1}{2}\|\mathbf{w}\|^2. \tag{3.6.4}$$

The quadratic programming with linear constraints as shown in Eq. 3.6.5 can be solved as Eq. 3.6.6 by the Lagrangian dual problem.

$$\text{minimize} \quad L_p(\mathbf{w}, b, \alpha_i) = \frac{1}{2}\|\mathbf{w}\|^2 - \sum_{i=1}^{n}\alpha_i\big(y_i(\mathbf{w}^T\mathbf{x}_i + b) - 1\big) \quad \textbf{s.t. } \alpha_i \geq 0 \tag{3.6.5}$$

$$g(\mathbf{x}) = \mathbf{w}^T\mathbf{x} + b = \sum_{i\in\text{SV}}\alpha_i\mathbf{x}_i^T\mathbf{x} + b. \tag{3.6.6}$$

However, due to the nature of EEG signals, there are some critical issues such as "noisy signal and outliers," "high dimensionality," "nonstationarity,", and "small training sets" (Lotte et al. 2007). Due to these issues, we cannot guarantee that the classifier will elicit good results most of the time (Polikar 2006; Sun et al. 2007). One approach to address this issue is ensemble methods in which the outputs of the multiple classifiers are combined to make a single decision that could improve performance of each classifier (Polikar 2006). Polikar defined two key components of an ensemble system as "*First, a strategy is needed to build an ensemble that is as diverse as possible. Some of the more popular ones, such as bagging, boosting, AdaBoost, ... A second strategy is needed to combine the outputs of individual classifiers that make up the ensemble in such a way that the correct decisions are amplified, and incorrect ones are cancelled out*" (Polikar 2006, p. 27). In this study, the second strategy is adopted by combining the class outputs of both LDA and SVM classifiers to improve classification accuracy and robustness. The author listed three combining methods which could be applied to this study including (1) majority voting, (2) weighted majority voting, and (3) behavior knowledge space. Among them, the weighted majority voting method was selected, because performance and robustness of each feature within the same classifier, as well as that of each classifier, could be varied from subject to subject. The final decision of the ensemble method with the weighted majority voting by combining the features of LDA and SVM can be calculated by Eq. 3.7.

$$\sum_{t=1}^{T} w_t d_{t,J} = \max_{j=1}^{C} \sum_{t=1}^{T} w_t d_{t,j}, \tag{3.7}$$

where
t = the type of classifiers (1 = LDA, 2 = SVM),
T = the number of classifiers ($T = 2$),
j = the type of classes (j = 1–2),
C = the number of classes ($C = 2$),
w_t = the predefined weight of classifier t,
$d_{t,J}$ = If tth classifier chooses class j, then $d_{t,j} = 1$, and = 0, otherwise.

The accuracy (p) of each classification method including LDA, SVM, and ensemble methods can be assessed by the significance of the accuracy (Noirhomme et al. 2014). The normal distribution can be used to approximate the binomial distribution, but this approximation can be valid only if the number of trial (N) is large enough such as $N \geq 5/p/(1-p)$ (Berrar et al. 2006; Rosner 2015). To estimate the binomial lower bound of the accuracy, Jeffreys' beta distribution is usually utilized (Berrar et al. 2006) and can be calculated by Eq. 3.8.

$$\lambda \approx \left\{ a + \frac{2(N-2m)z\sqrt{0.5}}{2N(N+3)} \right\} - z\sqrt{\frac{a(1-a)}{N+2.5}}, \tag{3.8}$$

where
λ = the true error rate of binomial classification,
α = the expected accuracy (0.5),
N = the number of trials,
m = the expected number of correct trials ($N/2$ with $\alpha = 0.5$), and
z = z-score based on the standard normal distribution (1.65; $p = 0.05$; one-sided).

For example, the true chance level, the binomial lower bound, can be estimated as 62.7% with 40 trails, and this lower bound is much higher than the conventional chance level 50%. Table 6 shows the true error rate of binomial classification for

Table 6 True chance level for different numbers of trials

# of trials	True chance level (%)
30	64.43
60	60.40
120	57.43
240	55.28

different numbers of trials, and these rates were utilized as lower bounds to set the true chance level for hypotheses in Sect. 3.3.

3.4 Limitations of Current Research

From the systematic literature review, five main limitations of current SMR-based BCI studies have been identified as follows: (1) the ambiguous MI instruction given during training which not only increases the subjective mental workload but also negatively affects SMR performance due to incorrect or inconsistent MI tasks; (2) the lack of research that investigated different MI tasks in the same limb, such as imagining opening and closing in a single hand; (3) the lack of studies utilizing SMR features as a criterion to either stop or keep FES by voluntary MI-evoked SMRs; and (4) the unclear frequency band selection procedure for extracting SMR features that varied from study to study.

These limitations might impede a wide application of SMR-based BCI-controlled FES systems due to (1) long training periods due to ambiguous instruction for MI tasks, (2) low classification accuracy of a 2-class MI task in a single hand, and (3) difficulties in applying SMR features to either stop or keep FES.

4 Summary

From the systematic literature review, four main limitations of current SMR-based BCI studies have been identified:

- The ambiguous MI instruction given during training which not only increases the subjective mental workload but also negatively affects SMR performance due to incorrect or inconsistent MI tasks,
- The lack of research that investigated different MI tasks in the same limb, such as imagining opening and closing in a single hand,
- The lack of studies utilizing SMR features as a criterion to either stop or keep FES by voluntary MI-evoked SMRs, and
- The unclear frequency band selection procedure for extracting SMR features that varied from study to study.

These limitations might impede a wide application of SMR-based BCI-controlled FES systems due to (1) long training periods due to ambiguous instruction for MI tasks, (2) low classification accuracy of a 2-class MI task in a single hand, and (3) difficulties in applying SMR features to either stop or keep FES.

References

Ang, K. K., Chin, Z. Y., Zhang, H., & Guan, C. (2008). Filter Bank Common Spatial Pattern (FBCSP) in brain-computer interface. In *IEEE International Joint Conference on Neural Networks* (IEEE World Congress on Computational Intelligence), 2390–2397.

Ang, K. K., Guan, C., Ang, Kai Keng, & Guan, Cuntai. (2015). Brain-computer interface for neurorehabilitation of upper limb after stroke. *Proceedings of the IEEE, 103*(6), 944–953.

Bashashati, A., Fatourechi, M., Ward, R. K., & Birch, G. E. (2007). A survey of signal processing algorithms in brain-computer interfaces based on electrical brain signals. *Journal of Neural Engineering, 4*(2), R32–R57.

Berrar, D., Bradbury, I., & Dubitzky, W. (2006). Avoiding model selection bias in small-sample genomic datasets. *Bioinformatics, 22*(10), 1245–1250. Oxford Univ Press.

Blanchard, G., & Blankertz, B. (2004). BCI competition 2003—Data set IIa: Spatial patterns of self-controlled brain rhythm modulations. *IEEE Transactions on Biomedical Engineering, 51*(6), 1062–1066.

Choi, I., Bond, K., Krusienski, D., & Nam, C. S. (2015). Comparison of stimulation patterns to elicit steady-state somatosensory evoked potentials (SSSEPs): Implications for hybrid and SSSEP-based BCIs. *Systems, Man, and Cybernetics (SMC), 2015 IEEE International Conference on* (pp. 3122–3127).

Choi, I., Bond, K., & Nam, C. S. (2016). A hybrid BCI-controlled FES system for hand-wrist motor function. *IEEE International Conference on Systems, Man, and Cybernetics*.

Daly, J. J., Cheng, R., Rogers, J., Litinas, K., Hrovat, K., & Dohring, M. (2009). Feasibility of a new application of noninvasive brain computer interface (BCI): A case study of training for recovery of volitional motor control after stroke. *Journal of Neurologic Physical Therapy, 33*(4), 203–211.

Delorme, A., & Makeig, S. (2004). EEGLAB: An open source toolbox for analysis of single-trial EEG dynamics including independent component analysis. *Journal of Neuroscience Methods, 134*(1), 9–21.

Delorme, A., Makeig, S., & Sejnowski, T. (2001). Automatic artifact rejection for EEG data using high-order statistics and independent component analysis. *Proceedings of the third international ICA conference* (pp. 9–12).

Doucet, B. M., Lam, A., & Griffin, L. (2012). Neuromuscular electrical stimulation for skeletal muscle function. *Yale J Biol Med, 85*(2), 201–215.

Elnady, A. M., Zhang, X., Xiao, Z. G., Yong, X., Randhawa, B. K., Boyd, L., & Menon, C. (2015). A single-session preliminary evaluat on of an affordable BCI-controlled arm exoskeleton and motor-proprioception platform. *Frontiers in Human Neuroscience, 9*, 168. Switzerland.

Ferree, T. C., Clay, M. T., & Tucker, D. M. (2001). The spatial resolution of scalp EEG. *Neurocomputing, 38–40*, 1209–1216.

Forrester, B. J., & Petrofsky, J. S. (2004). Effect of electrode size, shape, and placement during electrical stimulation. *Journal of Applied Research, 4*(2), 346–354.

Gordon, C. C., Churchill, T., Clauser, C. E., Bradtmiller, B., & McConville, J. T. (1989). *Anthropometric survey of US army personnel: methods and summary statistics 1988*.

Gu, Y., Dremstrup, K., & Farina, D. (2009). Single-trial discrimination of type and speed of wrist movements from EEG recordings. *Clinical Neurophysiology, 120*(8), 1596–1600. International Federation of Clinical Neurophysiology.

Hamedi, M., Salleh, S.-H., & Noor, A. M. (2016). Electroencephalographic motor imagery brain connectivity analysis for BCI: A review. *Neural Computation, 28*(6), 999–1041.

Hyvärinen, a, & Oja, E. (2000). Independent component analysis: algorithms and applications. *Neural networks : the official journal of the International Neural Network Society, 13*(4–5), 411–430.

Kayser, J., & Tenke, C. E. (2003). Optimizing PCA methodology for ERP component identification and measurement: Theoretical rationale and empirical evaluation. *Clinical Neurophysiology, 114*(12), 2307–2325.

Kim, T., Kim, S., & Lee, B. (2016). Effects of action observational training plus brain-computer interface-based functional electrical stimulation on paretic arm motor recovery in patient with stroke: A randomized controlled trial. *Occupational therapy international, 23*(1), 39–47. England.

Lawrence, M. (2009). Transcutaneous electrode technology for neuroprostheses, (18213).

Lee, H., & Choi, S. (2003). PCA + HMM + SVM for EEG pattern classification. *Seventh International Symposium on Signal Processing and Its Applications, 2003. Proceedings., 1*(2), 1–4.

Liu, Y., Li, M., Zhang, H., Wang, H., Li, J., Jia, J., Wu, Y., et al. (2014). A tensor-based scheme for stroke patients' motor imagery EEG analysis in BCI-FES rehabilitation training. *Journal of neuroscience methods, 222*, 238–249. Elsevier.

Looned, R., Webb, J., Xiao, Z. G., & Menon, C. (2014). Assisting drinking with an affordable BCI-controlled wearable robot and electrical stimulation: a preliminary investigation. *Journal of neuroengineering and rehabilitation, 11*, 51. England.

Lotte, F., Congedo, M., Lécuyer, A., Lamarche, F., & Arnaldi, B. (2007). A review of classification algorithms for EEG-based brain-computer interfaces. *Journal of Neural Engineering, 4*(2), R1–R13.

Lyons, G. M., Leane, G. E., Clarke-Moloney, M., O'Brien, J. V., & Grace, P. A. (2004). An investigation of the effect of electrode size and electrode location on comfort during stimulation of the gastrocnemius muscle. *Medical Engineering & Physics, 26*(10), 873–878.

McGie, S. C., Zariffa, J. J., Popovic, M. R., & Nagai, M. K. (2015). Short-term neuroplastic effects of brain-controlled and muscle-controlled electrical stimulation. *Neuromodulation, 18*(3), 233–240. United States.

Moher, D., Shamseer, L., Clarke, M., Ghersi, D., Liberati, A., Petticrew, M., et al. (2015). Preferred reporting items for systematic review and meta-analysis protocols (PRISMA-P) 2015 statement. *Systematic Reviews, 4*(1), 1.

Mukaino, M., Ono, T., Shindo, K., Fujiwara, T., Ota, T., Kimura, A., Liu, M., et al. (2014). Efficacy of brain-computer interface-driven neuromuscular electrical stimulation for chronic paresis after stroke. *Journal of rehabilitation medicine, 46*(4), 378–382. Sweden: Medical Journals Limited.

Müller, G. R. R., Neuper, C., Rupp, R., Keinrath, C., Gerner, H. J. J., & Pfurtscheller, G. (2003). Event-related beta EEG changes during wrist movements induced by functional electrical stimulation of forearm muscles in man. *Neuroscience Letters, 340*(2), 143–147.

Nam, C. S., Lee, J., Bahn, S., Li, Y., & Choi, I. (2014). Brain-computer interface supported collaborative work. *Proceedings of 5th International Brain-Computer Interface Meeting*.

Nam, C. S., Moore, M., Choi, I., & Li, Y. (2015). Designing better, cost-effective brain-computer interfaces. *Ergonomics in Design: The Quarterly of Human Factors Applications, 23*(4), 13–19. SAGE.

Nicolas-Alonso, L. F., & Gomez-Gil, J. (2012). Brain computer interfaces, a review. *Sensors*.

Noirhomme, Q., Lesenfants, D., Gomez, F., Soddu, A., Schrouff, J., Garraux, G., Luxen, A., et al. (2014). Biased binomial assessment of cross-validated estimation of classification accuracies illustrated in diagnosis predictions. *NeuroImage: Clinical, 4*, 687–694.

Nolan, H., Whelan, R., & Reilly, R. B. (2010). FASTER: fully automated statistical thresholding for EEG artifact rejection. *Journal of neuroscience methods, 192*(1), 152–162. Elsevier.

Novi, Q., Guan, C., Dat, T. H., & Xue, P. (2007). Sub-band common spatial pattern (SBCSP) for brain-computer interface. *Proceedings of the 3rd International IEEE EMBS Conference on Neural Engineering*, 204–207.

Pfurtscheller, G., & Lopes, F. H. (1999). Event-related EEG/MEG synchronization and desynchronization: basic principles. *Clinical Neurophysiology, 110*, 1842–1857.

Pfurtscheller, G., Müller-Putz, G. R., Pfurtscheller, J. J., Rupp, R. R., Muller-Putz, G. R., Pfurtscheller, J. J., Rupp, R. R., et al. (2005). EEG-based asynchronous BCI controls functional electrical stimulation in a tetraplegic patient. *EURASIP Journal on Applied Signal Processing, 2005*(19), 3152–3155. Hindawi, USA.

Pfurtscheller, G., Müller, G. R., Pfurtscheller, J. J., Gerner, H. J. J., Rupp, R. R., Muller, G. R., Pfurtscheller, J. J., et al. (2003). "Thought"—control of functional electrical stimulation to restore hand grasp in a patient with tetraplegia. *Neuroscience letters*, *351*(1), 33–36. Ireland.

Pfurtscheller, G., & Neuper, C. (2006). Future prospects of ERD/ERS in the context of brain—computer interface (BCI) developments. *Progress in Brain Research, 159*, 433–437.

Pfurtscheller, G., Solis-Escalante, T., Ortner, R., Linortner, P., & Muller-Putz, G. R. (2010). Self-paced operation of an SSVEP-based orthosis with and without an imagery-based "brain switch": A feasibility study towards a hybrid BCI. *IEEE Transactions on Neural Systems and Rehabilitation Engineering, 18*(4), 409–414.

Polikar, R. (2006). Ensemble based systems in decision making. *Circuits and Systems Magazine, IEEE, 6*(3), 21–45.

Powers, J. C., Bieliaieva, K., Wu, S., & Nam, C. S. (2015). The human factors and ergonomics of P300-based brain-computer interfaces. *Brain sciences*, *5*(3), 318–56. Switzerland.

Reynolds, C., Osuagwu, B. A., & Vuckovic, A. (2015). Influence of motor imagination on cortical activation during functional electrical stimulation. *Clinical neurophysiology : official journal of the International Federation of Clinical Neurophysiology*, *126*(7), 1360–1369. Netherlands.

Rohm, M., Muller-Putz, G. R., Kreilinger, A., von Ascheberg, A., & Rupp, R. (2010). A hybrid-Brain Computer Interface for control of a reaching and grasping neuroprosthesis. *Biomedizinische Technik*, *55*(suppl. 1). Fachverlag Schiele & Schon GmbH, Germany.

Rohm, M., Schneiders, M., Müller, C., Kreilinger, A., Kaiser, V., Müller-Putz, G. R., Rupp, R. R. R., et al. (2013). Hybrid brain-computer interfaces and hybrid neuroprostheses for restoration of upper limb functions in individuals with high-level spinal cord injury. *Artificial Intelligence in Medicine*, *59*(2), 133–142. Netherlands: Elsevier Science B.V., Netherlands.

Roset, S. A., Gant, K., Prasad, A., & Sanchez, J. C. (2014). An adaptive brain actuated system for augmenting rehabilitation. *Frontiers in neuroscience*, *8*, 415. Switzerland.

Rosner, B. (2015). *Fundamentals of biostatistics*. Nelson Education.

Schalk, G., & Mellinger, J. (2010). *A practical guide to brain–computer interfacing with BCI2000: General-purpose software for brain-computer interface research, data acquisition, stimulus presentation, and brain monitoring*. Springer Science & Business Media.

Sun, S., Zhang, C., & Zhang, D. (2007). An experimental evaluation of ensemble methods for EEG signal classification. *Pattern Recognition Letters, 28*(15), 2157–2163.

Tan, H. G., Shee, C. Y., Kong, K. H., Guan, C., Ang, W. T., et al. (2011). EEG controlled neuromuscular electrical stimulation of the upper limb for stroke patients. *Frontiers of Mechanical Engineering*, *6*(1), 71–81. SP Higher Education Press, Germany.

Vuckovic, A., Wallace, L., & Allan, D. B. (2015). Hybrid brain-computer interface and functional electrical stimulation for sensorimotor training in participants with tetraplegia: a proof-of-concept study. *Journal of neurologic physical therapy : JNPT*, *39*(1), 3–14. United States.

Wang, D., Miao, D., & Blohm, G. (2012). Multi-class motor imagery EEG decoding for brain-computer interfaces. *Frontiers in Neuroscience*, *6*(OCT), 1–13.

Wolpaw, J. R., Birbaumer, N., McFarland, D. J., Pfurtscheller, G., & Vaughan, T. M. (2002). Brain-computer interfaces for communication and control. *Clinical Neurophysiology, 113*(6), 767–791.

Young, B. M., Nigogosyan, Z., Walton, L. M., Remsik, A., Song, J., Nair, V. A., Tyler, M. E., et al. (2015). Dose-response relationships using brain-computer interface technology impact stroke rehabilitation. *Frontiers in human neuroscience*, *9*, 361. Switzerland.

Young, B. M., Nigogosyan, Z., Nair, V. A., Walton, L. M., Song, J., Tyler, M. E., Edwards, D. F., et al. (2014). Case report: post-stroke interventional BCI rehabilitation in an individual with preexisting sensorineural disability. *Frontiers in neuroengineering*, *7*, 18. Switzerland.

Zickler, C., Riccio, A., Leotta, F., Hillian-Tress, S., Halder, S., Holz, E., Staiger-Salzer, P., et al. (2011). A brain-computer interface as input channel for a standard assistive technology software. *Clinical EEG and neuroscience*, *42*(4), 236–244.

Everyday and Emerging Applications

Neuroergonomics Behind the Wheel: Neural Correlates of Driving

Macie Ware, Jing Feng and Chang S. Nam

Abstract Each year around the world, approximately 1.25 million lives are lost in fatal vehicle crashes. This statistic equates to 3,434 deaths per day. In addition, another 20–50 million people are injured or disabled in crashes (Association for Safe International Road Travel 2002). This chapter provides a review of studies that utilized neurocognitive methods to study how a driver's brain functions to support various tasks while driving. These changes in brain activity can show how secondary tasks, distractions, and substances degrade driver performance and increase crash risks. The neural techniques analyzed in this chapter include electroencephalography (EEG), functional magnetic resonance imaging (fMRI), functional near-infrared spectroscopy (fNIRS), and magnetoencephalography (MEG). Most participants for these studies were in their mid-20s, except for a few in which the study's purpose was to analyze driving performance between age groups. Given the difficulty in collecting neural activity data while a driver is on the road, most studies utilized driving simulation as opposed to on-road driving. Current findings suggest a critical role of frontal regions of the brain in driving. In addition, this chapter analyzes the effects that independent variables added to the driving task have on the organizational planning levels—strategical, tactical, and operational planning along with the brain activity that accompanies this planning. Lastly, the chapter provides a discussion of limitations of this literature and future directions.

M. Ware · C. S. Nam (✉)
Edward P. Fitts Department of Industrial & Systems Engineering, North Carolina State University, Raleigh, NC 27695, USA
e-mail: csnam@ncsu.edu

J. Feng
Department of Psychology, North Carolina State University, Raleigh, NC 27695, USA

C. S. Nam (ed.), *Neuroergonomics*, Cognitive Science and Technology,
https://doi.org/10.1007/978-3-030-34784-0_18

1 Introduction

While driving a car may seem like second nature to some, the task of driving requires motor skills, problem-solving, decision-making, planning, and memory. According to ergonomics-based car driving task analysis, more than 45 different tasks are associated with driving (Foy et al. 2016). These tasks are not of equal importance during the task of driving as they activate different parts of the brain. Once the strategical and tactical decisions are made, the operational level executes these decisions. In order to understand the neural correlates of driving, methods such as electroencephalography (EEG), functional magnetic resonance imaging (fMRI), functional near-infrared spectroscopy (fNIRS), and magnetoencephalography (MEG) have been used to analyze brain activity during the driving task (Navarro et al. 2018).

Figure 1 illustrates historical events in neuroergonomics driving research. This timeline begins in 1885 with the invention of the first gasoline automobile powered by an internal combustion engine (The Columbia Encyclopedia 2019). In 1929, German psychiatrist Hans Berger discovered EEG. The first experiments utilizing EEG technology were performed on Berger's own son, Klaus. By 1938, Berger's findings were accepted by the international scientific community, and thus EEG became a key research tool (Sorkin 2017). At first, EEGs were inspected visually, yet today there is digital technology to produce more accurate readings. An EEG evaluates electrical activity in the brain as brain activity is stimulated through electrical impulses. EEG readings, resembling wavy lines with peaks and valleys, are used to track brain wave patterns through electrodes attached to the scalp with wires.

Driving simulators date back to before World War II when flight simulators were used for military training purposes. The first highway simulator was used in the early 1960s, yet a lack of technology led to a decline of their use in the mid-60s (Blana 1996). Physicist David Cohen at the Massachusetts Institute of Technology invented the MEG in the early 1970s. In his discovery, Cohen was investigating small magnetic field that develop in electrically active tissues such as the brain (Doherty 2014). MEG detects, records, and analyzes the magnetic fields produced by electrical currents in the brain. The distribution of the magnetic fields is then superimposed with an anatomical visual of the brain to aid in detecting the source of brain activity.

Following MEG, fNIRS was soon after discovered in 1985 and the first studies of cerebral oxygenation were done by Ferrari et al. (1985). fNIRS is a neuroimaging technology that measures changes in near-infrared light to monitor blood flow in the front part of the brain. In the year of 1990, Siji Ogawa and a group from Bell Laboratories invented the fMRI. This group discovered that oxygen is used by active neurons; therefore, the fMRI technique could distinguish neurons that were functionally active from those that were inactive. Spatial resolution of fMRI has made much progress since its first discovery. Current images of fMRI have a spatial resolution of about one millimeter, yet the end goal is to reduce resolution down to the size of a neuron (Uttal 2002). In recent years, fNIRS technology has gained a lot of traction in the medical community, so much so that in 2010 the first fNIRS conference was

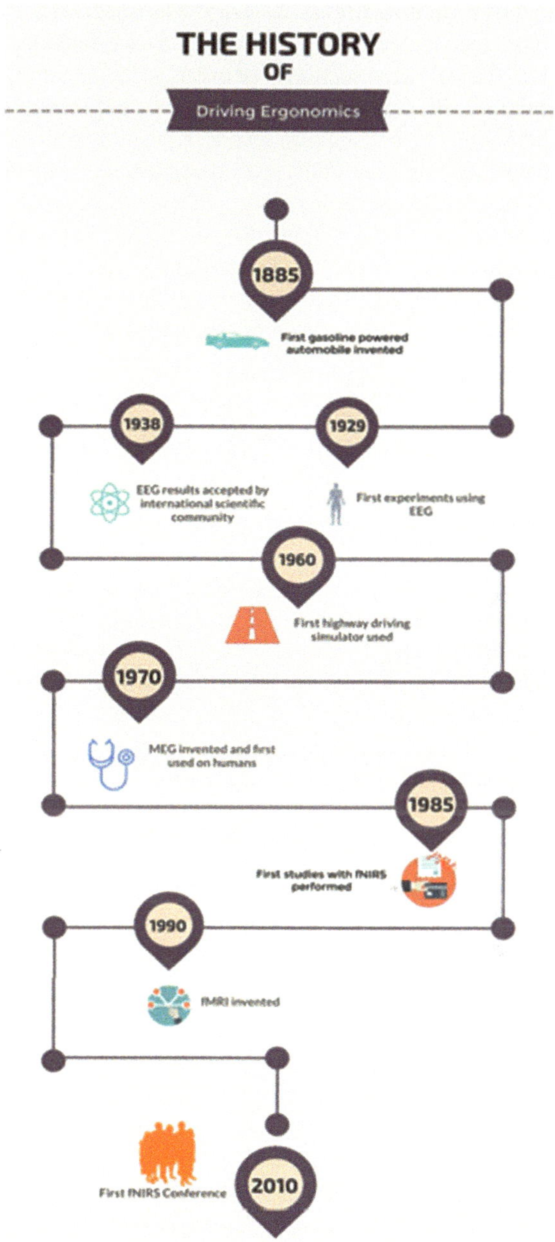

Fig. 1 Timeline of historical events in neuroergonomics driving research

held in Boston to discuss the expanding usage of the technology (Boas 2014). This conference has continued to grow with each meeting.

In the following sections, we organize the literature on neural correlates of driving based on the neural technique used in each study, including EEG, fMRI, fNIRS, and MEG, respectively. In Sect. 2, we summarize these studies that examined how various driving conditions, secondary tasks, and other factors have an impact on brain activity and driving performance. In Sect. 3, we will identify the shared brain regions in these tasks. In Sect. 4, the organizational planning levels, tactical, strategical, and operational will be analyzed to determine which driving factors affect which areas of planning along with the brain activity that corresponds to the planning. Lastly, Sect. 5 concludes with a discussion of limitations of this literature and future directions.

2 Neuroergonomics Research of Car Driving

In order to fully investigate brain activity while driving, neuroimaging techniques were needed to analyze said brain activity. In this chapter, fMRI, MEG, EEG, and fNIRS studies are analyzed to determine the effects certain independent variables have on the driving task. In using these different techniques, the task of driving can be analyzed from four different techniques while also incorporating many different variables to identify what factors affect driving. Through this thorough analysis, the brain regions stimulated are noted in the tables below. One can see that across the neuroimaging techniques, findings are consistent in mapping the brain regions that support driving.

2.1 fMRI-Based Driving Research

A functional magnetic resonance imaging, or fMRI, was developed to identify which level is activated during certain tasks. The fMRI investigates how the brain responds to changes within different driving scenarios. This technique is able to detect changes of blood flow and make an inference of brain activation as an increase in neural activity in a brain area is associated with an increase of local blood flow (Blana 1996).

The experiments given in Table 1 analyzed the brain activity with fMRI while driving in addition to secondary tasks. These experiments analyzed the effects of secondary tasks on brain activity while driving. For example, Choi et al. (2017) study on the activity of the brain while driving and performing a concurrent task showed decreased activity in the parietal area of the brain while increased activity in the inferior frontal gyrus and superior temporal gyrus. Just et al. (2008) study found that listening to someone speak while driving degraded driving performance with decreased brain activity in key brain areas, such as the frontal, parietal, and occipital regions that perform the driving task when compared to driving without a distraction. As noted in Table 1, many other studies incorporated the study of performing a

Table 1 Summary of exemplary fMRI-based driving studies

Author	Experimental task	Subject characteristics	Brain region stimulated	Key findings
Choi et al. (2017)	Driving while performing subtasks	• 15 male adults • M_{age}: 26 ± 1.4	• Frontal • Parietal, and • Occipital regions	• Decreased activity in parietal area • Increased activity in inferior frontal gyrus and superior temporal gyrus in comparison to driving without subtasks
Just et al. (2008)	Driving while listening to someone's speak	• 29 adults (15 M) • Age: 18–25	• Left superior frontal gyrus • Left orbital frontal gyrus • Right inferior frontal gyrus	• Degraded driving performance • Decreases in activation in key regions that perform the driving task
Bernardi et al. (2013)	Visuospatial tasks	• Experienced drivers: 11 male adults; $M_{age} = 24 \pm 4$ • Naive drivers: 11 male adults; $M_{age} = 28 \pm 4$	• Middle and inferior frontal cortex • Insula • Striatum • Cerebellum • Cingulate • Sensorimotor • Temporo-occipital • Parietal cortex	• Professional and naïve car drivers showed no differences in performance • Naïve subjects showed significantly greater response in supplementary motor area, left middle frontal and precentral cortex, bilateral inferior parietal lobule, right superior parietal, and postcentral cortex, cerebellum, and bilateral striatum

(continued)

Table 1 (continued)

Author	Experimental task	Subject characteristics	Brain region stimulated	Key findings
Schweizer et al. (2013)	Six right-hand turns and six left-hand turns with and without traffic	• 16 adults (9 M) • Age: 20–30	• Superior parietal lobule • Precuneus • Postcentral gyrus • Lingual gyrus • Cuneus • Precentral gyrus • Cerebellum • Superior frontal gyrus • Fusiform gyrus • Superior occipital gyrus • Middle temporal gyrus • Inferior temporal gyrus	• Left turn + traffic + audio showed larger activations of auditory, motor, somatosensory, visual, parietal, and cerebellar • Significant activation in anterior brain areas bilaterally • Driving performance showed speed differences in undistracted driving conditions (left-turn traffic being the slowest)
Calhoun et al. (2004)	10-min tasks of 1 min alterations between an asterisk fixation tasks	• 9 adults (7 M) • M_{age}: 24.2 ± 5.8	• Anterior cingulate cortex • Primary motor cortex • Parieto-occipital sulcus • Cerebellum	• Significant alcohol-related changes found in the anterior cingulate cortex and primary motor cortex • Consumption of alcohol-impaired error detection

(continued)

Table 1 (continued)

Author	Experimental task	Subject characteristics	Brain region stimulated	Key findings
Hung et al. (2014)	Driver's ability after brain damage	• Various drivers with and without brain damage	• Frontal region • Parietal region • Cerebellum	• Significant impairment of driving performance found in drivers with cerebellar damage • Drivers with brain damage turned with slower mean speeds • Drivers without cerebellar damage saw shorter gaps between vehicles and shorter estimated time to collision
Calhoun et al. (2004)	Alternated tasks between an asterisk fixation task	• 12 adults (2F) • M_{age}: 22.5	• Motor region • Cerebellar region • Anterior cingulate • Medial frontal regions • Frontoparietal	• Motor region and cerebellar networks were significantly different between the driving task and observing task • Visual/Cerebellar activity most active during driving and less during watching • Cerebellar and frontoparietal increased/decreased during driving • Anterior cingulate, medial frontal decreased during driving and increased during fixation • Visual areas activated during driving and watching

(continued)

Table 1 (continued)

Author	Experimental task	Subject characteristics	Brain region stimulated	Key findings
Jo et al. (2019)	Driving while performing subtasks	• 10 male adults; M_{age}: 23 ± 1.2 • Driving experience; • M_{age}: 2.6 ± 1.8	• Inferior frontal gyrus • Inferior parietal lobule • Inferior temporal gyrus • Insula • Lingual gyrus • Precentral gyrus • Precuneus • Superior temporal gyrus	• Insula region significantly large increase in activation during driving and performing additional task
Meda et al. (2009)	Tasks including fixation (30 s), drive (90 s), and observe (60 s)	• 40 adults (20 M) • M_{age}: 24.75 ± 4.7	• Superior frontal gyrus • Middle frontal gyrus • Precuneus • Parahippocampal gyrus • Inferior frontal gyrus • Medial frontal gyrus • Superior temporal gyrus • Uncus • Cingulate gyrus • Cuneus • Precentral gyrus • Anterior cingulate	• Increased crashes • Increased steering weave • More frequent crossing of passenger side white lines

(continued)

Table 1 (continued)

Author	Experimental task	Subject characteristics	Brain region stimulated	Key findings
Battistella et al. (2013)	Keeping car in the middle of the roadway despite lateral movement of vehicle	• 31 male adults • Age: 18–30	• Middle occipital gyrus • Inferior occipital gyrus • Lingual gyrus • Left motor cortex • Supplementary motor area • Central sulcus • Postcentral gyrus • Precentral gyrus • Superior parietal lobule • Intraparietal sulcus • Supramarginal gyrus • Thalamus • Insula • Left putamen	• Significant increase in blood oxygen level dependent (BOLD) response compared to placebo smoking covering the anterior cingulate cortex and ventromedial prefrontal cortex • Decrease of BOLD after cannabis smoking in anterior insula, dorsomedial thalamus, left middle frontal gyrus
Navarro et al. (2018)	A meta-analysis on driving fatigue utilizing a fMRI	• 131 participants from 9 studies	• Left posterior lobe and right anterior lobe of the cerebellum • Bilateral extrastriate cortex • Right middle temporal gyrus • Bilateral precuneus • Left precentral gyrus • Left superior frontal gyrus • Right middle frontal gyrus • Left insula • Right posterior cingulate cortex • Right dorsomedial nucleus of the thalamus	• Commonly activated areas ✓ Left posterior lobe and right anterior lobe of cerebellum ✓ Bilateral extrastriate cortex ✓ Right middle temporal gyrus ✓ Bilateral precuneus ✓ Left precentral gyrus ✓ Left superior frontal gyrus ✓ Right middle frontal gyrus ✓ Left insula, right posterior cingulate cortex ✓ Right dorsomedial nucleus of the thalamus

(continued)

Table 1 (continued)

Author	Experimental task	Subject characteristics	Brain region stimulated	Key findings
Li et al. (2010)	Fixation 30 s, driving 90 s, and watching 60 s	• 40 adults (20F) • M_{age}: 24.75 ± 4.7	• Frontal and parieto-occipital • Cerebellum • Medial frontal • Parietal lobe • Anterior and posterior cingulate	• Spatial pattern shows correlation at parietal lobe, anterior cingulate, and anti-correlation at motor cortex and posterior cingulate • Functional correlation is more consistent within a subject than across different subjects
Kim et al. (2014)	Driving video simulation recorded with four different driving speeds	• 16 adults (8 M) • M_{age}: 22.2 ± 1.7	• Occipital lobe • Parietal lobe • Frontal lobe	• Number of activated voxels in the frontal lobe increased as the speed differences got bigger which is associated with spatial attention • Tendency was consistent in both left and right hemispheres
Chung et al. (2014)	Driving without changing lanes at a constant speed of 110 km/h with and without a second task	• 16 male adults • M_{age}: 26.6 ± 2.1	• Frontal • Parietal • Temporal • Occipital • Limbic • Cerebellar regions	• Large increases were seen in driving with a task in the precuneus, inferior parietal lobule, supramarginal gyrus, middle frontal gyrus, cuneus, and declive brain areas
Tsunashima and Yanagisawa (2009)	Driving with and without adaptive cruise control	• 4 male adults • Age: 20	• Frontal lobe	• Oxygenated hemoglobin increased when subject drove without ACC • Both outer portions of frontal lobe became active during driving task

secondary task while driving, and these studies were consistent in finding degraded driving performance with increased brain activity in the inferior frontal gyrus. Hung et al. (2014) performed a study utilizing fMRI in order to determine the effects of distracted driving on brain activity and driving capability. As typically found in the other studies, there were large increases in brain activity in the frontal areas while driving with a secondary task. The frontal areas of the brain are those that are responsible for planning routes, decision-making, body movement, and making judgement calls.

fMRI studies also investigated the effects of substances on brain activity and driving performance. A study by Calhoun et al. (2004) analyzed the effects of alcohol on the brain in a driving task. This study found significant changes in the anterior cingulate cortex and primary motor cortex under the influence of alcohol. In addition, the consumption of alcohol-impaired error detection while driving. A similar study of the impacts of alcohol by Meda et al. (2009) found that higher alcohol doses were associated with increased crashes, steering weave, and more frequent crossing of passenger side white lines. Battistella et al. (2013) found that driving under the effects of cannabis caused a significant increase in blood oxygen level dependent (bold) response (as an indirect measure of change in neural activity) in comparison to placebo smoking. This increase was seen in the anterior cingulate cortex and ventromedial prefrontal cortex.

An advantage of using fMRI is that it does not use radiation like other techniques such as X-rays and computed tomography. fMRI is also a noninvasive technique that produces high-resolution images. While fMRI has its advantages, it also has its disadvantages. Obtaining the technology to perform an fMRI is expensive. In addition, the person being scanned must remain completely skill, which is difficult to do when analyzing a driving task. The participant also has to lay on a table that slides into the center of a magnet to produce the images needed, which is also difficult to accommodate when analyzing a driving activity (Watson and Mann 2018).

2.2 MEG-Based Driving Research

Magnetoencephalography, or MEG, identifies spatial distribution of cerebral oscillations using synthetic aperture magnetometry. In one study (Sakihara et al. 2014), participants' brain activities were recorded using MEG while driving using a simulator. Here is the setup and procedure of a typical MEG study examining brain activity while driving. During the study, participant's heads were stabilized so that no movement was allowed. This experiment consisted of an active car-following task. Using a 64-channel whole-head MEG system, neuromagnetic signals were recorded at a sampling rate of 625 Hz. In total, ten sets of passive viewing and active driving data were collected from each participant. During this experiment, MRI scans were obtained and the MEG data was then superimposed onto these scans for further evaluation.

Table 2 Summary of exemplary MEG-based driving studies

Author	Experimental task	Subject characteristics	Brain region stimulated	Key findings
Bowyer et al. (2008)	Watching a video of a driving scene and pressing a pedal when a red light appeared, while answering questions from a simulated cell phone conversation	• 16 adults (5 M) • M_{age}: 34 ± 11	• Lower central and left peripheral visual field • Wernicke's area	• When no speech production there was no activity in Wernicke's area • Overt speech showed Wernicke's area as highly coherent in all subjects • Covert speech condition, Wernicke's area was active while watching a driving scene • Broca's area is active in all three conditions
Sakihara et al. (2014)	Driving a freeway at night, two lanes, mostly straight with gentle curves, few oncoming cars	• 14 adults (2 M) • Age: 21–57	• Oscillatory activity	• Theta power identified in SFG and MFG • Theta band induced in MPC occurred during mental arithmetic and tracing a maze • Oscillatory differences in SFG, IPL, PoCG, MTG, and PCiG during simulated driving associated with cognitive function • Consistent with other studies done using EEG and fMRI

(continued)

Table 2 (continued)

Author	Experimental task	Subject characteristics	Brain region stimulated	Key findings
Bowyer et al. (2004)	Attending to surrounding vehicle activity in a driving video	• 5 adults (4F) • Age: 30–55	• Right motor and premotor cortex • Frontal gyrus • Occipital cortex • Superior frontal gyrus • Anterior cingulate gyrus • Precentral gyrus • Auditory cortex	• High frequency imaged in right motor and premotor context • Orbital frontal gyrus activation was active with peaks associated with light onset and pedal activation • Red light on activated occipital cortex, superior frontal gyrus, and anterior cingulate gyrus • Motor activity in the precentral gyrus correlated to foot and leg control followed the foot pedal response
Fort et al. (2010)	Driving in an urban environment, braking at yellow lights and following direction signs, while listening to radio	• 13 male adults • Age: 21–28	• Dorsal frontoparietal network • Ventral frontoparietal network	• 77% of participants gave correct answers about radio broadcasts • Strong activity observed around right temper parietal junction and spreads to posterior areas of the right frontal area • Bilateral medial frontal and bilateral superior parietal areas were activated
Bowyer et al. (2009)	Monitoring a driving video and pressing a foot pedal in response to a small red light at unpredictable times, with hands-free conversation	• 19 adults (14F) • Age: 18–65	• Right superior parietal lobe, occipital region	• Largest significant difference found in behavioral testing found overt conversation condition increased mean reaction time by 73 ms compared to no conversation • Largest effect sizes in behavioral lab for reaction times were for long conversation compared to no conversation

In reference to Table 2, one can see Fort et al. (2010) study using MEG to investigate a driver's attention and processing of visual information during simulated driving. During this task, drivers were to listen to a radio broadcast while driving and answer questions about said broadcast. 77% of participants in the study gave correct answers about the radio broadcasts. In addition, there was strong activity in the temporal–parietal junction and the posterior areas of the right frontal area. In another study by Bowyer et al. (2004), they analyze the effects on reaction time and brain activity due to visual events that occur on a driving scene. In addition, there was a secondary task of hands-free conversation (i.e., participants conversed using a Bluetooth device rather than holding a phone). Participants showed significantly delayed response time when carrying out a conversation compared to no conversation. This impairment in reaction time further intensified with increased duration of conversation.

MEG technology is highly precise as it is accurate down to the millimeter. In addition, MEG has high temporal resolution and is adaptable to many functions as it is noninvasive and easy to use. One of the main drawbacks to MEG is the large cost that comes along with it. Also, MEG's interest signals are on such a small scale that it is easy to pick up noise when analyzing the signals; thus, specialized shielding is required to eliminate interference. For the purpose of this study, it is hard to perform a driving simulation task without having too much noise to perform an MEG (MIT, n.d.).

2.3 EEG-Based Driving Research

An electroencephalogram, or EEG, allows for the exploration of information within the human body by processing cognitive function at a high resolution in situations where the performance cannot be measured. With this technology, the analysis of oscillatory brain activity is analyzed using frequency bands of the EEG that reflect different mental processes of perception and cognition. The three bands of interest are alpha (8–3 Hz), theta (4–7 Hz), and beta (>13 Hz). Alpha bands are most frequently found in the posterior regions of the head associated with mental fatigue. Theta bands are associated with cognitive control and are increased during monotonous tasks. Lastly, beta bands are associated with sensorimotor functions, which are related to mental workload (Karthaus et al. 2018).

As given in Table 3, a study performed by Mardi et al. (2011) analyzed brain activity changes due to varying levels of fatigue by using EEG technology. This study found that the brain is in its lowest level of activity and complexity when the driver is fatigued and drowsy. Fatigued drivers lose their concentration and control and cannot respond quickly to risky situations. Another study performed by Haak et al. (2009) used EEG signals and the tracking of blinking eyes when presented with visual distractions to determine the effects of stress while driving. In this study, the presence of billboards on a straight track has an effect on the emotional state of the participant, as the mean time between blinks is short. On curvy tracks, billboards

Table 3 Summary of exemplary EEG-based driving studies

Author	Experimental task	Subject characteristics	Brain region stimulated	Key findings
Lin et al. (2005)	Lane-keeping driving task	• 10 adults • Age: 20–40	• Anterior and posterior regions	• Two EEG channels (Cz and Pz) found to have the highest correlation between EEG power spectrum and driving performance • Using all channels caused too much noise for the experiment
Mardi et al. (2011)	Diving game	• 10 adults (3F) • M_{age}: 27.67	• Frontal • Occipital • Prefrontal	• Brain in lowest level of activity and complexity when driver is fatigued and drowsy • Fatigued drivers lose concentration and control and cannot respond quickly
Lin et al. (2011)	Freeway scene built on a 6-DOF platform, with five different additional tasks	• 15 male adults • Age: 20–28	• Frontal • Left motor	• Significant power increased in frontal cluster related to the math task • Theta power increases in three out of five dual-task cases • Power in dual-task is stronger than math task • Frontal area deals with impulse, judgment, language, memory, motor function, and problem-solving
Haak et al. (2009)	Driving a race car simulator	• N/A	• Frontal • Occipital • Prefrontal	• Presence of billboards on a straight track has effect on emotional state of participant; mean time between blinks is short • On curvy track, billboards are linked to higher count of blink and lower mean time between blinks due to stress of having to turn the vehicle • Sharp curves require more cognitive processing • Brain much busier when eyes are open

(continued)

Table 3 (continued)

Author	Experimental task	Subject characteristics	Brain region stimulated	Key findings
Karthaus et al. (2018)	Driving 31 mph on a monotonous straight two-lane road through grassland	• 14 adults; • M_{age}: 25.1 ± 2.7 • 28 elderly; • M_{age}: 64.6 ± 3.7	• Fronto-central • Posterior region	• Fronto-central alpha power was slightly stronger in young than old group and decreased with increasing crosswind • Posterior and fronto-central theta power was stronger in young than old • Fronto-central beta power decreased with increasing crosswind, but no significant differences between groups
Brown et al. (2013)	Three types of roadways (urban, freeway, and rural) at three times of day (9 a.m.–1 p.m., 10 p.m.–2 a.m., and 2 a.m.–6 a.m.)	• 72 adults: • 3 age groups (21–34, 38–51, and 55–68 years)	• Central • Lateral • Parietal	• Power ranges great in early and late night in lateral sites for both alpha and theta waveforms, and parietal area for alpha • Distraction differences found later in drive
Gharagozlou et al. (2015)	Driving in an overnight study first 90-km-long monotonous road and last 20 km a winding mountain out road	• 12 male adults • Age: 20–30	• Parietal	• Strong correlation found on the fatigue scale in final ten minutes of driving • Final ten minutes of driving alpha power saw differences in parietal area of brain
Touryan et al. (2014)	Maintaining vehicle position within the current cruising lane by correcting drift	• 25 adults (10 M) • M_{age}: 34.6	• Frontal • Parietal	• Self-reported fatigue increased during driving tasks • Driving task perceived as boring by participants

(continued)

Table 3 (continued)

Author	Experimental task	Subject characteristics	Brain region stimulated	Key findings
Brooks et al. (2018)	Driving straight four-lane highway with a speed limit sign (25/45 mph)	• 10 male adults • Age: 27–39	• Parietal • Left posterior • Right anterior	• Strongest performance variability was captured by reaction time and heading error • All 32 channels showed significant increases in alpha power with increasing heading errors (strongest relationships in left posterior and right anterior locations)
Perrier et al. (2016)	Two-day simulations: one night of sleep deprivation and a normal night of sleep; maintain a constant speed of 58mph and a steady lateral position	• 24 adults (12 M) • M_{age}: 26.9 ± 3.4	• Frontal • Central • Parieto-occipital	• Standard deviation of lane position overall higher after sleep deprivation compared to normal sleep • SDLP remain constant with time on task under normal sleep but varies more after sleep deprivation
Ahn et al. (2016)	Driving tasks when well-rested and sleep-deprived	• 11 adults (10 M) • M_{age}: 26.6 ± 1.4	• Fronto-central	• Relative power level from EEG study did not differ in delta, theta, and gamma band but did in alpha and beta band • Less oxygen consumed when fatigued; therefore, brain activity is suppressed
Nguyen et al. (2017)	Driving an oval shape with four curves, with two short and two long paths	• 11 adults (10 M) • Age: 24–28	• Parietal • Frontal • Temporal	• Beta band power is high when eye closure is zero and gets lower as eye closure increases • When drowsy alpha band power level values are higher in parietal lobes, beta band is lower in frontal lobe and gamma band is lower in both temporal lobes

(continued)

Table 3 (continued)

Author	Experimental task	Subject characteristics	Brain region stimulated	Key findings
Kim et al. (2013)	Varying driving/road conditions	• 30 adults • Age: 20–60	• Frontal	• Left turn found to have 12% increase in driving workload compared to straight driving
Getzmann et al. (2018)	Keeping a car on track on a one-lane road with varying curves	• 16 Young (8 M); M_{age}: 24.1 • 16 Older (8 M); M_{age}: 63.3	• Fronto-central • Anterior • Posterior	• Time off track increased with higher task load but not on time on task or age • Steering variability higher with higher task load decreased with time on task • Total power from EEG readings stronger in younger age group • Frontal relative theta power slightly stronger in younger than older group • Older group showed overall decrease in relative alpha power with lower task load and younger group demonstrates a more pronounced increase in alpha power

were linked to a higher number of blinks and lower mean time between blinks due to the stress of having to turn the vehicle. Sharper curves require more cognitive processing. This study also found that the brain is much busier when the eyes are open. To further analyze curves while driving, Kim et al. (2013) examined the effects of increasing driving workload and found that left turns would have a 12% increase in driving workload compared to traveling straight.

Many of the studies analyzed in Table 3 investigated fatigue and its effect on brain activity and driving capability. In Perrier et al. (2016) study, they did just this. They found that the standard deviation of lane position was higher overall after sleep deprivation. Time on task during this study had higher variation with increased sleep deprivation as well. Getzmann et al. (2018) performed another study on analyzing the age-related differences in proactive driving utilizing EEG measurements. They found that time off track increased with a higher task load but not on time on task or age. In addition, steering variability was higher with a higher task load and decreased with time on task. Total brain power from the EEG measurements was stronger in the younger age group.

For brain research, EEG has clear advantages. EEG is highly precise as it captures changes in the brain's electrical activity that occurs quickly. In addition, EEG is a noninvasive procedure, which makes it a more viable option for study purposes. In comparison to other imaging technologies, EEG equipment is relatively inexpensive and simple to operate. However, EEG is not useful for pinpointing the exact source of the brain activity. More advanced techniques are currently available for analyzing EEG which allows for more accurate detection of brain area activation (e.g., Low-resolution brain electromagnetic tomography or LORETA, effective connectivity, etc.).

2.4 *fNIRS-Based Driving Research*

A functional near-infrared spectroscopy is a portable technique which can be used for simulated and real-world driving. The fNIRS approach is more suitable for motion and has a higher temporal resolution. This technique measures the changes in the concentration of oxygenated and deoxygenated hemoglobin to analyze brain activity. fNIRS operates at a frequency of 2 Hz and is placed behind the participant in the driving simulator. Various filters are then added to filter out noise from other occurrences not focused on within the study (Grohol 2018).

In reference to Table 4, one can see a study performed by Sasai et al. (2016) where they analyzed the participants' brain activity using fNIR and performance in three driving tasks. This study employed listening to GPS instructions and listening to the radio as auditory secondary tasks. Contrary to what may be expected, there was no significant difference when adding an auditory secondary task. In addition, task difficulty, arousal level, and driving performance did not significantly differ between the two tasks. Another study by Tsunashima and Yanagisawa (2009) utilized fNIRS to analyze changes in brain activity when using cruise control was added as a secondary task to driving. This study found that the level of oxygenated hemoglobin in the frontal region of the brain increased when driving without adaptive cruise control. Unni et al. (2018) investigated adding additional tasks to the driving task to see its effect on the brain while driving. During this study, participants performed a typical working memory task in which a series of numbers were presented and then participants would have to recite the numbers in the same order. This added distraction caused an increase in time in which the participants needed to achieve the correct speed. In addition, it increased the variability of brake pedal position and response. The increased memory load also had a significant effect on heart rate as it increased with difficulty. Herff et al. (2014) performed a similar experiment to Unni et al. (2018) where the n-back test was performed. This study found that when the n-back test went from a 1-back to a 2-back to a 3-back test the percentage of targets missed by participants went from 5.7% to 16.7% to 33.7%. The proportion of incorrect responses while driving was significantly influenced by the working memory task as wrong reactions increased from 1.4 to 1.9 to 4.5 (2014).

Other studies used fNIRS to analyze the development of the brain by comparing younger drivers to older drivers. A study by Foy et al. (2016) analyzes changes in the prefrontal cortex in younger drivers as opposed to older drivers. In this study, it

Table 4 Summary of exemplary fNIRS-based driving studies

Author	Experimental task	Subject characteristics	Brain region stimulated	Key findings
Sasai et al. (2016)	• Task 1: Drive following GPS instructions • Task 2: Drive while listening to a radio • Task 3: Drive the simulator alone	• 13 male adults • Age: 22–34	• Frontal • Parietal • Occipital	• Task difficulty, arousal level, and driving performance did not significantly differ between integrated and split tasks • No significant difference in listening comprehension between listening to the radio and listening to GPS
Tsunashima and Yanagisawa (2009)	• Scenario 1: Driving with adaptive cruise control and driving without • Scenario 2: Driving twice under each condition	• 4 male adults • M_{age}: 20	• Frontal lobe	• Oxygenated hemoglobin increased when subject drove without ACC • Both outer portions of frontal lobe became active during driving task
Foy et al. (2016)	60 MPH stretch of single roadway	• 32 students	• Prefrontal Cortex	• There were greater increases in the activity of the right hemisphere of the prefrontal cortex than the left for all tasks • Younger drivers overestimated driving skill, underestimate accident risk, and have poorer situational awareness

(continued)

Table 4 (continued)

Author	Experimental task	Subject characteristics	Brain region stimulated	Key findings
Ahn et al. (2016)	Simulated driving when well-rested and when sleep-deprived	• 11 adults (10 M) • M_{age}: 26.6 ± 1.4	• Fronto-central region, alpha, beta, delta, theta, and gamma band activity	• The relative power level from EEG study did not differ in delta, theta, and gamma band but did in alpha and beta band • Less oxygen may be consumed when fatigued; therefore, brain activity is likely suppressed
FakhrHosseini et al. (2015)	Driving along an urban road with road signs, vehicles, traffic signals, and pedestrians	• 10 adult (4F) • M_{age}: 24 ± 5.7	• Compares oxygenation between left and right hemisphere	• Right hemisphere shows higher oxygen concentration than left hemisphere after writing • Right hemisphere also increases as hazardous events occurred • Overall oxygen levels in emotional driving were higher than baseline driving

(continued)

Table 4 (continued)

Author	Experimental task	Subject characteristics	Brain region stimulated	Key findings
Pradhan et al. (2015)	• Solo driving used as a control • Ten events integrated into the drive that would require decisions regarding risks	• 12 male adults • Age: 25–29	• Medial prefrontal cortex	• There were no significant differences in stopping behaviors between solo and passenger-drive conditions • When light turned yellow drivers showed higher brain activity in left medial prefrontal cortex and right medial prefrontal cortex while driving with a passenger
Nguyen et al. (2017)	Driving oval shape track with four curves, with two short and two long	• 11 adults (10 M) • Age: 24–28	• Parietal • Frontal • Temporal	• In most cases, beta band power is high when eye closure is zero and gets lower as eye closure increases • When drowsy, alpha band power level values are higher in parietal lobes, beta band is lower in frontal lobe, and gamma band is lower in both temporal lobes

(continued)

Table 4 (continued)

Author	Experimental task	Subject characteristics	Brain region stimulated	Key findings
Unni et al. (2018)	Driving a round course with a slightly curvy three-lane highway with standard lane widths	• 19 adults (17 M) • M_{age}: 25.2 ± 3.7	• Inferior frontal areas: ventral and dorsal prefrontal cortices • Temporo-occipital areas: inferior parietal areas	• The added distraction while driving cause an increase in the time in which participants needed to achieve the correct speed and the variability of the brake pedal position increased as well • The working memory load had a significant effect on heart rate as it increased with difficulty • Highest correlations of brain activation for working memory load are found bilaterally in the inferior frontal areas reflecting activation changes in ventral and dorsal prefrontal cortices

(continued)

Table 4 (continued)

Author	Experimental task	Subject characteristics	Brain region stimulated	Key findings
Oka et al. (2015)	Driving an expressway with a straight line	• 15 adults (8 M) • M_{age}: 31.4 ± 4.4	• Right premotor cortex • Right frontal eye field • Bilateral prefrontal cortex	• Cerebral oxygen exchange increased significantly more during left curves in right premotor cortex, right frontal eye field, and the bilateral prefrontal cortex • In non-driving conditions, increases were significantly greater during left curves in the right frontal eye field
Herff et al. (2014)	Driving while performing a n-back test	• 10 adults (6 M) • M_{age}: 22.0	• Parietal • Frontal • Temporal	• Percentage of targets missed by subjects increased from 5.7% for 1-back to 16.7% for 2-back and 33.7% for 3-back test • Amount of wrong reactions significantly influenced by the n-back test; Wrong reactions increased from 1.4 to 1.9 to 4.5
Khan and Hong (2015)	Driving a simulated car for an hour while keeping the car within 40–60 km/h with and without sleep deprivation	• 13 male adults • M_{age}: 28.5 ± 4.8	• Region A: Left hemisphere of brain • Region B: Prefrontal cortex • Region C: Right hemisphere of brain	• Region A accounts for the left hemisphere of the brain • Region A is most active during drowsy state • Most significant variations of hemodynamic changes area seen in region A

(continued)

Table 4 (continued)

Author	Experimental task	Subject characteristics	Brain region stimulated	Key findings
Li et al. (2018)	Driver's reaction to 3 traffic signs (red light, green light, and random sign)	• 13 adults (8 M) • Age: 21–24	• Prefrontal cortex	• First 1–2 h showed obvious spatial variation for all participants • Brain functional activation map in the prefrontal cortex varied significantly with increase of driving time • Accuracy and reaction time dropped quickly after the first hour • Fatigue affected the left hemisphere more than the right

(continued)

Table 4 (continued)

Author	Experimental task	Subject characteristics	Brain region stimulated	Key findings
Foy and Chapman (2018)	12 mile virtual loop of 4 road types with 2 examples of each, 40–70 mph, traffic lights, signs, pedestrians, roundabouts included	• 26 adults • Age: 18–24	• Prefrontal	• Road type had significant effect for oxygenated hemoglobin • Deoxygenated hemoglobin had no significant main effect of road type • Heart rate and respiration rate showed no significant main effect of road type • Mean skin conductance had significant main effect from road type and number of skin conductance responses per minute • Mean fixation duration showed a significant effect from road type • Horizontal spread of search showed a significant effect of road type

(continued)

Table 4 (continued)

Author	Experimental task	Subject characteristics	Brain region stimulated	Key findings
Yoshino et al. (2013)	Driving in a real expressway	• 12 adults (8 M) • M_{age}: 33.3 ± 4.5	• Prefrontal Cortex	• fNIRS images showed areas activated changed according to the driving operation and not due to the time of day • Bilateral and medial regions of the brain increased significantly more during acceleration than deceleration • Change in cerebral blood volume increased significantly during U-turns than parking in the right hemisphere and medial cortex
Bruno et al. (2018)	• Section 1: drive down center lane • Section 2: turning portion • Section 3: autonomous driving	• 21 adults (11 M) • M_{age}: 23.48	• Prefrontal cortex • Parietal cortex	• Vehicle lateral position deviation during turning maneuver was significantly different • Pupil diameter was significantly greater during turning portion and straight portion as oppose to observing • Turning pupil diameter was higher than straight • During the turning portion of the trial, cortical activation was greater for left prefrontal cortex, right prefrontal cortex, right parietal cortex, and left parietal cortex

was found that younger drivers overestimate their driving skills and underestimated accident risk. Younger drivers also had poorer situational awareness. Similarly, while Foy and his colleagues investigated age and its effects on the brain, Oka et al. (2015) investigated the effects left versus right turns have on the prefrontal cortex. This study showed that cerebral oxygen exchange significantly increased more during left curves in the right premotor cortex and right frontal eye field. In non-driving conditions, increases were significantly greater during left curves in the right frontal eye field.

FNIRS, like the other imaging technologies, offers advantages and disadvantages as well. FNIRS may seem uncomfortable to participants, as it requires the attachment of a set of probes on the scalp. On the other hand, it does not expose participants to high magnetic fields. In addition, fNIRS uses two near-IR wavelengths to provide separate measures of oxygenated and deoxygenated hemoglobin as opposed to fMRI that provides only a measure of deoxygenated hemoglobin. A big advantage for this study's purposes is the fact that fNIRS does not require severe head movement constraint like fMRI (Aslin and Mehler 2005).

3 Common Brain Areas and Activation Patterns While Driving

When utilizing the fMRI technology, a "core circuit" in the brain was found during driving activities. The cerebellum was found to be connected to motor error correction, the tracking of a visual target, and coordinated limb movements. In addition, the bilateral extrastriate cortex is utilized when processing visual images and identifying objects. Another region widely used while driving is the right middle temporal gyrus which is linked to visual motion processing. As given in Table 3, driving with the addition of subtasks typically activates the frontal region of the brain such as the left superior frontal gyrus, left orbital frontal gyrus, and the inferior frontal cortex. There is also significant brain activity in the parietal region of the brain.

Findings showed that the bilateral precuneus which aids in the coordination of motor behavior. The left motor and bilateral premotor cortex execute voluntary movements and allow driving at the motor preparation level. Also, the left anterior part of the insula has recently shown a role in high-level cognitive control which aids in responding to stimuli and prioritizing them. The right posterior cingulate gyrus is used for topographical memory which is needed for mental routing while driving. Lastly, the right dorsomedial part of the thalamus usually relays sensory information to the cortex which is key in alertness (Navarro et al. 2018). With the MEG technology along with the synthetic aperture magnetometry (SAM) analysis (Vrba and Robinson, 2001), brain areas involved in concentration, attention, motor control, and visual images were able to be analyzed. In Table 2 much like the fMRI studies, when analyzing MEG studies, there was significant brain activity seen in the frontal region of the brain. When introducing visual stimulants such as a driving scene, there is significant brain activity in the occipital cortex.

The study utilizing fNIRS found that there are greater increases in brain activity in the right hemisphere of the prefrontal cortex than the left for all tasks analyzed during the study. This study found that younger drivers lack development within the prefrontal cortex as it is not fully developed until around the age of 25; therefore, they are more likely to take risks while driving. In addition, young drivers overestimate their driving skills, while underestimating the risk of an accident, and have poorer situational awareness than older drivers (Joy et al. 2001). As seen before, when analyzing fNIRS studies, it also found most brain stimulation in the frontal region of the brain such as the prefrontal cortex, parietal region, medial prefrontal cortex, and temporal lobes.

As given in Table 3, EEG studies did not stray from what has been seen so far in the other neuroimaging techniques and their relation to areas of the brain activated by driving studies. These studies like all the others saw significant brain activity mainly in the frontal region of the brain.

4 Organizational Planning Levels

With the three-level planning model (Navarro et al. 2018; strategical, tactical and operational), brain regions that are stimulated during certain tasks while driving are identified. The planning of a driving trip occurs in the strategical level. This step is the overall big picture of the trip, while the important details soon follow. For example, the driver knowing that they must turn in a mile would be a strategical level of planning. Those plans are then made into movements at the tactical level. The tactical level of planning occurs before the execution of the movement. In the driving scenario, secondary tasks and traffic levels affect the tactical level. Finally, the movements are executed into movements during the operational level.

4.1 Strategical Level

The strategical level of the brain is responsible for the planning tasks while driving. Using the fMRI analysis with strategical tasks showed that the cerebellar posterior lobes, used for motor control, right precuneus, responsible for motor coordination, and the bilateral extrastriate cortex, used in visual representation, are highly utilized with strategical level tasks. Also, the right superior temporal gyrus has recently come under a lot of investigation as it potentially involves mentalizing others' thoughts. During any time of driving, the strategical level is engaged first (Navarro et al. 2018).

Throughout the MEG study, oscillatory differences were detected during the active driving portion of the experiment. A power increase in the theta band was observed in the superior frontal gyrus along with the medial frontal gyrus which are involved in the decision-making process of driving (Sakihara et al. 2014). Therefore, active driving requires more of the strategical level of the brain than passive viewing. The conducted MEG study was an hour long; therefore, drowsiness and fatigue were a factor. During passive viewing, the alpha value in the middle temporal gyrus showed a positive correlation with drowsiness and fatigue which affected the participant's focus on the task at hand.

In Choi et al. (2017) fMRI study, they analyzed a driver's ability to perform subtasks in addition to driving. When performing subtasks, the reaction time of driving is delayed, therefore making the strategical task of planning delayed as well. Any sort of secondary task such as additional conversation, additional visual distractions, or additional difficult maneuvers while driving will also affect the planning process of driving. Along with having to overcome these new distractions, drivers still have to plan their next course of action while driving. With added tasks while driving, it is apparent that the planning task will see some delay. There were other experiments where the speed limit was altered changing the time in which participants had to plan their routes. Driver fatigue was another experimental area that slowed down brain activity and thus slowed down the driver's ability to plan their routes.

Pradhan et al. (2015) analyzed the effects the presence of passengers can have on the prefrontal cortex activity of male drivers. In regards to the strategical level of planning, most would think that the presence of passengers would always affect one's ability to effectively plan out driving actions, but this study found that there were no significant differences in stopping behaviors between solo and passenger driving conditions. On the other hand, changes were seen between these two scenarios in a yellow light situation, so one cannot claim that a passenger never affects the strategical level of planning when driving.

4.2 Tactical Level

The tactical level of the brain is utilized while driving when determining which course of action to take. While the strategical level of planning plans a route, the tactical level determines what movements will be used in order to achieve that goal. During the fMRI study, only two brain regions were found to be associated to tactical tasks—the bilateral middle frontal areas and the right middle temporal gyrus. The bilateral middle frontal areas plan voluntary complex movements, while the right middle temporal gyrus is used for visual recognition. The tactical level is only used when deemed necessary after a strategical level is engaged (Navarro et al. 2018).

Schweizer et al. (2013) fMRI study investigated the difference in brain activity when turning with no traffic, with traffic, and while having audio distractions. This study showed brain activity in the frontal gyrus along with the middle temporal gyrus which correlates to the tactical level of planning. In this study, tactical level of planning is used to judge when a turn should be executed when presented with traffic obstacles and distractions such as audio. Unni et al. (2018) performed a study using fNIRS to analyze the effect varying road conditions have on brain activity. In addition to varying road conditions such as curves, the participant had to perform the n-back test that was discussed in previous studies. These added distractions while driving caused an increase in the time in which participants needed to achieve the correct speed and variability of the brake pedal position. In other words, the added distractions made it difficult to determine a proper course of action. Lin et al. (2011) performed a similar study using EEG technology. This study simulated a freeway but incorporated five different additional tasks such as the car drifting off the road along with presenting two-digit addition equations. The addition equations of this study affect the strategical level of planning as it adds in a delay in the driving task. On the other hand, the additional tasks such as drifting off the road requires the driver to correct the car's lane position, thus affecting the tactical level of planning.

4.3 *Operational Level*

The operational level executes the decisions made in the strategical and tactical levels of the brain. Three brain areas are used during operational tasks—the right extrastriate cortex, the anterior lobe of the right cerebellum, and the right medial dorsal nucleus of the thalamus (Navarro et al. 2018). The conducted MEG study was an hour long; therefore, drowsiness and fatigue were a factor. During the active driving of this experiment, the theta power in the superior frontal gyrus decreased toward the last quarter of the experiment (Sakihara et al. 2014). Therefore, the driver had less brain activity for executing his decisions as time progressed and drowsiness and fatigue increased.

In Li et al. (2018) study, they analyzed the changes of hemoglobin levels using fNIRS technology through cognitive tests of identifying street signs. This study lasted a total of eight hours so fatigue became a key factor in this study. The first one to two hours showed spatial variation for all participants. In terms of length of the drive, activity of the prefrontal cortex varied significantly with the increase of driving time. After the first hour of driving, reaction time and accuracy of driving dropped quickly. The left hemisphere of the brain was affected more than the right. This study proved that fatigue can and will affect how the body operates. This observation is key when analyzing driving as when one performs a lengthy drive they are more likely to make mistakes and react more slowly than usual.

Utilizing EEG technologies, Karthaus et al. (2018) analyzed how varying ages of driving affected driving performance. In this study, participants drove around thirty miles per hour for sixty minutes on a monotonous two-lane road with no bends of visual distractions. This study analyzed changes in alpha, beta, and theta band activity within the front, central, and posterior regions of the brain. From the study, it was found that front and center brain region had a slightly stronger alpha band power in the young age group when compared to the older age group, and posterior and fronto-central theta power was stronger in the young as well. Fronto-central beta power saw no significant differences between the age groups. The low frontal theta and alpha activity could be an age-related decline in activity which has been found in other studies as well. While the study found less brain activity and slower reaction times in older drivers, which may affect driver safety, the study also found that young drivers are more likely to take risks while driving. Therefore, both age groups have different operational risks when driving.

5 Summary and Future Directions

Due to the physical requirements of neuroimaging techniques, almost all of the studies analyzed were required to use a driving simulator. For instance, measuring brain activity while a driver is behind the wheel on the road is difficult. Although some studies have used EEG to measure brain activity as indices of arousal and workload during on-road driving (Lin et al. 2011), these attempts of on-road brain recording are limited likely due to the technical difficulties in the setup and signal processing during data analysis. While driving simulator seems an apparent easier alternative, significant challenges still exist. For example, in order to collect data with fMRI, the participant must lie in a supine position within a scanner. This protocol restricts the participants' personal space while interacting with a simulated environment while also allowing very minimal head movement. Therefore, this simulated driving is much different than a real-life driving situation which could skew results.

It is also hypothesized that the areas of the brain engaged could fluctuate with a driver's experience level due to the fact that drivers in the studies are experienced drivers (Navarro et al. 2018). While the MEG system uses a t-statistic for oscillatory differences, the variance in power increases and decreases between subjects suggests that the t-test might not be optimal (Sakihara et al. 2014). fMRI offers room for further exploration as it is a noninvasive technique and offers a cross between spatial and temporal resolutions in regards to driving. Technical difficulties with fNIRS led to some participants scenarios containing no data; therefore, further development of the technology to make it easier to use would be preferable. Although fNIRS had its flaws, fNIRS could be used in the future to detect the overloading of the threshold of young drivers prefrontal cortex to determine how much of a risk driving could be due to their lack of development (Foy et al. 2016).

The vast majority of the studies analyzed used a driving simulator. The reason for this could varies. As discussed previously, many of the neuroimaging techniques are infeasible to incorporate into a real-life driving situation; therefore, the only option is some sort of simulation. Driving simulators can also be a safer way to perform studies. For example, some of the studies analyzed in this paper analyze the effects of driving while under the influence of alcohol and driving while fatigued. These studies would not be able to be performed in a real-life scenario without putting both the driver and the analyst at risk. In all simulated studies, the participant is put in a safer risk, as there are no repercussions for crashing, yet this may also skew results, as the participant knows there are no repercussions as well. In addition, many of these studies did not clearly define what planning level was affected; therefore, some speculation had to be made between tactical, strategical, and operational levels being affected.

There are many future direction researches in neural activity when driving can be taken. One could further investigate these neuroimaging techniques and their ability to detect brain activity while driving. On the other hand, one could also expand and incorporate new methods for analyzing brain activity while driving in order to gain a new perspective. There could also be a comparison made between driving and a similar task to show that these brain areas seen while driving also appear in other similar tasks.

References

Ahn, S., Nguyen, T., Jang, H., Kim, J. G., & Jun, S. C. (2016). Exploring neuro-physiological correlates of drivers mental fatigue caused by sleep deprivation using simultaneous EEG, ECG, and fNIRS data. *Frontiers in Human Neuroscience*, 10.

Aslin, R. N., & Mehler, J. (2005). Near-infrared spectroscopy for functional studies of brain activity in human infants: promise, prospects, and challenges. *Journal of biomedical optics, 10*(1), 011009.

Association for Safe International Road Travel. (2002). Road Safety Facts. (n.d.). Retrieved March 7, 2019, from https://www.asirt.org/safe-travel/road-safety-facts/.

Battistella, G., Fornari, E., Thomas, A., Mall, J. F., Chtioui, H., Appenzeller, M., ... & Giroud, C. (2013). Weed or wheel! FMRI, behavioural, and toxicological investigations of how cannabis smoking affects skills necessary for driving. *PLoS one*, *8*(1), e52545.

Bernardi, G., Ricciardi, E., Sani, L., Gaglianese, A., Papasogli, A., Ceccarelli, R., ... Pietrini, P. (2013). How Skill Expertise Shapes the Brain Functional Architecture: An fMRI Study of Visuo-Spatial and Motor Processing in Professional Racing-Car and Naïve Drivers. *PLoS ONE*, *8*(10).

Blana, E. (1996). A survey of driving research simulators around the world. Working Paper. Institute of Transport Studies, University of Leeds, Leeds, UK.

Boas, G. (2014, October 28). The birth of an academic society. Retrieved March 25, 2019, from https://www.nmr.mgh.harvard.edu/news/141028/the-birth-of-an-academic-society.

Brooks, J. R., Passaro, A. D., Kerick, S. E., Garcia, J. O., Franaszczuk, P. J., & Vettel, J. M. (2018). Overlapping brain network and alpha power changes suggest visuospatial attention effects on driving performance. *Behavioural Neuroscience, 132*(1), 23–33.

Brown, T., Johnson, R., & Milavetz, G. (2013). Identifying periods of drowsy driving using EEG. *Annals of Advances in Automotive Medicine, 57*, 99.

Bowyer, S. M., Hsieh, L., Moran, J. E., Chiang, Y. R., Young, R. A., & Tepley, N. (2004, August 8–12). Neural correlates of event related distractions during a driving task using MEG. In *Biomag 2004: Proceedings of the 14th International Conference on Biomagnetism: Boston, Massachusetts, USA* (p. 215). Biomag 2004 Ltd.

Bowyer, S. M., Hsieh, L., Moran, J. E., Young, R. A., Manoharan, A., Liao, C. J., et al. (2009). Conversation effects on neural mechanisms underlying reaction time to visual events while viewing a driving scene using MEG. *Brain Research, 1251*, 151–161.

Bowyer, S. M., Moran, J. E., Seaman, S., Young, R. A., Sullivan, J., Farjam, R., ... & Hsieh, L. (2008). Language processes during overt and covert speech in a simulated driving task. In *Proceedings of the 16th International Conference on Biomagnetism, Sapporo, Japan* (pp. 25–29).

Bruno, J. L., Baker, J. M., Gundran, A., Harbott, L. K., Stuart, Z., Piccirilli, A. M., ... Reiss, A. L. (2018). *Mind over motor mapping: driver response to changing vehicle dynamics*. Wiley.

Calhoun, V. D., Pekar, J. J., & Pearlson, G. D. (2004). Alcohol intoxication effects on simulated driving: Exploring alcohol-dose effects on Brain activation using functional MRI. *Neuropsychopharmacology, 29*(11), 2097–2107.

Choi, M. H., Kim, H. S., Yoon, H. J., Lee, J. C., Baek, J. H., Choi, J. S., … & Chung, S. C. (2017). Increase in brain activation due to sub-tasks during driving: fMRI study using new MR-compatible driving simulator. *Journal of physiological anthropology*, *36*(1), 11.

Chung, S., Choi, M., Kim, H., You, N., Hong, S., Lee, J., … Kim, H. (2014). Effects of distraction task on driving: A functional magnetic resonance imaging study. *Bio-Medical Materials and Engineering*, 2971–2977.

Doherty, E. (2014). *MEG matters: MIT's new magnetoencephalography lab delivers results.* McGovern Institute for Brain Research at MIT.

FakhrHosseini, M., Jeon, M., & Bose, R. (2015, September). Estimation of drivers' emotional states based on neuroergonmic equipment: An exploratory study using fNIRS. In *Adjunct Proceedings of the 7th International Conference on Automotive User Interfaces and Interactive Vehicular Applications* (pp. 38–43). ACM.

Ferrari, M., Giannini, I., Sideri, G., & Zanette, E. (1985). Continuous non invasive monitoring of human brain by near infrared spectroscopy. *Advances in Experimental Medicine and Biology Oxygen Transport to Tissue VII*, 873–882.

Foy, H. J., & Chapman, P. (2018). Mental workload is reflected in driver behaviour, physiology, eye movements and prefrontal cortex activation. *Applied Ergonomics, 73,* 90–99.

Foy, H. J., Runham, P., & Chapman, P. (2016). Prefrontal cortex activation and young driver behaviour: a fNIRS study. *PLoS ONE, 11*(5), e0156512.

Fort, A., Martin, R., Jacquet-Andrieu, A., Combe-Pangaud, C., Daligault, S., Foliot, G., & Delpuech, C. (2010). Attention and processing of relevant visual information while simulated driving: A MEG study. *Brain Research*, 117–127.

Getzmann, S., Arnau, S., Karthaus, M., Reiser, J. E., & Wascher, E. (2018). Age-related differences in pro-active driving behaviour revealed by EEG measures. *Frontiers in Human Neuroscience, 12*, 321.

Gharagozlou, F., Saraji, G. N., Mazloumi, A., Nahvi, A., Nasrabadi, A. M., Foroushani, A. R., … Samavati, M. (2015). Detecting driver mental fatigue based on EEG alpha power changes during simulated driving. *Iranian Journal of Public Health*, 1693–1700.

Grohol, J. M., Psy, D. (2018). What is functional near-infrared spectroscopy? *PsychCentral.*

Haak, M., Bos, S., Panic, S., & Rothkrantz, L. (2009) Detecting stress using eye blinks and brain activity from eeg signals. In *Proceeding of the 1st driver car interaction and interface (DCII2008)* (pp. 35–60).

Herff, C., Heger, D., Fortmann, O., Hennrich, J., Putze, F., & Schultz, T. (2014). Mental workload during n-back task—quantified in the prefrontal cortex using fNIRS. *Frontiers in Human Neuroscience, 7.*

Hung, Y., Vetivelu, A., Hird, M. A., Yan, M., Tam, F., Graham, S. J., et al. (2014). Using fMRI virtual-reality technology to predict driving ability after brain damage: A preliminary report. *Neuroscience Letters, 558,* 41–46.

Jo, J., Kim, H., Chung, S., & Choi, M. (2019). BOLD signal change during driving with addition task using fMRI. *Research of Institute of Biomedical Engineering.*

Joy, S., Fein, D., Kaplan, E., & Freedman, M. (2001). Quantifying qualitative features of block design performance among healthy older adults. *Archives of Clinical Neuropsychology, 16*(2), 157–170.

Just, M. A., Keller, T. A., & Cynkar, J. (2008). A decrease in brain activation associated with driving when listening to someone speak. *Brain Research, 1205,* 70–80.

Karthaus, M., Wascher, E., & Getzmann, S. (2018). Proactive vs. reactive car driving: EEG evidence for different driving strategies of older drivers. *Plos One*, *13*(1).

Khan, M. J., & Hong, K. (2015). Passive BCI based on drowsiness detection: An fNIRS study. *School of Mechanical Engineering, Pusan National University*. Retrieved March 6, 2019.

Kim, H., Choi, M., Yoon, H., Kim, H., Jeoung, U., Park, S., … Lee, B. (2014). Cerebral activation and lateralization due to the cognition of a various driving speed difference: An fMRI study. *Bio-Medical Materials and Engineering*, 1133–1139. Retrieved March 6, 2019.

Kim, J. Y., Jeong, C. H., Jung, M. J., Park, J. H., & Jung, D. H. (2013). Highly reliable driving workload analysis using driver electroencephalogram (EEG) activities during driving. *International Journal of Automotive Technology, 14*(6), 965–970.

Li, D. H., Liu, Q., Yuan, W., & Liu, H. X. (2010). Relationship between fatigue driving and traffic accident. *Journal of traffic and transportation engineering (Xi'an, Shaanxi), 10*(2), 104–109.

Li, T., Lin, Y., Gao, Y., & Zhong, F. (2018). Longtime driving induced cerebral hemodynamic elevation and behavior degradation as assessed by functional near-infrared spectroscopy and a voluntary attention test. *Journal of Biophotonics, 11*(12), e201800160.

Lin, C., Chen, S., Chiu, T., Lin, H., & Ko, L. (2011). Spatial and temporal EEG dynamics of dual-task driving performance. *Journal of NeuroEngineering and Rehabilitation, 8*(1), 11.

Lin, C., Wu, R., Jung, T., Liang, S., & Huang, T. (2005). Estimating driving performance based on EEG spectrum analysis. *EURASIP Journal on Advances in Signal Processing, 2005*(19), 521368.

Mardi, Z., Ashtiani, S. N., & Mikaili, M. (2011). EEG-based drowsiness detection for safe driving using chaotic features and statistical tests. *Journal of Medical Signals and Sensors*, 130–137.

Meda, S. A., Calhoun, V. D., Astur, R. S., Turner, B. M., Ruopp, K., & Pearlson, G. D. (2009). Alcohol dose effects on brain circuits during simulated driving: An fMRI study. *Human Brain Mapping, 30*(4), 1257–1270.

Navarro, J., Osiurak, F., & Reynaud, E. (2018). Neuroergonomics of car driving: A critical meta-analysis of neuroimaging data on the human brain behind the wheel. *Neuroscience & Biobehavioral Reviews*.

Nguyen, T., Ahn, S., Jang, H., Jun, S. C., & Kim, J. G. (2017). Utilization of a combined EEG/NIRS system to predict driver drowsiness. *Scientific Reports, 7*(1).

Oka, N., Yoshino, K., Yamamoto, K., Takahashi, H., Li, S., Sugimachi, T., … Kato, T. (2015). Greater activity in the frontal cortex on left curves: a vector-based fNIRS study of left and right curve driving. *PLos One, 10*(5).

Perrier, J., Jongen, S., Vuurman, E., Bocca, M., Ramaekers, J., & Vermeeren, A. (2016). Driving performance and EEG fluctuations during on-the-road driving following sleep deprivation. *Biological Psychology, 121,* 1–11.

Pradhan, A. K., Hu, X. S. F., Buckley, L., & Bingham, C. R. (2015). Pre-frontal cortex activity of male drivers in the presence of passengers during simulated driving: an exploratory functional near-infrared spectroscopy (fNIRS) study.

Sakihara, K., Hirata, M., Ebe, K., Kimura, K., Ryu, S. Y., Kono, Y., … Yorifuji, S. (2014). Cerebral oscillatory activity during simulated driving using MEG. *Frontiers in Human Neuroscience, 8.*

Sasai, S., Boly, M., Mensen, A., & Tononi, G. (2016). Functional split brain in a driving/listening paradigm. *Proceedings of the National Academy of Sciences.*

Schweizer, T. A., Kan, K., Hung, Y., Tam, F., Naglie, G., & Graham, S. J. (2013). Brain activity during driving with distraction: An immersive fMRI study. *Frontiers in Human Neuroscience, 7.*

Sorkin, J. (2017). *Getting into brain waves: History and resources.* Retrieved March 25, 2019 from https://neuroscience.stanford.edu/news/getting-brainwaves-history-and-resources.

The Columbia Encyclopedia. (2019). Internal-combustion engine. Retrieved from https://www.encyclopedia.com/science-and-technology/technology/technology-terms-and-concepts/internal-combustion-engine.

Touryan, J., Apker, G., Lance, B. J., Kerick, S. E., Ries, A. J., & Mcdowell, K. (2014). Estimating endogenous changes in task performance from EEG. *Frontiers in Neuroscience, 8.*

Tsunashima, H., & Yanagisawa, K. (2009). Measurement of Brain Function of Car Driver Using Functional Near-Infrared Spectroscopy (fNIRS). *Computational Intelligence and Neuroscience,* 1–12.

Unni, A., Ihme, K., Jipp, M., & Rieger, J. (2018). Corrigendum: Assessing the drivers current level of working memory load with high density functional near-infrared spectroscopy: A realistic driving simulator study. *Frontiers in Human Neuroscience, 12.*

Uttal, W. R. (2002). Précis of the new phrenology: The limits of localizing cognitive processes in the brain. *Brain and mind, 3*(2), 221–228.

Vrba, J., & Robinson, S. E. (2001). Signal processing in magnetoencephalography. *Methods, 25*(2), 249–271.

Watson, T. M., & Mann, R. E. (2018). Harm reduction and drug-impaired driving: Sharing the road?

Yoshino, K., Oka, N., Yamamoto, K., Takahashi, H., & Kato, T. (2013). Functional brain imaging using near-infrared spectroscopy during actual driving on an expressway. *Frontiers in Human Neuroscience, 7.*

Fundamentals and Emerging Trends of Neuroergonomic Applications to Driving and Navigation

Inki Kim, Erfan Pakdamanian and Vishesh Hiremath

Abstract The state-of-the-art vehicle automation and navigation technologies promise to augment or even replace diverse human functions in driving. For safety assurance on roadways, vehicles need to be informed about the humans (not only their presence but also mental states) in and around the vehicle. Yet, the uncertainties of drivers' mental states under varying traffic situations make it difficult to provide such information. In this regard, neuroergonomics have potential to help bridge human and automation toward the next level of integration. In this chapter, we aim to provide an appreciation of neuroergonomic application to driving and navigation, with an emphasis on drivers' cognitive tasks and performance. Particularly, the four main cognitive constructs associated with driving—attention, situation awareness (SA), intent, and mental workload—are reviewed in terms of their theoretical foundations and recent applications in neuroergonomic studies. The effects of special demographic population and environmental selection for such study are also discussed. To the readers who are interested in understanding the effects of next-generation vehicle technologies on humans or who aspire for the breakthrough in neuroergonomics for new driving and navigation technology, this chapter may provide a useful source.

1 Introduction: Cognitive Functions in Driving and Navigation

Driving involves complicated information processing that demands a range of cognitive resources or functions. Inadequate functions or scant resource may negatively impact the driver's abilities to make appropriate decision under certain traffic situations and operate vehicle in a safe manner. The scientific structure, mechanism, and capacity of cognitive functions are common topics both in neuroergonomics and cognitive science. In particular, neuroergonomics is the study of human brain and

I. Kim (✉)
University of Illinois at Urbana-Champaign, Urbana, IL, USA
e-mail: inkikim@illinois.edu

E. Pakdamanian · V. Hiremath
University of Virginia, Charlottesville, VA, USA

C. S. Nam (ed.), *Neuroergonomics*, Cognitive Science and Technology,
https://doi.org/10.1007/978-3-030-34784-0_19

behaviors at work that integrate scientific principles and practices of neuroscience and ergonomics, for improved function and performance. Contrary to the theories in cognitive science and psychology that may help understand how certain human functions work in driving tasks, neuroergonomics contributes to the evaluation and augmentation of such functions in completion of driving tasks.

Just as modern navigation systems have changed a driver's wayfinding task into a perceptual follow-through, increased vehicle automation can reshape the nature of cognitive functions in driving. For example, the set of dominant cognitive tasks in a vehicle to be operating under *cruise control* is different from those under *adaptive cruise control*. It is an obvious cognitive gain to relieve oneself from the attentional demand of longitudinal distance monitoring. However, this gain is accompanied by the loss of situation awareness associated with both longitudinal control and monitoring the road to find critical obstacles and events.

In general, automation will transform driving tasks from physical control to mental supervision, and neuroergonomics are suitable for supporting this rapid transition. Neuroergonomics could provide unique opportunities to investigate the neural bases of perceptual and cognitive functions, in order to shed lights on performance improvement and technological designs dedicated to driving automation. In the long run, automation in driving is expected to accompany two contradictory effects: While automation is capable of compensating for driver errors, and thus the drivers grow less sensitive to the maintenance of situation awareness (Merat et al., 2019), the drivers still need to intervene in the face of system failures or when the automation reaches its operational limit. Finding the balance between legitimate situation awareness and cognitive performance of drivers is likely to be an important issue to be investigated by neuroergonomics.

This chapter primarily aims to provide a foundation of important cognitive constructs that are common to driving and navigation, regardless of automation level, technical properties, or driving conditions in traffic ecosystems. Scholarly articles, technical reports, and seminal works in the fields of cognitive science and engineering, cognitive neuroscience and neurophysiology, and transportation engineering and human–machine interaction studies were reviewed for this foundation. Based on an initial review of over 150 articles, we could parse them out into the four topical areas of attention, situation awareness, intent, and mental workload. Certainly, these topics are closely interconnected, but separate treatment of each topic helps comprehend their relations in detail with neuroergonomic establishments. In addition, this chapter attempts to provide an overview of most recent trends in neuroergronomic applications associated with each one of the topics. Especially, current applications for driver state inference, intent recognition, and action prediction necessitates thorough literature search, because of their potential impacts for breakthrough in vehicle automation. In the following sections, we also summarized the methods of processing and analysis of the brain signals associated with driving performance.

2 Attention in Driving and Navigation

2.1 Fundamentals of Attention

Attention is one of the most fundamental human functions, common to all cognitive tasks. As a metaphor, attention may serve as a filter to selectively constrain attentional resources to a meaningful target, or as a resource that facilitates information processing in a given work environment (Wickens & McCarley, 2008; Wiegmann & Kirlik, 2007). The former concept of *selective attention* (Broadbent, 1970) elaborates the attentional processing of sensory registration and perceptual encoding in terms of early- and late-selection. In other words, selective attention prioritizes some of the critical environmental components from the range of perceived information on which to make a proper decision response. Traditionally, *event-related potentials* (ERPs) have provided a classic framework in neurophysiology to relate electric signals with the stimuli and its processing in the brain (Gazzaniga, Ivry, & Mangun, 2002). In the controlled experiments (Eason, Harter, & White, 1969; Mangun, Hillyard, & Luck, 1993), an ERP wave known as *P1* (first major positive wave) can be observed over lateral occipital regions of the scalp, which amplifies itself in a latency of 70–90 ms when a visual stimuli appear at a location that the subject attends. Furthermore, succeeding evidence confirms that the signal amplification on the occipital scalp reflects the location of the stimulus in the visual field (Hopfinger, Buonocore, & Mangun, 2000).

Contrary to such theoretical establishment, the feasibility of ERP waves for the estimation of voluntary (or reflexive) attention requires further validation in driving studies, due to the inherent complexity associated with naturalistic driving tasks (Fastenmeier & Gstalter, 2007). The related methodological issues of subject and platform choice will be discussed in Sects. 5 and 6, respectively.

The latter concept of shared resource for dividing attention between multiple tasks was developed (Navon & Gopher, 1979; Wickens, 2008). As the complex nature of driving requires significant amount of road scanning—either as an active controller or a passive monitor once the automation is prompted— the driver needs to constantly perceive and make decisions. According to *multiple resource theory*, humans are less likely to make invalid decisions due to separately engaged cognitive resources. This *multiple resource theory* assumes that such resource-sharing process is determined by three factors that encompass both human cognition and task environment: (1) the attentional demand of the tasks, (2) the degree to which related mental resources are shared within the specific human information processing, and (3) allocation policy set forth by the individuals at work (Lee, Kirlik, & Dainoff, 2013).

Since this divided attention in driving relies on the factors outside the human information processing (for instance, the attentional demands may be externally imposed, or allocation policy can be altered with education), only limited neurophysiological evidence exists to support the theory (Just et al., 2001). Besides, the fact that both cognitive states of under- and overload could deteriorate driver performance emphasizes moderate utilization of mental resources. Advancement in onboard neurophysiological measurement could assist in this. At a macro-level, attentional resource in a driving task is known to have systemic relations with the level of vehicle automation and workload (Kaber & Endsley, 2004), and this topic will be further discussed in the Sect. 5.

2.2 Neuroergonomic Application to Driver Attention, Attention Failures, and Fatigue

Attention involves allocating cognitive resources to target according to the predefined structure of priority. Practically, most driver attention studies in neuroergonomics revolve around accident prevention and safety regulations. Improper attention allocation and distraction has been identified as the most common crash factors (Klauer, Dingus, Neale, Sudweeks, & Ramsey, 2006). Attentional problems are a dominant type of driver errors, accounting for approximately 40% of 340,278 road accidents in the United Kingdom (Brown, 1990). Reallocating attentional resources by shifting from the most important location for safe driving to another area-of-interest could cause one to lose situational awareness and to become involved in an accident (Singh, 2015; Victor et al., 2015). A closer look at attentional failures explicates faulty mechanisms in the driver's attentional process; information is not processed due to competition for attention (i.e., a failure of divided attention); certain features in the visual field of view were given undue priority of attention or obscured (i.e., a failure of selective attention) (Brown, 2002).

Taxonomy of driver attention provides a foundation for neuroergonomic application with clearly defined mental states. According to the American Automobile Association Foundation for Traffic Safety (AAAFTS), a driver's attentional state can belong to one of the five categories: attentive, distracted, looked-but-did-not-see, sleepy, and unknown (Stutts, Reinfurt, Staplin, & Rodgman, 2001). For distraction, in particular, the National Highway Transportation Safety Association (NHTSA) presents the four categories of visual distraction, cognitive distraction, auditory distraction, and biomechanical distraction, based on the driver's function (Ranney, Garrott, & Goodman, 2001).

So far, neuroergonomic applications have shown substantial evidence of accurately inferring attentional states and detecting attentional failures. In a review of inattention monitoring systems (Dong, Hu, Uchimura, & Murayama, 2011), the authors examined literature that used electroencephalography (EEG) signals to successfully identify and detect attentional states, mostly for the state of sleepiness or fatigue (Berka et al., 2007; Chouvarda et al., 2007; Jap, Lal, Fischer, & Bekiaris, 2009; Lin et al., 2005, 2008; Skinner, Nguyen, & Liu, 2007; Yang, Lin, & Bhattacharya, 2010; Yeo, Li, Shen, & Wilder-Smith, 2009). Despite their demonstrated viability, however, the authors recommended a hybrid combination of driver's physical measures (e.g., eye movements) and driving performance measures (e.g., vehicle parameters) for a more acceptable solution. Obstructive EEG sensors and heavy signal processing were pointed out as technical hurdles, although neurophysiological measures are still useful for ground-truth markers in controlled driving environment (Dong et al., 2011).

More recently, advances in Deep Learning and data analytics are substantially improving the classification performance of EEG time-series data into appropriate attentional states. Table 1 summarizes the input EEG data and feature extraction technique, classifier algorithm, and classification performance for the inference of driver attentional state in the past 3 years. To date, the classification performance is not predictable, subject to the changes of input data. Besides, the performance variance across different feature extraction and classifiers implies room for further improvement.

Although advanced analytics and computational power can offload the burden of signal processing to some degree, the poor usability of sensing in EEG still remains a challenge. The choice of neuroergonomic measurement and computational method needs to be based on the extensive analysis of cost and benefit. For instance, a recent study that used electrocardiogram (ECG) in place of EEG showed improved detection of driver distraction up to 88.45% in a naturalistic driving environment (Deshmukh & Dehzangi, 2017).

Another trend of monitoring driver attention is a pursuit of realism. A field driving test together with long duration of driving is suitable to track the time-dependent vigilance of drivers. In (Schmidt et al., 2009), the alpha spindle rate derived from alpha band (6–12 Hz) and the stimulus-induced P3 ERP wave confirmed continuously decreased vigilance over 3-hour-long driving on a monotonous road environment. Enhanced realism for the test and validation of neuroergonomic application will become critical with vehicle automation. A new norm of desirable driver attention and cases of attentional failures will be redefined under different level-of-automation (LOA) (Banks & Stanton, 2017), and their detection also needs to incorporate the changes of vehicle features and road infrastructure.

3 Situation Awareness (SA) in Driving and Navigation

3.1 Fundamentals of SA

For a wide variety of human–machine interaction, situation awareness (SA) provides a useful construct as a basis of human performance. It enables to explain how humans overcome the limits of attentional resources and working memory in practice, by focusing on the "content" of the perceived elements in the task environment (Kihlstrom, 1984), contrary to the "content-less" aspect of human information processing (Jacko, 2012). By definition, SA is involved in the perception of specific cues in task environment, comprehension of current situation, and projection of future status (Endsley, 1995).

Table 1 Recent Progress in Driver State Inference based on EEG

Article	Purpose of study	Input data and feature extraction method	Classifier algorithm	Inference performance
Zeng et al. (2018)	To infer the driver mental state predefined in terms of the eye blink and heart rate	15-Channel EEG/none	Convolutional neural network (CNN) combined with deep residual learning	92.68
Hu and Min (2018)	To detect driver fatigue	32-Channel EEG, EOG/entropy	Gradient boost decision tree (GBDT)	91.90
Nguyen, Ahn, Jang, Jun, and Kim (2017)	To classify into drowsy or awake state	64-Channel EEG, NIRS/beta relative power level (RPL), hemoglobin concentration change	Fisher's linear discriminant analysis (FLDA)	79.20
Hu (2017)	To detect driver fatigue	32-Channel EEG/entropy	Decision tree classifier	91.80
Hajinoroozi, Zhang, and Huang, (2017)	To classify into drowsy or alert state	64-Channel EEG/riemann distances of EEG covariance matrices	Convolutional neural network (CNN) with deep Riemannian model	70.96
Gao et al. (2018)	To classify into fatigue or alert state	32-Channel EEG/Wavelet Entropy	Relative wavelet entropy complex network (RWECN)	95.75
Cui and Wu (2017)	To detect driver drowsiness	32-Channel EEG/Power spectral density in theta and alpha bands	3-layer Convolutional neural network (CNN), regression	64.46*
Li (2018)	To detect driver fatigue	32-Channel EEG	Particle swarm optimization (PSO) with SVM	93.20
Fu, Wang, and Zhao (2016)	To model dynamic driver fatigue	O1- and O2-channels of EEG, EMG, Respiration/Feature optimization with relative operating characteristic (ROC) curve	Hidden Markov Model (HMM)	–

* Indicates a correlation coefficient in percentage (%), significant at 5%, as a measure of performance, while other values in the same column indicate the average percentage of accurate classification across all subjects

The mismatch between the current allocation of cognitive resources and attention demand set forth by activity requirements could possibly place humans in critical situations (Engström, Victor, & Markkula, 2017). Such a broad relevance of SA to key cognitive constructs, including schema, mental model, goal-oriented processes, and team cognition, has made it a widely accepted trait of quality and expert performance (Adams, Tenney, & Pew, 1995; Brown, 1990; Endsley, 1988, 1995; Fracker, 1988; Smith & Hancock, 1995; Taylor & Selcon, 1994; Tenney, Adams, Pew, Huggins, & Rogers, 1992).

Driver SA is no exception: it requires efficient and timely attentional allocation to properly maneuver the vehicle and successfully respond to a critical event. Driver SA represents the perception of multimodal vehicle display, navigation information, and visual cues in the driving scene for *Level 1 SA* (Ma & Kaber, 2007). At *Level 2*, SA includes the driver's understanding of significance, constraints, changes, risk, priorities, timing and confidence about the perceived element (Sarter & Woods, 1991). At *Level 3*, SA represents proactive judgment and prediction (Amalberti & Deblon, 1992).

Despite its popularity and meaningfulness to driving tasks (Baumann & Krems, 2007), a critical quest after the underlying cognitive basis has persisted from early on (Sarter & Woods, 1991), and relatively limited neuroergonomic evidence confirms the basis. For instance, in (French, Clarke, Pomeroy, Seymour, & Clark, 2007), continuous EEG data from the 19 tin scalp electrodes was recorded from the 12 participants while they played a simulation game, and the raw EEG records were time-segmented according to their semantic functions in the game play, into Level 1, 2, and 3 SA, respectively. As a result, discriminant analysis of the raw EEG data could correctly classify 89.4% of Level 2 SA, indicating an increased involvement of the frontal lobe activities in the theta and gamma bands during Level 2 SA events. To our knowledge, such partial findings were not validated on a greater number of trials or with time-locked EEG data with ERPs.

3.2 *Neuroergonomic Application to SA and Judgment*

To date, literature continues to seek the neurocognitive basis of SA and situation-driven judgment over a variety of interactive tasks. In (Catherwood et al., 2014), an extensive review of *Brodmann Areas* (Amunts, Schleicher, & Zilles, 2007) associated with an individual or all levels of SA was followed by 128-channel EEG recording of the participants over two different tasks (to identify a target pattern and to identify "threat" associated with a target in a scene) in which the target changed without warning to incur SA loss in the experiment. The key regions of the brain activity were identified by using standardized low-resolution electrical tomography (sLORETA) (Pascual-Marqui, 2002). The results showed that significant loss of SA was followed by concurrent engagement of visual regions, as well as rapid memory operations, suggesting a top-down processing in perception to rebuild Level 1 SA. In (Johnson et al., 2014), the situational judgment of law-enforcement personnel regarding the use of deadly force with different level of expertise was recorded in a simulation environment. The recordings of 10-channel EEG data for 25 officers in response to 27 scenarios in simulation revealed the significant increase of the alpha and frontal theta as partial evidence of SA construction, while other aspects of SA in the frontal and right-parietal gamma was not evident. The application of discriminant analysis to the raw EEG data could discriminate between experts and novices with 72.6% accuracy.

With an increased level of automation in vehicle control and navigation, the importance of SA will continue to grow, together with other related psychological constructs, including mental model, trust, workload, and risk perception (De Winter, Happee, Martens, & Stanton, 2014; Horswill & McKenna, 2004; Ma & Kaber, 2007; Ma & Kaber, 2005; Stanton & Young, 2000). Reliable and reproducible recognition of these situationally-driven states remains a big challenge to the multidisciplinary fields of neuroergonomics and transportation engineering.

4 Intent in Driving and Navigation

4.1 Action Plan and Intent in Driving Under Uncertainty

The innovation toward vehicle-to-vehicle communication and advanced crash avoidance in vehicle-pedestrian or vehicle–bicycle interaction (Balk, Bertola, Shurbutt, & Do, 2014; Ben-Chaim, Lappan, & Houang, 1988; Richard et al., 2015; Sayed, Zaki, & Autey, 2013), necessitates that all entities on the roadway "share" their plan of operation to some degree. For effective safety intervention, this information sharing needs to occur before each entity, either vehicle or human, executes its own action plan.

Due to this time gap, handling the uncertainty of on-road interactions in a predictive manner becomes an essential quality of intelligent vehicle technology. For preemptive safety responses, accurate recognition and prediction of the other entity's intent are critical. For trust, it is also important for autonomous vehicles not to operate against the passengers' intent. In essence, the fear of manual drivers in mixed traffic environment who will encounter self-driving vehicles on roadways will be caused by the inability to understand and communicate intent with automation (Sarter, Woods, & Billings, 1997).

Humans learn to read others' intent and communicate their own by using certain set of behavioral cues. Infants as young as 15 months learn to notice the difference between intentional and unintentional behaviors, and at the age of two, they can encode intent from a sequence of actions. Adults are even better at encoding action sequences, with an aid of refined *hierarchy of intentions* (Baldwin & Baird, 2001). The granularity of the intention hierarchy depends on the importance of an action in a certain context, or the level of experience or knowledge one has about the action. The uncertainty in human intent recognition associated with contextual circumstances or an individual's knowledge can often be compensated by inferences.

At the lower level of neurophysiology, how motion plans are formed, internally represented, and translated into actions has been a topic of fundamental research (Keele, 1986). The scientific accounts of representing motor plans vary by studies, including representation in terms of the final position (location coding) (Bizzi, Accornero, Chapple, & Hogan, 1984), trajectory, or distance. Another issue is how different motion representations are structured together to form a sequence of actions. Some experiments support separate processing (Abrams, Stemmons, & Vandillen, 1991), while others in favor of hierarchical structure (MacKay, 2012).

The difficulties with seeking the neural basis of motor control continue when identifying motor attributes as a sign of specific motor plans or intent. Partly, such difficulties are due to the variability of movements. Ironically, the motion variability itself appears to be a clue to the process of motion plan: according to the neuromuscular analyses of reaching tasks performed by monkeys, the movement variability arises before movements begin and is determined by the variability of neural activities (neuromuscular noise) at the motion plan stages (Churchland, Afshar, & Shenoy, 2006). More recent evidence explores which aspects of motor plan or intent increases or decreases the motion variability, such as competition between multiple motion plans (Wijdenes, Ivry, & Bays, 2016) or sensory-motor control policies (Gallivan, Logan, Wolpert, & Flanagan, 2016).

At the high level of goal-directed actions (Hutchins, Hollan, & Norman, 1985; Norman, 2013), Norman presumes that an individual interacts with the world (and machines) by specifying and executing an action plan that best implement their intent, subject to the goal, and constraints in the environment. The neural basis of intent as a precursor to action has found some evidence. In (Gallivan, McLean, Valyear, & Culham, 2013), brain imaging (f-MRI) identified the parieto-frontal regions of the brain to become activated before the onset of purposeful behaviors. With EEG data, intent can be detected prior to movements as early as 500 ms (Lew, Chavarriaga, Silvoni, & Millán, 2012). The feasibility of intent recognition for action prediction in short time period has motivated application to assistive technologies for prosthetic design and robot-assisted rehabilitation for motion-impaired patients (Cincotti et al., 2008; Lenzi, De Rossi, Vitiello, & Carrozza, 2012).

5 Neuroergonomics Applied to Intent Recognition and Action Prediction

Neuroergonomic application has great potential to satisfy the needs to recognize intent and predict action for human–vehicle interaction. Substantial number of recent studies have attempted to accomplish this goal with demonstrated feasibility. For example, a simulation study (Kim, Kim, Haufe, & Lee, 2013) investigated the feasibility of detecting braking intent across three driving scenarios and emergency situations. The drivers were instructed to follow a lead vehicle within a specified distance without passing through it and then induced to an immediate brake in order to avoid a crash. The result showed a significant positive deflection at around 300 ms after the stimulus onset and before emergency braking. Many neuroergonomic studies investigated braking intent and braking behaviors (Haufe et al., 2014; Kim et al. 2013; Teng, Bi, & Liu, 2018); steering (Calderó-Bardají et al., 2016; Zhang et al., 2015); head rotation on a head-mounted display (Brouwer, van der Waa, Hogervorst, Cacace, & Stokking, 2017); or a broader set of control actions (Khaliliardali, Chavarriaga, Gheorghe, & del Millán, 2015). Especially, this study presented one of the earliest attempts to predict voluntary action intended to traffic events in driving by focusing on anticipatory ERP. Besides the confirmation of the anticipatory ERPs with an average peak value of -8 μV at the Cz electrode, the authors could successfully distinguish between the go and no-go intent from the traffic condition. This prediction accuracy was over 83% for braking and 79% for acceleration. Also, the centro-medial anticipatory potentials before action were observed at 320 ± 200 ms, with prediction accuracy of over 77%.

To date, most studies are confined to simulation setting or monotonous driving environment, and the number of target actions is limited to one (for detection) or a few (for classification). Table 2 summarizes the recent relevant applications with methods and findings.

6 Mental Workload and Special Driving Population

Mental workload is a well-investigated topic in human–machine interaction with rich foundation for assessment, mitigation, and system integration (Schmorrow, Kruse, Reeves, & Bolton, 2007). Over the past decades, several funding agencies have supported neuroergonomic projects that aim to measure dynamic workload under various task environment, by using EEG, magnetoencephalography (MEG), functional magnetic resonance imaging (fMRI), and functional near-infrared imaging (fNIR). In driving, neuroergonomic research of workload may be aspired by different purposes and context, and lead to different contributions. First, it can enable the objective assessment of cognitive demand for certain driving tasks, including the primary and the secondary (Kohlmorgen et al., 2007; Strayer & Drews, 2007), or driving environment (Kim, Jeong, Jung, Park, & Jung, 2013). In (Lei & Roetting, 2011), the frontal theta and the parietal-alpha activation in EEG data were correlated with perceived mental workload.

Second, it can focus on the real-time monitoring and prediction of dynamic workload during driving, with the ultimate goal of mitigation. In (Borghini et al., 2012), the powers of theta band in the prefrontal cortex and alpha band in the parietal areas were effective predictors of dynamic workload. In (Kim, Yoon, Shin, & Park, 2018), the repetitive patterns of EEG activation and driving behaviors were machine-learned for workload prediction, which can go without the sensing electrode after adequate training.

Table 2 Neuroergonomic application to driver intent recognition and action prediction

Article	Purpose of study	Methods	Key findings
Kim et al. (2013)	To recognize intent of emergency braking in a simulation setting	Post hoc analysis of ERPs	A significant, positive deflection at about 300 ms after the stimulus onset in emergency braking
Kim, Kim, and Lee (2014)	To predict braking in a simulation setting	Post hoc application of kernel ridge regression (KRR) applied to event-related potential (ERP), event-related desynchronization (ERD), and readiness potential (RP)	Superior prediction accuracy of voluntary braking over emergency braking Improved accuracy by the combination of ERP, ERD, and RP
Haufe et al. (2011)	To recognize intent of emergency braking in a simulation setting	Post hoc analysis of RP and movement-related potential from EMG	Braking intent precedes the pedal response by approximately 130 ms
Zhang et al. (2015)	To recognize and classify the intent of turn in a simulation and on-road driving environment	Both Post hoc and real-time application of discriminant analysis to EEG	Classification accuracy was as high as 70% in post hoc/simulation setting. The real-time analysis was about 2% lower in accuracy
Teng et al. (2018)	To recognize intent of emergency braking in a simulation setting	Post hoc application of discriminant analysis and sequential forward floating search (SFFS) to the optimized power spectra of EEG	Accurate recognition becomes feasible 420 ms after the onset of emergency situation. The recognition accuracy was at least 94%
Khaliliardali et al. (2015)	To predict an action dichotomy of braking and acceleration in a simulation setting	Application of discriminant analysis to ERPs with the contingent negative variation (CNV) paradigm	Confirmed the presence of the anticipatory ERPs with an average peak value of $-8\ \mu$Vat on electrode Cz. Classification accuracy was over 83% for braking and 79% for acceleration
Calderó-Bardají et al. (2016)	To predict an action dichotomy of steering direction in on-road driving environment	Application of principal component analysis (PCA) and support-vector machine (SVM) with a linear kernel to EEG	Sample entropy provides a superior classification accuracy (73.5%), compared with those with the estimated powers of the theta and alpha bands (62.6%)

Third, it can help compare different interface designs in vehicle display and navigation systems (Heger, Putze, & Schultz, 2010). Through such comparative evaluation, vehicle interface can be balanced between information overload and negligence. Fourth, it helps test new vehicle technology and identify emergent cognitive tasks with the technology. As illustrated with emergence of adaptive cruise control, higher level of automation tends to transfer direct control into monitoring tasks while the driver is potentially faced with a demand explosion in emergency (Stanton, Young, & McCaulder, 1997). Therefore, as the driver's role changes, the control authority may shift back and forth on a regular basis from the human driver to the automated system. During the transition period, the automated system will prompt the driver about the vehicle status and ask for temporary intervention, takeover request. In this case "readiness potential" could indicate the capability of driver to intervene (Kim, Kim, Kim, Lee, & Yoon, 2018; Kim, Hwang, Yoon, Choi, & Park, 2014). As the driver is expected to effectively reengage situation awareness for safe driving in a matter of seconds, this readiness potential can be useful to investigate the optimized states of driver alert and/or the design of feedback systems (S. Fu & Parasuraman, 2006). The driver readiness potentials obtained from ERP components could be used to alarm the driver to take over properly via various modalities such as auditory (Forster, Naujoks, Neukum, & Huestegge, 2017); visual (Liu et al., 2014); haptic (Pakdamanian, Feng, & Kim, 2018); or multimodal stimuli, which has shown superior to unimodal alarms in prompt attention orientation (Lee, McGehee, Brown, & Marshall, 2006).

Finally, it can provide cost-effective triage and customized support for special driving population. In (Fan, Wade, Key, Warren, & Sarkar, 2018), the features of fractal dimension, higher order crossings, power of frequency bands could detect the drivers of *autistic spectrum disorder* (ASD) at an accuracy of over 86%. At a glance, older population has always been on a higher risk of automobile crashes. Rather than age itself, neurodegenerative disorders and treatment are likely to be the factor. A neurophysiological testing of senior drivers could associate a particular set of driver errors that originate from the lack of judgment about their own driving competence, with dementia (Dobbs, 1997).

7 Discussion

The choice of experimental setting for neuroergonomic studies determines the degree to which the findings can be generalized. The most common types are interactive simulation in a controlled setting, mixed-reality simulation (Augmented/Virtual Reality) for immersive experiences with a medium level of control (Brown et al., 2018; Charissis & Naef, 2007; Weidner, Hoesch, Poeschl, & Broll, 2017), and on-road field test (Akita et al., 2014; Schmidt et al., 2009; Sisiopiku, 2001), with or without naturalistic setting. Certainly, one type of setting cannot entirely cover for the problem space (e.g., risk factors and hazards in driving environment) that threaten safety. For example, studies that rely on simulation, epidemiological, and controlled environments are not able to identify hidden factors in driving that lead to accidents. A large-scale naturalistic data is a better study design for this purpose. Likewise, a study of vigilance would not be accurate in a simulator, because the level of vigilance, engagement, and fatigue might be different in a simulation than in real driving conditions. However, there is not explicit and standardized guidance as to the choice of experimental settings and protocols. More studies are needed to compare the benefits and cost for the choice of experimental setting, as well as statistically valid guidance for the choice of experimental parameters, given the expected level of noise in environment, variability in measurement, and required precision of the findings.

References

Abrams, R. A., Stemmons, V. A., & Vandillen, L. (1991). Multiple sources of spatial information for aimed limb movements. In *Bulletin of the Psychonomic Society* (Vol. 29, p. 520). Psychonomic Soc Inc 1710 Fortview Rd, Austin, TX 78704.

Adams, M. J., Tenney, Y. J., & Pew, R. W. (1995). Situation awareness and the cognitive management of complex systems. *Human Factors, 37*(1), 85–104. https://doi.org/10.1518/001872095779049462.

Akita, T., Shirahige, H., Seunghee, H., Hayashi, J., Suzuki, K., & Doi, S. (2014). Research of stop assistance considering visibility of Intersection. *International Journal of Automotive Engineering, 5*(2), 65–71. https://doi.org/10.20485/jsaeijae.5.2_65.

Amalberti, R., & Deblon, F. (1992). Cognitive modelling of fighter aircraft process control: a step towards an intelligent on-board assistance system. *International Journal of Man-Machine Studies, 36*(5), 639–671.

Amunts, K., Schleicher, A., & Zilles, K. (2007). Cytoarchitecture of the cerebral cortex—more than localization. *Neuroimage, 37*(4), 1061–1065.

Baldwin, D. A., & Baird, J. A. (2001). Discerning intentions in dynamic human action. *Trends in Cognitive Sciences, 5*(4), 171–178.

Balk, S. A., Bertola, M. A., Shurbutt, J., & Do, A. (2014). *Human factors assessment of pedestrian roadway crossing behavior.*

Banks, V. A., & Stanton, N. A. (2017). *Automobile automation: distributed cognition on the road.* CRC Press.

Baumann, M., & Krems, J. F. (2007). Situation awareness and driving: A cognitive model. In *Modelling driver behaviour in automotive environments* (pp. 253–265). Springer.

Ben-Chaim, D., Lappan, G., & Houang, R. T. (1988). The effect of instruction on spatial visualization skills of middle school boys and girls. *American Educational Research Journal, 25*(1), 51–71.

Berka, C., Levendowski, D. J., Lumicao, M. N., Yau, A., Davis, G., Zivkovic, V. T., … Craven, P. L. (2007). EEG correlates of task engagement and mental workload in vigilance, learning, and memory tasks. *Aviation, Space, and Environmental Medicine*, *78*(5), B231–B244.

Bizzi, E., Accornero, N., Chapple, W., & Hogan, N. (1984). Posture control and trajectory formation during arm movement. *Journal of Neuroscience, 4*(11), 2738–2744.

Borghini, G., Vecchiato, G., Toppi, J., Astolfi, L., Maglione, A., Isabella, R., … Zhou, Z. (2012). Assessment of mental fatigue during car driving by using high resolution EEG activity and neurophysiologic indices. In *2012 Annual International Conference of the IEEE Engineering in Medicine and Biology Society* (pp. 6442–6445). IEEE.

Broadbent, D. E. (1970). Stimulus set and response set: Two kinds of selective attention.

Brouwer, A.-M., van der Waa, J. S., Hogervorst, M. A., Cacace, A., & Stokking, H. (2017). A feasible BCI in real life: Using predicted head rotation to improve HMD imaging. In *Proceedings of the 2017 ACM Workshop on An Application-oriented Approach to BCI out of the laboratory* (pp. 35–38). ACM.

Brown, B., Park, D., Sheehan, B., Shikoff, S., Solomon, J., Yang, J., & Kim, I. (2018). Assessment of human driver safety at Dilemma Zones with automated vehicles through a virtual reality environment. In *Systems and Information Engineering Design Symposium (SIEDS), 2018* (pp. 185–190). IEEE.

Brown, I. D. (1990). Drivers' margins of safety considered as a focus for research on error. *Ergonomics, 33*(10–11), 1307–1314.

Brown, I. D. (2002). A review of the'looked but failed to see'accident causation factor. In *behavioural research in road safety: Eleventh Seminar.*

Calderó-Bardají, P., Longfei, X., Jaschke, S., Reermann, J., Mideska, K. G., Schmidt, G., … Muthuraman, M. (2016). Detection of steering direction using EEG recordings based on sample entropy and time-frequency analysis. In *Engineering in Medicine and Biology Society (EMBC), 2016 IEEE 38th Annual International Conference of the* (pp. 833–836). IEEE.

Catherwood, D., Edgar, G. K., Nikolla, D., Alford, C., Brookes, D., Baker, S., et al. (2014). Mapping brain activity during loss of situation awareness: An EEG investigation of a basis for top-down influence on perception. *Human Factors, 56*(8), 1428–1452.

Charissis, V., & Naef, M. (2007). Evaluation of prototype automotive head-up display interface: testing driver's focusing ability through a VR simulation. In *2007 IEEE Intelligent Vehicles Symposium* (pp. 560–565). IEEE.

Chouvarda, I., Papadelis, C., Kourtidou-Papadeli, C., Bamidis, P. D., Koufogiannis, D., Bekiaris, E., & Maglaveras, N. (2007). Non-linear analysis for the sleepy drivers problem. In *Medinfo 2007: Proceedings of the 12th World Congress on Health (Medical) Informatics; Building Sustainable Health Systems* (p. 1294). IOS Press.

Churchland, M. M., Afshar, A., & Shenoy, K. V. (2006). A central source of movement variability. *Neuron, 52*(6), 1085–1096.

Cincotti, F., Mattia, D., Aloise, F., Bufalari, S., Schalk, G., Oriolo, G., … Babiloni, F. (2008). Non-invasive brain–computer interface system: towards its application as assistive technology. *Brain Research Bulletin, 75*(6), 796–803.

Cui, Y., & Wu, D. (2017). EEG-based driver drowsiness estimation using convolutional neural networks. In *International Conference on Neural Information Processing* (pp. 822–832). Springer.

De Winter, J. C. F., Happee, R., Martens, M. H., & Stanton, N. A. (2014). Effects of adaptive cruise control and highly automated driving on workload and situation awareness: A review of the empirical evidence. *Transportation Research Part F: Traffic Psychology and Behaviour, 27,* 196–217.

Deshmukh, S. V, & Dehzangi, O. (2017). ECG-based driver distraction identification using wavelet packet transform and discriminative kernel-based features. In *Smart Computing (SMARTCOMP), 2017 IEEE International Conference on* (pp. 1–7). IEEE.

Dobbs, A. R. (1997). Evaluating the driving competence of dementia patients. *Alzheimer Disease and Associated Disorders, 11,* 8–12.

Dong, Y., Hu, Z., Uchimura, K., & Murayama, N. (2011). Driver inattention monitoring system for intelligent vehicles: A review. *IEEE Transactions on Intelligent Transportation Systems, 12*(2), 596–614.

Eason, R. G., Harter, M. R., & White, C. T. (1969). Effects of attention and arousal on visually evoked cortical potentials and reaction time in man. *Physiology & Behavior, 4*(3), 283–289.

Endsley, M. R. (1988). Design and evaluation for situation awareness enhancement. In *Proceedings of the Human Factors Society annual meeting* (Vol. 32, pp. 97–101). SAGE Publications Sage CA: Los Angeles, CA.

Endsley, M. R. (1995). Toward a theory of situation awareness in dynamic systems. *Human Factors, 37*(1), 32–64.

Engström, J., Victor, T., & Markkula, G. (2017). Attention selection and multitasking in everyday driving: A conceptual model. In *Driver Distraction and Inattention* (pp. 27–54). CRC Press.

Fan, J., Wade, J. W., Key, A. P., Warren, Z. E., & Sarkar, N. (2018). EEG-based affect and workload recognition in a virtual driving environment for ASD intervention. *IEEE Transactions on Biomedical Engineering, 65*(1), 43–51.
Fastenmeier, W., & Gstalter, H. (2007). Driving task analysis as a tool in traffic safety research and practice. *Safety Science, 45*(9), 952–979.
Forster, Y., Naujoks, F., Neukum, A., & Huestegge, L. (2017). Driver compliance to take-over requests with different auditory outputs in conditional automation. *Accident Analysis and Prevention, 109,* 18–28.
Fracker, M. L. (1988). A theory of situation assessment: Implications for measuring situation awareness. In *Proceedings of the Human Factors Society Annual Meeting* (Vol. 32, pp. 102–106). SAGE Publications Sage CA: Los Angeles, CA.
French, H. T., Clarke, E., Pomeroy, D., Seymour, M., & Clark, C. R. (2007). Psycho-physiological measures of situation awareness. *Decision Making in Complex Environments, 291.*
Fu, R., Wang, H., & Zhao, W. (2016). Dynamic driver fatigue detection using hidden Markov model in real driving condition. *Expert Systems with Applications, 63,* 397–411.
Fu, S., & Parasuraman, R. (2006). Event-related potentials (ERPs) in neuroergonomics. *Oxford Series in Human-Technology Interaction, 32.*
Gallivan, J. P., Logan, L., Wolpert, D. M., & Flanagan, J. R. (2016). Parallel specification of competing sensorimotor control policies for alternative action options. *Nature Neuroscience, 19*(2), 320.
Gallivan, J. P., McLean, D. A., Valyear, K. F., & Culham, J. C. (2013). Decoding the neural mechanisms of human tool use. *Elife, 2,* e00425.
Gao, Z., Li, S., Cai, Q., Dang, W., Yang, Y., Mu, C., et al. (2018). Relative wavelet entropy complex network for improving EEG-based fatigue driving classification. *IEEE Transactions on Instrumentation and Measurement, 99,* 1–7.
Gazzaniga, M. S., Ivry, R. B., & Mangun, G. R. (2002). *Cognitive neuroscience.* New York: W. W. Norton & Company.
Hajinoroozi, M., Zhang, J., & Huang, Y. (2017). Prediction of fatigue-related driver performance from EEG data by deep Riemannian model. In *Engineering in Medicine and Biology Society (EMBC), 2017 39th Annual International Conference of the IEEE* (pp. 4167–4170). IEEE.
Haufe, S., Kim, J.-W., Kim, I.-H., Sonnleitner, A., Schrauf, M., Curio, G., et al. (2014). Electrophysiology-based detection of emergency braking intention in real-world driving. *Journal of Neural Engineering, 11*(5), 56011.
Haufe, S., Treder, M. S., Gugler, M. F., Sagebaum, M., Curio, G., & Blankertz, B. (2011). EEG potentials predict upcoming emergency brakings during simulated driving. *Journal of Neural Engineering, 8*(5), 56001.
Heger, D., Putze, F., & Schultz, T. (2010). Online workload recognition from EEG data during cognitive tests and human-machine interaction. In *Annual Conference on Artificial Intelligence* (pp. 410–417). Springer.
Hopfinger, J. B., Buonocore, M. H., & Mangun, G. R. (2000). The neural mechanisms of top-down attentional control. *Nature Neuroscience, 3*(3), 284.
Horswill, M. S., & McKenna, F. P. (2004). Drivers' hazard perception ability: Situation awareness on the road. *A Cognitive Approach to Situation Awareness: Theory and Application*, 155–175.
Hu, J. (2017). Comparison of different features and classifiers for driver fatigue detection based on a single EEG channel. *Computational and Mathematical Methods in Medicine, 2017.*
Hu, J., & Min, J. (2018). Automated detection of driver fatigue based on EEG signals using gradient boosting decision tree model. *Cognitive Neurodynamics*, 1–10.
Hutchins, E. L., Hollan, J. D., & Norman, D. A. (1985). Direct manipulation interfaces. *Human-Computer Interaction, 1*(4), 311–338.
Jacko, J. A. (2012). *Human computer interaction handbook: Fundamentals, evolving technologies, and emerging applications.* CRC press.
Jap, B. T., Lal, S., Fischer, P., & Bekiaris, E. (2009). Using EEG spectral components to assess algorithms for detecting fatigue. *Expert Systems with Applications, 36*(2), 2352–2359.

Johnson, R. R., Stone, B. T., Miranda, C. M., Vila, B., James, L., James, S. M., … Berka, C. (2014). Identifying psychophysiological indices of expert vs. novice performance in deadly force judgment and decision making. *Frontiers in Human Neuroscience.* Retrieved from https://www.frontiersin.org/article/10.3389/fnhum.2014.00512.
Just, M. A., Carpenter, P. A., Keller, T. A., Emery, L., Zajac, H., & Thulborn, K. R. (2001). Interdependence of nonoverlapping cortical systems in dual cognitive tasks. *Neuroimage, 14*(2), 417–426.
Kaber, D. B., & Endsley, M. R. (2004). The effects of level of automation and adaptive automation on human performance, situation awareness and workload in a dynamic control task. *Theoretical Issues in Ergonomics Science, 5*(2), 113–153.
Keele, S. W. (1986). Motor control.
Khaliliardali, Z., Chavarriaga, R., Gheorghe, L. A., & del Millán, J. R. (2015). Action prediction based on anticipatory brain potentials during simulated driving. *Journal of Neural Engineering, 12*(6), 66006.
Kihlstrom, J. F. (1984). Conscious, subconscious, unconscious: A cognitive perspective.
Kim, H., Kim, W., Kim, J., Lee, S.-J., & Yoon, D. (2018). Design of driver readiness evaluation system in automated driving environment. In *2018 International Conference on Information and Communication Technology Convergence (ICTC)* (pp. 300–302). IEEE.
Kim, H. S., Hwang, Y., Yoon, D., Choi, W., & Park, C. H. (2014a). Driver workload characteristics analysis using EEG data from an urban road. *IEEE Transactions on Intelligent Transportation Systems, 15*(4), 1844–1849.
Kim, H. S., Yoon, D., Shin, H. S., & Park, C. H. (2018b). Predicting the EEG level of a driver based on driving information. *IEEE Transactions on Intelligent Transportation Systems, 99,* 1–11.
Kim, I.-H., Kim, J.-W., Haufe, S., & Lee, S.-W. (2013). Detection of multi-class emergency situations during simulated driving from ERP. In *Brain-Computer Interface (BCI), 2013 International Winter Workshop on* (pp. 49–51). IEEE.
Kim, J.-W., Kim, I.-H., & Lee, S.-W. (2014). Decision of braking intensity during simulated driving based on analysis of neural correlates. In *Systems, Man and Cybernetics (SMC), 2014 IEEE International Conference on* (pp. 4129–4132). IEEE.
Kim, J. Y., Jeong, C. H., Jung, M. J., Park, J. H., & Jung, D. H. (2013b). Highly reliable driving workload analysis using driver electroencephalogram (EEG) activities during driving. *International Journal of Automotive Technology, 14*(6), 965–970.
Klauer, S. G., Dingus, T. A., Neale, V. L., Sudweeks, J. D., & Ramsey, D. J. (2006). The impact of driver inattention on near-crash/crash risk: An analysis using the 100-car naturalistic driving study data.
Kohlmorgen, J., Dornhege, G., Braun, M., Blankertz, B., Müller, K.-R., Curio, G., … Kincses, W. (2007). Improving human performance in a real operating environment through real-time mental workload detection. *Toward Brain-Computer Interfacing*, 409–422.
Lee, J. D., Kirlik, A., & Dainoff, M. J. (2013). *The Oxford handbook of cognitive engineering.* Oxford University Press.
Lee, J., McGehee, D., Brown, T., & Marshall, D. (2006). Effects of adaptive cruise control and alert modality on driver performance. *Transportation Research Record: Journal of the Transportation Research Board, 1980,* 49–56.
Lei, S., & Roetting, M. (2011). Influence of task combination on EEG spectrum modulation for driver workload estimation. *Human Factors, 53*(2), 168–179.
Lenzi, T., De Rossi, S. M. M., Vitiello, N., & Carrozza, M. C. (2012). Intention-based EMG control for powered exoskeletons. *IEEE Transactions on Biomedical Engineering, 59*(8), 2180–2190.
Lew, E., Chavarriaga, R., Silvoni, S., & del Millán, J. R. (2012). Detection of self-paced reaching movement intention from EEG signals. *Frontiers in Neuroengineering, 5,* 13.
Li, Y. (2018). Recognition algorithm of driving fatigue related problems based on EEG signals. *NeuroQuantology, 16*(6).
Lin, C.-T., Chen, Y.-C., Huang, T.-Y., Chiu, T.-T., Ko, L.-W., Liang, S.-F., … Duann, J.-R. (2008). Development of wireless brain computer interface with embedded multitask scheduling and

its application on real-time driver's drowsiness detection and warning. *IEEE Transactions on Biomedical Engineering, 55*(5), 1582–1591.
Lin, C.-T., Wu, R.-C., Liang, S.-F., Chao, W.-H., Chen, Y.-J., & Jung, T.-P. (2005). EEG-based drowsiness estimation for safety driving using independent component analysis. *IEEE Transactions on Circuits and Systems I: Regular Papers, 52*(12), 2726–2738.
Liu, Y.-T., Lin, Y.-Y., Wu, S.-L., Chuang, C.-H., Prasad, M., & Lin, C.-T. (2014). EEG-based driving fatigue prediction system using functional-link-based fuzzy neural network. In *2014 International Joint Conference on Neural Networks (IJCNN)* (pp. 4109–4113). IEEE.
Ma, R., & Kaber, D. B. (2005). Situation awareness and workload in driving while using adaptive cruise control and a cell phone. *International Journal of Industrial Ergonomics, 35*(10), 939–953.
Ma, R., & Kaber, D. B. (2007). Situation awareness and driving performance in a simulated navigation task. *Ergonomics, 50*(8), 1351–1364.
MacKay, D. G. (2012). *The organization of perception and action: A theory for language and other cognitive skills*. Springer Science & Business Media.
Mangun, G. R., Hillyard, S. A., & Luck, S. J. (1993). IQ electrocortical substrates of visual selective attention. *Attention and Performance XIV: Synergies in Experimental Psychology, Artificial Intelligence, and Cognitive Neuroscience, 14,* 219.
Merat, N., Seppelt, B., Louw, T., Engström, J., Lee, J. D., Johansson, E., … Itoh, M. (2019). The "out-of-the-loop" concept in automated driving: Proposed definition, measures and implications. *Cognition, Technology & Work, 21*(1), 87–98.
Navon, D., & Gopher, D. (1979). On the economy of the human-processing system. *Psychological Review, 86*(3), 214.
Nguyen, T., Ahn, S., Jang, H., Jun, S. C., & Kim, J. G. (2017). Utilization of a combined EEG/NIRS system to predict driver drowsiness. *Scientific Reports, 7,* 43933.
Norman, D. (2013). *The design of everyday things: Revised and expanded edition*. Constellation.
Pakdamanian, E., Feng, L., & Kim, I. (2018). The effect of whole-body haptic feedback on driver's perception in negotiating a curve. In *Proceedings of the Human Factors and Ergonomics Society Annual Meeting* (Vol. 62, pp. 19–23). SAGE Publications Sage CA: Los Angeles, CA.
Pascual-Marqui, R. D. (2002). Standardized low-resolution brain electromagnetic tomography (sLORETA): technical details. *Methods and Findings in Experimental and Clinical Pharmacology, 24*(Suppl D), 5–12.
Ranney, T. A., Garrott, W. R., & Goodman, M. J. (2001). *NHTSA driver distraction research: Past, present, and future*. SAE Technical Paper.
Richard, C. M., Morgan, J. F., Bacon, L. P., Graving, J. S., Divekar, G., & Lichty, M. G. (2015). *Multiple sources of safety information from v2v and v2i: Redundancy, decision making, and trust—safety message design report*.
Sarter, N. B., & Woods, D. D. (1991). Situation awareness: A critical but ill-defined phenomenon. *The International Journal of Aviation Psychology, 1*(1), 45–57.
Sarter, N. B., Woods, D. D., & Billings, C. E. (1997). Automation surprises. In *Handbook of human factors and ergonomics* (Vol. 2, pp. 1926–1943). New York: Wiley.
Sayed, T., Zaki, M. H., & Autey, J. (2013). Automated safety diagnosis of vehicle–bicycle interactions using computer vision analysis. *Safety Science, 59,* 163–172.
Schmidt, E. A., Schrauf, M., Simon, M., Fritzsche, M., Buchner, A., & Kincses, W. E. (2009). Drivers' misjudgement of vigilance state during prolonged monotonous daytime driving. *Accident Analysis and Prevention, 41*(5), 1087–1093.
Schmorrow, D., Kruse, A., Reeves, L., & Bolton, A. (2007). Augmenting cognition in HCI: 21st century adaptive system science and technology. In *The Human-Computer Interaction Handbook* (pp. 1247–1266). CRC Press.
Singh, S. (2015). *Critical reasons for crashes investigated in the national motor vehicle crash causation survey*.
Sisiopiku, V. P. (2001). *Implications of cellular telephony to traffic safety*. Michigan: Michigan State University.

Skinner, B. T., Nguyen, H. T., & Liu, D. K. (2007). Classification of EEG signals using a genetic-based machine learning classifier. In *Engineering in Medicine and Biology Society, 2007. EMBS 2007. 29th Annual International Conference of the IEEE* (pp. 3120–3123). IEEE.

Smith, K., & Hancock, P. A. (1995). Situation awareness is adaptive, externally directed consciousness. *Human Factors, 37*(1), 137–148.

Stanton, N. A., Young, M., & McCaulder, B. (1997). Drive-by-wire: the case of driver workload and reclaiming control with adaptive cruise control. *Safety Science, 27*(2–3), 149–159.

Stanton, N. A., & Young, M. S. (2000). A proposed psychological model of driving automation. *Theoretical Issues in Ergonomics Science, 1*(4), 315–331.

Strayer, D. L., & Drews, F. A. (2007). Cell-phone–induced driver distraction. *Current Directions in Psychological Science, 16*(3), 128–131.

Stutts, J. C., Reinfurt, D. W., Staplin, L., & Rodgman, E. A. (2001). *The role of driver distraction in traffic crashes*. Washington, DC: Report prepared for AAA Foundation for Traffic Safety.

Taylor, R. M., & Selcon, S. J. (1994). Situation in mind: Theory, application and measurement of situational awareness. *Situational Awareness in Complex Settings*, 69–78.

Teng, T., Bi, L., & Liu, Y. (2018). EEG-based detection of driver emergency braking intention for brain-controlled vehicles. *IEEE Transactions on Intelligent Transportation Systems, 19*(6), 1766–1773.

Tenney, Y. J., Adams, M. J., Pew, R. W., Huggins, A. W. F., & Rogers, W. H. (1992). A principled approach to the measurement of situation awareness in commercial aviation.

Victor, T., Dozza, M., Bärgman, J., Boda, C.-N., Engström, J., Flannagan, C., … Markkula, G. (2015). *Analysis of naturalistic driving study data: Safer glances, driver inattention, and crash risk*.

Weidner, F., Hoesch, A., Poeschl, S., & Broll, W. (2017). Comparing VR and non-VR driving simulations: An experimental user study. In *2017 IEEE Virtual Reality (VR)* (pp. 281–282). https://doi.org/10.1109/VR.2017.7892286.

Wickens, C. D. (2008). Multiple resources and mental workload. *Human Factors, 50*(3), 449–455.

Wickens, C. D., & McCarley, J. S. (2008). Applied attention theory.

Wiegmann, A. F. K. D. A., & Kirlik, A. (2007). *Attention: From theory to practice* (Vol. 4). Oxford University Press.

Wijdenes, L. O., Ivry, R. B., & Bays, P. M. (2016). Competition between movement plans increases motor variability: evidence of a shared resource for movement planning. *American Journal of Physiology-Heart and Circulatory Physiology*.

Yang, G., Lin, Y., & Bhattacharya, P. (2010). A driver fatigue recognition model based on information fusion and dynamic Bayesian network. *Information Sciences, 180*(10), 1942–1954.

Yeo, M. V. M., Li, X., Shen, K., & Wilder-Smith, E. P. V. (2009). Can SVM be used for automatic EEG detection of drowsiness during car driving? *Safety Science, 47*(1), 115–124.

Zeng, H., Yang, C., Dai, G., Qin, F., Zhang, J., & Kong, W. (2018). EEG classification of driver mental states by deep learning. *Cognitive Neurodynamics, 12*(6), 597–606.

Zhang, H., Chavarriaga, R., Khaliliardali, Z., Gheorghe, L., Iturrate, I., & del Millán, J. R. (2015). EEG-based decoding of error-related brain activity in a real-world driving task. *Journal of Neural Engineering, 12*(6), 66028.

Neuroergonomic Solutions in AR and VR Applications

Paruthi Pradhapan, Jolanda Witteveen, Navid Shahriari, Alessio Meroni and Vojkan Mihajlović

Abstract Although wearable sensors are omnipresent in consumer market nowadays, due to wearability and comfort issues they often avoid human head. Reliable and robust estimation of human cognitive states and traits, unfortunately, requires positioning electrodes at this sensitive location to capture electrical activity of the brain, eyes, and/or (facial) muscles. Novel augmented and virtual reality (AR/VR) use cases provide a new area that could lower the acceptance threshold of users toward head-worn devices and open new application spaces. To foster this new trend, head-mounted solutions have to offer clear benefits to users and also provide convenience and wearing comfort. Two prototype solutions are discussed in more detail, an AR-enhanced glasses and a VR-compatible electroencephalography (EEG) headset, along with the methods used to assess and improve data quality such that reliable information on brain and eye activity can be extracted. Systems and methods are evaluated through large-scale user experiments during a music festival (Lowlands) in case of AR glasses and a simulated real-life scenario for VR EEG solution. Improvements achieved in ergonomics and signal integrity are clear over state of the art, however, further adaptations of the solutions toward final use case could lead to increased end-user benefits.

1 Introduction

The first augmented/virtual reality (AR/VR) solutions are linked to stereoscope creation almost two centuries ago, however, the AR/VR solutions as we know them today started with the first headsets that included head-mounted displays, developed in the 1960s. Most of the solutions were a part of simulator setups and only toward the end of the last century those started to include actuators such as gloves or motion sensors to interact with AR/VR content. The applications were mainly in training and entertainment domain. Only recently, with proliferation of low-cost consumer AR/VR devices, such solutions have reached wider user groups. We witness modern

P. Pradhapan (✉) · J. Witteveen · N. Shahriari · A. Meroni · V. Mihajlović
Holst Centre, IMEC, Eindhoven, The Netherlands
e-mail: paruthi.pradhapan@imec.nl

C. S. Nam (ed.), *Neuroergonomics*, Cognitive Science and Technology,
https://doi.org/10.1007/978-3-030-34784-0_20

AR/VR solutions being used in applications that range from health care and military to retail and education. Furthermore, they start to move away from wired to wireless solutions, enabling more natural interaction with the environment and/or immersion in the VR content.

While a lot of development in the AR/VR space is focused on improving the quality of visually rendered content, either within the VR glasses display or overlaid on the AR glasses, less emphasis is given on methods that capture users' states or intentions. Examples of those include VR solutions utilizing gloves to facilitate interaction and motions sensors placed over different body parts to render users in the VR space, as well as VR/AR solutions that include cameras that either capture users' environment or track his/her eye activity. One must wear additional pieces of clothing (e.g., gloves or tight suits) or substantial adaptations on the device mounted on a user's head has to be done to fit such sensing modalities. This renders most such sensor-enhanced VR/AR devices impractical for many applications.

Only a handful of AR/VR solutions exist where head-mounted devices include sensors that are a natural extension of the device and neither introduce additional complexity and clutter on the device nor discomfort or inconvenience to the user. The typical examples of such devices include AR glasses incorporating biopotential sensors that can capture electrical movement of the eyes—electrooculography (EOG) (Jins Meme, 2019) and VR headsets that capture EOG, electromyography (EMG), or electroencephalography (EEG) from the head (Looxid Labs, 2019; Neurable, 2019). The former can be used to provide a window to the human mind, given that information about the eye activity (gaze, focus, blinking patterns) can provide valuable information about both healthy individuals (Bulling, Roggen, & Tröster, 2009; Ishimaru et al., 2014) but also patient populations (MacAskill & Anderson, 2016; Mani, Asper, & Khuu, 2018; Pretegiani & Optican, 2017; Sabatos-DeVito, Schipul, Bulluck, Belger, & Baranek, 2016; Zargari Marandi, Madeleine, Omland, Vuillerme, & Samani, 2018). The latter enables new interaction modalities (Jantz, Molnar, & Alcaide, 2017) and is an essential tool for assessing the impact of AR/VR content on the user, which can help in understanding cognitive processes (Tromp, Peeters, Meyer, & Hagoort, 2017) but also in monitoring and treating certain health conditions (Lupu, Irimia, Ungureanu, Poboroniuc, & Moldoveanu, 2018).

The main advantage of such devices is that if they are designed in a smart way, they can be considered as a truly ergonomic AR/VR solutions that could add a new dimension to the AR/VR experience. One of the main challenges for such solutions is to provide reliable information that can be used to enhance user interaction (Barbara & Camilleri, 2016; Iáñez, Azorin, & Perez-Vidal, 2013). This calls for specific technological breakthroughs and data analysis methods to preserve signal integrity and facilitate the information extraction process (Mihajlović, Grundlehner, Vullers, & Penders, 2015; Witteveen, Pradhapan, & Mihajlović, 2019).

The development of ergonomic wearable AR/VR solutions that captures reliable information about the user's state is the approach that was followed at imec. Two prototype solutions have been developed:

- A VR-compatible EEG headset,
- An EOG glasses for AR applications.

They both introduce minimal adaptations or additions to the existing VR/AR devices and enable the use of such devices in real-world setting, without introducing additional burden to the user. Specific design solutions, as well as integration of low-power and miniaturized electronics in user-friendly prototypes is discussed in this chapter. The acceptance and functionality of the solutions are detailed. In addition, methods used to assess signal quality and reduce artefacts for both EOG and EEG signals are discussed. The performance of those devices was evaluated in two scenarios, one involving users wearing a VR and VR-compatible EEG headset while performing simulated daily life activities, and the other investigating the ergonomic and signal reliability aspects of users wearing EOG glasses during a music festival.

2 Ergonomic Devices

2.1 VR-Compatible EEG Headset

Imec's VR-compatible EEG headset is designed to cover the most relevant electrode positions for cognitive state monitoring, as well as audio and visual stimulus recognition. Monitoring cognitive state requires electrodes positioned at the front of the scalp (i.e., frontal lobe of the brain hemispheres; electrode positions F7, F3, F4 and F8), which is known to be the cognitive and emotions center of the brain (McEvoy, Smith, & Gevins, 2000; Papousek & Schulter, 2004; Shallice & Evans, 1978). Electrodes at the temporal lobes (T3 and T4) are positioned to detect EEG patterns during auditory perception (Grillon & Buchsbaum, 1986; Huotilainen et al., 1998), whereas the electrodes at the occipital lobes (O1 and O2) are predominantly used to pick up visually evoked responses (Makeig et al., 2002; Sharon, Hamalainen, Tootell, Halgren, & Belliveau, 2007). The measurement reference and subject bias are placed at the frontal midline (Fz and F2), respectively. By measuring activity at different brain regions during immersive engagement in the VR environment, the cognitive and emotional state of a person can be estimated. The measurement locations and the headset design of the imec's EEG headset are shown in Fig. 1.

The headset is "one-size-fits all" structure by means of a flexible mechanism incorporated in the mechanical design, such that the headset can be fastened or loosened as required. It can be attached to most of the commercially available VR headsets by means of strap mechanism and can provide optimal user comfort. The headset is designed such that the weight of the VR systems is compensated, and the headset fits the curvature of the scalp to ensure optimal weight distribution and electrode contact at the measurement positions. For this purpose, the miniaturized electronics, which includes the analogue active electrode printed circuit board (PCB) and the extension PCB, are placed at the back of the EEG headset (Xu et al., 2014). The active electrode PCB acquires analog signals from the measurement positions, while the extension board digitizes it and performs certain basic signal processing steps (i.e., high-pass filtering) and transmits the data using Bluetooth at a sampling

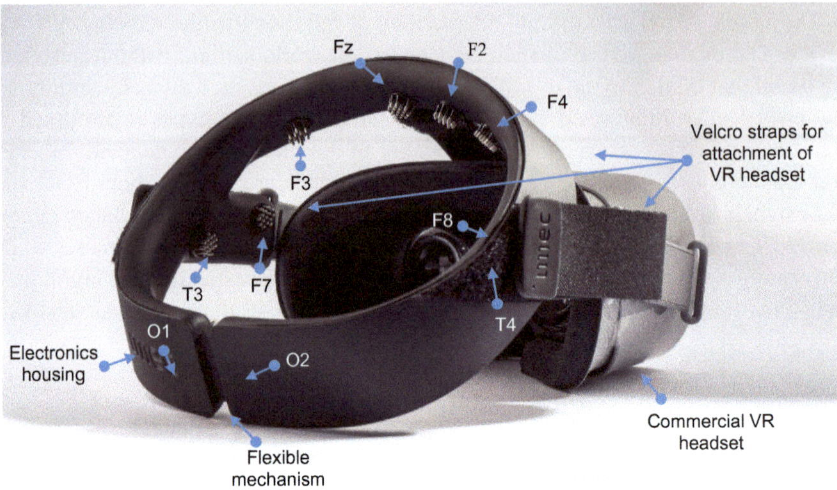

Fig. 1 Imec's 8-channel VR-compatible headset

frequency of 256 Hz. The active electrode board continuously captures the electrode-tissue contact impedance aside from the EEG. Furthermore, the extension board includes a motion sensor capturing 3D acceleration and rotation information. The electronics housing also accommodates a 300 mAh battery that powers the system for data acquisition and transmission, enabling for more than 8 h of continuous use.

The measurement positions are connected to the electronics by means of a flex-PCB, that is neatly enclosed within the permanent insert. All electrodes are attached using a spring mechanism to increase user comfort by decreasing the pressure exerted on the surface of the scalp. The electrodes are attached using an ECG snap mechanism, thereby allowing easy cleaning and replacement of the sensing elements. State of the art, dry electrodes made of a conductive polymer base (Chen et al., 2014) with silver/silver-chloride coating are used to measure brain activity (shown in Fig. 2). These electrodes are of two types. The first type consists of flexible pins to penetrate through hairy regions of the scalp while minimizing discomfort or skin damage even during long hours of use. The second type is a flat electrode, i.e., without pin structures, that can be used at non-hairy regions of the scalp. Furthermore, these electrodes can be easily cleaned using alcohol wipes so that risk of cross contamination among users is eliminated.

2.2 *EOG Glasses*

Imec's EOG glasses are designed to capture eye movements swiftly and unobtrusively. The design aims at usability for large population, needing minimal support

Fig. 2 Flat- and pin-type conductive polymer electrodes with silver/silver-chloride coating

for placement on the user's head. They can record eye movements for long periods of time such that users perceive them as normal glasses, as illustrated in Fig. 3.

The classical electrodes for EOG measurements are placed near the outer canti of each eye for the horizontal eye movements, and above and below one eye for the vertical eye movements. The positions of electrodes are modified in the Imec's glasses offering three main advantages:

Fig. 3 Imec's EOG glasses are designed to capture eye movements unobtrusively

1. The electrodes are unobtrusive,
2. The positions are natural pressing points of a pair of glasses,
3. These positions allow more flexibility in deriving the horizontal and vertical eye movements.

The glasses contain a total of six electrodes, four recording electrodes, a reference electrode, and a ground electrode. Two of the recording electrodes are positioned on the temples and two are positioned on each side of the nose, as shown in Fig. 4. The reference electrode is placed strategically in the middle of the frame, just above the nose bridge. The ground electrode is placed on the left leg of the glasses, touching the ear. The four symmetrically placed recording electrodes within the glasses allow the derivation of the vertical and horizontal eye movements in at least six different ways. This enables choosing the best electrode combination in terms of signal quality and analyzing each eye separately. This flexibility of deriving the horizontal and vertical EOG not only makes a variety of applications possible but also enables a more robust design in terms of signal quality.

The electrodes used are dry conductive polymer electrodes with silver/silver-chloride coating, similar to the one used for the VR headset. The shape of the electrodes is specifically designed for each position to enable easy setup and one-size-fits-all design. The electrodes on the temples have a cylindrical shape pinned onto a stainless steel spring wire. The spring wire gives the electrode a constant soft pressure onto the skin.

The nose measurement electrodes have an oval disk shape similar to what can be found in regular glasses. The electrodes are pinned onto stainless steel wires which allow the electrodes one degree of freedom to pivot around the wire and follow the curvature of the nose. The reference electrode is also a cylindrically shaped electrode,

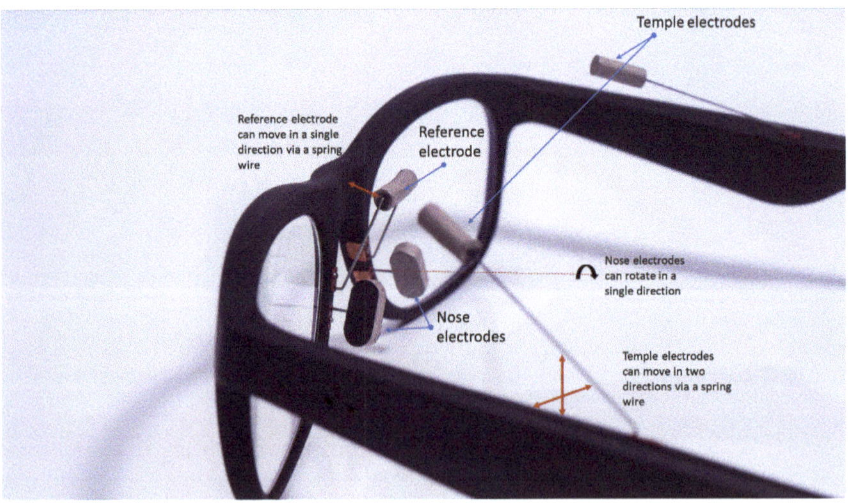

Fig. 4 Electrode configuration of imec's EOG glasses

pinned on a stainless steel wire from two sides, applying a small pressure to the skin when wearing the glasses. The wires introduced form a spring mechanism that is required for good contact between the electrode and the skin (Fig. 4). The electrodes with silver/silver-chloride coating are specifically designed for low electrode-skin impedance. The design allows for the electrodes to be replaceable and to be cleaned with alcohol.

The miniaturized electronics, which includes the analogue signal acquisition PCB and the extension board, similar to the one integrated in the EEG headset, are placed at the back of the glasses. The PCB acquires analog EOG and contact impedance signals from the measurement positions at a sampling rate of 256 Hz, while the extension board digitizes it and performs basic signal processing steps and transmits the data using Bluetooth. EOG and impedance data, acceleration and rotation information are sent as the extension board contains a motion sensor. The electronics housing accommodates a 300 mAh battery that powers the system for data acquisition and transmission.

3 Measuring EEG While Exploring VR

3.1 Data Collection

3.1.1 Aim

The aim of this experiment was to assess usability and signal integrity while simulating common daily activities that might be performed while engaged in the interaction within the VR environment. The measurements were used to assess whether common head or body movements affect the EEG measurements and if so, the possibility to treat and retrieve useful information using state-of-the-art signal quality and artefact handling algorithms.

3.1.2 Subjects

Data was collected on 15 subjects (32.91 $\pm$ 6.68 years; 4 females) using imec's VR-compatible EEG headset, described in Sect. 2.1. All subjects were recruited on a voluntary basis and were informed about the experimental protocol beforehand. Prior to the start of the experiments, signed informed consent was obtained from all participants and they were given the choice to stop the experiments at any point during the measurements.

Subject exclusion criteria were defined as follows:

- Subjects with known neurological disorders (Alzheimer's, dementia, epilepsy, encephalopathy, etc.)

- Subjects with other medical conditions (Parkinson's disease, central nervous system disorders, stroke, head injuries, skin allergies, etc.).

3.1.3 Measurement Protocol

The experimental setup and measurement protocol were reviewed and approved by imec's internal ethical committee. The experiments were performed in a laboratory setting. Participants were seated comfortably about 1 m away from a computer screen that was used to play the stimuli of the measurement protocol. The measurement protocol consisted of three sections, performed one after the other, in the following order:

1. Baseline (eyes open—EO/eyes closed—EC) measurements
2. Electromyography (EMG) artefacts
3. Head motion artefacts
4. Baseline (EO/EC) measurements
5. Body motion artefacts
6. Baseline (EO/EC) measurements.

The different measurement sections are discussed below.

Baseline: Baseline measurements were performed to record reference data absent of activity-specific EEG or artefacts. Baseline measurements included eyes closed and eyes open activities, each lasting 1 min, during which the subject sat in a relaxed position and was instructed to keep his/her eyes open or closed for the duration of measurement. The baseline measurements (both eyes closed and eyes open) were repeated three times at different phases of measurement, as shown above.

EMG artefacts: To simulate EMG artefacts that affect EEG measurements, a number of activities were considered:

- *Talking*: Talking is a main source of artefacts in EEG and EOG streams as the movement of facial muscles and natural facial gestures can contaminate these measurement streams, either with physiological or other motion artefacts. To simulate this situation, subjects were asked to read a text that was displayed on the screen aloud. This activity lasted for 1 min.
- *Chewing*: Chewing is also considered a major source of artefact due to the influence of high-frequency activity in the EEG data stream, due to the movement of facial muscles. In order to effectively produce chewing action, subjects were provided with gum which they were asked to chew upon for 1 min.
- *Facial expressions*: Cues were pictures of four different actors who performed the following four expressions: fear, happy, neutral, and sad. One measurement set comprised 16 different cues (4 actors showing 4 expressions each). The facial expression measurements were repeated twice per measurement session, i.e., two iterations were performed.

Each epoch lasted 5 s in duration, during which the facial expression cue was displayed on the screen for 4 s. During this period of 4 s, the subject was asked to perform and hold the expression that was indicated in the cue. The last second of each epoch was the relax phase, were the subject returns to baseline, i.e., neutral expression position. Prior to protocol execution, the researcher informed the subjects to mimic the expressions displayed on-screen as accurately as possible.

Head motion artefacts: Various head movements were included that are a common source of motion artefacts. Among several movements considered, periodical movements such as vertical, horizontal, tilt, rotation, and sequence of horizontal and vertical movements were included in a fixed sequence for the current evaluation. Periodic movements were set at 0.5 Hz frequency of movements and angle at approximately 60° to simulate normal viewing angle in the field of vision. All activities were executed for 1 min each by the subjects.

Body motion artefacts: Similar to head movements, common body movements expected to be exercised during a VR experience were chosen as part of the experiment. Body rotation, on-the-spot walking and squatting/ducking, in a fixed sequence, were chosen for the current protocol. Although several other movements were considered, they were discarded keeping in mind the safety of the participants, as they would be completely blinded from their surrounding environment while wearing the EEG headset with VR goggles attached. As in the case of head movements, all activities were performed for 1 min each.

3.2 Assessment of Signal Quality

The data acquired during the protocol described above is expected to contain artefactual data as this was the aim of the experiment. The first step in assessing the usability of the data acquired is to determine the percentage of data that remains within the dynamic range of the imec's EEG acquisition system. EEG data from various activities were precisely segmented based on timing information of cues that was recorded in the data file. For the current dataset, once the activities were segmented, the percentage of data that is clipping, i.e., saturated data due to the limits of dynamic range of the measurement system, is determined based on a modified version of the SQI algorithm described in (Witteveen et al., 2019). The results of the evaluation are presented in Table 1.

The next step is to determine the percentage of data that is within the dynamic range (i.e., not clipping) but corrupt due to artefacts in the signal. The data was analyzed using data quality indicator (SQI) algorithms developed within IMEC for the purpose of evaluating EEG signal quality and integrity (Witteveen et al., 2019). The SQI algorithm is based on a pragmatic approach, where various statistical parameters are evaluated, and each segment of EEG data assessed as clean or contaminated based on data-driven thresholds. The mean percentage of clean data during each activity performed in the protocol is shown in Table 2.

From Tables 1 and 2, it is evident that significant portions of data is lost due to artefacts in the EEG stream. The proportion of data, i.e., classified as corrupt EEG but not clipping, that can be recovered by adept artefact handling methods per activity

Table 1 Percentage data that is not clipping (i.e., within dynamic range) while performing different activities using the VR-compatible EEG headset

Activities		T3	F7	F8	T4	F4	O2	F3	O1	Activity mean	Activity group mean
B	Baseline	79.35	94.07	93.06	83.61	97.50	89.26	90.46	97.50	90.60	90.60
EMG	Talking	67.78	82.78	80	68.89	96.67	89.44	87.78	97.78	83.89	70.6
	Chewing	25	16.11	10	22.22	85.56	89.44	86.11	97.78	54.03	
	Facial expressions	55	78.13	64.38	58.96	82.71	82.10	79.58	90.21	73.88	
Head	Vertical	27.78	78.89	72.22	49.44	75.56	53.33	77.22	76.11	63.82	62.88
	Horizontal	67.22	93.89	87.78	63.89	99.44	90.56	91.67	98.89	86.67	
	Tilt	08.89	27.22	11.11	18.33	80.56	50	71.67	62.22	41.25	
	Rotation	08.33	48.89	35.56	30.56	80	37.78	74.44	53.89	46.18	
	Sequence	42.78	88.33	76.67	53.33	94.44	78.33	87.22	90.56	76.46	
Body	Body rotation	76.11	93.89	95	85.56	99.44	91.11	88.33	99.44	91.11	81.94
	Spot walking	54.44	84.44	88.89	78.33	97.78	91.11	90	97.22	85.28	
	Squatting	38.33	63.33	63.33	58.89	87.78	80.56	77.22	86.11	69.44	
Channel mean		45.92	70.83	64.83	56.00	89.79	76.92	83.48	87.31		

Table 2 Percentage artefact-free EEG data while performing different activities using the VR-compatible EEG headset. Epochs that are clipping are considered artefact contaminated

Activities		T3	F7	F8	T4	F4	O2	F3	O1	Activity mean	Activity group mean
B	Baseline	51.68	85.34	79.59	56.97	92.69	54.53	84.50	65.55	71.36	71.36
EMG	Talking	22.47	48.60	46.30	12.32	82.89	38.24	68.96	44.96	45.59	30.28
	Chewing	01.53	01.88	0	02.17	15.26	08.88	09.13	09.90	06.09	
	Facial expressions	22.13	47.13	36.26	17.42	59.40	38.27	50.32	42.40	39.16	
Head	Vertical	13.24	53.08	48.13	23.19	48.16	12.58	61.23	15.71	34.42	33.05
	Horizontal	22.95	50.08	45.64	26.99	78.85	45.33	77.55	44.87	49.03	
	Tilt	07.05	09.44	04.02	08.69	43.81	18.58	40.99	27.59	20.02	
	Rotation	03.02	26.75	16.13	12.47	52.40	04.09	42.76	11.85	21.18	
	Sequence	14.90	52.04	33.29	28.67	67.39	29.90	53.38	45.32	40.61	
Body	Body rotation	34.96	51.48	63.51	59.80	92.94	48.55	69.43	42.24	57.86	40.68
	Spot walking	10.56	49.81	38.17	36.32	63.44	26.61	51.36	26.29	37.82	
	Squatting	9.81	41.98	28.76	16.75	55.74	5.12	49.15	03.49	26.35	
Channel mean		17.86	43.13	36.65	25.15	62.75	27.56	54.90	31.68		

is as follows: baseline (19%), EMG artefacts (40%), head movements (29%), and body movements (41%), respectively. These numbers indicate the benefits of artefact handling methods are profound in any application space.

3.3 *Handling Artefacts*

In order to extract meaningful information from the corrupted data, adequate artefact handling methods need to be implemented. Artefact reduction is most often performed using various filterings and blind source separation methods, such as independent component analyses (ICA) and principal component analyses (PCA) (Jung et al., 2000). However, the amount of data that can be recovered is dependent on the type of artefact and availability of reference sources of noise signals. Therefore, for effective artefact reduction, it is essential to know the source of the artefact and the reference signals that closely correspond to the noise signals. Our approach to artefact handling involved two steps discussed in more detail in the remainder of the section:

- Artefact classification and
- Artefact reduction.

3.3.1 Artefact Classification

Artefact classification was performed to determine the type of artefact that is corrupting the EEG signals as this helps identify the reference sources that might be useful in reducing the artefact from the EEG stream. For this purpose, an artefact classification scheme was formulated and is shown in Fig. 5. Features derived from EEG, accelerometer, gyroscope, and impedance data streams were used in artefact classification. In total, 394 features, comprising statistical (i.e., mean, standard deviation, mode, 25th percentile, 75th percentile, kurtosis, skewness, etc.) and time/frequency (mean frequency, median frequency, spectral deformation, mean amplitude, median amplitude, etc.) were extracted for this purpose. Relevant features were then chosen based on sequential forward selection which works by selecting a subset of features until there is no improvement in prediction. Models were created based on leave-one-session out, where data from one session (i.e., one subject) is left out while the model is trained using all the other sessions. The created model is then validated using the left out session. The process is repeated until the models are validated in every session. Different models are created for each step in the artefact classification scheme and the aggregated results are presented in Table 3.

Overall, motion artefacts (i.e., head and body movements) could be well differentiated from non-motion artefacts (i.e., talking, chewing, and facial expressions) with accuracies up to 96%. This is substantial since the reference channels for artefact reduction may vary significantly between the two classes. Non-motion artefacts were

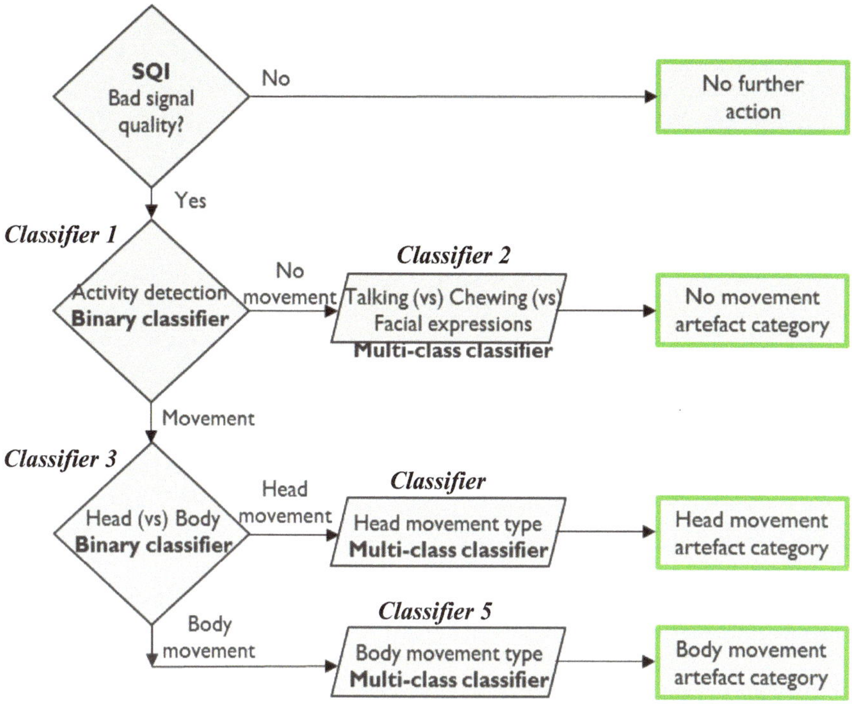

Fig. 5 Artefact classification pipeline to determine the type of artefact corrupting the signals

Table 3 Artefact classification performance in the pipeline depicted in Fig. 5

Classifier	Accuracy	Sensitivity	Specificity	F1 score	Kappa
Classifier 1 (binary)	96.11	96.16	96.03	95.50	92.27
Classifier 2 (multi-class)	43.14	51.27	72.00	–	19.39
Classifier 3 (binary)	62.72	71.41	49.67	66.29	21.91
Classifier 4 (multi-class)	56.74	56.75	78.39	–	13.52
Classifier 5 (multi-class)	66.78	68.81	83.99	–	37.16

harder to distinguish, with classification accuracies reaching 43%. This can mainly be attributed to the magnitude of data that was clipping making it harder to distinguish. Further, the features of facial expressions overlapped with features computed from talking and chewing, thereby making it harder to achieve higher classification performance. Artefacts from head and body movements could be distinguished with an accuracy of 63%. The classification accuracy of various head movements was estimated at 57% and those of body movements at 67%, respectively. In this way, by determining the type of activity to its finest detail, a better selection of the most suitable methods for artefact reduction, hence contributing to better overall artefact reduction performance.

3.3.2 Artefact Reduction

The method used to reduce EEG artefacts is based on a multichannel adaptive filtering (MCAF) algorithm (Mihajlović, Patki, & Grundlehner, 2014). MCAF has been shown useful in reducing movement artefacts when using electrode-tissue impedance (ETI) as a reference signal capturing artefact components (Mihajlović, Li, Grundlehner, Penders, & Schouten, 2013). We extended the same framework to handle not only head and body movement artefacts but also talking, chewing, and facial expressions. Relative movement of the electrode on the scalp and deformation of the skin underneath due to those artefacts induces changes in the electrical properties of the contact interface. Given that movements can also be characterized by acceleration and rotation, we included these signals, being available in our headset, as potential reference signals. Hence, the reference signal candidates evaluated for each EEG channel include the following: ETI magnitude from the same measurement channel, acceleration and rotation magnitude, and acceleration rotation in each of the three directions. Before applying, MCAF, the signal is band-pass filtered to the frequency band containing relevant artefact components, which is empirically determined to be 0.1–5 Hz. The best combination of reference signals are selected per artefact type based on performance metrics of the MCAF and applied in the artefact reduction step.

MCAF uses subsequent signal (vector) observations, where the t-th observation (i.e., at sample time t) can be expressed as shown in Eq. (1). Here, $\mathbf{r}[t]$ is the observation of the reference signal r, observed at sampling time t. The reference signal in our case is ETI magnitude, any of the components of acceleration or rotation, or any combination of those. In Eq. (1), D represents the time lags, or the number of filter coefficients used, and T is the matrix transpose operator.

$$\boldsymbol{I}[t] = [r[t]^T r[t-1]^T r[t-2]^T \ldots r[t-D]^T]^T \tag{1}$$

$$\boldsymbol{w}[t+1] = (1-\mu\alpha) \times \boldsymbol{w}[t] + \mu \times \varepsilon[t]\boldsymbol{I}[t]/||\boldsymbol{I}[t]|| \tag{2}$$

$$\varepsilon[t] = d[t] - \boldsymbol{w}^T[t]\boldsymbol{I}[t] \tag{3}$$

The coefficient values of the adaptive filter ($\boldsymbol{w}[t]$) are updated for each observation time instance t, using an iterative algorithm, as shown in Eq. (2). In the equation, α is the forgetting factor (such that the expression $b = 1-\mu\alpha$ is between 0 and 1), μ is the step size of the algorithm (between 0 and 2), and $\varepsilon[t]$ is the estimation error, computed as shown in Eq. (3). The "cleaned" EEG signal at the time instant t is then equal to the estimation error, as we assume that the EEG signal is uncorrelated with the reference signals.

To estimate the impact of artefact reduction we applied the tailored artefact reduction MCAF algorithm per artefact type on the complete dataset and then evaluate the performance using the estimation of the signal quality of corrected EEG signals. Table 4 illustrates the amount of data assessed as artefact-free after the application

Table 4 Percentage artefact-free EEG data after artefact reduction

Activities		T3	F7	F8	T4	F4	O2	F3	O1	Activity mean	Activity group mean
B	Baseline	53.53	87.39	81.22	57.88	93.54	53.86	87.28	62.00	**72.09 (0.73 ↑)**	**72.09 (0.73 ↑)**
EMG	Talking	17.91	58.43	57.38	16.43	78.22	46.09	65.49	54.16	**49.26 (3.67 ↑)**	**33.03 (2.75 ↑)**
	Chewing	1.09	2.08	0.00	1.48	10.83	8.96	9.64	12.91	***05.87 (0.22↓)***	
	Facial expressions	18.82	53.83	43.50	22.02	59.81	45.65	55.39	52.29	**43.91 (4.75 ↑)**	
Head	Vertical	10.72	57.96	59.07	21.38	58.50	15.38	67.08	19.24	**38.67 (4.25 ↑)**	**39.75 (6.70 ↑)**
	Horizontal	18.15	63.25	60.49	29.40	79.73	54.82	81.94	59.35	**55.89 (6.86 ↑)**	
	Tilt	5.47	17.80	11.76	11.91	57.71	29.57	57.03	48.45	**29.96 (9.94 ↑)**	
	Rotation	2.44	36.09	20.78	19.78	66.61	6.05	47.77	11.38	**26.36 (5.18 ↑)**	
	Sequence	12.97	64.40	49.38	33.81	71.97	39.50	61.64	49.24	**47.86 (7.25 ↑)**	
Body	Body rotation	30.34	64.67	71.94	68.01	92.35	55.87	72.20	49.31	**63.09 (5.23 ↑)**	**46.47 (5.79 ↑)**
	Spot walking	9.58	58.43	46.15	48.54	62.65	25.64	54.44	23.25	**41.08 (3.26 ↑)**	
	Squatting	9.58	53.36	49.62	27.53	66.83	6.40	60.07	8.49	**35.24 (8.89 ↑)**	
Channel mean		15.88	51.48	45.94	29.85	66.56	32.32	60.00	37.51		

Bold value indicates the percentage change in artefact free EEG data
Bold & Italics value indicates the negative change (i.e. deterioration of signal quality after artefact reduction)

of artefact classification, artefact reduction, and signal quality indication algorithms. Similar variability can be observed across channels as in the case of original data (see Table 2) but overall signal quality is improved after applying artefact reduction. The improvements are substantial in case of head and body movements and minor in case of EMG and baseline, as illustrated in the activity and group mean columns. Given that impedance and acceleration are better suited to capture movement artefacts than EMG ones, such results are expected. One may argue that the improvements still leave a lot of corrupted data but reaching higher signal quality might require adaptations of the complete signal pipeline from data acquisition modules to more sophisticated artefact handling methods.

4 Capturing Eye Activity

4.1 Data Collection

4.1.1 Aim

The aim of this experiment was to assess the usability of the imec's glasses and its fit on random subjects as well as to investigate the signal quality. The study was set up in an unusual environment, a music festival. Prior to the start of the experiments, signed informed consent was obtained from all participants and they were given the choice to stop the experiments at any point during the measurement. The study was reviewed and granted exemption from being a clinical study by the Maxima Medisch Centrum ethical committee (Ref. N18.066). The study was part of imec's internal research program.

4.1.2 Subjects

The number of subjects was only limited by the number of hours available for the study and the duration of a single test. The festival attracts around 20,000 people so a large number of participants were expected, biased toward young Caucasians. The data collection, which was spread over three subsequent days, was envisioned as an ultimate field test for the glasses. Subjects were not actively recruited, they needed to visit the "science" stand on their own account. They were able to read the content of the study in a separate leaflet. All participants were above 18 years old and were able to sign a consent form. Exclusion criteria were set to protect those who were too intoxicated, which was left to the judgement of the investigators, typically persons who were not able to understand or reliably answer basic questions about the study. Furthermore, people with vision problems (based on questionnaires) were excluded, and wearing glasses was not seen as a problem.

4.1.3 Protocol

For the study exemption, ethical approval was granted by an external ethical committee. Prior to the start of the study, all research assistants were trained by the sponsor on the proper use of the devices and execution of the protocol. During the entire study, the research coordinator was present for technical support and assessment of the correct execution of the measurements. Participants were required to fill in a questionnaire with basic information including age, weight, height, gender, ethnicity, wearing glasses, alcohol used, and drugs used. Each participant was asked to use a breathalyzer to measure alcohol consumption. Thereafter, the participant started with the EOG calibration measurement and a game. In the short calibration session subjects were asked to look left, right, up, down, and blink several times. This is used to distinguish different eye states and it took 1 min. After this, the subject played a game for around 8 min. The game used was a commercially available platform game in which a monkey runs and jumps to collect coins. The entire session per subject took 15–25 min. All sessions were performed at a single site, the Lowlands festival, in Biddinghuizen. Subjects were measured only once at a time.

4.1.4 Analysis

The EOG data was annotated for signal quality by means of visual inspection of the signals. Criteria for contaminated signal was the presence of high amplitude noise, 50 Hz noise and the absence of eye movements or blinks (vertical only). When bad signal quality was assigned, the presence of artefact was prominent, and no eye movements or blinks could be observed in the signal. In case of good signals, although they could contain a small artefact, the eye movements and/or blinks were present and clearly identifiable. For some analysis the distinction was made within the good signals to good and mediocre, the former containing no artefacts whatsoever and the latter containing small artefacts. The vertical and horizontal EOG channels were assessed independently. The vertical EOG channel is derived only from the nose electrodes. The horizontal EOG channel is derived as a combination of the nose and temple electrodes.

4.2 Signal Quality and Fit

4.2.1 Study Population and Device Use

The study population consisted of 153 participants, out of which 62% were males ($N = 95$). The age was between 18 and 64 with mean at 28.5 years old. Five different EOG devices were used in the study. They were tested on a different number of participants, 19 for EOG01, 27 for EOG02, 34 for EOG03, 39 for EOG04, and 37 participants used EOG05.

4.2.2 Fit of the Glasses

The fit of the glasses was assessed on all participants by the investigator. It was judged both on the physical attachment of the electrodes to the skin and the quality of the signal. The judgement was done when fitting the glasses to the participant, just before the start of the recording. The fit was assessed based on the pair of electrodes on the temple or nose position and on the ear for the ground electrode. Criteria to judge the fit were as follows:

- Is the electrode touching the skin?
- Is it touching the skin at the intended location?
- Is most of the electrode surface touching the skin?

If all the above questions could be answered with "yes" the fit of the assessed pair of electrodes was considered as good. If one of the questions was answered with "no" the fit would be judged as poor.

As shown in Fig. 6, in 62% of the participant all electrodes had a good fit. The ear electrode had a good fit in 93% of the participants, being the most well-fitted position. The temples were well fitted in 86% of the participants while the nose in 73% of the participants. Out of five prototypes used in the experiment, two prototypes had worse fit compared to the three others (see Fig. 6). For EOG04 prototype this is caused by a poor fit of the temple electrodes, while for EOG02 it seems that all positions had problems with the fit but typically not at the same time. The fit of the glasses can be improved by using different sizes. A smaller size in which distance between the temple electrodes is smaller could potentially improve the fit for those with a smaller head.

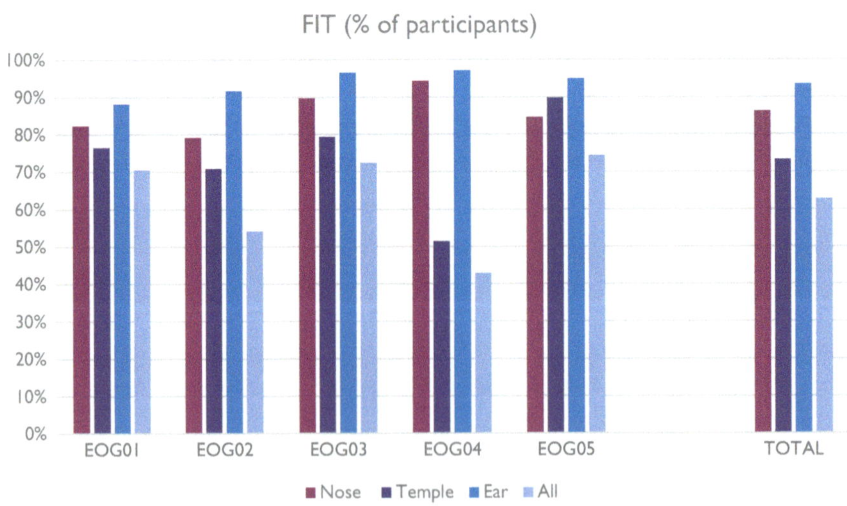

Fig. 6 Percentage of participants with a good fit of the different positions for each prototype glass

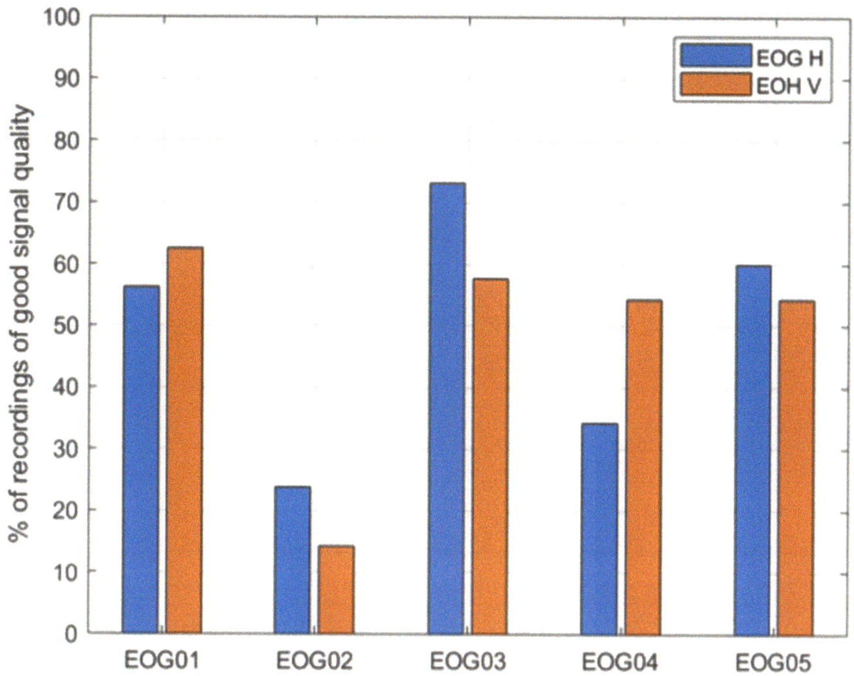

Fig. 7 Percentage of the recordings for each prototype with good signal quality for the vertical and horizontal EOG channels. EOG0# refers to different glasses prototype

4.2.3 Signal Quality

Based on manual annotation, overall the signal quality was assessed as good in 46% of the participants for the vertical EOG and in 49% for the horizontal EOG. Signal quality was not equal for all prototypes. As can be seen in Fig. 7, 3 EOG prototypes show good signal quality in 55–75% of the participants. This implies a relation between the fit of the glasses and the signal quality.

4.2.4 Comparing Fit and Signal Quality

Although poor fit of the glasses implies also bad signal quality, there could be other causes for bad signal quality as well. Tables 5 and 6 show the percentage of good and bad quality signals for participants with a certain fit. For both horizontal and vertical EOG, the signal quality is good about half of the time when the fit is good. It is evident from the results that fit, visually judged at the start of the data collection, is not the only predictor for signal quality. Poor fit seems to be a better predictor for bad signal quality. The ear electrode is the ground, so assumed to be vital for good signal quality. A poor fit on the ear electrode only occurred in four participants. If all the electrodes would fit, especially for the horizontal EOG, the chances are higher to

Table 5 Signal quality for the vertical EOG channel compared to fit. The first cell indicates that 52% of the participants with good fit also had good signal quality

Fit	SQ EOGv	Nose fit (%)	Temples fit (%)	Ear fit (%)	All fit (%)
Good	Good	52	54	50	59
Good	Poor	48	46	50	41
Bad	Good	20	33	0	30
Bad	Poor	80	67	100	70

Table 6 Signal quality for the horizontal EOG channel compared to fit. The first cell indicates 51% of the participants with good fit also had good signal quality

Fit	SQ EOGv	Nose fit (%)	Temples fit (%)	Ear fit (%)	All fit (%)
Good	Good	51	59	50	62
Good	Poor	49	41	50	38
Bad	Good	40	22	50	28
Bad	Poor	60	78	50	72

get good signal quality. Other artefacts seen in the data are caused by environmental noise (50 Hz) and data loss, due to poor Bluetooth connection.

4.2.5 Ergonomic Differences Based on Gender

The EOG glasses were made as one-size-fits-all; therefore, the fit could be different for males and females, which in turn could influence the signal quality. As indicated in Table 7, the fit is worse for females in all positions. The difference is most prominent across all electrodes combined. This suggests that it is possible to get some of the positions right, but it is harder to fit all electrodes on females compared to males. Apart from the combination of electrodes, the temple electrode seems to be the least well fitted in females. This can be explained by typically smaller head sizes of females compared to males. The ear electrodes do not seem to pose much fitting problems for both males and females and are well chosen to be the ground electrode.

Table 7 Signal quality and fit across male and female subjects

	Male (%)	Female (%)
SQ EOGv	47	49
Fit nose	89	81
Fit temples	77	67
Fit ear	95	91
Fit all	70	50

Table 8 Signal quality and fit across male and female subjects

	FP I	FP II	FP III	FP IV
Number of participants	11	117	23	2
SQ good	45%	52%	39%	0%

4.2.6 Signal Quality Differences Based on Skin Type

In order to record a signal of good quality, a low skin-electrode impedance is required. This can be achieved by making good contact between the two. However, differences between different skin types also exist. Skin type was assessed based on Fitzpatrick scale for each participant (Fitzpatrick, 1988). Most of the participants were light skinned, FP I and II. Only 15% reported an FP scale of III and only two participants reported FP IV. There is substantial difference in signal quality across skin type, as shown in Table 8. However, additional research is required to establish the relationship between skin type, contact impedance, and signal quality.

4.2.7 Signal Quality Differences Based on Age

Age is known to have an effect on the skin (Nicander, Nyren, Emtestam, & Ollmar, 2006) and possibly on the electrode-skin impedance and signal quality. The impact of age on signal quality is shown in Fig. 8. It shows the relation between age and signal quality (noted as good, mediocre, or bad) on the vertical EOG channel. There is a significant correlation between the age and signal quality (correlation of 0.17, $p = 0.04$) based on t-test.

4.3 *Identifying Eyeblinks*

Eyeblinks are visible in the EOG as spikes with larger amplitude compared to the fast-saccadic eye movements. The eyeblinks are typically seen in the vertical EOG. To perform blink identification a multilayer feedforward artificial neural network was employed. The neural network uses one hidden layer of size 3 and an output layer of size 1, with relu and sigmoid activation functions, respectively. The data used for training and validation was taken randomly from a manually annotated set of blinks ($N = 468$) and non-blinks ($N - 881$) from the Lowlands database. Each blink or non-blink was filtered EOG data of 0.5 s window length, downsampled by a factor 4. Eighty percent of the blinks and non-blinks were randomly selected for training and 20% for testing. The neural network was trained using adam optimization, with batch size of 10 and epoch size of 100. Evaluation of the test dataset reveals an accuracy of 96%, having 98% specificity and 96% sensitivity. Evaluation of the blink detector was also performed using a subset of the participants. The algorithm was given the

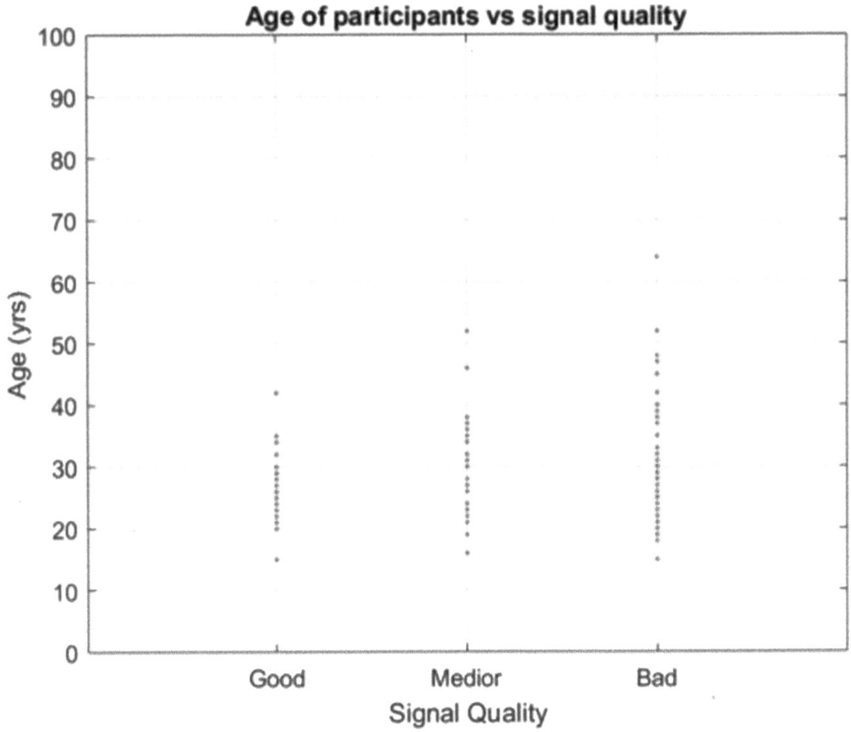

Fig. 8 Signal quality versus age. Each dot represents a participant which had certain signal quality and age

data stream of 28 participants with good quality signals. The validation set contained 429 blinks within 28 min of EOG recording of a subset of participants with good quality signals. The correctly identified blinks (TP), as compared to the annotated set, wrongly marked blinks (FP), and the missed blinks (FN) were counted. This results in a total sensitivity of 79% and F1 score of 81%.

Figure 9a shows an example of a signal containing blinks which are identified correctly by the algorithm. The artefactual spikes between the blinks are correctly not detected as blinks. Figure 9b shows an example of signal with more noise and one missed blink (red star) and one wrongly marked blink (green circle).

5 Discussion

Head-mounted augmented and virtual reality systems to a certain degree compromise comfort and convenience of the user. Additional sensing modalities must be blended within AR/VR solutions such that they do not result in additional burden to users but at the same time provide identifiable benefits. In this chapter, we discussed two such

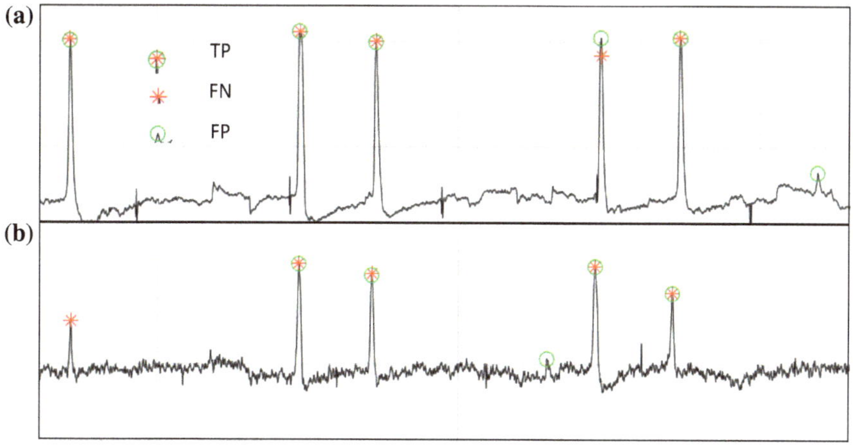

Fig. 9 Example of blinks detected by the annotator (red stars) and algorithm (green circles). TP, true positive; FN, false negative, FP, false positive

solutions—a wearable EEG headset solution designed as an add-on to an off-the-shelf VR headset and pair of glasses that can capture EOG activity. Specific designs concepts have been introduced to minimize the impact of sensing systems on the typical AR and VR interaction. Those designs have been described in detail, and the required miniaturized and low-power electronics integration is discussed.

Both solutions have been tested in an environment that introduces typical AR/VR interaction activities. The EEG solution is deployed in a simulated VR use case, while EOG glasses have been tested on users during a music festival in AR like interaction scenario. The tests were designed to assess not only wearability of such devices but also possibility to extract useful information from the recorded signals. In the EEG case, our investigation was focused on signal quality and its relationship to different head and body movements, as well as the impact of artefact handling methods in recovering useful EEG data. Thorough investigation of signal quality was performed using EOG glasses, focusing on impact of the electrode fit, demographics, and skin properties of users. Finally, an assessment of how well eyeblinks can be identified was performed.

5.1 VR Solution That Captures EEG

The EEG headset is designed such that it can easily be attached to most commercially available VR headsets using stretchable Velcro straps. Due to the usage of dry electrodes, it can be mounted and dismounted without hustle—almost like a normal VR headset. It has a shape that resembles standard straps used to attach VR headsets and yet enables capturing EEG from frontal locations involved in cognitive processing and occipital locations capturing visual aspects. The electronic solution is neatly integrated in a semi-rigid structure such that it provides equal weight distribution. Due to the low-power design, EEG headset can be in operation for more than 8 h without recharging. During the experiment performed, the users reported more discomfort due weight of the VR headset at the front (i.e., nose) than the EEG setup.

The EEG signal obtained in a simulated VR usage scenario demonstrated large impact of noise. The signal was clipping in less than 10% of epochs of baseline data, while clipping was present in almost 30% of facial muscle activity, 37% of head movement activity, and 18% of body movement activity. Nevertheless, 71% of baseline, 30% of EMG contaminated, 33% of head movement contaminated, and 41% of body movement contaminated data was assessed as artefact-free. After the application of motion artefact removal handling, consisting of artefact classification and artefact reduction using adaptive filtering, these numbers increased to 72%, 33%, 40%, and 47%, respectively.

Limited increase in clean data percentage can be explained by several factors. Limited capability of adaptive filtering based motion artefact detection algorithm (across different movement artefacts) and absence of EMG specific artefact correction method are some of those. Also, the poor performance of the artefact classification method. Although the classification performance was satisfactory for distinguishing movement and no-movement artefacts (accuracy higher than 96%), identifying no-movement artefacts (43%), distinguishing head and body movements (63%), identifying head movement type (57%), and identifying body movement type (67%) were shown to be challenging. Finally, limited capability of contact impedance and acceleration information used as a reference in adaptive filtering to capture artefactual components is also one of the reasons. Performance improvements of classification algorithms, as well as better tailoring of artefact type-specific algorithms could substantially increase the amount of clean EEG data recovered.

5.2 AR Glasses That Capture EOG

The design of EOG glasses is such that aside from the two temple electrodes, they look like regular glasses. This also holds for the comfort of users as no complaints were raised during the data collection at the music festival. In general, the size and shape of the EOG glasses seem to be adequate given that they are designed as one-size-fits-all but could fit well a large sample population (all electrodes combined had a good fit on 62% of subjects). There was difference in the fit (and signal quality) observed across different EOG prototype samples, indicating that the production process could be improved.

The overall level of EOG signal quality is also above expectations, given that the EOG glasses are deployed in a field test with low level of control over subjects and how they used the glasses. Across different glasses, 55–75% of the EOG data was assessed as good quality data. The goodness of fit seemed to be a good predictor of signal quality but was not the only factor, considering also different electrode locations. Although the fit was better in males (70%) compared to females (50%), this was not reflected in signal quality: 47% versus 49%. There were differences observed in terms of signal quality related to skin type and age, but full characterization of those requires dedicated exploration. Several improvement possibilities have been identified for adapting the design, one being the fit of the temple electrodes for females. Furthermore, different sizes or a more adaptable design for the nose electrodes should be considered.

The percentage of correctly identified eyeblinks in a selected database was 79% on a real data stream. Typical mistakes of the neural network algorithm introduced include spurious detection of small peaks, including overshoot of a saccade and blinks of longer duration, as well as exaggerated blinks. These misjudgments also illustrate the shortcomings of the training set of blinks and non-blinks. The opposite is also true, typical non-blinks that were part of the training set, artefactual peaks of different amplitudes and spiky noise data were successfully detected as non-blink. The current blink detector was deliberately made independent of signal amplitude. Not only does blink amplitude differ between persons but this also allows flexibility in the distance between the electrodes, the different leads which can be chosen to compose the vertical and horizontal EOG channels and improves applicability to other systems. The use of small, low sampling rate, windows for blink detection also permits the possibility of real-time implementation with low latency. To demonstrate this, the blink detector was implemented in a game in which the player uses the blinks to navigate a maze, applied during the tests.

6 Summary

EEG and EOG are well-known physiological signals which are most frequently acquired by inconvenient and often wired solutions. Imec's EEG headset and EOG glasses demonstrate the state of the art of wearable wireless recording of brain activity and eye movements. Individually, these technologies are a powerful tool to get more insights in diseases and improve therapies. Integration of both technologies with VR and AR is shown to be feasible, enabling as a direct feedback loop between the AR/VR content and the perception of it by the user. Taking the technology one step further, both can be combined with AR/VR enabling feedback on the user's state of mind and their focus and at the same time facilitating intuitive AR/VR content control. Applications in AR/VR that range from healthcare and military to retail and education can benefit from this greatly. Combining attentional level and point of focus to adapt therapies to the learning rate of the user in case of healthcare and educational applications or tracking the emotional status and enabling touchless controls for military and retail applications are just a few options. Imec's high tech EEG and EOG solutions hold the promise to enable wireless, user-friendly, unobtrusive solutions not only taking control of AR/VR content but moreover providing an adaptive feedback loop, taking personalization of the AR/VR content to the next level.

References

Barbara, N., & Camilleri, T. (2016). Interfacing with a speller using EOG glasses. In *IEEE International Conference on Systems, Man, and Cybernetics (SMC)* (pp. 1069–1074). Budapest: IEEE.

Bulling, A., Roggen, D., & Tröster, G. (2009). Wearable EOG goggles: Eye-based interaction in everyday environments. In *Extended Abstracts on Human Factors in Computing Systems (CHI EA '09)* (pp. 3259–3264). New York: ACM.

Chen, Y.-H., Op de Beeck, M., Vanderheyden, L., Carrette, E., Mihajlović, V., Vanstreels, K., ... Hoof, C. (2014). Soft, comfortable polymer dry electrodes for high quality ECG and EEG recording. *Sensors*, 23758–23780.

Fitzpatrick, T. (1988). The validity and practicality of sun-reactive skin types I through VI. *Archives of Dermatology, 124*(6), 869–871.

Grillon, C., & Buchsbaum, M. (1986). Computed EEG topogaphy of response to visual and auditory stimuli. *Electroencephalography and Clinical Neurophysiology*, 42–53.

Huotilainen, M., Winkler, I., Alho, K., Escera, C., Virtanen, J., Ilmoniemi, R., ... Naatanen, R. (1998). Combined mapping of human auditory EEG and MEG responses. *Electroencephalography and Clinical Neurophysiology*, 370–379.

Iáñez, E., Azorin, J., & Perez-Vidal, C. (2013). Using eye movement to control a computer: A design for a lightweight electro-oculogram electrode array and computer interface. *PLoS One*, e6709.

Ishimaru, S., Kunze, K., Uema, Y., Kise, K., Inam, M., & Tanaka, K. (2014). Smarter eyewear: Using commercial EOG glasses for activity recognition. In *ACM International Joint Conference on Pervasive and Ubiquitous Computing: Adjunct Publication (UbiComp '14 Adjunct)* (pp. 239–242). New York: ACM.

Jantz, J., Molnar, A., & Alcaide, R. (2017). A brain-computer interface for extended reality interfaces. In *ACM SIGGRAPH 2017 VR Village (SIGGRAPH '17)* (p. 2). New York: ACM.

Jins Meme. (2019, February 2). Retrieved from Jins Meme: https://jins-meme.com.

Jung, T.-P., Makeig, S., Humphries, C., Lee, T.-W., McKeown, M., Iragui, V., & Sejnowski, T. (2000). Removing electroencephalographic artifacts by blind source separation. *Psychophysiology*, 163–178.

Looxid Labs. (2019, February 2). Retrieved from Looxid Labs: https://looxidlabs.com/.

Lupu, R., Irimia, D., Ungureanu, F., Poboroniuc, M., & Moldoveanu, A. (2018). BCI and FES based therapy for stroke rehabilitation using VR facilities. *Wireless Communications and Mobile Computing*, *8*.

MacAskill, M. R., & Anderson, T. J. (2016). Eye movements in neurodegenerative diseases. *Current Opinion in Neurology*, 61–68.

Makeig, S., Westerfield, M., Jung, T., Enghoff, S., Townsend, J., Courchesne, E., & Sejnowski, T. (2002). Dynamic brain sources of visual evoked responses. *Science*, 690–694.

Mani, R., Asper, L., & Khuu, S. K. (2018). Deficits in saccades and smooth-pursuit eye movements in adults with traumatic brain injury: A systematic review and meta-analysis. *Brain Injury*, 1–22.

McEvoy, L., Smith, M., & Gevins, A. (2000). Test-retest reliability of cognitive EEG. *Clinical Neurophysiology*, 457–463.

Mihajlović, V., Grundlehner, B., Vullers, R., & Penders, J. (2015). Wearable, wireless EEG solutions in daily life applications: What are we missing? *IEEE Journal of Biomedical and Health Informatics*, 6–21.

Mihajlović, V., Li, H., Grundlehner, B., Penders, J., & Schouten, A. (2013). Investigating the impact of force and movements on impedance magnitude and EEG. *Engineering in Medicine and Biology Society (EMBC)* (pp. 1466–1469). IEEE.

Mihajlović, V., Patki, S., & Grundlehner, B. (2014). The impact of head movements in EEG and contact impedance: An adaptive filtering solution for motion artifact reduction. *Engineering in Medicine and Biology Society (EMBC)* (pp. 5064–5067). IEEE.

Neurable. (2019, February 2). Retrieved from Neurable: http://neurable.com/.

Nicander, I., Nyren, M., Emtestam, L., & Ollmar, S. (2006). Baseline electrical impedance measurements at various skin sites—related to age and sex. *Skin Research & Technology*, 252–258.

Papousek, I., & Schulter, G. (2004). Manipulation of frontal brain asymmetry by cognitive tasks. *Brain and Cognition*, 43–51.

Pretegiani, E., & Optican, L. M. (2017). Eye movements in Parkinson's disease and inherited parkinsonian syndromes. *Frontiers in Neurology*, 1–7.

Sabatos-DeVito, M., Schipul, S., Bulluck, J., Belger, A., & Baranek, G. (2016). Eye tracking reveals impaired attentional disengagement associated with sensory response patterns in children with autism. *Journal of Autism and Developmental Disorders*, 1319–1333.

Shallice, T., & Evans M E. (1978). The involvement of the frontal lobes in cognitive estimation. *Cortex*, 294–303.

Sharon, D., Hamalainen, M., Tootell, R., Halgren, E., & Belliveau, J. (2007). The advantage of combining MEG and EEG: Comparison to fMRI in focally-stimulated visual cortex. *Neuroimage*, 1225–1235.

Tromp, J., Peeters, D., Meyer, A., & Hagoort, P. (2017). The combined use of virtual reality and EEG to study language processing in naturalistic environments. *Behavior Research Methods*, 862–869.

Witteveen, J., Pradhapan, P., & Mihajlović, V. (2019). Comparison of a Pragmatic and Regression Approach for Wearable EEG Signal Quality Assessment. *IEEE Journal of Biomedical and Health Informatics*, 1–1.

Xu, J., Mitra, S., Matsumoto, A., Patki, S., Van Hoof, C., Makinwa, K., & Yazicioglu, R. (2014). A wearable 8-channel active-electrode EEG/ETI acquisition system for body area networks. *IEEE Journal of Solid-State Circuits*, 2005–2016.

Zargari Marandi, R., Madeleine, P., Omland, Ø., Vuillerme, N., & Samani, A. (2018). Eye movement characteristics reflected fatigue development in both young and elderly individuals. *Scientific Reports*, 1–10.

Neuroergonomic Applications in Information Visualization

Joseph K. Nuamah and Ranjana K. Mehta

Abstract Information visualization affords us the ability to reason, understand, and gain insight into data. Traditional behavioral and subjective methods of evaluating visualizations are inadequate. Neuroergonomic techniques can be used to complement traditional evaluation methods by assessing noninvasively and unobtrusively participants' working memory as they engage with the information visualization. In this chapter, we discuss information visualization techniques with an emphasis on evaluating visualizations from a cognitive load perspective. In particular, we review noninvasive functional brain monitoring techniques—electroencephalography (EEG) and functional near-infrared spectroscopy (fNIRS)—and how they have been used in information visualization research. Both measures of neural activity and task-related performance measures can be combined in a manner that enables the relationship between neural activity and performance to be quantified in terms of efficiency. In this chapter, we discuss visual efficiency, a neuroergonomic metric of information visualization. Finally, we review neuroadaptive interfaces—interfaces that modify their functionality in response to changes in the operator's cognitive state and needs.

1 Introduction

The term information visualization (InfoVis) may refer to a process or a type of display. As a process, InfoVis refers to representing information in a visual and meaningful way to enhance understanding and provide insight (e.g., Ward, Grinstein, & Keim, 2015). It leverages the working of the human visual system to enable humans to reason, understand, and gain insight into data (Nuamah, 2018). As a type of display, InfoVis refers to external visuospatial displays which can be categorized, as (1) iconic—to represent visual–spatial entities, (2) relational—to represent abstract

---

J. K. Nuamah · R. K. Mehta (✉)
Department of Industrial and Systems Engineering, Texas A&M University, College Station, TX 77843, USA
e-mail: rmehta@tamu.edu

C. S. Nam (ed.), *Neuroergonomics*, Cognitive Science and Technology,
https://doi.org/10.1007/978-3-030-34784-0_21

relationships, or (3) complex displays—based on the relation between the representation and the represented world, and the complexity (i.e., number of data points, configuration of data points, regularity) of the data represented (Hegarty, 2011, 2018; Meyer, Shinar, & Leiser, 1997). The data to be represented may be static—does not change or streaming—continuous, rapid, and time-varying (Babcock, Babu, Datar, Motwani, & Widom, 2002; Dasgupta, Arendt, Franklin, Wong, & Cook, 2018).

InfoVis (i.e., visuospatial displays) augments cognition (Card, Mackinlay, & Schneiderman, 1999; Larkin & Simon, 1987; Tory & Moller, 2004) by (1) acting as an external storage of information and freeing up working memory (Card et al., 1999; Fekete, Van Wijk, Stasko, & North, 2008; Scaife & Rogers, 1996), (2) organizing information by spatial indexing (Chambers & Reisberg, 1985; Larkin & Simon, 1987), (3) shifting cognitive burden to the perceptual system (Hegarty, 2018; Scaife & Rogers, 1996), and (4) offloading cognition on action (Card et al., 1999). InfoVis is needed in many disciplines including health care (e.g., Widanagamaachchi, Livnat, Bremer, Duvall, & Pascucci, 2017), computer network traffic monitoring (e.g., Cao et al., 2018), cybersecurity (e.g., Shi, Zhao, Zhou, Shi, & Zhang, 2018), and aviation (e.g., Lim et al., 2018; Wickens & Ward, 2017) to support decision-making, and to conduct exploratory and confirmation analyses (Marty, 2009). It is worth noting that task performance can be very different with different representations of the same information (e.g., Breslow, Trafton, & Ratwani, 2009; Hegarty, Canham, & Fabrikant, 2010).

Evaluating the effectiveness of InfoVis is challenging. Traditional behavioral and subjective approaches are inadequate because they are obtrusive and suffer from recall bias (Frey, Mühl, Lotte, & Hachet, 2013; Hill & Bohil, 2016). Physiological methods, especially brain monitoring methods, have shown promise in continuously and objectively assessing cognitive states of users as they interact with systems. In this chapter, we will first discuss how these methods, especially noninvasive functional brain monitoring techniques, may be used to evaluate InfoVis.

Understanding of both cognitive and neural underpinnings of task performance is crucial in evaluating InfoVis. Measures of neural activity obtained from noninvasive functional brain monitoring techniques and task-related performance measures can be combined in a manner that enables the relationship between neural activity and performance to be quantified in terms of efficiency. We will present how newer neuroergonomic metrics of InfoVis may potentially provide a better understanding and evaluation of visual efficiency.

Traditionally, interactions between the human and the computer have been asymmetrical—the computer sends status information to humans but has no means of independently assessing the human's status (Hettinger, Branco, Encarnacao, & Bonato, 2003). Thus, computers are unable to modify their behavior in real time to adapt to the needs of the human until the human initiates some sort of command. Advances in neuroergonomics have allowed for the development of neuroadaptive interfaces, closed-loop human–computer systems that are able to modify their functionality in response to changes in the human's cognitive state and needs (Hettinger et al., 2003; Parasuraman, 2011). Thus, the remainder of this chapter will expand the utility of neuroadaptive interfaces for improved information visualization.

2 Evaluation of InfoVis

Evaluation, though a challenge (Elmqvist & Yi, 2015; Plaisant, 2004), is crucial in ensuring the effectiveness of InfoVis (Borgo, Micallef, Bach, McGee, & Lee, 2018; Huang, Eades, & Hong, 2009). Traditionally, performance measures such as response time (or completion time) and accuracy are used to evaluate InfoVis (Herman, Melançon, & Marshall, 2000; Zagermann, Pfeil, & Reiterer, 2016). However, these measures discount the neural cost in achieving and/or maintaining the same level of performance for two different visualizations—say, one good and the other bad—of the same data. This means that performance-based measures are unable to differentiate between the amount of cognitive effort required for two different visualizations representing the same data (Paas, Tuovinen, Tabbers, & Van Gerven, 2003). This issue becomes particularly critical when operators are working under non-normal scenarios (e.g., emergency response) or for prolonged periods of time (e.g., drillers). As such, there is a need for other measurement techniques to complement performance-based measures.

Since InfoVis seeks to minimize cognitive burden, it is expedient to also evaluate visualizations from a cognitive load perspective (Huang et al., 2009; Zagermann et al., 2016). Techniques employed to evaluate InfoVis cognitive load include the subjective NASA Task Load Index (NASA-TLX; Hart & Staveland, 1988) questionnaire, eye tracking, and noninvasive functional brain imaging. The NASA-TLX has been used to evaluate working memory demand of different types of InfoVis on older adults (Price, Crumley-Branyon, Leidheiser, & Pak, 2016), measure cognitive load of participants while they used two different electronic health record user interfaces (Ahmed, Chandra, Herasevich, Gajic, & Pickering, 2011), and to evaluate preattentive visualization techniques (Krekhov & Krüger, 2019). Eye tracking measures, including voluntary (e.g., fixations, saccades) and involuntary (e.g., pupil dilation, blinking) eye movements, have been used in previous research to evaluate simple information graphics (Goldberg & Helfman, 2010), measure the distribution of attention on objects with different video visualizations (Kurzhals, Höferlin, & Weiskopf, 2013), evaluate ontology visualization techniques (Fu, Noy, & Storey, 2017), improve the design of surgical interfaces (Barkana & Açık, 2014), and evaluate drug label designs (Bojko, Gaddy, Lew, Quinn, & Israelski, 2005). Techniques of interest in this chapter, i.e., functional brain imaging techniques, make it possible for researchers to identify roles that specific regions of the brain play as participants engage with any visualization. The next section presents these functional imaging techniques, and how they have been used in InfoVis.

3 Noninvasive Functional Brain Monitoring Techniques in InfoVis

Functional neuroimaging techniques allow us to capture and/or monitor brain activities while participants perform various perceptual, motor, and/or cognitive tasks (Gui, Chuansheng, Zhong-Lin, & Qi, 2010). These techniques include functional magnetic resonance imaging (fMRI), functional near-infrared spectroscopy (fNIRS), electroencephalography (EEG), and magnetoencephalography (MEG). Each technique has its own advantages and limitations. The choice of measurement modality for a particular application is influenced by cost, portability, temporal resolution, and spatial resolution (Buccino, Keles, & Omurtag, 2016; Mehta & Parasuraman, 2013). EEG and fNIRS are used more often in visualization research than other modalities because they are relatively cheaper, and offer monitoring in seated/standing postures. Furthermore, EEG, which has been thoroughly researched (Zander & Kothe, 2011), measures summated postsynaptic electrical activity, has moderate portability, relatively low spatial resolution but high temporal resolution. Standard EEG has a source localization problem that can be addressed when a large number of electrodes are used (Grech et al., 2008). High-density EEG (hdEEG) with up to 256 electrodes combines the high temporal resolution of EEG recordings with high spatial resolution, allowing localization of sources (Liu, Farahibozorg, Porcaro, Wenderoth, & Mantini, 2017). On the other hand, fNIRS measures oxyhemoglobin and deoxyhemoglobin responses, has relatively high portability, relatively good spatial resolution but lower temporal resolution (Mehta & Parasuraman, 2013).

EEG has been used to assess cognitive load associated with information presented visually in data plots (Anderson et al., 2011), detect insight and emotion in visualization applications (Cernea, Kerren, & Ebert, 2011), evaluate graphic icons (Cherng, Lin, King, & Lee, 2016), evaluate 3-D user interfaces (Frey, Hachet, & Lotte, 2017), assess stereoscopic images (Avarvand et al., 2017), estimate visual comfort in head-mounted displays (Mai, Hassib, & Königbauer, 2017), estimate visual comfort in stereoscopic visualization (Frey, Appriou, Lotte, & Hachet, 2016), and evaluate cognitive load imposed by intuition-inducing versus analysis-inducing user interfaces (Nuamah, 2018). For example, Anderson et al. (2011) evaluated the effectiveness of different box plots in visualizing distribution of data. They collected EEG data derived from electrodes placed around the prefrontal cortex and segmented the data into baseline and stimulus epochs. They focused on the theta (4–7.5 Hz) and alpha (7.5–12.5 Hz) frequency bands, and used the S-Transform, a variable window of the short-time Fourier Transform, to decompose epochs into time–frequency representations. They computed mean frequencies in the theta and alpha bands at each epoch and combined them into a single time series at each electrode placement. Finally, they combined the time series through spatially aware averaging using Gaussian weights to form each trial's overall cognitive load.

Nuamah (2018) simulated an experiment in which an automated water sampling station with six different sensors measured six water quality parameters. Data from all six sensors were presented in two visual formats: table format (see Fig. 1a) and

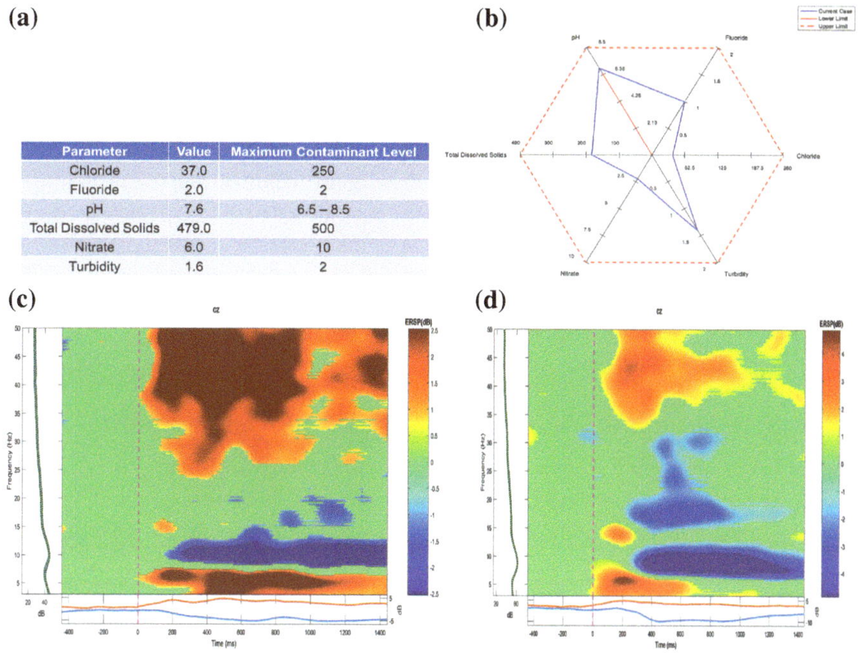

Parameter	Value	Maximum Contaminant Level
Chloride	37.0	250
Fluoride	2.0	2
pH	7.6	6.5 – 8.5
Total Dissolved Solids	479.0	500
Nitrate	6.0	10
Turbidity	1.6	2

Fig. 1 InfoVis in (**a**) Table format and (**b**) Graphical format. Event-related spectral perturbation at Cz in (**c**) table format and (**d**) graphical format

graphical format (Fig. 1b). The objective was to investigate neural correlates of different visualizations of the same data. EEG data was collected from participants as they judged water quality (as good or bad) based on the two visualizations. To visualize changes in spectral power across time within frequency ranges of interest, event-related spectral perturbation (ERSP; Makeig, Debener, Onton, & Delorme, 2004) analysis was conducted. Power spectrum over a sliding window was calculated and averaged across data trials. Results (see Fig. 1c, d) showed differences in spectral power of theta, alpha, and beta, suggesting that event-related brain activity could be useful markers for measuring whether two InfoVis differ.

fNIRS has been used to distinguish differences in cognitive states that derive from InfoVis (Peck, Yuksel, Ottley, Jacob, & Chang, 2013), evaluate different types of web form filling interfaces (Lukanov, Maior, & Wilson, 2016), and automatically adjust the difficulty of an unmanned aerial vehicle (UAV) task in real time (Afergan et al., 2014; Yuksel et al., 2015). For example, Peck et al. (2013) used two fNIRS probes—each probe had four light sources and one light detector for each hemisphere—to record light intensity data as participants compared bar graphs and pie charts. Using the Beer–Lambert Law (Delpy et al., 1988), they transformed the filtered raw data into oxygenated hemoglobin and deoxygenated hemoglobin concentrations. They used deoxygenated hemoglobin for their analysis and found changes in deoxygenated hemoglobin during the use of bar graphs to be statistically different from those observed during use of pie charts. Their findings suggest that deoxygenated hemoglobin changes in the prefrontal cortex (PFC) is indicative of the amount of mental effort associated with the different visualizations.

EEG and fNIRS are complementary brain monitoring methods (Mandrick, Chua, Causse, Perrey, & Dehais, 2016). Previous studies (e.g., Banville, Parent, Tremblay, & Falk, 2018; Hong, Khan, & Hong, 2018; Sargent, Heiman-Patterson, Feldman, Shewokis, & Ayaz, 2018) have combined both modalities to obtain a more complete picture of brain activity. Although these modalities can be combined to evaluate InfoVis, there are no existing studies that have evaluated InfoVis with a hybrid fNIRS–EEG.

4 Neuroergonomic Metrics of Information Visualization

The Cognitive Load Theory (CLT; Chandler & Sweller, 1991), which relates working memory capacity to cognitive demands of a task, provides a theoretical underpinning for developing neuroergonomic measures of InfoVis. The CLT categorizes cognitive load into (1) intrinsic load—generated by intrinsic complexity of the task, (2) germane load—generated from effort needed to learn new cognitive schema, and (3) extraneous load—imposed by the design of the task (Chandler & Sweller, 1991; Paas et al., 2003). Overall cognitive load is characterized by the combination of these three constructs (Anderson et al., 2011; Seufert, Jänen, & Brünken, 2007). Both extraneous load (see Anderson et al., 2011; Zagermann et al., 2016) and germane load (see Debue & Van De Leemput, 2014) can be controlled by the format and the way information is presented. Consequently, InfoVis should aim to minimize both germane load and extraneous load.

Mental effort, the assigned cognitive capacity to accommodate task demands, reflects actual cognitive load (Huang et al., 2009), and can be measured physiologically. EEG measures of mental effort include the task load index—ratio of theta activity at frontal midline sites to alpha at parietal sites (Gevins & Smith, 2003; Kamzanova, Kustubayeva, & Matthews, 2014; Nuamah & Seong, 2017), and amplitude of the P300 event-related potential (ERP; Kirchner et al., 2013; Kok, 2001; Polich, 2007). fNIRS measures of mental effort include changes in concentration of oxygenated and deoxygenated hemoglobin (Causse, Chua, Peysakhovich, Campo, & Matton, 2017; Peck et al., 2013; Vassena, Gerrits, Demanet, Verguts, & Siugzdaite, 2018).

Measured cognitive load is meaningful only to the extent that it is interpreted in relation with its corresponding performance level (Huang et al., 2009; Paas et al., 2003). Similar to Paas and Van Merriënboer (1993), Huang et al. (2009) combined normalized measures of mental effort (ME) and performance (P) into one measure referred to as visualization efficiency (VE) defined by Eq. 1. Performance and neural measures are normalized across all task conditions and participants.

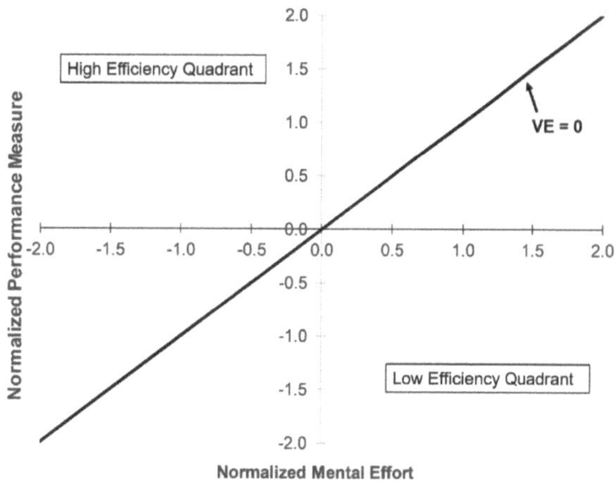

Fig. 2 Relating visual efficiency to performance and mental effort. High-efficiency quadrant is marked by relatively low neural demand and relatively high performance. Low-efficiency quadrant is marked by relatively high neural demand and lower-than-expected performance

$$VE = \frac{z(P) - z(ME)}{\sqrt{2}} \tag{1}$$

where $z(P)$ and $z(ME)$ are z-score measures of performance and mental effort, respectively.

High visual efficiency is achieved when performance (e.g., accuracy) is high and the corresponding mental effort (e.g., level of concentration of oxygenated hemoglobin) is low, and vice versa. Equation 1 affords researchers the opportunity to compare relative visual efficiencies between experimental conditions and assess neurocognitive effectiveness of different visualizations for the same type of information. When $VE = 0$, performance and mental effort are balanced. If two axes are drawn with the vertical axis representing the performance measure, *P*, and the horizontal axis representing the mental effort measure, *ME*, (see Fig. 2), then *VE* represents the perpendicular distance from a point in the Cartesian space to the plane whose equation is by Eq. 2 (Huang et al., 2009).

$$z(P) - z(ME) = 0 \tag{2}$$

where $z(P)$ and $z(ME)$ are z-score measures of performance and mental effort, respectively.

A limitation of employing neuroergonomic metrics is their reliance on the sensitivity and utility of both neural and performance measures for the specific InfoVis evaluation. An ambiguity in any of the measures will present itself with the analysis of the metric. Thus, future work is warranted that tests the efficacy of this metric on different InfoVis applications.

5 Neuroadaptive Interfaces for InfoVis

Brain–computer interfaces (BCIs) are communication and control systems that take biosignals measured from a person and translate them into messages and commands for an interactive application (Lotte & Roy, 2019). BCIs may be classified as active, reactive, or passive (Zander & Kothe, 2011). Active BCIs translate consciously imagined mental tasks into specific commands. Reactive BCIs detect and translate biosignals resulting from stimuli presented to the user into commands associated with the stimuli. Passive BCIs monitor the user's mental states and automatically modify their functionality accordingly without the user sending voluntary command (Zander & Kothe, 2011). Neuroadaptive interfaces, also referred to as adaptive interface technology (e.g., Mulgund, Rinkus, & Zacharias, 2003) or adaptive automation (e.g., Scerbo, Freeman, & Mikulka, 2003), are passive BCIs which adapt to brain activity (Krol & Zander, 2017). They are closed-loop systems comprised of controls and displays that modify their functionality in response to changes in the operator's cognitive state and needs (Hettinger et al., 2003; Parasuraman, 2011). Changes in functionality can be initiated by either the human only—"static" adaptable interface, or the human and/or system—"dynamic" adaptive system (Scerbo et al., 2001). Figure 3 below shows a schematic of neuroadaptive interface for improved InfoVis. Changes in operator states are indexed by brain activity which, in turn, controls adaptive modifications to the interface (Hettinger et al., 2003). The main purpose of neuroadaptive interfaces is not only to widen the communication bottleneck between operators and computer-based systems (Zander, Krol, Birbaumer, & Gramann, 2016), but also in real time to enhance human–computer system performance (Hettinger et al., 2003). This enables the human–computer system to automatically adapt to estimates of its operator's cognitive state thereby maintaining a relatively constant level of performance (Hettinger et al., 2003; Zander et al., 2016).

Brain signals acquired with an EEG system have high temporal resolution (~1 ms) from which frequency and time–frequency features can be generated. EEG features

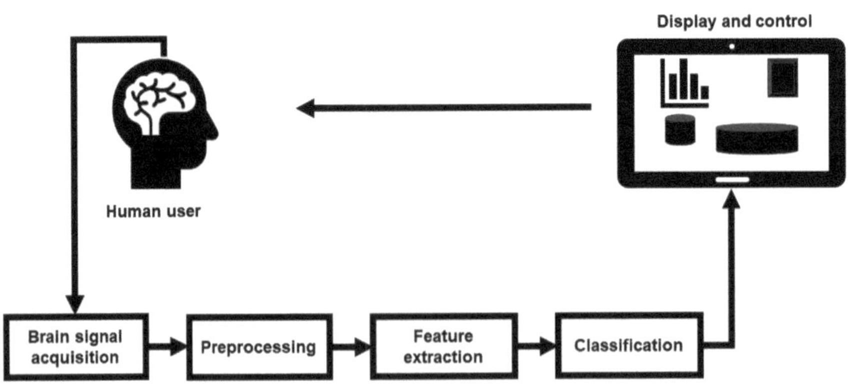

Fig. 3 Neuroadaptive interface schematic

include amplitudes derived from event-related potentials (ERPs), measurements of power spectral densities, and common spatial pattern (CSP). On the other hand, hemodynamic response (~4–5 s), measured by fNIRS, is not as quick as electrophysiological response (Treacy Solovey, Afergan, Peck, Hincks, & Jacob, 2015). Consequently, fNIRS is usually analyzed in the time domain. Statistics of oxygenated hemoglobin and deoxygenated hemoglobin are commonly used as fNIRS features (Naseer, Noori, Qureshi, & Hong, 2016). Common features include mean, variance, peak value, slope, skewness, and kurtosis of preprocessed hemodynamic response signals (Naseer & Hong, 2015). With respect to neuroadaptive interfaces (see Fig. 3), brain signals are acquired using the appropriate neuroimaging modality, including EEG (e.g., Freeman, Mikulka, Prinzel, & Scerbo, 1999; Freeman, Mikulka, Scerbo, Prinzel, & Clouatre, 2000; Prinzel, Freeman, Scerbo, Mikulka, & Pope, 2000; Zander et al., 2016) and fNIRS (e.g., Afergan et al., 2014; Girouard, Solovey, & Jacob, 2013; Solovey et al., 2012). Identifying correct features, which may be classified as temporal, spatial, or spatiotemporal, from acquired brain signals is crucial to the proper functioning of neuroadaptive interfaces (Hong et al., 2018). Indeed, brain patterns are characterized by certain features (Graimann, Allison, & Pfurtscheller, 2009). For example, features for EEG-based neuroadaptive interfaces include event-related potentials to detect intentions when fixating an object (Protzak, Ihme, & Zander, 2013), spectral power of frontal theta and parietal alpha band activity to detect cognitive workload (Gerjets, Walter, Rosenstiel, Bogdan, & Zander, 2014), and index of engagement ratio given by beta/(alpha +theta) to detect changes in cognitive workload demands (Pope, Bogart, & Bartolome, 1995). Features for fNIRS-based systems include signal mean to classify motor imagery versus rest (Noori, Naseer, Qureshi, Nazeer, & Khan, 2017), signal peak to differentiate between motor imagery and arithmetic tasks (Stangl, Bauernfeind, Kurzmann, Scherer, & Neuper, 2013), and mean oxyhemoglobin, signal peak, and sum of peaks (Khan & Hong, 2015).

Classifiers use the aforementioned features to differentiate between brain signals to quantify different levels of operator states (such as workload). Commonly used classifiers include linear discriminant analysis and support vector machines (Aghajani, Garbey, & Omurtag, 2017; Khan & Hong, 2015; Yin et al., 2015). These studies have shown that it is possible to detect indices of brain activities to cognitive states, and in some instances, use these indices to drive the activity of a human–machine interface. Some limitations that impede the realization of neuroadaptive interfaces to real application include: (1) the difficulty in determining optimal input features for a particular user, considering the variability within individuals and among users, and (2) the difficulty in capturing brain dynamics in untethered and natural environments.

6 Conclusion

Neuroergonomic techniques can be used to complement traditional evaluation methods by assessing noninvasively and unobtrusively participants' working memory as they engage with the InfoVis. This allows researchers and system designers to determine whether an InfoVis is more or less effective, whether it reduces working memory demand and overall cognitive load. Neuroergomic metrics, particularly visual efficiency, may help researchers determine whether an InfoVis evokes relatively low neural demand and relatively high performance (high efficiency) versus relatively high neural demand and lower-than-expected performance (low efficiency).

References

Afergan, D., Peck, E. M., Solovey, E. T., Jenkins, A., Hincks, S. W., Brown, E. T., … & Jacob, R. J. (2014, April). Dynamic difficulty using brain metrics of workload. In *Proceedings of the 32nd Annual ACM Conference on Human Factors in Computing Systems* (pp. 3797–3806). ACM.

Aghajani, H., Garbey, M., & Omurtag, A. (2017). Measuring mental workload with EEG + fNIRS. *Frontiers in Human Neuroscience, 11,* 359.

Ahmed, A., Chandra, S., Herasevich, V., Gajic, O., & Pickering, B. W. (2011). The effect of two different electronic health record user interfaces on intensive care provider task load, errors of cognition, and performance. *Critical Care Medicine, 39*(7), 1626–1634.

Anderson, E. W., Potter, K. C., Matzen, L. E., Shepherd, J. F., Preston, G. A., & Silva, C. T. (2011, June). A user study of visualization effectiveness using EEG and cognitive load. In *Computer graphics forum* (Vol. 30, No. 3, pp. 791–800). Oxford, UK: Blackwell Publishing Ltd.

Avarvand, F. S., Bosse, S., Müller, K. R., Schäfer, R., Nolte, G., Wiegand, T., … & Samek, W. (2017). Objective quality assessment of stereoscopic images with vertical disparity using EEG. *Journal of Neural Engineering, 14*(4), 046009.

Babcock, B., Babu, S., Datar, M., Motwani, R., & Widom, J. (2002, June). Models and issues in data stream systems. In *Proceedings of the Twenty-First ACM SIGMOD-SIGACT-SIGART Symposium on Principles of Database Systems* (pp. 1–16). ACM.

Banville, H., Parent, M., Tremblay, S., & Falk, T. H. (2018). Toward mental workload measurement using multimodal EEG–fNIRS monitoring. In *Neuroergonomics* (pp. 245–246). Academic Press.

Barkana, D. E., & Açık, A. (2014). Improvement of design of a surgical interface using an eye tracking device. *Theoretical Biology and Medical Modelling, 11*(1), S4.

Bojko, A., Gaddy, C., Lew, G., Quinn, A., & Israelski, E. (2005, September). Evaluation of drug label designs using eye tracking. In *Proceedings of the Human Factors and Ergonomics Society Annual Meeting* (Vol. 49, No. 11, pp. 1033–1037). Sage, CA; Los Angeles, CA: SAGE Publications.

Borgo, R., Micallef, L., Bach, B., McGee, F., & Lee, B. (2018, June). Information visualization evaluation using crowdsourcing. *Computer Graphics Forum, 37*(3), 573–595.

Breslow, L. A., Trafton, J. G., & Ratwani, R. M. (2009). A perceptual process approach to selecting color scales for complex visualizations. *Journal of Experimental Psychology: Applied, 15*(1), 25.

Buccino, A. P., Keles, H. O., & Omurtag, A. (2016). Hybrid EEG-fNIRS asynchronous brain-computer interface for multiple motor tasks. *PLoS ONE, 11*(1), e0146610.

Cao, N., Lin, C., Zhu, Q., Lin, Y. R., Teng, X., & Wen, X. (2018). Voila: Visual anomaly detection and monitoring with streaming spatiotemporal data. *IEEE Transactions on Visualization and Computer Graphics, 24*(1), 23–33.

Card, S. K., Mackinlay, J. D., & Schneiderman, B. (1999). *Readings in information visualization: Using vision to think.* San Francisco: Morgan Kaufmann Publishers.

Causse, M., Chua, Z., Peysakhovich, V., Campo, N., & Matton, N. (2017). Mental workload and neural efficiency quantified in the prefrontal cortex using fNIRS. *Scientific Reports, 7*(1), 5222.

Cernea, D., Kerren, A., & Ebert, A. (2011, November). Detecting insight and emotion in visualization applications with a commercial EEG headset. In *Proceedings of SIGRAD 2011. Evaluations of graphics and visualization—Efficiency; Usefulness; Accessibility; Usability; November 17–18; 2011; KTH; Stockholm; Sweden* (No. 65, pp. 53–60). Linköping University Electronic Press.

Chambers, D., & Reisberg, D. (1985). Can mental images be ambiguous? *Journal of Experimental Psychology: Human Perception and Performance, 11*(3), 317.

Chandler, P., & Sweller, J. (1991). Cognitive load theory and the format of instruction. *Cognition and Instruction, 8*(4), 293–332.

Cherng, F. Y., Lin, W. C., King, J. T., & Lee, Y. C. (2016, May). An eeg-based approach for evaluating graphic icons from the perspective of semantic distance. In *Proceedings of the 2016 Chi Conference on Human Factors in Computing Systems* (pp. 4378–438). ACM.

Dasgupta, A., Arendt, D. L., Franklin, L. R., Wong, P. C., & Cook, K. A. (2018, February). Human factors in streaming data analysis: Challenges and opportunities for information visualization. *Computer Graphics Forum,* 37(1), 254–272.

Debue, N., & Van De Leemput, C. (2014). What does germane load mean? An empirical contribution to the cognitive load theory. *Frontiers in Psychology, 5,* 1099.

Delpy, D. T., Cope, M., van der Zee, P., Arridge, S. R., Wray, S., & Wyatt, J. S. (1988). Estimation of optical pathlength through tissue from direct time of flight measurement. *Physics in Medicine & Biology, 33*(12), 1433.

Elmqvist, N., & Yi, J. S. (2015). Patterns for visualization evaluation. *Information Visualization, 14*(3), 250–269.

Fekete, J. D., Van Wijk, J. J., Stasko, J. T., & North, C. (2008). The value of information visualization. In *Information visualization* (pp. 1–18). Berlin, Heidelberg: Springer.

Freeman, F. G., Mikulka, P. J., Prinzel, L. J., & Scerbo, M. W. (1999). Evaluation of an adaptive automation system using three EEG indices with a visual tracking task. *Biological Psychology, 50*(1), 61–76.

Freeman, F. G., Mikulka, P. J., Scerbo, M. W., Prinzel, L. J., & Clouatre, K. (2000). Evaluation of a psychophysiologically controlled adaptive automation system, using performance on a tracking task. *Applied Psychophysiology and Biofeedback, 25*(2), 103–115.

Frey, J., Appriou, A., Lotte, F., & Hachet, M. (2016). Classifying EEG signals during stereoscopic visualization to estimate visual comfort. *Computational Intelligence and Neuroscience, 2016,* 7.

Frey, J., Hachet, M., & Lotte, F. (2017). EEG-based neuroergonomics for 3D user interfaces: Opportunities and challenges. *Le travail humain, 80*(1), 73–92.

Frey, J., Mühl, C., Lotte, F., & Hachet, M. (2013). Review of the use of electroencephalography as an evaluation method for human-computer interaction. arXiv preprint. arXiv:1311.2222.

Fu, B., Noy, N. F., & Storey, M. A. (2017). Eye tracking the user experience–An evaluation of ontology visualization techniques. *Semantic Web, 8*(1), 23–41.

Gerjets, P., Walter, C., Rosenstiel, W., Bogdan, M., & Zander, T. O. (2014). Cognitive state monitoring and the design of adaptive instruction in digital environments: Lessons learned from cognitive workload assessment using a passive brain-computer interface approach. *Frontiers in Neuroscience, 8,* 385.

Gevins, A., & Smith, M. E. (2003). Neurophysiological measures of cognitive workload during human-computer interaction. *Theoretical Issues in Ergonomics Science, 4*(1–2), 113–131.

Girouard, A., Solovey, E. T., & Jacob, R. J. (2013). Designing a passive brain computer interface using real time classification of functional near-infrared spectroscopy. *International Journal of Autonomous and Adaptive Communications Systems, 6*(1), 26–44.

Goldberg, J. H., & Helfman, J. I. (2010, April). Comparing information graphics: A critical look at eye tracking. In *Proceedings of the 3rd BELIV'10 Workshop: Beyond Time and Errors: Novel Evaluation Methods for Information Visualization* (pp. 71–78). ACM.

Graimann, B., Allison, B., & Pfurtscheller, G. (2009). Brain–computer interfaces: A gentle introduction. In *Brain-computer interfaces* (pp. 1–27). Berlin, Heidelberg: Springer.

Grech, R., Cassar, T., Muscat, J., Camilleri, K. P., Fabri, S. G., Zervakis, M., … & Vanrumste, B. (2008). Review on solving the inverse problem in EEG source analysis. *Journal of Neuroengineering and Rehabilitation, 5*(1), 25.

Gui, X. U. E., Chuansheng, C. H. E. N., Zhong-Lin, L. U., & Qi, D. O. N. G. (2010). Brain imaging techniques and their applications in decision-making research. *Xin li xue bao. Acta psychologica Sinica, 42*(1), 120.

Hart, S. G., & Staveland, L. E. (1988). Development of NASA-TLX (Task Load Index): Results of empirical and theoretical research. In *Advances in psychology* (Vol. 52, pp. 139–183). North-Holland.

Hegarty, M. (2018). Advances in cognitive science and information visualization. In *Score reporting research and applications* (pp. 19–34). Routledge.

Hegarty, M. (2011). The cognitive science of visual-spatial displays: Implications for design. *Topics in Cognitive Science, 3*(3), 446–474.

Hegarty, M., Canham, M. S., & Fabrikant, S. I. (2010). Thinking about the weather: How display salience and knowledge affect performance in a graphic inference task. *Journal of Experimental Psychology. Learning, Memory, and Cognition, 36*(1), 37.

Herman, I., Melançon, G., & Marshall, M. S. (2000). Graph visualization and navigation in information visualization: A survey. *IEEE Transactions on Visualization and Computer Graphics, 6*(1), 24–43.

Hettinger, L. J., Branco, P., Encarnacao, L. M., & Bonato, P. (2003). Neuroadaptive technologies: Applying neuroergonomics to the design of advanced interfaces. *Theoretical Issues in Ergonomics Science, 4*(1–2), 220–237.

Hill, A. P., & Bohil, C. J. (2016). Applications of optical neuroimaging in usability research. *Ergonomics in Design, 24*(2), 4–9.

Hong, K. S., Khan, M. J., & Hong, M. J. (2018). Feature extraction and classification methods for hybrid fNIRS-EEG brain-computer interfaces. *Frontiers in Human Neuroscience, 12.*

Huang, W., Eades, P., & Hong, S. H. (2009). Measuring effectiveness of graph visualizations: A cognitive load perspective. *Information Visualization, 8*(3), 139–152.

Kamzanova, A. T., Kustubayeva, A. M., & Matthews, G. (2014). Use of EEG workload indices for diagnostic monitoring of vigilance decrement. *Human Factors, 56*(6), 1136–1149.

Khan, M. J., & Hong, K. S. (2015). Passive BCI based on drowsiness detection: An fNIRS study. *Biomedical Optics Express, 6*(10), 4063–4078.

Kirchner, E. A., Kim, S. K., Straube, S., Seeland, A., Wöhrle, H., Krell, M. M., … & Fahle, M. (2013). On the applicability of brain reading for predictive human-machine interfaces in robotics. *PLoS One, 8*(12), e81732.

Kok, A. (2001). On the utility of P3 amplitude as a measure of processing capacity. *Psychophysiology, 38*(3), 557–577.

Krekhov, A., & Krüger, J. (2019). Deadeye: A novel preattentive visualization technique based on dichoptic presentation. *IEEE Transactions on Visualization and Computer Graphics, 25*(1), 936–945.

Krol, L. R., & Zander, T. O. (2017). Passive BCI-based neuroadaptive systems. In *Proceedings of the 7th Graz Brain-Computer Interface Conference* (Vol. 2017, pp. 248–253).

Kurzhals, K., Höferlin, M., & Weiskopf, D. (2013, June). Evaluation of attention-guiding video visualization. In *Computer graphics forum* (Vol. 32, No. 3pt1, pp. 51–60). Oxford, UK: Blackwell Publishing Ltd.

Larkin, J. H., & Simon, H. A. (1987). Why a diagram is (sometimes) worth ten thousand words. *Cognitive Science, 11*(1), 65–100.

Lim, Y., Gardi, A., Sabatini, R., Ramasamy, S., Kistan, T., Ezer, N., … & Bolia, R. (2018). Avionics human-machine interfaces and interactions for manned and unmanned aircraft. *Progress in Aerospace Sciences.*

Liu, Q., Farahibozorg, S., Porcaro, C., Wenderoth, N., & Mantini, D. (2017). Detecting large-scale networks in the human brain using high-density electroencephalography. *Human Brain Mapping, 38*(9), 4631–4643.

Lotte, F., & Roy, R. N. (2019). Brain–computer interface contributions to neuroergonomics. In *Neuroergonomics* (pp. 43–48). Academic Press.

Lukanov, K., Maior, H. A., & Wilson, M. L. (2016, May). Using fNIRS in usability testing: Understanding the effect of web form layout on mental workload. In *Proceedings of the 2016 CHI Conference on Human Factors in Computing Systems* (pp. 4011–4016). ACM.

Mai, C., Hassib, M., & Königbauer, R. (2017, September). Estimating visual discomfort in head-mounted displays using electroencephalography. In *IFIP Conference on Human-Computer Interaction* (pp. 243–252). Cham: Springer.

Makeig, S., Debener, S., Onton, J., & Delorme, A. (2004). Mining event-related brain dynamics. *Trends in Cognitive Sciences, 8*(5), 204–210.

Mandrick, K., Chua, Z., Causse, M., Perrey, S., & Dehais, F. (2016). Why a comprehensive understanding of mental workload through the measurement of neurovascular coupling is a key issue for neuroergonomics? *Frontiers in Human Neuroscience, 10,* 250.

Marty, R. (2009). *Applied security visualization* (p. 552). Upper Saddle River: Addison-Wesley.

Mehta, R. K., & Parasuraman, R. (2013). Neuroergonomics: A review of applications to physical and cognitive work. *Frontiers in Human Neuroscience, 7,* 889.

Meyer, J., Shinar, D., & Leiser, D. (1997). Multiple factors that determine performance with tables and graphs. *Human Factors, 39*(2), 268–286.

Mulgund, S. S., Rinkus, G., & Zacharias, G. (2003). Adaptive pilot–vehicle interfaces for the tactical air environment. *Virtual and Adaptive Environments, 483.*

Naseer, N., Noori, F. M., Qureshi, N. K., & Hong, K. S. (2016). Determining optimal feature-combination for LDA classification of functional near-infrared spectroscopy signals in brain-computer interface application. *Frontiers in Human Neuroscience, 10,* 237.

Naseer, N., & Hong, K. S. (2015). fNIRS-based brain-computer interfaces: A review. *Frontiers in Human Neuroscience, 9,* 3.

Noori, F. M., Naseer, N., Qureshi, N. K., Nazeer, H., & Khan, R. A. (2017). Optimal feature selection from fNIRS signals using genetic algorithms for BCI. *Neuroscience Letters, 647,* 61–66.

Nuamah, J. K. G. (2018). *Effects of information visualization and task type on cognition and judgment in human-system interaction: A neuroergonomic approach*. Doctoral dissertation, North Carolina Agricultural and Technical State University.

Nuamah, J. K., & Seong, Y. (2017, September). Neural correspondence to human cognition from analysis to intuition–implications of display design for cognition. In *Proceedings of the Human Factors and Ergonomics Society Annual Meeting* (Vol. 61, No. 1, pp. 51–55). Sage CA: Los Angeles, CA: SAGE Publications.

Paas, F., Tuovinen, J. E., Tabbers, H., & Van Gerven, P. W. (2003). Cognitive load measurement as a means to advance cognitive load theory. *Educational Psychologist, 38*(1), 63–71.

Paas, F. G., & Van Merriënboer, J. J. (1993). The efficiency of instructional conditions: An approach to combine mental effort and performance measures. *Human Factors, 35*(4), 737–743.

Parasuraman, R. (2011). Neuroergonomics: Brain, cognition, and performance at work. *Current Directions in Psychological Science, 20*(3), 181–186.

Peck, E. M. M., Yuksel, B. F., Ottley, A., Jacob, R. J., & Chang, R. (2013, April). Using fNIRS brain sensing to evaluate information visualization interfaces. In *Proceedings of the SIGCHI Conference on Human Factors in Computing Systems* (pp. 473–482). ACM.

Plaisant, C. (2004, May). The challenge of information visualization evaluation. In *Proceedings of the Working Conference on Advanced Visual Interfaces* (pp. 109–116). ACM.

Protzak, J., Ihme, K., & Zander, T. O. (2013). A passive brain-computer interface for supporting gaze-based human-machine interaction. In *International Conference on Universal Access in Human-Computer Interaction* (pp. 662–671). Berlin, Heidelberg: Springer.

Polich, J. (2007). Updating P300: An integrative theory of P3a and P3b. *Clinical Neurophysiology, 118*(10), 2128–2148.

Pope, A. T., Bogart, E. H., & Bartolome, D. S. (1995). Biocybernetic system evaluates indices of operator engagement in automated task. *Biological Psychology, 40*(1–2), 187–195.

Price, M. M., Crumley-Branyon, J. J., Leidheiser, W. R., & Pak, R. (2016). Effects of information visualization on older adults' decision-making performance in a medicare plan selection task: A comparative usability study. *JMIR Human Factors, 3*(1).

Prinzel, L. J., Freeman, F. G., Scerbo, M. W., Mikulka, P. J., & Pope, A. T. (2000). A closed-loop system for examining psychophysiological measures for adaptive task allocation. *The International Journal of Aviation Psychology, 10*(4), 393–410.

Sargent, A., Heiman-Patterson, T., Feldman, S., Shewokis, P. A., & Ayaz, H. (2018). Mental fatigue assessment in prolonged BCI use through EEG and fNIRS. In *Neuroergonomics* (pp. 315–316). Academic Press.

Scaife, M., & Rogers, Y. (1996). External cognition: How do graphical representations work? *International Journal of Human-Computer Studies, 45*(2), 185–213.

Scerbo, M. W., Freeman, F. G., & Mikulka, P. J. (2003). A brain-based system for adaptive automation. *Theoretical Issues in Ergonomics Science, 4*(1–2), 200–219.

Scerbo, M. W., Freeman, F. F., Mikulka, P. J., Parasuraman, R., Di Nocera, F. & Prinzel, L. J. (2001). The efficacy of psychophysiological measures for implementing adaptive technology, NASA TP-2001-211018. Hampton, VA: NASA Langley Research Center.

Seufert, T., Jänen, I., & Brünken, R. (2007). The impact of intrinsic cognitive load on the effectiveness of graphical help for coherence formation. *Computers in Human Behavior, 23*(3), 1055–1071.

Shi, Y., Zhao, Y., Zhou, F., Shi, R., & Zhang, Y. (2018). A novel radial visualization of intrusion detection alerts. *IEEE Computer Graphics and Applications, 38*(6), 83–95.

Solovey, E., Schermerhorn, P., Scheutz, M., Sassaroli, A., Fantini, S., & Jacob, R. (2012, May). Brainput: Enhancing interactive systems with streaming fnirs brain input. In *Proceedings of the SIGCHI Conference on Human Factors in Computing Systems* (pp. 2193–2202). ACM.

Stangl, M., Bauernfeind, G., Kurzmann, J., Scherer, R., & Neuper, C. (2013). A haemodynamic brain–computer interface based on real-time classification of near infrared spectroscopy signals during motor imagery and mental arithmetic. *Journal of Near Infrared Spectroscopy, 21*(3), 157–171.

Tory, M., & Moller, T. (2004). Human factors in visualization research. *IEEE Transactions on Visualization and Computer Graphics, 10*(1), 72–84.

Treacy Solovey, E., Afergan, D., Peck, E. M., Hincks, S. W., & Jacob, R. J. (2015). Designing implicit interfaces for physiological computing: Guidelines and lessons learned using fNIRS. *ACM Transactions on Computer-Human Interaction (TOCHI), 21*(6), 35.

Vassena, E., Gerrits, R., Demanet, J., Verguts, T., & Siugzdaite, R. (2018). Anticipation of a mentally effortful task recruits Dorsolateral Prefrontal Cortex: An fNIRS validation study. *Neuropsychologia*.

Ward, M. O., Grinstein, G., & Keim, D. (2015). *Interactive data visualization: Foundations, techniques, and applications*. AK Peters/CRC Press.

Wickens, C. D., & Ward, J. (2017). Cockpit displays of traffic and weather information: Effects of 3D perspective versus 2D coplanar rendering and database integration. *The International Journal of Aerospace Psychology, 27*(1–2), 44–56.

Widanagamaachchi, W., Livnat, Y., Bremer, P. T., Duvall, S., & Pascucci, V. (2017). Interactive visualization and exploration of patient progression in a hospital setting. In *AMIA Annual Symposium Proceedings* (Vol. 2017, p. 1773). American Medical Informatics Association.

Yin, X., Xu, B., Jiang, C., Fu, Y., Wang, Z., Li, H., & Shi, G. (2015). Classification of hemodynamic responses associated with force and speed imagery for a brain-computer interface. Journal of medical systems, *39*(5), 53.

Yuksel, B. F., Peck, E. M., Afergan, D., Hincks, S. W., Shibata, T., Kainerstorfer, J., … & Jacob, R. J. (2015, March). Functional near-infrared spectroscopy for adaptive human-computer interfaces. In *Optical tomography and spectroscopy of tissue XI* (Vol. 9319, p. 93190R). International Society for Optics and Photonics.

Zagermann, J., Pfeil, U., & Reiterer, H. (2016, October). Measuring cognitive load using eye tracking technology in visual computing. In *Proceedings of the Sixth Workshop on Beyond Time and Errors on Novel Evaluation Methods for Visualization* (pp. 78–85). ACM.

Zander, T. O., & Kothe, C. (2011). Towards passive brain–computer interfaces: Applying brain–computer interface technology to human–machine systems in general. *Journal of Neural Engineering, 8*(2), 025005.

Zander, T. O., Krol, L. R., Birbaumer, N. P., & Gramann, K. (2016). Neuroadaptive technology enables implicit cursor control based on medial prefrontal cortex activity. *Proceedings of the National Academy of Sciences, 113*(52), 14898–14903.

Neural Correlates and Mechanisms of Trust

Elizabeth Eskander, Nathan Sanders and Chang S. Nam

Abstract The decision to trust is the foundation for proper functioning in technology, work, and social environments. For this reason, it is of critical interest to understand and investigate the brain mechanisms involved in human trust for the neuroergonomics community. This chapter seeks to review the current understanding of the neurotransmitters and different neural correlates involved in the decision to trust compared to the decision not to trust. The neurotransmitters with these multifaceted behaviors are Oxytocin and Testosterone. While previously dubbed, the "love hormone" oxytocin is now believed to have an impact on the brain's encoding of prediction error, and therefore its ability to modify preexisting beliefs. Testosterone, which was thought of as an inhibitor of trust, only shows decreases of trust only in men and a subgroup of women who displayed social naivety. The structural correlates involved in trust/distrust identified are the prefrontal cortex (PFC) (specifically the ventromedial PFC and orbitofrontal cortex), the insula, caudate, amygdala, and anterior paracingulate cortex (PCC). The caudate and anterior paracingulate cortex are currently thought to be involved with the positive decision to trust, whereas the prefrontal cortex, insula, and amygdala seem to have higher correlations with the negative decision to distrust. The academic literature used to determine these correlates methodology included lesion studies, intranasal administration, fMRI imaging of the trust game, and transcranial direct stimulation studies.

1 Introduction

By definition, trust is the assured reliance on the character, ability, strength, or truth of someone or something (Merriam-Webster's Collegiate Dictionary, 1993). The foundations of our complex human social lives and interactions are driven by trust,

E. Eskander
Department of Psychology, North Carolina State University, Raleigh, NC 27695, USA

N. Sanders · C. S. Nam (✉)
Edward P. Fitts Department of Industrial & Systems Engineering, North Carolina State University, Raleigh, NC 27695, USA
e-mail: csnam@ncsu.edu

C. S. Nam (ed.), *Neuroergonomics*, Cognitive Science and Technology,
https://doi.org/10.1007/978-3-030-34784-0_22

and its pervasive necessity in relationships between humans and society has led to much research in the fields of neuroscience, psychology, and economics (Krueger & Meyer-Lindenberg, 2019). Trust allows for a more open and inclusive society, economic stability and development, and overall feelings of well-being. Modern society has grown ever more complicated with the introduction of easily accessible mass media, global stock markets, and increased immigration rates expanding ethnodiversity in previously culturally uniform areas. As our societies, relationships, and interactions become more complex, researchers have become all the more curious to understand the neural correlates and mechanisms behind our foundations of trust.

It is important to give a distinction between trust and distrust as these are two separate constructs that arise through different determinations in the brain. Distrust is not defined as a lack of trust, but rather the expectation that the truster will have his/her interest violated by the trustee (Dimoka, 2010). Establishing this dichotomy allows for deeper introspection into the separate neural processes that occur prior to our behavioral decisions between trust and distrust. Before continuing, it is also critical to note that the following brain structures and neurobiological chemicals should cautiously only be regarded as correlates of trust and should not be overstated as causes of trust. While the field of neuroscience and use of neuroimaging has exploded in the past decade, the methodology and technology used are still in their infancy and must be tested both critically and rigorously before the data yielded can be accepted as scientific fact. Structural imaging and studies on subjects with brain abnormalities currently can be considered more reliable than functional imaging (Vidal, 2015), but this does not imply that researchers should not continue to examine and critically analyze all the data provided.

Understanding the behavioral decision to trust has also become an endeavor taken up in an emerging field of neuroergonomics. Its goal is to use the existing/emerging knowledge of human performance and brain function to design systems for safer and more efficient operation, and to advance this understanding of the relationship between brain function and performance in real-world tasks (Parasuraman & Rizzo, 2008). As previously established, trust is an important foundation for proper functioning in all of these environments. This chapter seeks to understand the specific brain mechanisms involved in human trust by first exploring the current research on the neurobiological correlates, and then conclude by reviewing the spatial neurodynamics of trust.

The remainder of this chapter is organized as follows. Section 2 reviews the research of the two current neurobiological correlates implicated in trust/distrust: oxytocin and testosterone. Section 3 presents the currently understood structural neural correlates of trust/distrust: Prefrontal cortex, insula, caudate, amygdala, and anterior paracingulate cortex. Section 4 concludes with a few suggestions for future research.

2 Neurobiological Correlates of Trust

This section will explore the current understanding and state of research on the neurobiological correlates of oxytocin and testosterone and their relationship with the decision to trust. Social neuroendocrinology has recently displayed evidence that considers two hormones that interact in a complex balance to promote interpersonal trust from experience (Bos, Hermans, Ramsey, & Van Honk, 2012). Oxytocin, which is thought of as a promoter of trust, and testosterone, the inhibitor of trust (Kosfeld, Heinrichs, Zak, Fischbacher, & Fehr, 2005). While this may seem like a parsimonious explanation, the realities are far more complex and neuroscientists are only now beginning to understand the nuances of the roles of these neurotransmitters in the brain. Expression and effects of these hormones have several influencing factors such as age, sex, and psychopathology.

2.1 Oxytocin

Oxytocin is a nine-amino acid peptide hormone that acts as a neurotransmitter in the brain. It is released from the magnocellular cells of the hypothalamus and is accumulated and stored in the pituitary gland (Standring, 2015). It is prevalent in the female body during childbirth where it stimulates the uterus to contract, and during breast-feeding when it functions to stimulate milk secretion during lactation (Chiras, 2012; Venes, 2017). Interest in understanding the role of oxytocin in human social life began when it was discovered that mothers who breastfed experienced less stress than mothers who bottle-fed (Mezzacappa & Katkin, 2002). Behavioral scientists had also shown that oxytocin plays a key role in social attachment among nonhumans, increasing enthusiasm for research (e.g., Nave, Camerer, & McCullough, 2015).

Pop science has dubbed oxytocin the "love hormone" because of its associations with pleasant feeling and emotions (van IJzendoorn & Bakermans-Kranenburg, 2012). Until 2015, researchers themselves believed that oxytocin was causally related to increasing interpersonal trust (Nave et al., 2015). What is now more widely accepted is that oxytocin has an impact on the brain's encoding of prediction error and therefore its ability to modify preexisting beliefs. Specifically, increase of trust under administration of intranasal oxytocin is associated with a decrease of activation in the amygdala (Ide et al., 2018), consistent with studies showing individuals with damage to the amygdala damage display increased rates of trust (Koscik & Tranel, 2011; van Honk, Eisenegger, Terburg, Stein, & Morgan, 2013).

While many intranasal studies (Nave et al., 2015) have failed to replicate findings of increasing trust with administration of oxytocin, studies on subjects with psychiatric abnormalities seem to offer researchers more profound insight into the neurotransmitters' role. A 2017 study examining the effects of intranasal oxytocin administration on adolescents found that oxytocin increases trust in individuals with moderate-to-low attachment security, but paradoxically decreased trust in their healthy controls (Venta et al., 2017). Researchers are also studying possible treatment methods using oxytocin for schizophrenic patients who suffer from social deficits, most notably trust (e.g., Gul, Gul, & Godil, 2018). Self-report placebo-controlled studies show that schizophrenic patients administered with intranasal oxytocin have increased capacity for empathetic perspective taking as well as introspective accuracy (Halverson, Jarskog, Pedersen, & Penn, 2019).

The history of research on oxytocin's relationship to trust has been complex and controversial. Three main methods have been used, each with their own contentions. The first method of study is done by examining the effects of oxytocin on behavioral measures through the administration of placebo-controlled intranasal oxytocin. Unlike other hormones, the pharmacokinetics of oxytocin's movement through the body is not yet well understood and direct evidence is still urgently needed (Carson, Yuan, & Labuschagne, 2016; Ide et al., 2018). While evidence does exist of intranasal oxytocin crossing the blood–brain barrier, enhancing resting state amygdala–prefrontal connectivity and increasing cerebral blood flow to areas containing oxytocin receptors (Sripada et al., 2012), direct evidence that it reaches its target tissues is very limited (Nave et al., 2015). There is, however, strong evidence that other large molecules similar to oxytocin delivered intranasally penetrate the olfactory bulb and rostral brain regions via the peripheral olfactory system (Lochhead & Thorne, 2012).

The second method of study is to correlate plasma oxytocin levels with behavioral measures of trust among individuals. The first publications claiming a correlation between oxytocin plasma levels and trust were released in 2005. Studies published since then have often failed to replicate results and are also plagued with methodological issues (Nave et al., 2015). These studies cannot establish causation, and unlike intranasal studies, there is an even larger gap in assuming that peripheral oxytocin crosses the blood–brain barrier. There is also a concern with how oxytocin levels are measured within the body. Accurate immunoassay requires an extraction step for all extraneous substances to be removed. Many studies have failed to complete this step and therefore grossly overestimate their findings (Nave et al., 2015).

The last method is correlating oxytocin-related genetic polymorphisms with behavioral measures among individuals. A single study in 2012 reported an association between a common variation in the oxytocin receptor (OXTR) gene and trust game investment (Krueger et al., 2012). Other studies claimed different variations of this gene correlated with a range of social behaviors including prosocial, empathy, and autism (e.g., Nishina et al., 2018). Meta-analyses have shown that none of these studies have a combined effect greater than zero, and a large-scale genome-wide association study has defended the notion that no single SNP across the whole genome can explain more than 1.25% variation of any trait (Nave et al., 2015).

2.2 *Testosterone*

Testosterone is a steroid sex hormone responsible for the growth and development of masculine characteristics in addition to being linked to aggressive and predatory behaviors (Venes, 2017). A qualitative review by Heany, van Honk, Stein and Brooks (2016) showed adult testosterone levels are associated with larger limbic, insula and occipital lobe volumes, and smaller parietal lobe and opercular left inferior frontal gyrus volumes.

Studies that examined the effects of testosterone administration (TA) sublingually have yielded results showing that the neurotransmitter has behavioral effects on subjects considered to be more socially naive (Silver, 2010). Specifically, subjects who had a higher tendency for initial trust showed lower trust levels compared to their placebo controls. This effect was not seen in the subgroup that was administered testosterone but already had lower levels of initial trust (Bos et al., 2012). It must be mentioned that the mentioned studies were conducted using only female subjects, but a 2014 study conducted by Carre and colleagues, comparing the behavioral effects of TA between men and women, found that increases in testosterone do decrease trusting behavior in men but had no such effect on women (Carré et al., 2014).

3 Spatial Neural Dynamics of Trust

According to the NPE (neuropsychoeconomic) model of trust proposed in 2019 by Krueger et al., the decision to trust arises from the psychological systems of motivation, affect, and cognition, interacting and engaging with key neural networks: the reward network (associated with motivation), salience network (affective system), central-executive network (cognitive control), and default mode network (social cognition system) (Krueger & Meyer-Lindenberg, 2019). The brain regions associated with these networks are the prefrontal cortex (associated with executive functions) and the limbic system (associated with emotional processes) (Dimoka, 2010). The following correlates chosen to be discussed were the brain structures identified in Dimoka's (2010) publication after her extensive review of the current literature inspired her own fMRI study to validate her hypothesis. In this article, Dimoka hypothesized that trust is associated with higher brain activation in the caudate nucleus and anterior paracingulate cortex (PCC) but lower activation in the orbitofrontal cortex (OFC). Dimoka identified the insula as well as the amygdala to have higher activations for expressions of distrust.

3.1 Prefrontal Cortex (PFC)

Orbitofrontal (OFC): The orbitofrontal cortex (OFC) is a PFC region in the frontal lobes of the brain, which consists of Brodmann area (BA) 10, 11, and 47 (Kringelbach, 2005). Findings by Dimoka (2010) correlate the decision to increase trust with lower activation of the orbitofrontal cortex, which is consistent with the assumption that it is the primary area for assessing trustee's characteristics, credibility, and calculating uncertainty by determining risk value. Damage to the orbitofrontal cortex has shown to impair decision-making (Manes et al., 2002), a crucial component necessary to form the decision to trust. fMRI studies have concluded that neurons located in the orbitofrontal cortex are responsible for assessing risk versus reward evaluations (Drnec, Marathe, Lukos, & Metcalfe, 2016). More specifically, studies have determined it to encode the value of a choice relative to the recent history of choice values (Wallis & Kennerley, 2011). Animal studies show that damage causes an increase in the variability of successive decisions (Noonan et al., 2010). This is potentially caused by the override of other neural networks responsible for decision-making leading the OFC to explore new options without the risk assessment of known values (Noonan, Kolling, Walton, & Rushworth, 2012) The results of a transcranial direct current stimulation study suggest that the activity of the right orbitofrontal cortex has direct causal effects on cooperative behavior (Wang, Li, Yin, Li, & Wei, 2016).

Ventromedial Prefrontal Cortex (vmPFC): The ventral medial prefrontal is located in the frontal lobe at the bottom of the cerebral hemispheres, which consists of BA 10, 14, 25, and 32, as well as portions of BA 11, 12, and 13 (Finger et al., 2008). Damage to the vmPFC is associated with abnormal decision-making as well as a reduction in trust decisions during the trust game (Zheng et al., 2016). Activity has also been observed during personal and interpersonal decision-making (Haas, Ishak, Anderson, & Filkowski, 2015). Voxel-based morphometry analyses of high-resolution structural images have associated increased regional gray matter volume of the ventromedial cortex with increased propensity for individual trust (Haas et al., 2015). Lesion studies have shown that patients with a damaged vmPFC show less trust and altruism than their controls (Zheng et al., 2016), as well as increased risk-taking but reduced generosity (Moretto, Sellitto, & di Pellegrino, 2013). While conducting a trust game study on vmPFC lesion patients, Moretto et al. (2013) found an increased initial endowment separate from the trust game. vmPFC group had overall significantly higher transfer levels compared to their control and healthy control patients. Results from a transcranial direct stimulation study showed an increase in trustworthiness following the excitement of the vmPFC (Zheng et al., 2016). It is also implicated in the reinforcement learning process due to its association with reward sensitivity (Parasuraman, de Visser, Wiese, & Madhavan, 2014).

3.2 *Insula*

The insula, which is hidden deep within the lateral sulcus of the brain, is a part of the limbic system and is associated with the specific fear of loss and coding of trust and distrust (Dimoka, 2010). fMRI studies and multi-round trust games conclude that the insula responds to violations of social norms during interpersonal reactions (Belfi, Koscik, & Tranel, 2015; Walter et al., 2004). It is associated with the salience network where it codes the subjective and aversive feelings of unpredictable events (Krueger & Meyer-Lindenberg, 2019). Activation of the insular cortex leads to the avoidance of future negative interactions. Activation of this area has also been associated with discredibility as well as malevolence (Dimoka, 2010). Increased activation is associated with evaluating untrustworthy faces, with levels peaking during mid-adolescence (Kragel, Zucker, Covington, & LaBar, 2014; Santos, Almeida, Oliveiros, & Castelo-Branco, 2016).

Increased gray matter of the insula has also been associated with increased levels of trust (Zheng et al., 2016). Damage to the insula results in abnormal interpersonal trust during the trust game. Patients displayed misplaced trust when playing the part of the investor and increased malevolence when acting as the role of the trustee (Belfi et al., 2015).

3.3 *Caudate*

The caudate nucleus makes up the dorsal striatum, which is a component of the basal ganglia. The caudate nucleus has been identified as a key area associated with subjects anticipating positive rewards, being innervated by dopamine receptors which fire when reward is received (Dimoka, 2010). fMRI studies suggest that the caudate is involved in signaling and integrating reputations gained through experience into trust decisions (Wardle et al., 2013). Behavioral findings suggest that the caudate also selectively encodes the transfer of social values during moral learning (FeldmanHall et al., 2018). Caudate nucleus activation has also been shown when subjects acted cooperatively to obtain an anticipated reward and that activation can be used to predict the decision to reciprocate cooperation for mutual gain (Dimoka, 2010).

The caudate, along with the putamen, also shows activation for benevolence and credibility, implicating benevolence as being associated with higher rewards (Dimoka, 2010). This has been contrasted with studies showing increased activation of the caudate during interaction with indifferent and unfair partners over fair partners. Because the caudate is a reward-sensitive area, it is suggested to be a correlate of the reinforcement learning process which aids in establishing trust decisions (Wardle et al., 2013).

3.4 Amygdala

The vast majority of research concerning the amygdala's correlation with interpersonal trust shows that it has a significant role in processing facial trustworthiness (Santos et al., 2016). Meta-analysis specifically implicates bilateral amygdala activations to be greater in negative facial trustworthiness evaluations versus positive evaluations.

Studies conducted on patients with damage to the amygdala also display abnormal interpersonal behavior (Koscik & Tranel, 2011; van Honk et al., 2013). In a 2011 study, Koscik and Tranel used the trust game to show that amygdala damage is associated with increased benevolent behavior. More notably they saw this increase in benevolence in response to betrayal whereas neurologically normal adults show decreases in benevolence in response to breaches of trust. A 2013 study later conducted by Honk et al. showed that subjects with basolateral amygdala damage invest nearly 100 percent more money in unfamiliar partners than their healthy controls (van Honk et al., 2013). Their data suggest that the behavior exhibited by these subjects is the result of impairment in instrumental choice behaviors.

3.5 Anterior Paracingulate Cortex (PCC)

As previously mentioned, Dimoka's (2010) study found higher activation of the anterior paracingulate cortex associated with the decision to trust. The decision to trust involves mental processes that require the trustee to determine their partner's intentions and whether their trust will be reciprocated. Neuroimaging has correlated the anterior paracingulate cortex as this "social brain" area that is activated when predicting the actions of others (Gallagher & Frith, 2003). This possible "prediction" function of the paracingulate cortex seems to be unique to the decision for cooperative behaviors. This is suggested due to the fact that studies consistently only found higher activation of this area when subjects decide to engage in more social and cooperative behaviors (Dimoka, 2010; Krueger et al., 2012; Walter et al., 2004).

4 Summary

The decision to trust is the foundation of our complex societies and is pertinent to proper societal functioning. For this reason, understanding the underlying neural mechanisms and correlates involved in this decision has been of interest to the fields of neuroscience, psychology, economics as well as neuroergonomics.

Research in these fields has found that trust is a multifaceted decision and requires the neural processing of prediction, anticipation, reflection, and encoding prior interactions. Studies have also identified two key neurotransmitters that have a role in regulating between the decision to trust and distrust: Oxytocin and Testosterone. The functional and neurobiological correlates mentioned in this chapter are not the only correlates that have been implicated in trust, but more research must be conducted before making any certain conclusions. This also applies to the abovementioned correlates for which more research and reflection of methodologies, data, and conclusions must be conducted in order to conclusively reduce the mental process of trust to neurological functions.

References

Belfi, A. M., Koscik, T. R., & Tranel, D. (2015). Damage to the insula is associated with abnormal interpersonal trust. *Neuropsychologia, 71,* 165–172.

Bos, P. A., Hermans, E. J., Ramsey, N. F., & Van Honk, J. (2012). The neural mechanisms by which testosterone acts on interpersonal trust. *Neuroimage, 61*(3), 730–737.

Carré, J. M., Baird-Rowe, C. D., & Hariri, A. R. (2014). Testosterone responses to competition predict decreased trust ratings of emotionally neutral faces. *Psychoneuroendocrinology, 49,* 79–83.

Carson, D. S., Yuan, H., & Labuschagne, I. (2016). Improving research standards to restore trust in intranasal oxytocin. *Biological Psychiatry, 79*(8), e53–e54.

Chiras, D. D. (2012). *Human biology* (7th ed.). Sudbury, MA: Jones & Bartlett Learning. p. 262. ISBN 978-0-7637-8345-7.

Dimoka, A. (2010). What does the brain tell us about trust and distrust? Evidence from a functional neuroimaging study. *Mis Quarterly*, 373–396.

Drnec, K., Marathe, A. R., Lukos, J. R., & Metcalfe, J. S. (2016). From trust in automation to decision neuroscience: Applying cognitive neuroscience methods to understand and improve interaction decisions involved in human automation interaction. *Frontiers in Human Neuroscience, 10,* 290.

FeldmanHall, O., Dunsmoor, J. E., Tompary, A., Hunter, L. E., Todorov, A., & Phelps, E. A. (2018). Stimulus generalization as a mechanism for learning to trust. *Proceedings of the National Academy of Sciences, 115*(7), E1690–E1697.

Finger, E. C., Marsh, A. A., Mitchell, D. G., Reid, M. E., Sims, C., Budhani, S., … Pine, D. S. (2008). Abnormal ventromedial prefrontal cortex function in children with psychopathic traits during reversal learning. *Archives of General Psychiatry, 65*(5), 586–594.

Gallagher, H. L., & Frith, C. D. (2003). Functional imaging of 'theory of mind'. *Trends in Cognitive Sciences, 7*(2), 77–83.

Gul, O., Gul, S., & Godil, A. A. (2018). Posterior pituitary neurohormonal disturbances in schizophrenia and role of oxytocin in treatment–need for more short-and long-term studies. *Neuropsychiatric Disease and Treatment, 14,* 2579.

Haas, B. W., Ishak, A., Anderson, I. W., & Filkowski, M. M. (2015). The tendency to trust is reflected in human brain structure. *Neuroimage, 107,* 175–181.

Halverson, T. L., Jarskog, F., Pedersen, C., & Penn, D. (2019). Effects of oxytocin on empathy, introspective accuracy, and social symptoms in schizophrenia: A 12-week twice-daily randomized controlled trial. *Schizophrenia Research, 204,* 178–182. ISSN 0920-9964.

Heany, S. J., van Honk, J., Stein, D. J., & Brooks, S. J. (2016). A quantitative and qualitative review of the effects of testosterone on the function and structure of the human social-emotional brain. *Metabolic Brain Disease, 31*(1), 157–167.

van Honk, J., Eisenegger, C., Terburg, D., Stein, D. J., & Morgan, B. (2013). Generous economic investments after basolateral amygdala damage. *Proceedings of the National Academy of Sciences, 110*(7), 2506–2510.

Ide, J. S., Nedic, S., Wong, K. F., Strey, S. L., Lawson, E. A., Dickerson, B. C., … Mujica-Parodi, L. R. (2018). Oxytocin attenuates trust as a subset of more general reinforcement learning, with altered reward circuit functional connectivity in males. *Neuroimage, 174*, 35–43.

Koscik, T. R., & Tranel, D. (2011). The human amygdala is necessary for developing and expressing normal interpersonal trust. *Neuropsychologia, 49*(4), 602–611.

Kragel, P. A., Zucker, N. L., Covington, V. E., & LaBar, K. S. (2014). Developmental trajectories of cortical–subcortical interactions underlying the evaluation of trust in adolescence. *Social Cognitive and Affective Neuroscience, 10*(2), 240–247.

Krueger, F., & Meyer-Lindenberg, A. (2019). Toward a model of interpersonal trust drawn from neuroscience, psychology, and economics. *Trends in Neurosciences, 42*(2), 92–101.

Krueger, F., Parasuraman, R., Iyengar, V., Thornburg, M., Weel, J., Lin, M., … Lipsky, R. (2012). Oxytocin receptor genetic variation promotes human trust behavior. *Frontiers in Human Neuroscience, 6*, 4.

Kosfeld, M., Heinrichs, M., Zak, P. J., Fischbacher, U., & Fehr, E. (2005). Oxytocin increases trust in humans. *Nature, 435*(7042), 673.

Kringelbach, M. L. (2005). The human orbitofrontal cortex: Linking reward to hedonic experience. *Nature Reviews Neuroscience, 6*(9), 691.

Lochhead, J. J., & Thorne, R. G. (2012). Intranasal delivery of biologics to the central nervous system. *Advanced Drug Delivery Reviews, 64*(7), 614–628.

Manes, F., Sahakian, B., Clark, L., Rogers, R., Antoun, N., Aitken, M., & Robbins, T. (2002). Decision-making processes following damage to the prefrontal cortex. *Brain, 125*(3), 624–639.

Merriam-Webster. (1993). *Webster's third new international dictionary of the English language unabridged*. Merriam-Webster Inc.

Mezzacappa, E. S., & Katkin, E. S. (2002). Breast-feeding is associated with reduced perceived stress and negative mood in mothers. *Health Psychology, 21*(2), 187.

Moretto, G., Sellitto, M., & di Pellegrino, G. (2013). Investment and repayment in a trust game after ventromedial prefrontal damage. *Frontiers in Human Neuroscience, 7*, 593.

Nave, G., Camerer, C., & McCullough, M. (2015). Does oxytocin increase trust in humans? A critical review of research. *Perspectives on Psychological Science, 10*(6), 772–789.

Nishina, K., Takagishi, H., Fermin, A. S. R., Inoue-Murayama, M., Takahashi, H., Sakagami, M., & Yamagishi, T. (2018). Association of the oxytocin receptor gene with attitudinal trust: Role of amygdala volume. *Social Cognitive and Affective Neuroscience, 13*(10), 1091–1097.

Noonan, M. P., Kolling, N., Walton, M. E., & Rushworth, M. F. S. (2012). Re-evaluating the role of the orbitofrontal cortex in reward and reinforcement. *European Journal of Neuroscience, 35*(7), 997–1010.

Noonan, M. P., Walton, M. E., Behrens, T. E. J., Sallet, J., Buckley, M. J., & Rushworth, M. F. S. (2010). Separate value comparison and learning mechanisms in macaque medial and lateral orbitofrontal cortex. *Proceedings of the National Academy of Sciences, 107*(47), 20547–20552.

Parasuraman, R., & Rizzo, M. (Eds.). (2008). *Neuroergonomics: The brain at work*. Oxford University Press.

Parasuraman, R., de Visser, E., Wiese, E., & Madhavan, P. (2014, September). Human trust in other humans, automation, robots, and cognitive agents: Neural correlates and design implications. In *Proceedings of the Human Factors and Ergonomics Society Annual Meeting* (Vol. 58, No. 1, pp. 340–344). Sage, CA: Los Angeles, CA: SAGE Publications.

Santos, S., Almeida, I., Oliveiros, B., & Castelo-Branco, M. (2016). The role of the amygdala in facial trustworthiness processing: A systematic review and meta-analyses of fMRI studies. *PLoS One, 11*(11), e0167276.

Silver, J. (2010). Don't be deceived: Testosterone decreases trust. *Journal Watch Psychiatry*. http://dx.doi.org.prox.lib.ncsu.edu/10.1056/JP201006140000002.

Standring, S. (Ed.). (2015). *Gray's anatomy international edition: The anatomical basis of clinical practice*. Elsevier Health Sciences.

Sripada, C. S., Phan, K. L., Labuschagne, I., Welsh, R., Nathan, P. J., & Wood, A. G. (2012). Oxytocin enhances resting-state connectivity between amygdala and medial frontal cortex. *International Journal of Neuropsychopharmacology, 16*(2), 255–260.

Venes, D. (2017). *Taber's cyclopedic medical dictionary*. FA Davis.

Venta, A., Ha, C., Vanwoerden, S., Newlin, E., Strathearn, L., & Sharp, C. (2017). Paradoxical effects of intranasal oxytocin on trust in inpatient and community adolescents. *Journal of Clinical Child & Adolescent Psychology*, 1–10.

Vidal, F. (2015). Historical and ethical perspectives of modern neuroimaging. *Handbook of Neuroethics*, 535–550.

Wallis, J. D., & Kennerley, S. W. (2011). Contrasting reward signals in the orbitofrontal cortex and anterior cingulate cortex. *Annals of the New York Academy of Sciences, 1239*(1), 33–42.

Walter, H., Adenzato, M., Ciaramidaro, A., Enrici, I., Pia, L., & Bara, B. G. (2004). Understanding intentions in social interaction: The role of the anterior paracingulate cortex. *Journal of Cognitive Neuroscience, 16*(10), 1854–1863.

Wang, G., Li, J., Yin, X., Li, S., & Wei, M. (2016). Modulating activity in the orbitofrontal cortex changes trustees' cooperation: A transcranial direct current stimulation study. *Behavioural Brain Research, 303,* 71–75.

Wardle, M. C., Fitzgerald, D. A., Angstadt, M., Sripada, C. S., McCabe, K., & Phan, K. L. (2013). The caudate signals bad reputation during trust decisions. *PLoS One, 8*(6), e68884.

Van IJzendoorn, M. H., & Bakermans-Kranenburg, M. J. (2012). A sniff of trust: Meta-analysis of the effects of intranasal oxytocin administration on face recognition, trust to in-group, and trust to out-group. *Psychoneuroendocrinology,* 37(3), 438–443.

Zheng, H., Huang, D., Chen, S., Wang, S., Guo, W., Luo, J., … Chen, Y. (2016). Modulating the activity of ventromedial prefrontal cortex by anodal tDCS enhances the trustee's repayment through altruism. *Frontiers in Psychology, 7*, 1437.

Conclusion

Conclusion: Moving Forward in Neuroergonomics

Chang S. Nam

This book was motivated to provide experienced researchers and practitioners, including those with no formal training in neuroergonomics, with a synopsis of key findings and theoretical and technical advances in neuroergonomics. Fortunately, we could deliver 22 chapters, including an introduction to the evolution of neuroergonomics, in the following five main themes which supported and drove the important opportunities for neuroergonomics research: neuroimaging methods (Part I), computational approaches (Part II), neuroergonomics assessments of cognitive and physical performance (Part III), and brain–computer interfaces (Part IV), and everyday emerging applications (Part V).

As evidenced by the contributed chapters in this book, a great success has been achieved in studying the human brain at work and great strides are currently being made to extend neuroergonomics research. However, it is also true that neuroergonomics still faces many challenges neuroergonomists need to address in order to advance even further. Exemplary future directions in neuroergonomics would include

- advances in neuroimaging technologies that can directly or indirectly visualize the structure, function, and connectivity of the brain while humans interact with the world,
- the development and validation of brain-inspired computational models or frameworks that can lead to novel predictions of the operator's human performance in dynamic contexts,
- investigation into new neural mechanisms and correlates of human cognition and behavior under various work contexts at a greater level of detail,
- the design, development, implementation, and validation of neuroadaptive technology that can monitor and infer the cognitive state of an individual in real time during task performance, and enable intelligent adaptation,

C. S. Nam (✉)
Edward P. Fitts Department of Industrial & Systems Engineering, North Carolina State University, Raleigh, NC 27695, USA
e-mail: csnam@ncsu.edu

C. S. Nam (ed.), *Neuroergonomics*, Cognitive Science and Technology,
https://doi.org/10.1007/978-3-030-34784-0_23

- the identification of emerging applications, barriers, and actions to advance neuroergonomics, and
- the development of advanced research methodologies through enhanced inter and multidisciplinary collaboration with researchers in other fields.

To sum up, we hope that the collection of chapters in this book will be able to serve as the next step in the journey of examining the human brain function and behavior in relation to behavioral performance from multiple perspectives in the context of changing work environments and everyday settings. This book can be used as an additional source of reference in a broader study of neuroergonomics.

Author Index

C. S. Nam (ed.), *Neuroergonomics*, Cognitive Science and Technology,
https://doi.org/10.1007/978-3-030-34784-0

Subject Index

C. S. Nam (ed.), *Neuroergonomics*, Cognitive Science and Technology,
https://doi.org/10.1007/978-3-030-34784-0

Kolibri 360 GmbH
Ferdinand-Jühlke-Straße 7
99095 Erfurt, Deutschland
produktsicherheit@kolibri360.de